Catalytic Hydroarylation of Carbon-Carbon Multiple Bonds

Catalytic Hydroarylation of Carbon-Carbon Multiple Bonds

Edited by

Lutz Ackermann

T. Brent Gunnoe

Laurel Goj Habgood

WILEY-VCH

The Editors

Prof. Lutz Ackermann
Georg-August-Universität
Institut für Organische Chemie
Tammannstr. 2
37077 Göttingen
Germany

Prof. T. Brent Gunnoe
University of Virginia
Department of Chemistry
McCormick Road
PO Box 400319
Charlottesville, VA
United States

Prof. Laurel Goj Habgood
Rollins College
Department of Chemistry
1000 Holt Avenue
Box 2743
Winter Park, FL
United States

Cover
fotolia_©zolnierek

Library of Congress Card No.:
applied for

British Library Cataloguing-in-Publication Data
A catalogue record for this book is available from the British Library.

Bibliographic information published by the Deutsche Nationalbibliothek
The Deutsche Nationalbibliothek lists this publication in the Deutsche Nationalbibliografie; detailed bibliographic data are available on the Internet at <http://dnb.d-nb.de>.

Print ISBN: 978-3-527-34013-2
ePDF ISBN: 978-3-527-69763-2
ePub ISBN: 978-3-527-69765-6
Mobi ISBN: 978-3-527-69766-3
oBook ISBN: 978-3-527-69764-9

Cover Design Adam-Design, Weinheim, Germany
Typesetting SPi Global, Chennai, India

Printed by CPI Group (UK) Ltd, Croydon, CR0 4YY

Printed on acid-free paper

Contents

Introduction and Preface

Laurel G. Habgood[1], *Lutz Ackermann*[2] *and T. Brent Gunnoe*[3]

[1] *Rollins College, Department of Chemistry, 1000 Holt Avenue, Winter Park, FL 32789-4499, USA*
[2] *Georg-August-Universität, Institut für Organische und Biomolekulare Chemie, Tammannstrasse 2, 37077, Göttingen, Germany*
[3] *Department of Chemistry, University of Virginia, McCormick Road, Charlottesville, VA 22904-4319, USA*

Whether the target is a biologically active compound or a chemical feedstock, the ability to selectively install alkyl or alkenyl groups at a position of an aromatic or heteroaromatic compound is of central importance to synthetic chemistry. Friedel–Crafts catalysis and transition metal-catalyzed carbon–carbon bond coupling using aryl halides and organometallic reagents (e.g., Suzuki, Stille, and Sonogashira reactions) have been developed over time to install C—C bonds into aromatic substrates. These reactions are powerful tools for synthetic chemists, yet they are often hampered by one or more issues including harsh reaction conditions, lack of regioselectivity and/or stereoselectivity, use of toxic organometallic reagents, and the use of halogenated substrates. The direct addition of aromatic C—H bonds to unsaturated substrates (e.g., olefins or alkynes) provides an atom-economical strategy that is complementary to Friedel–Crafts and traditional C—C coupling reactions. Thus, it is not surprising that after the first published reports of hydroarylation of olefins and alkynes, the number of groups interested in the synthetic and mechanistic aspects of transition metal-mediated addition of C—H bonds across C—C double and triple bonds has steadily grown.

To our knowledge, the first review article to include olefin or alkyne hydroarylation chemistry was published in 1990 [1], the first focused review appeared in 2002 [2], and the first book was published in 2009 [3]. The extensive advancements in catalytic addition of C—H bonds across C—C multiple bonds and their rising impact on the methods used in synthetic organic chemistry warrants a comprehensive textbook that can be a single source providing a broad overview of the state of the art in the field. Skilled chemists may look to utilize the chemistry in synthetic applications while graduate students and novices to the field may discover fundamentally interesting chemistry. The authors of each chapter have presented a detailed review of their topic supplementing areas not covered with references for the interested reader, including both experimental and theoretical data. The book provides an overview of what has been accomplished, and also includes commentary on existing challenges and opportunities.

Catalytic Hydroarylation of Carbon-Carbon Multiple Bonds, First Edition.
Edited by Lutz Ackermann, T. Brent Gunnoe and Laurel Goj Habgood.

In Chapter 1, Villuendas, Ruiz, and Urriolabeitia detail the use of group 9 and 10 catalysts to functionalize heteroarenes. Hydroarylation reactions with aryl halides or arylboronic acids are not covered as emphasis is placed on the cleavage of heteroaryl C—H bonds. The synthetic utility and versatility of these reactions is conveyed by organizing the sections by type of heterocycle (thiophene, furan, indoles, etc.) rather than by type of catalyst. Future challenges and directions in the field are discussed in the summary.

Chapter 2 includes the alkylation of functionalized arene and heteroaromatic substrates utilizing ruthenium catalysts as detailed by Burns, Kozhushkov, and Ackermann. Electrophilic alkyl halides are omitted as chelation-assisted *ortho*-C—H activation is the focus. The sections of the chapter are organized based on the mechanistic modes of C—H cyclometalation via C—H bond cleavage, C—H oxidative addition, and carboxylate-assisted C—H activation. The summary highlights the broad applicability of the reactivity for a variety of chemical industries while identifying the need for further improvements to develop milder reaction conditions and asymmetric C—H bond functionalizations.

Chapters 3 and 4 examine olefin hydroarylation as organized by catalyst identity. In Chapter 3, McKeown, Habgood, Cundari, and Gunnoe examine d^6 transition metal catalysts for the alkylation of arenes without chelation assistance while in Chapter 4 Suslick and Tilley detail the use of d^8 transition metal catalysts. The industrial relevance of the reactions is emphasized as alkyl benzenes are precursors to large-scale chemicals such as polymer precursors and surfactants. The synergistic use of experimental and theoretical experiments for mechanistic information to further catalyst development is highlighted.

Nakao focuses on nickel-catalyzed hydroarylation of carbon–carbon multiple bonds in Chapter 5, which is an area that has experienced dramatic advancement in recent years. The characteristic features of nickel catalysis that are highlighted include reactivity with electron-deficient arenes including heteroarenes. This chapter is organized by reactions of alkenes and alkynes with fluorobenzenes, five-membered heteroarenes, and azines.

While ruthenium, rhodium, and iridium complexes are the most prevalent for directing group-assisted hydroarylation reactions, there is a growing body of work demonstrating the utility of catalysts from the first row transition metals. In Chapter 6, Yoshikai builds on the nickel catalysis discussed in Chapter 5 by examining the hydroarylation of alkynes and alkenes catalyzed by first row transition metals in groups 7–9. The majority of the chapter consists of hydroarylation reactions of alkynes and alkenes utilizing both high- and low-valent cobalt catalysts. Examples of iron complexes for the hydroarylation of alkenes and alkynes, as well as the use of a Lewis acidic iron salt for the hydroarylation of alkenes, are included. Notable at the end is a discussion on low-valent manganese complexes for the hydroarylation of alkynes.

In Chapter 7 Kirillova, Miloserdov, and Echavarren review copper, silver, and gold catalysts utilized for alkyne hydroarylation. The content is organized by the reactivity, selectivity, and mechanistic aspects of intra- and intermolecular reactions with emphasis on electron-rich heterarenes and unactivated arenes. Alkyne hydroarylation using arylboron, aryl halide, and related congeners are reviewed by Yamamoto in Chapter 8. The organization of the sections is first by substrate, and then by catalyst identity. Alkyne hydroarylations as both singular reactions and as part of sequential processes are discussed. Both Chapters 7 and 8 feature the inclusion of synthetic applications to biologically active compounds.

In the final chapter, Chapter 9, Widenhoefer reviews the hydroarylation of allenes. In comparison with the hydroarylations of alkenes and alkynes, which often involve directed C—H bond activation, allenes undergo a π-activation followed by an arene outer-sphere addition. Organized by nucleophile with emphasis on electron-rich arenes and heteroaromatics, both intramolecular and intermolecular reactions are presented with discussion on the mechanistic details related to the modes of ring closure.

Whether read in its entirety or as a specific chapter, the reader is provided with the historical development of catalytic olefin and alkyne hydroarylation chemistry including scope, mechanistic details, and areas for future development. Examples of industrial relevance and synthetic targets are included where appropriate. It is our hope that the readers find the information useful for their endeavors in the laboratory.

References

1 Cacchi, S. (1990) The palladium-catalyzed hydroarylation and hydrovinylation of carbon-carbon multiple bonds: new perspectives in organic synthesis. *Pure Appl. Chem.*, **62**, 713–722.

2 Kakiuchi, F. and Murai, S. (2002) Catalytic C-H/olefin coupling. *Acc. Chem. Res.*, **35**, 826–834.

3 Bandini, M. and Umani-Ronchi, A. (eds) (2009) *Catalytic Asymmetric Friedel-Crafts Alkylations*, Wiley-VCH Verlag GmbH & Co. KGaA, Weinheim, Germany.

1

Functionalization of Heteroaromatic Substrates using Groups 9 and 10 Catalysts

Pedro Villuendas, Sara Ruiz and Esteban P. Urriolabeitia

CSIC-Universidad de Zaragoza, Instituto de Síntesis Química y Catálisis Homogénea (ISQCH), Department of Activation of Bonds by Metallic Complexes, Pedro Cerbuna 12, E-50009, Zaragoza, Spain

1.1 Introduction

A hydroarylation reaction is the formal addition of aromatic or heteroaromatic C—H bonds across an olefin C=C or an alkyne C≡C bond, as represented in Figure 1.1b,d, respectively. This C—C bond forming reaction, catalyzed by transition metals, is one of the most popular synthetic tools in metal-mediated organic synthesis to introduce alkyl or alkenyl groups at given positions of aromatic or heteroaromatic compounds. It combines a perfect atom economy, the use of simple, non-prefunctionalized reagents, and an environmentally benign design. From the point of view of the synthesis shown in Figure 1.1, it is evident that hydroarylation of olefins is an alternative route to the Friedel–Crafts alkylation (Figure 1.1a), while hydroarylation of alkynes can be considered complementary to the alkenylation of (hetero)aromatic rings (i.e., Heck and Fujiwara–Moritani reactions, Figure 1.1c). A quick comparison shows that Friedel–Crafts alkylation needs halogenated precursors, strongly acidic reagents, usually high temperatures and long reaction times, shows moderate to poor selectivity, and generates stoichiometric amounts of waste products, while Heck (or Suzuki, Sonogashira, and other couplings) also needs halogenated substrates and produces large amounts of residue. It is clear that hydroarylation provides additional simple and advantageous pathways to landmark C—C bond forming reactions.

The processes shown in Figure 1.1 are general examples of intermolecular couplings. The corresponding intramolecular versions, where the heteroaromatic ring and the olefin or the alkyne are linked by a tether, are also well known. Both processes, intra- and intermolecular, involving alkenes and alkynes, have been used as main synthetic tools for the synthesis and functionalization of a myriad of heterocycles, whose industrial and academic importance resides in the fact that they are basic scaffolds of products with biological and pharmacological activity, new optical materials, or important synthetic precursors and intermediates [1–3]. Due to the importance and the widespread use of these reactions, several reviews covering this area have been published along the years [4–28].

Catalytic Hydroarylation of Carbon-Carbon Multiple Bonds, First Edition.
Edited by Lutz Ackermann, T. Brent Gunnoe and Laurel Goj Habgood.

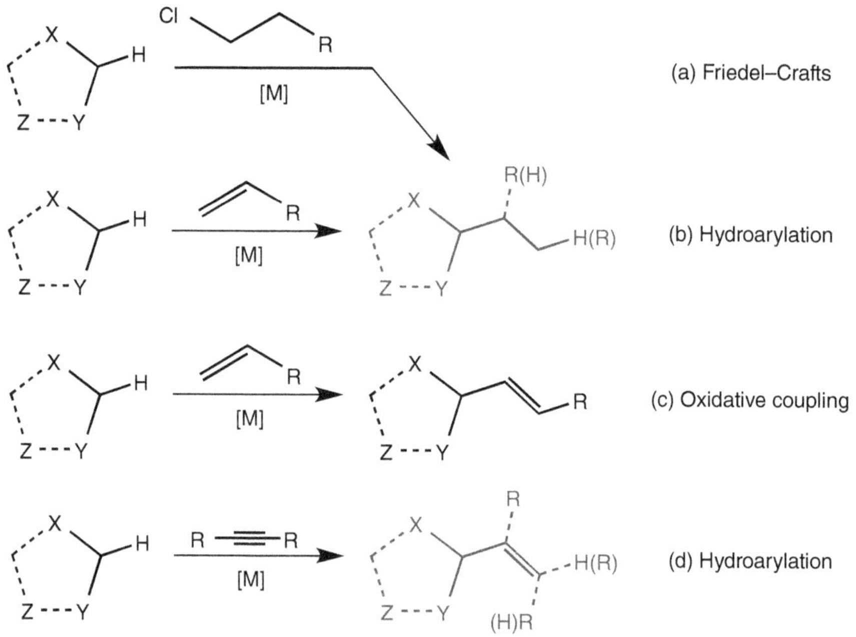

Figure 1.1 General intermolecular hydroarylation of C—C multiple bonds with (hetero)aromatic substrates and comparison to Friedel-Crafts alkylation.

An additional important aspect of the hydroarylation reaction is the selectivity of the reaction, which is closely related to the mechanism through which it takes place. Figure 1.2 exemplifies the most representative cases found for heteroarene–alkyne coupling, and a very similar mechanism scheme can be drawn for reactions involving olefins.

The reaction can take place either through alkyne activation or heteroarene activation. In the former case, vinylidene or π-complexes are formed as intermediates, and subsequent reaction with electron-rich arenes results in the formation of the vinylated derivatives, usually as a mixture of cis and trans stereoisomers. The reaction can also occur through metalation of the arene through C—H bond activation, either by oxidative addition or by concerted-metallation deprotonation. The resulting intermediates undergo migratory insertion of the alkyne into the M—C bond or the M—H bond, respectively. Protodemetalation or reductive elimination by C—C coupling afford selectively the *cis*-adducts.

The potential of this reaction was very clear from the first examples of hydroarylation of alkenes and alkynes, which were reported during 1978–1980 by Hong and Yamazaki [29–34]. In these works, the reaction of benzene (and other arenes) as solvents with $Ph_2C{=}C{=}O$ [29], ethylene [30, 34], or alkynes [31] under Rh catalysis and CO atmosphere afforded $Ph_2CHC(O)Ar$ ($Ar = C_6H_5$ in 68% yield based on ketene; other aryl groups in 53–57% yield), styrene (yields up to 9170% based on Rh atom), and stilbenes (around 45% yield based on alkyne), respectively, among other byproducts [33]. The processes are shown in Figure 1.3a–c. While the formation of the substituted acetophenone and

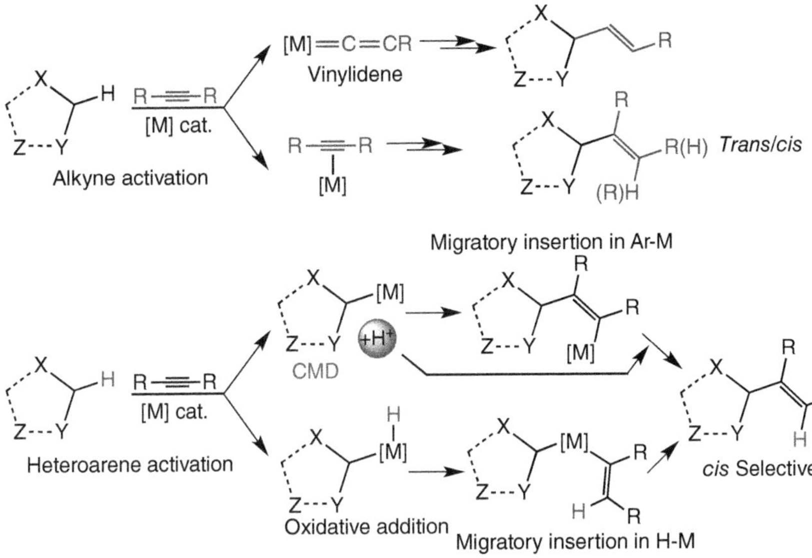

Figure 1.2 Hydroarylation of alkynes: mechanisms and selectivity of the resulting compounds.

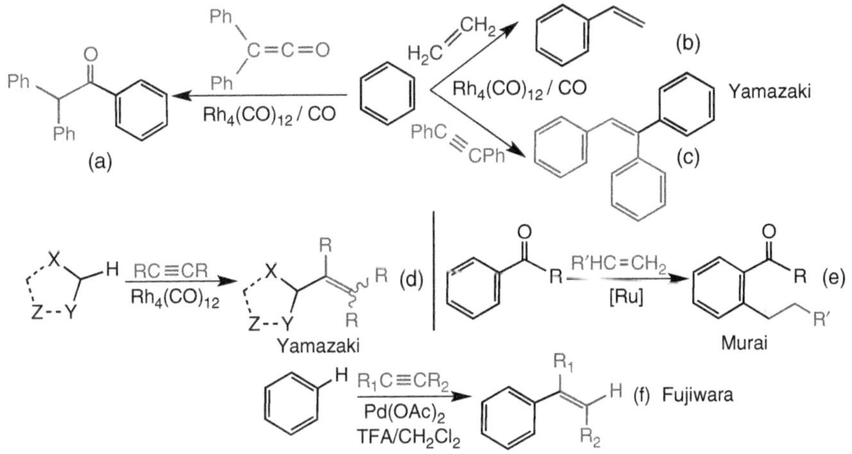

Figure 1.3 Examples of seminal hydroarylation reactions.

stilbenes are true hydroarylations, the production of styrene is formally a Fujiwara–Moritani oxidative coupling. The coupling with alkynes was extended to heteroarenes such as furan (80% yield; 41–86% for substituted furanes), thiophene (48%) and *N*-methylpyrrole (31%), as shown in Figure 1.3d [32]. In 1993, Murai and coworkers described the regioselective *ortho*-alkylation of acetophenones with different alkenes (Figure 1.3e), catalyzed by Ru-complexes, a milestone reaction considered a paradigm of atom- and step-economy [35]. This work was also one of the former examples of chelation-assisted functionalization, and paved the way for future research in the area. It is also worth mentioning the work of Fujiwara and coworkers, who in 2000 reported a very efficient addition of simple arenes to alkynes catalyzed by Pd(II), Pt(II), or other electrophilic metals. The reaction takes place in a mixture of CF_3CO_2H (which increases the electrophilicity of the catalyst) and other solvents, and affords unusual *trans*-hydroarylated compounds under kinetic control (Figure 1.3f) [36].

This chapter aims to cover the most relevant literature on hydroarylation reactions, catalyzed by transition metals from groups 9 and 10, involving heteroaromatic substrates. In particular, only hydroarylation reactions involving challenging cleavage of heteroaryl C—H bonds will be considered, excluding most of those dealing with aryl halides and/or arylboronic acids. The chapter has been organized taking into account the nature of the heterocycle to be functionalized, since this type of classification gives to the reader an overview of how many different structural motifs are accesible starting from each individual ring; that is, the versatility of each substrate. Therefore, furans, thiophenes, indoles, pyrroles, pyridines, and other miscellaneous heterocycles will be described separately.

1.2 Thiophenes, furans, and Related Heterocycles

Hong *et al.* [32] reported in 1980 that under an atmosphere of CO the catalyst $Rh_4(CO)_{12}$ is able to achieve the activation of aromatic C—H bonds in five-membered heteroarenes and, in this way, promote the hydroarylation of alkynes. Both unsubstituted and 2-substituted furans react at the α-position (Figure 1.4a). When the reaction is performed with an unsymmetrical alkyne (1-phenylpropyne), the process is regioselective, obtaining the isomer with the phenyl group attached to the same C of the alkene as the furyl ring. The CO pressure must be higher than 10 kg/cm^2 in order to avoid cyclotrimerization of the alkynes, and furan is added in great excess (acting as the solvent). If both α-positions are occupied by substituents (2,5-dimethylfuran), then the functionalization takes place at the ß-position, although the yield (40%) is lower than that of mono-α-substituted furans. All these reactions yield vinyl derivatives as a mixture of Z and E isomers, enriched in the Z isomer in all cases. The authors propose that the E isomer is first formed, but after some time the Z isomer becomes predominant in the mixture since it is thermodynamically more stable. The same catalytic system was applied to thiophene to obtain the corresponding 2-vinylated heterocycle (Figure 1.4b). Competitive experiments were carried out in order to determine the relative reactivity of various heterocycles. Furan was found to be more reactive than thiophene, which in turn is more reactive than

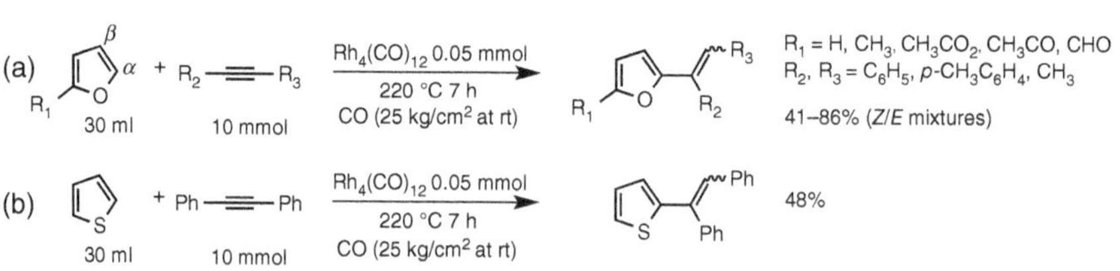

Figure 1.4 Hydroarylation of alkynes under Rh catalysis (yields are based on alkyne).

N-methylpyrrole. All of these heteroaromatic substrates are more reactive than benzene toward acetylenes [32].

As previously mentioned, the group of Fujiwara was pioneer in developing hydroarylation reactions using alkynes [36]. The year 2000 saw the publication of several seminal papers dealing with this topic, which describe the efficient hydroarylation of alkynes and alkenes with electron-rich aromatic substrates using catalytic amounts of Pd(II) or Pt(II) compounds, in solvent mixtures containing trifluoroacetic acid (HTFA), and both inter- and intramolecular transformations were reported [36, 37]. These reactions are proposed to proceed through alkyne-activation pathways by coordination to cationic and electrophilic complexes of the metals. In the same year the Fujiwara group dedicated another work to heterocycles, making use of the same catalytic process [38]. From a number of reports of detailed exploration of the reactivity of pyrrole and indole derivatives, a single example of the functionalization of a furan derivative is presented (Figure 1.5a): 2-methylfuran adds to an alkynoate at room temperature in the presence of catalytic $Pd(OAc)_2$ (5%) in acetic acid, affording exclusively the *Z*-heteroarylalkene. The addition of heteroaromatic compunds to alkynoates likely follows the mechanism outlined in Figure 1.5b. The formation of intermediate **A** proceeds through electrophilic metalation of the aromatic C—H bond with the cationic Pd(II) species $[PdOAc]^+$, and, after that, the coordination of the alkyne affords **B**. The *trans* insertion of the C—C triple bond to the σ-aryl-Pd

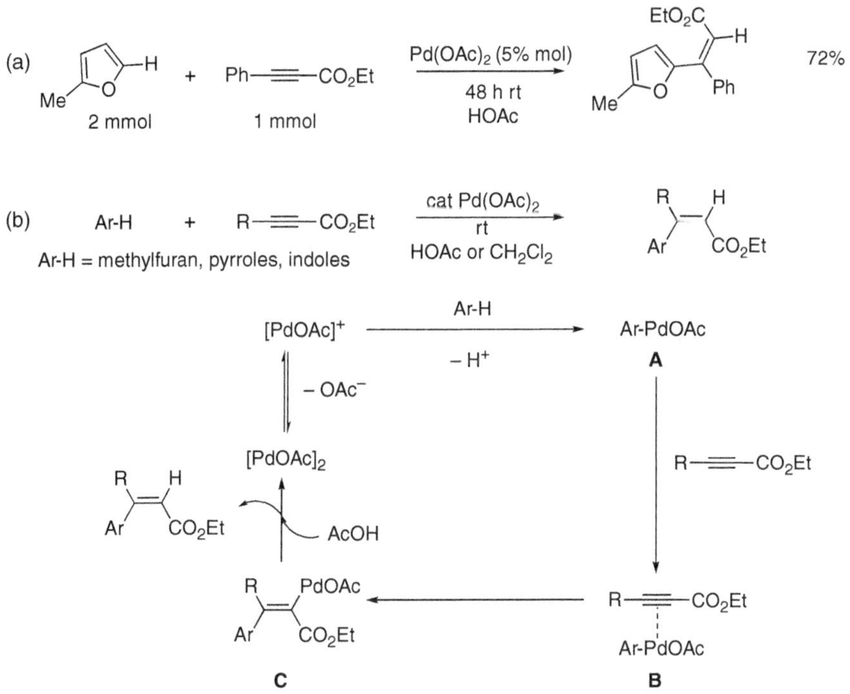

Figure 1.5 Hydroarylation of ethylphenylpropiolate with 2-methylfuran (a) and possible mechanism (b).

bond results in the vinyl complex **C**, which upon protonation by AcOH releases the *cis*-heteroarylalkenes.

Several years later Kitamura revisited the reactions between heteroarenes and propiolates, this time using a Pt(II) catalyst instead of Pd(II), following the assumption that Pt(II) was more active than Pd(II). Both thiophenes [39] and furans [40] react with ethyl propiolates or propiolic acids using K_2PtCl_4 and AgOTf under strong acidic conditions: the former in HTFA, and the later in acetic acid (AcOH). The reactions take place preferentially at C2, but when this position is already substituted the addition of the alkene occurs at C3. In contrast to the reaction catalyzed by $Pd(OAc)_2$ [38], in the reaction of 2-methylfuran with ethyl phenylpropiolate two molecules of furan are added to the triple bound by means of two consecutive hydroarylation transformations (Figure 1.6a). However, in the case of 2,5-dimethylfuran, the mono-adduct is obtained as an *E*/*Z* mixture (Figure 1.6b). When adding terminal ethyl propiolate to 2,5-dimethylfuran, only one hydroarylation takes place yielding the *Z* isomer. The second addition of the heterocycle to the double bond requires higher temperatures in order to take place (50 °C vs 30 °C) (Figure 1.6c).

In the reactions with thiophenes, the alkynes undergo a double hydroarylation (Figure 1.7a–c). The hydroarylation possibly follows an electrophilic aromatic substitution mechanism, as illustrated in Figure 1.7d. The process starts with the interaction between the alkyne and a cationic platinum species **A** (formed from the platinum precatalyst and AgOTf by ion exchange), which activates the alkyne (**B**). Then, the heteroarene attacks the triple C—C bond, forming a Wheland intermediate **C**. Proton release affords vinyl platinum complex **D**, which after protonation by TFA or HOAc produces the heteroarylacrylate **E**. A second hydroarylation can then take place by subsequent activation of the alkene fragment in **E**

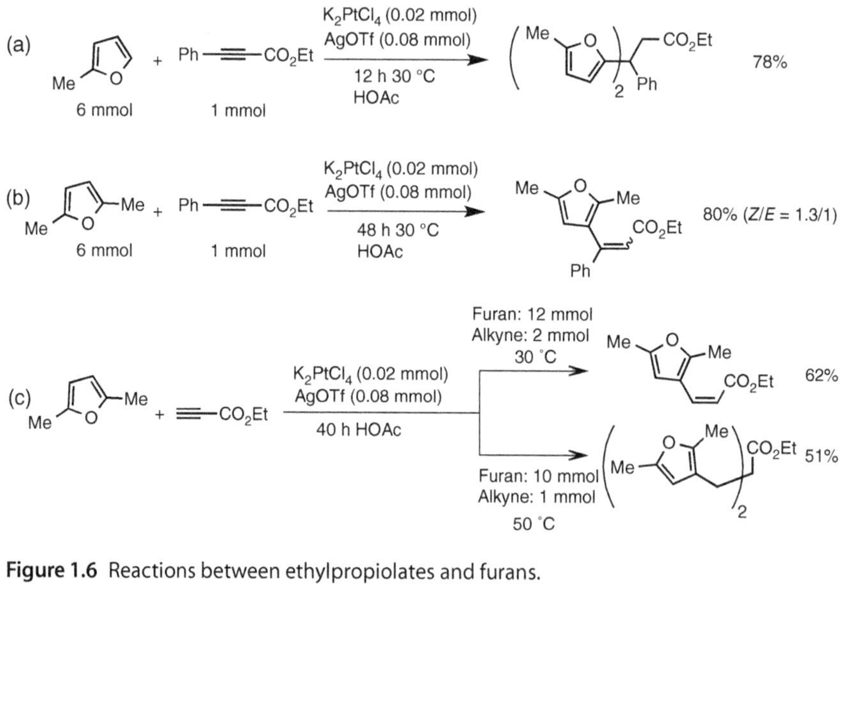

Figure 1.6 Reactions between ethylpropiolates and furans.

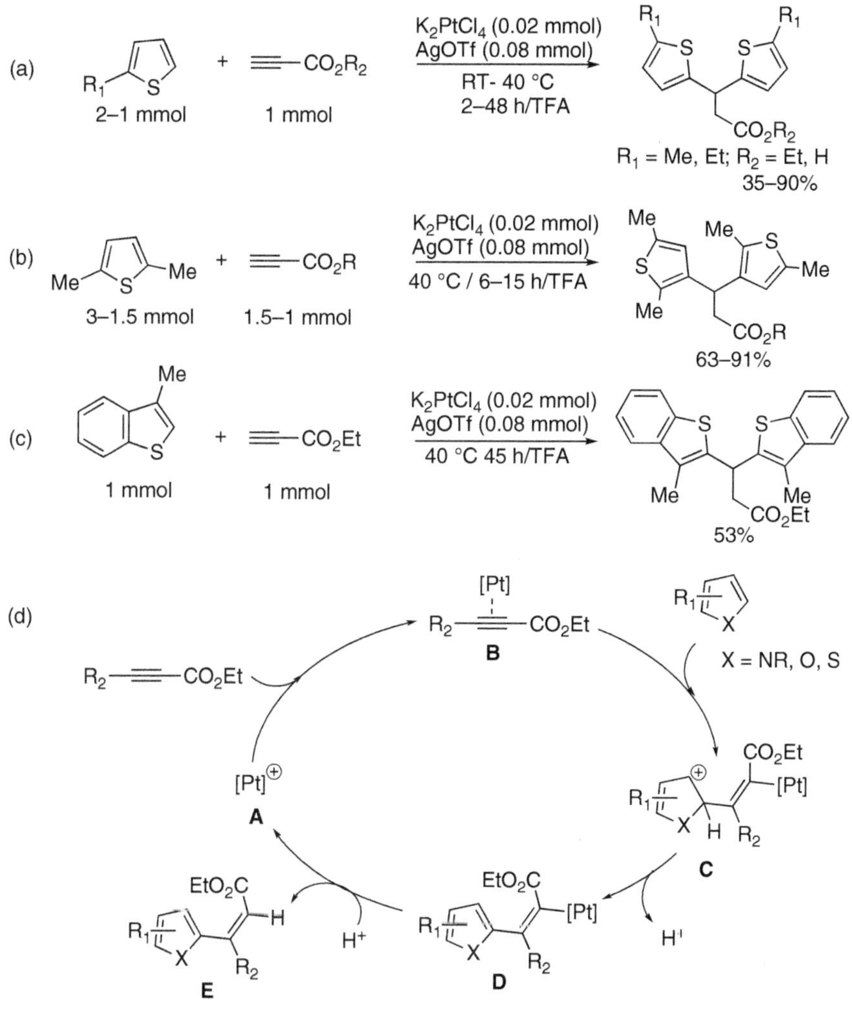

Figure 1.7 Reactions between ethylpropiolates and thiophenes, and mechanistic proposal.

by interaction with platinum cationic species, followed by heteroarene attack and protonation, yielding the final double-hydroarylation product.

As part of a study on hydroarylation reactions of ethyl phenylpropiolate with heterocycles, catalyzed by the chelating dicarbene Pd(II) complex shown in Figure 1.8 (**I** and $AgBF_4$ in HOAc), Biffis and coworkers [41] applied their optimized catalytic system to the same substrates explored by Fujiwara in order to compare the effectiveness of the two different catalytic systems. 2-Methylfuran reacted with the alkyne yielding a mixture of the Z-vinylated derivative and the diaddition product. Conversion of thiophene derivatives using the dicarbene Pd catalyst precursor **I** was not successful, showing that this system is clearly less effective than that reported by Fujiwara.

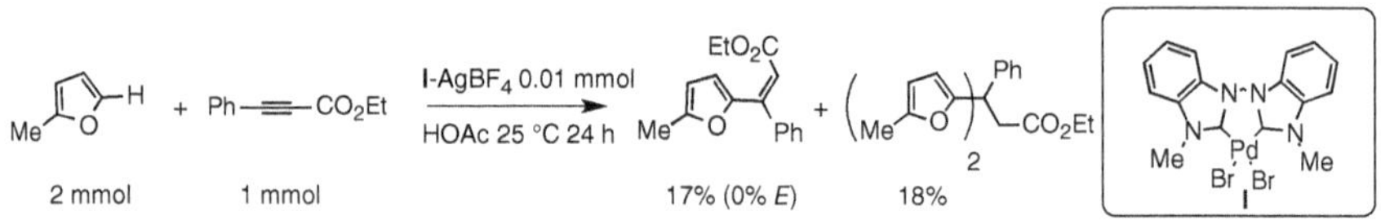

Figure 1.8 Hydroarylation of ethyl phenylpropiolate using a dicarbene Pd(II) complex.

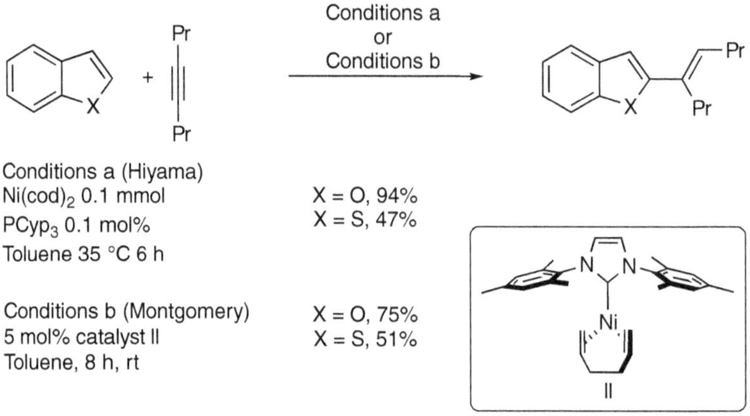

Figure 1.9 Ni-catalyzed hydroarylation of alkynes.

The group of Hiyama was interested in the development of nickel catalysts for hydroarylation reactions of unactivated alkynes under mild catalysis. In 2006, they reported the hydroarylation of 4-octyne with benzofuran and benzothiophene, which exclusively took place at the C2 position using $Ni(cod)_2$ and tricyclopentylphosphine ($PCyp_3$) as a ligand (Figure 1.9, conditions a) [42]. Later, in 2015, the group of Montgomery offered an alternative procedure using Ni(0)-NHC complexes as pre-catalysts (Figure 1.9, conditions b) [43].

The group of Yoshikai explored the use of a Co(II) catalyst in conjunction with a Grignard reagent, triarylphosphine ligand, and pyridine in order to achieve the hydroarylation of internal alkynes with heteroaromatic imines (Figure 1.10) [44]. Their strategy consisted in making use of an imine as a directing group for the C—H functionalization, improving its effectiveness by exploiting the chelation-assistance effect, and also allowing for the consecutive transformation of the initial product of hydroarylation. The alkenylation of a benzofuran derivative with 4-octyne afforded exclusively the *E* isomer, while the isomerization of

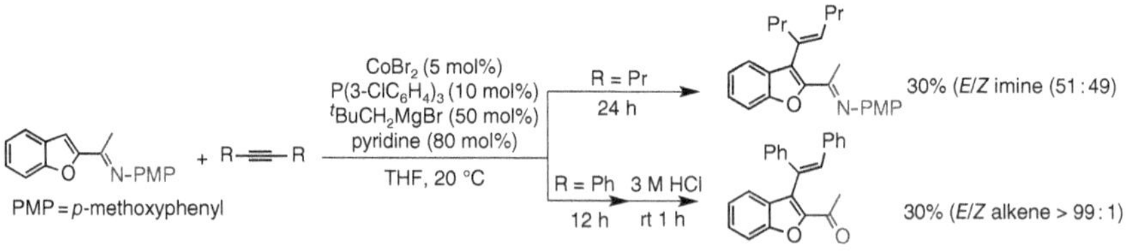

Figure 1.10 Imine as an effective directing group for the functionalization of benzofuran.

the imine moiety (*E*/*Z* 51 : 49) took place. The coupling with diphenylacetylene afforded the corresponding ketone by imine hydrolysis under acidic conditions, predominantly as the *E* isomer.

Thiophenes have also been subjected to hydroarylation reactions of unactivated internal alkynes, such as 3-hexyne and diphenylacetylene. Inoue reported in 2005 the use of a dinuclear palladium complex that enabled high stereo- and regioselectivities, producing *E*-2-alkenylthiophenes [45]. The catalyst of choice is $[Pd_2Me_2(\mu\text{-}OH)(\mu\text{-dpfam})]$ (dpfam = *N*,*N′*-bis[2-(diphenylphosphino)phenyl] formamidinate), shown in Figure 1.11. The procedure tolerates the presence of ketone and ester R_1 groups, but not of aldehyde.

Fujiwara suggested that intramolecular reactions could be more efficient than the corresponding intermolecular processes [36, 46]. The intramolecular hydroarylation of the triple bond of dibenzofurane alkynoates yielded selectively the kinetically favored six-membered rings by *endo*-cyclization (instead of the five-membered ones by *exo*-cyclization), as a mixture of regioisomers (Figure 1.12).

Sames and coworkers [47, 48] discovered that $PtCl_4$ was a better catalyst for intramolecular hydroarylation reactions of arene-alkyne (arene-yne) substrates than those previously described by Fujiwara using Pd(II) and Pt(II) [36, 46]. Alkynoate esters formed a fused furo-dihydropyran (the *exo*-cyclization product) in good yield (57% *Z*/*E* = 1 : 1) with $PtCl_4$ (Figure 1.13). $PtCl_2$ was ineffective and $PtCl_4$ was superior to $Pd(OAc_2)$ in HTFA/AcOH.

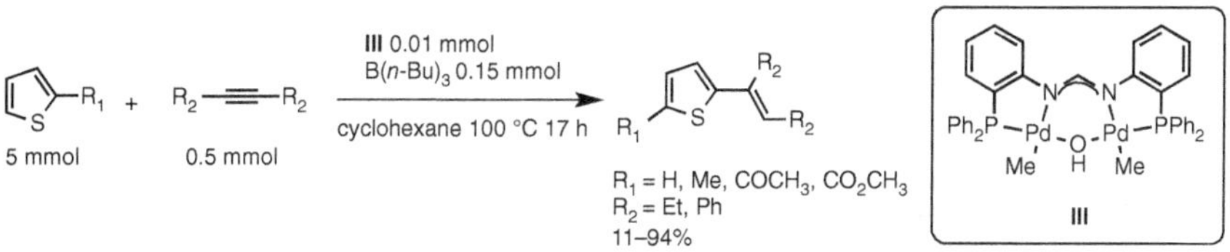

Figure 1.11 Hydroarylation of unactivated alkynes catalyzed by a Pd(II) complex.

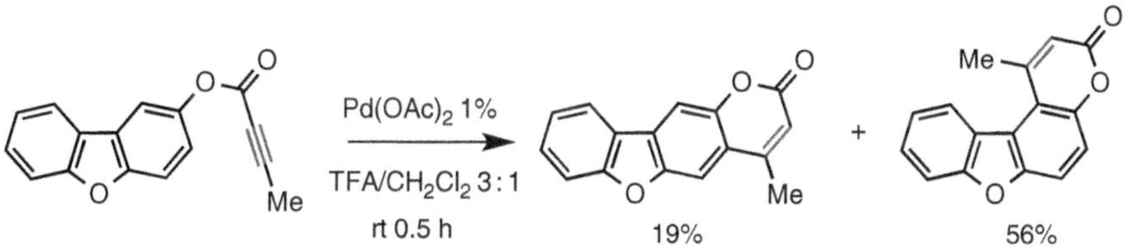

Figure 1.12 Intramolecular hydroarylation of alkynoates using $Pd(OAc)_2$ as the catalyst precursor.

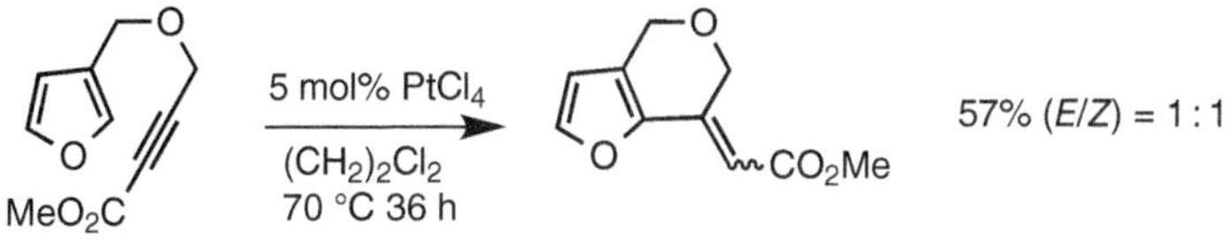

Figure 1.13 Intramolecular hydroarylation of a furan-yne substrate.

Kitamura and Otsubo applied this strategy to the functionalization of benzofurans in order to obtain angelicins, valuable materials for photobiological applications [49]. They reported the intramolecular hydroarylation of 4-benzofuranyl alkynoates containing different substituents R_1 in the presence of $Pd(OAc)_2$ in $HTFA/CH_2Cl_2$ (Figure 1.14). The angelicin derivatives were obtained in more than 70% yield, and HTFA was essential for the reaction to occur probably because it promotes the formation of highly reactive $[Pd(TFA)]^+$ species.

Unactivated terminal alkenes can also take part in intramolecular hydroarylations. In 2002, Fürstner and Mamane explored the formation of polycyclic structures by intramolecular hydroarylation catalyzed by $PtCl_2$ [50]. They reported the formation of naphthothiophene by 6-*endo*-dig cyclization of 2-(2-ethynylphenyl)thiophene (Figure 1.15a). In 2004, formation of the same structural core using other electrophilic metal salts as catalysts was explored [51]. Surprisingly, $GaCl_3$ and $InCl_3$ proved to be very effective and superior in their performance to $PtCl_2$ (Figure 1.15b). Lee's group reported the platinum-catalyzed synthesis of naphthalenes from 2-alkynyl cinnamates by 6-*endo* cyclization. Among them, 5-ethoxycarbonylbenzofuran was obtained from the hydroarylation of a terminal enyne (Figure 1.16) [52].

Furylalkynes are valuable starting materials that can sometimes exhibit divergent reactivity. The group of Echavarren has extensively studied this topic and the mechanisms involved in the formation of the different products. 5-(2-Furyl)-1-alkynes, containing either ether or malonate functionality, react in acetone and with $PtCl_2$ as a catalyst to afford mixtures of phenols (Figure 1.17a)

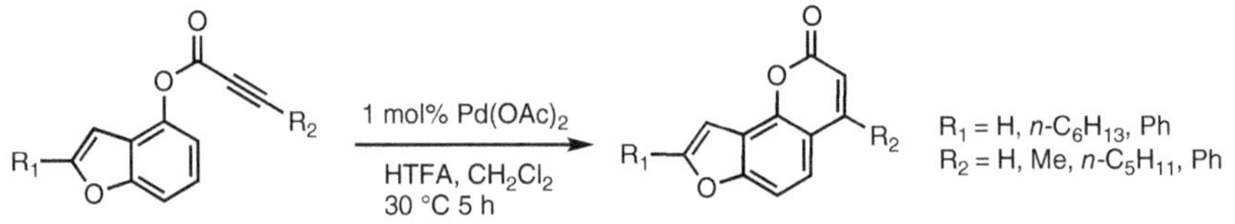

Figure 1.14 Synthesis of angelicin derivatives by intramolecular hydroarylation of benzofuran.

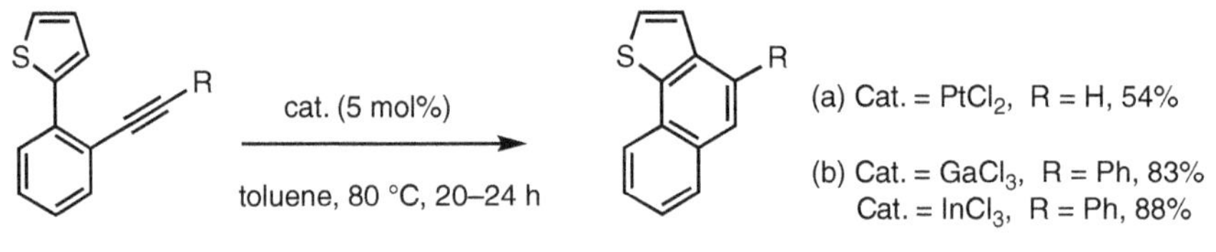

Figure 1.15 Intramolecular hydroarylation of 2-(2-ethynylphenyl)thiophene.

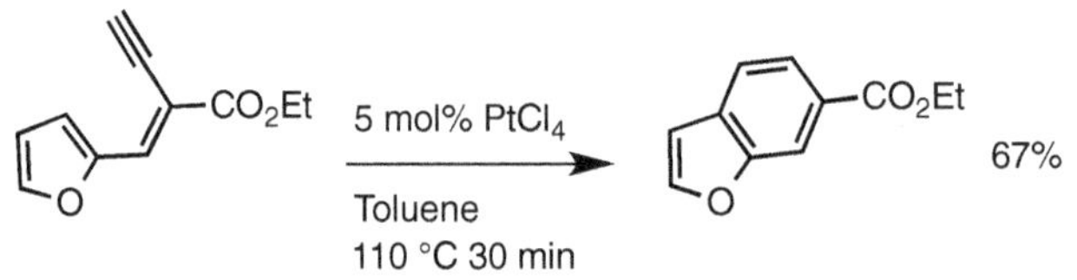

Figure 1.16 Intramolecular hydroarylation of terminal enynes catalyzed by $PtCl_4$.

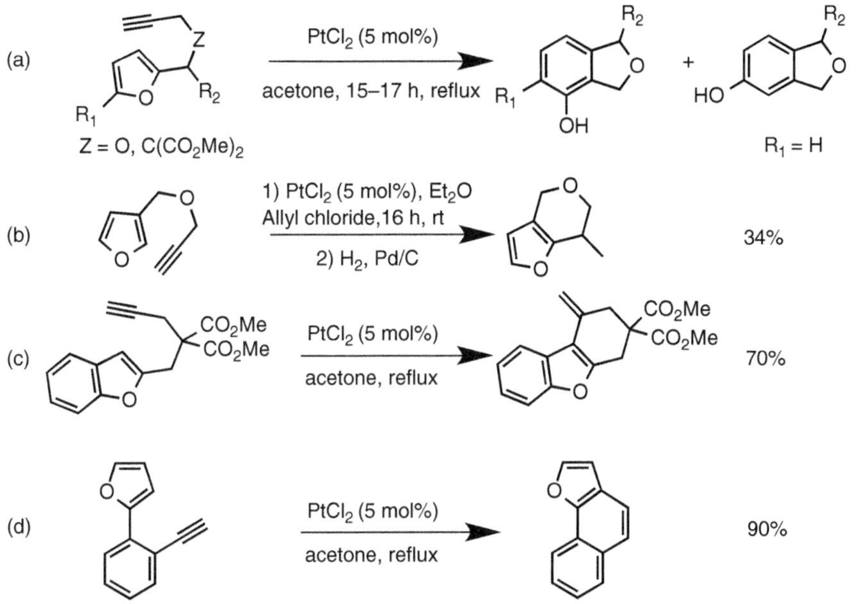

Figure 1.17 Intramolecular reactions of furylalkynes catalyzed by $PtCl_2$.

[53, 54]. These reactions involve intramolecular cyclization and further complex structural rearrangements, whereas 3-furylmethyl propargyl ether undergoes cyclization (Figure 1.17b). The product is hydrogenated to avoid polymerization. The role of the additive allyl chloride is not clear. A benzofuran malonate derivative also undergoes *exo*-cyclization (Figure 1.17c). Finally, the intramolecular cyclization of ethynylphenylfuran to yield naphthofuran was also reported (Figure 1.17d).

Gunnoe and coworkers reported in 2008 the hydroarylation of simple ethylene using a bipyridine Pt(II) complex (Figure 1.18), enabling the regioselective formation of 2-ethylfuran with 76 turnovers after 16 h [55]. Traces of 2,5-dialkylated products were also observed.

The hydroarylation of styrene and its derivatives has been reported. The group of Hiyama [56] made use, once again, of a Ni(0) catalyst precursor in combination with the NHC-ligand IMes (IMes = 1,3-bis(2,4,6-trimethylphenyl)imidazol-2-ylidene) to develop the hydroarylation of styrene with benzofuran (Figure 1.19a). Sigman and coworker reported the hydroheteroarylation of vinyl phenols catalyzed by Pd(0) with phosphines as ancillary ligands, and base and butyl

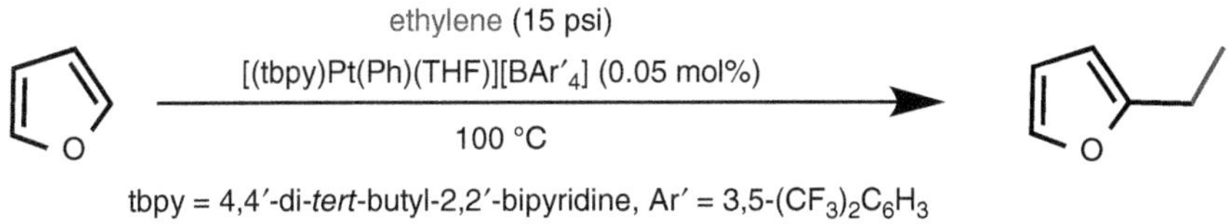

Figure 1.18 Hydroarylation of ethylene catalyzed by a bipyridine Pt(II) complex.

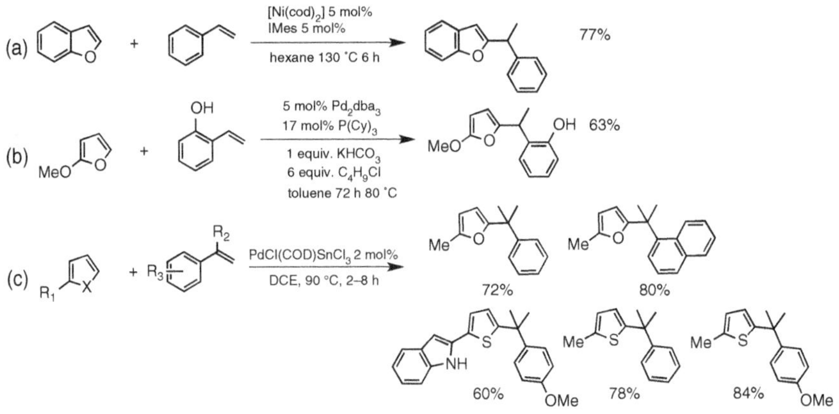

Figure 1.19 Examples of hydroarylation of styrene and styrene derivatives.

chloride as the H source (Figure 1.19b) [57]. The hydroarylation of α-methyl substituted aryl alkenes is challenging because they are prone to dimerize or polymerize in the presence of Lewis acids [58]. But, bimetallic catalysis successfully allowed the functionalization of heteroarenes with α-methyl substituted aryl alkenes using the catalyst $PdCl(SnCl_3)(COD)$ (COD = 1,5-cyclooctadiene), which is air- and moisture-stable (Figure 1.19c). This high catalytic performance is remarkable because, individually, Pd(II) and Sn(II) species were ineffective. However, no explanation was provided for this feature. In the case of 2-(thiophen-2-yl)-1*H*-indole, only alkylation on the thiophene ring was observed. All reactions shown in Figure 1.19 were regioselective: only Markovnikov products were obtained with selectivity at the C2 position of furan, benzofuran, and thiophene.

Hartwig and coworker developed an asymmetric Ir-catalyzed intermolecular hydroarylation of bicycloalkenes [59]. Furans and thiophenes with different substituents reacted with norbornene in the presence of $[IrCl(coe)_2]_2$ (coe = cyclooctene) and a chiral bisphosphine ligand in good yields (Figure 1.20). Reactions of thiophenes proceeded with high enantioselectivity, while the enantiomeric excess only reaches 78% in the case of furans. Under the same reaction conditions, couplings with norbornadiene afforded mixtures of products. Apart from the desired alkylated product, oxidative homocouplings as well as the reduction of the double bond of the incorporated fragment were detected. By increasing the concentration of the alkene (up to 2.5 equiv. instead of 1.2), the amount of product formed by homocoupling of the heteroarene was reduced, and thus the yields were increased (doubled for the benzothiophene), although no further explanations were provided about this fact.

Inspired by the hydroarylation of styrenes and benzofurans reported by Hiyama and coworkers [56], the group of Sames developed the catalytic intramolecular alkylation of benzofurans, aiming to obtain structural analogs to *Iboga* alkaloids, of important pharmaceutical activity [60]. Two examples, depicted in Figure 1.21, were successful, with isolated yields of 74% and 38%. The lower yield for the latter example may result from greater steric hindrance of the

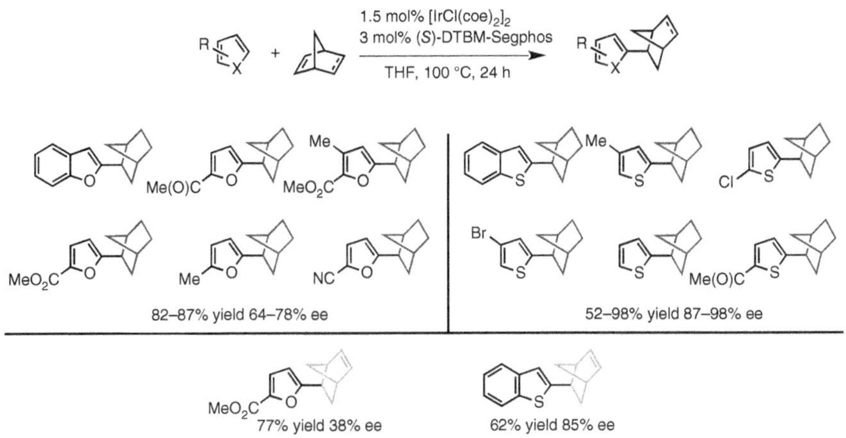

Figure 1.20 Enantioselective hydroarylation of bicycloalkenes with heteroarenes using an Ir catalyst.

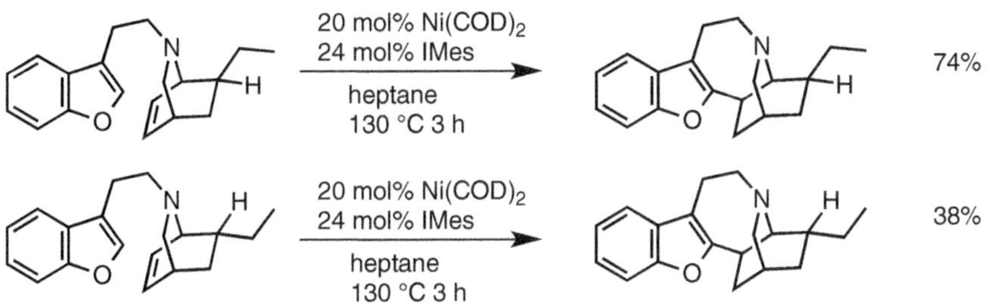

Figure 1.21 Intramolecular hydroarylations affording *Iboga* alkaloid benzofuran analogs using a Ni catalyst.

ethyl group toward the amine in the *exo*-epimer, thus preventing the amine from obstructing the catalytic cycle. Moreover, the ethyl group in the *exo*-epimer could favor the most suitable configuration of the N atom of the amine for the cyclization. The corresponding intermolecular alkylation does not take place under these conditions, which proves the suitability of this method.

The group of Hartwig explored the hydroarylation of unactivated terminal alkenes with heteroarenes, to afford the linear alkylation products, using the Ni-NHC catalytic system shown in Figure 1.22 (IPrMe = 4,5-dimethyl-*N*,*N*′-bis (2,6-diisopropylphenyl)imidazol-2-ylidene; IPrOMe = 4,5-dimethoxy-*N*,*N*′-bis (2,6-diisopropylphenyl)imidazol-2-ylidene) [61]. The production of linear products with non-functionalized olefins is uncommon. With benzofurans, the reaction was regioselective toward the C2 position. Furan afforded a 1 : 1 mixture of C2-mono- and C2,C5-dialkylation products. On the other hand, C2-substituted rings afforded selectively the C5-alkylated derivatives. The presence of potential directing groups (bearing a coordinating O atom) at C2 or C3 did not alter the regioselectivity. After fine-tuning the reaction time, temperature, and catalyst loading, substrates containing functional groups such as carbonyl groups, boronate esters, or siloxysilanes were efficiently functionalized.

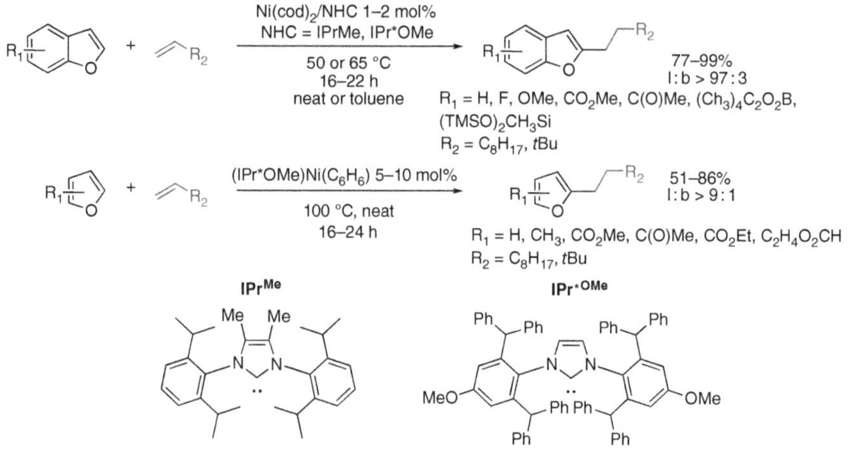

Figure 1.22 Hydroarylation of alkenes with furan and benzofuran derivatives.

When tolerated by the functional group, the addition of NaO^{t}Bu increases the yields because it prevents isomerization of the terminal alkene to the internal alkene.

The formation of branched alkylated products opens the door for the analysis of asymmetric hydroarylations. It was recently reported that a hydroxoiridium complex with a chiral diene ligand based on the tetrafluorobenzobarrelene (tfb) framework is able to catalyze the asymmetric hydroarylation of vinylethers with heteroarenes (Figure 1.23) [62]. Making use of a sulfonylamide as a directing group, branched alkylated furan and thiophene derivatives were obtained in good yield and excellent enantiomeric excess.

Chelation assistance has enabled the alkylation (using olefins) and alkenylation (using alkynes) of thiophenes at the C3 position. Several works in recent years make use of pyridine as a directing group, and thus describe the functionalization of 2-(thiophen-2-yl)pyridine (Figure 1.24). Yoshikai and coworkers reported the hydroarylation of 4-octyne with thiophene using a catalytic system comprising $CoBr_2$, $PMePh_2$, and the Grignard reagent MeMgCl, as shown in Figure 1.24a. The E/Z ratio was higher than 99 : 1 [63]. A better yield was achieved by the Chang group using the rollover cyclometalation strategy using $Rh(acac)_3$ in

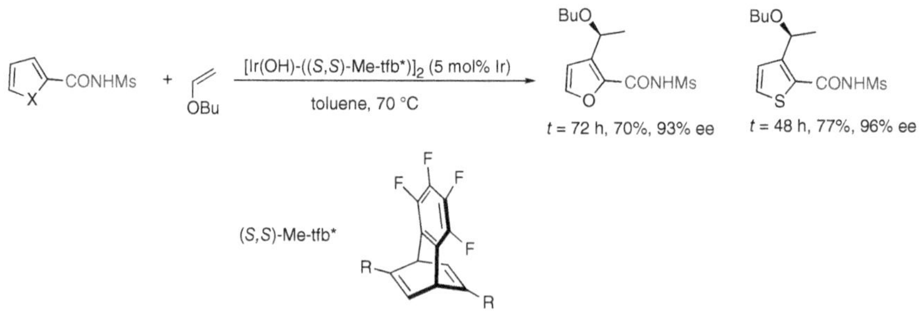

Figure 1.23 Asymmetric hydroarylation of vinyl ethers using an Ir catalyst precursor.

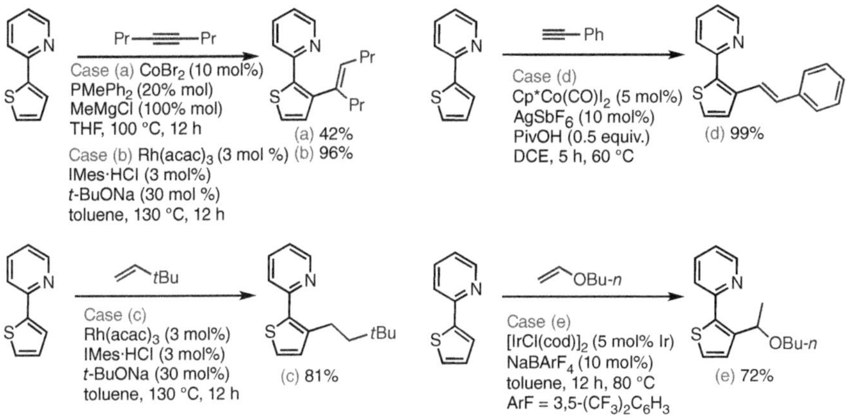

Figure 1.24 Hydroarylation reactions of alkenes and alkynes with 2-(thiophen-2-yl)pyridine.

combination with IMes·HCl ligand and *t*-BuONa (Figure 1.24b) [64]. The strong *trans*-effect exerted by the carbene ligand on the Rh center is proposed for the partial decoordination of the bidentate ligand, thus leading to cyclometalation and subsequent C—H activation selectively on the thiophene ring. The same system is effective for the reaction with alkenes (Figure 1.24c) [64]. 2-(Thiophen-2-yl)pyridine reacts almost quantitatively with phenylacetylene yielding exclusively the *E* isomer using a Cp*Co(III) catalyst under mild conditions (Figure 1.24d) [65]. Moreover, the Ir-catalyzed reaction with vinyl ethers takes place with high selectivity to the branched derivatives (Figure 1.24e) [66].

The group of Castarlenas explored the use of the Rh(III)-NHC complex $[Rh(\mu\text{-}Cl)(H)_2(IPr)]_2$ (IPr= 1,3-bis(2,6-diisopropylphenyl)imidazol-2-ylidene) as a catalyst for the hydroarylation of alkynes and alkenes with 2-(thiophen-2-yl) pyridine (Figure 1.25) [67]. This work included a detailed mechanistic study, involving the isolation and identification of intermediates as well as the study of their reactivity by NMR experiments and DFT calculations. The catalyst proved to be efficient for the hydroarylation of terminal alkenes, affording only linear products, and of internal alkynes, where only *E*-isomers were formed. However, terminal alkynes afforded homocoupling adducts [67].

The mechanistic study suggests that the $[Rh(\mu\text{-}Cl)(H)_2(IPr)]_2$ dimer is broken down upon coordination of thienylpyridine, and through loss of H_2 the active species **3** is generated (Figure 1.26a). The C—H bond on the thiophene ring is

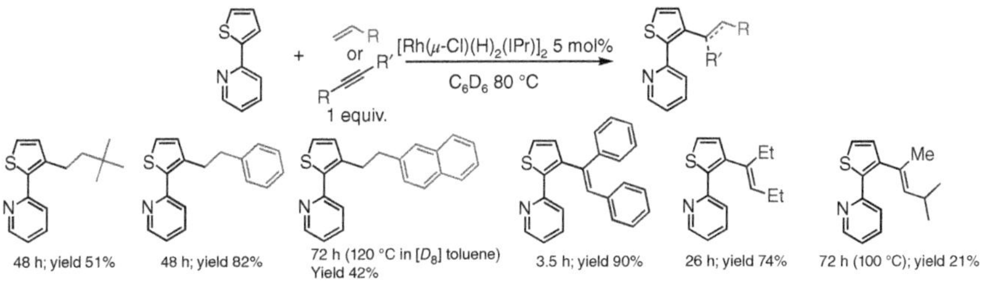

Figure 1.25 Rh-catalyzed functionalization of 2-(thiophen-2-yl)pyridine.

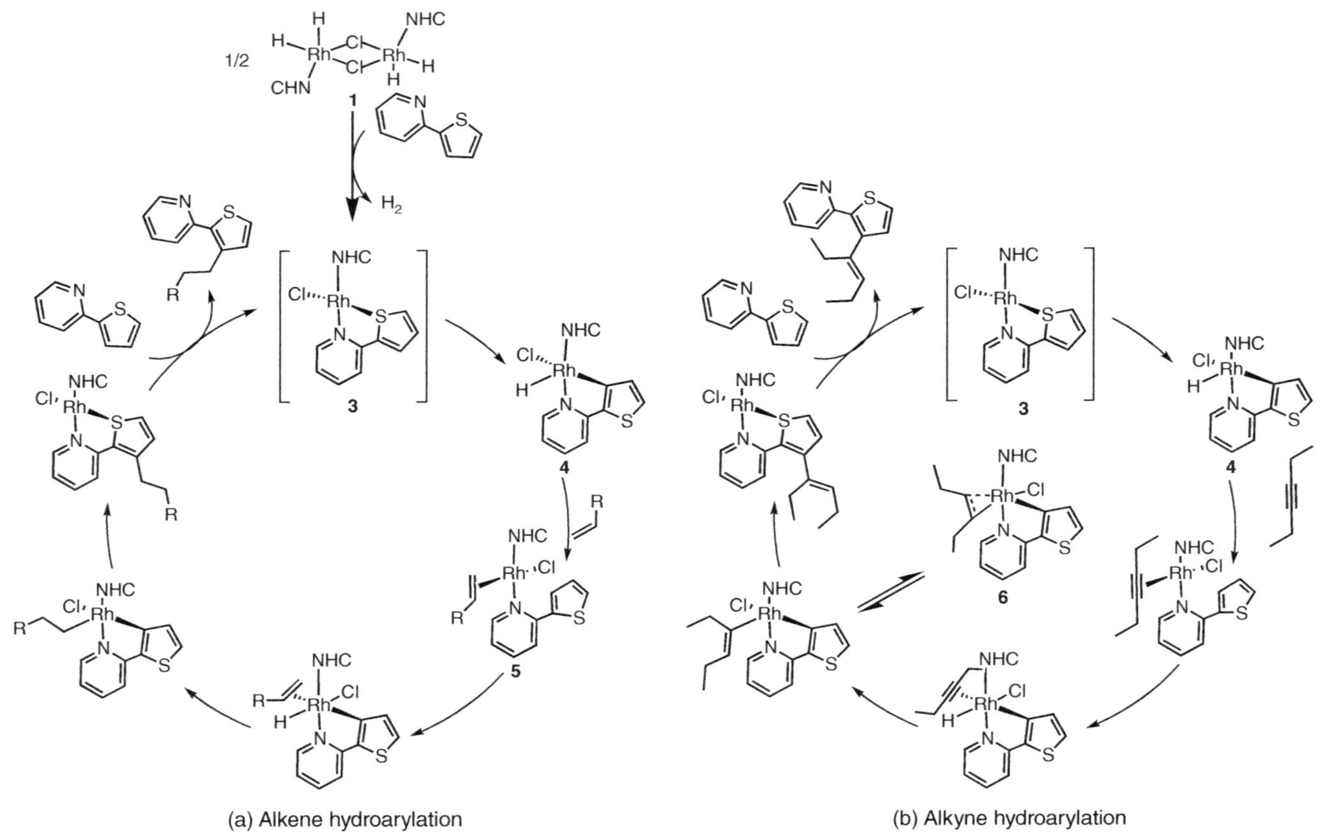

Figure 1.26 Mechanistic proposal for hydroarylation of alkynes and alkenes catalyzed by an NHC-Rh complex.

activated by oxidative addition, affording **4**. Complex **4** was prepared from the Rh dimer **1** by addition of two equivalents of 2-(2-thienyl)pyridine and was identified by NMR spectroscopy. The mode of C—H activation is also supported by DFT calculations, which show that σ-bond metathesis or concerted pathways have much higher activation barriers compared to the loss of H_2. The coordination of the olefin is followed by reductive elimination of the thiophene and the hydrido ligand to give intermediate [Rh(Cl)(IPr)(η^2-PhCH=CH_2)(*N*-PyTh)] (**5**), which was also characterized by NMR. The alkene coordinates at the free position *trans* to the H ligand, so the migratory insertion of the alkene into the Rh—H bond is discarded. Subsequently, the authors proposed a second cyclometalation of the thiophene ring, which affords a hydride *cis* to the alkene moiety and thus the insertion of the olefin into the C—H bond takes place. Eventually, reductive elimination liberates the functionalized product and the active species is regenerated by coordination of another molecule of 2-(thiophen-2-yl)pyridine. When 3-hexyne was added to complex **4**, intermediate **6** was obtained as the only reaction product. The structure of **6** was assigned by means of a detailed NMR study. The authors proposed an analogous catalytic cycle for the hydroarylation of alkynes to the one described for the reaction of alkenes. (Figure 1.26b). Complex **6** (a metallacyclopropene species) is in equilibrium with an alkenyl complex, which undergoes reductive elimination of the alkenyl ligand and the cyclometalated substrate to form the functionalized product.

1.3 Pyrroles, Indoles, Pyridines, and Imidazopyridines

There are a few examples of the hydroarylation using pyrroles with Pd and Pt complexes as catalysts. In most cases the pyrroles are protected, but cases of reactivity of free pyrroles are known. Early examples, published by Fujiwara and coworkers involve the use of $Pd(OAc)_2$ as catalyst in the hydroarylation of unprotected pyrroles and electron-poor internal alkynes, obtaining reactivity at position 2 and the *Z*-regioisomer in good yield (Figure 1.27a) [38]. Free pyrroles have also been functionalized in position 2 using a Pt catalyst by Kitamura and coworker [40], although a mixture of products functionalized in positions 2 and 3 can be obtained when electron-poor internal alkynes, such as ethyl phenylpropiolate, are used (Figure 1.27b). This contribution also reports a double pyrrole insertion into an alkyne to yield the bifunctionalized product (Figure 1.27c) [40]. A double insertion was also reported by Biffis and coworkers using a bis-NHC Pd catalyst (Figure 1.27d) [41].

There are two ways for the less active position 3 of a pyrrole to be selectively functionalized. The first, proposed by Tsukada, is by *N*-protection of the pyrrole with a bulky Boc group (Boc = *tert*-butyloxycarbonyl) using a Pd(II) catalyst (Figure 1.28a) [45]. This method allows the selective synthesis of modified pyrroles using electron-rich internal alkynes. The second is reported by the Fujiwara group and involves the protection of the positions 2 and 5 of free pyrroles in order to be functionalized at C3 by a more reactive ethyl phenylpropionate (Figure 1.28b) [68].

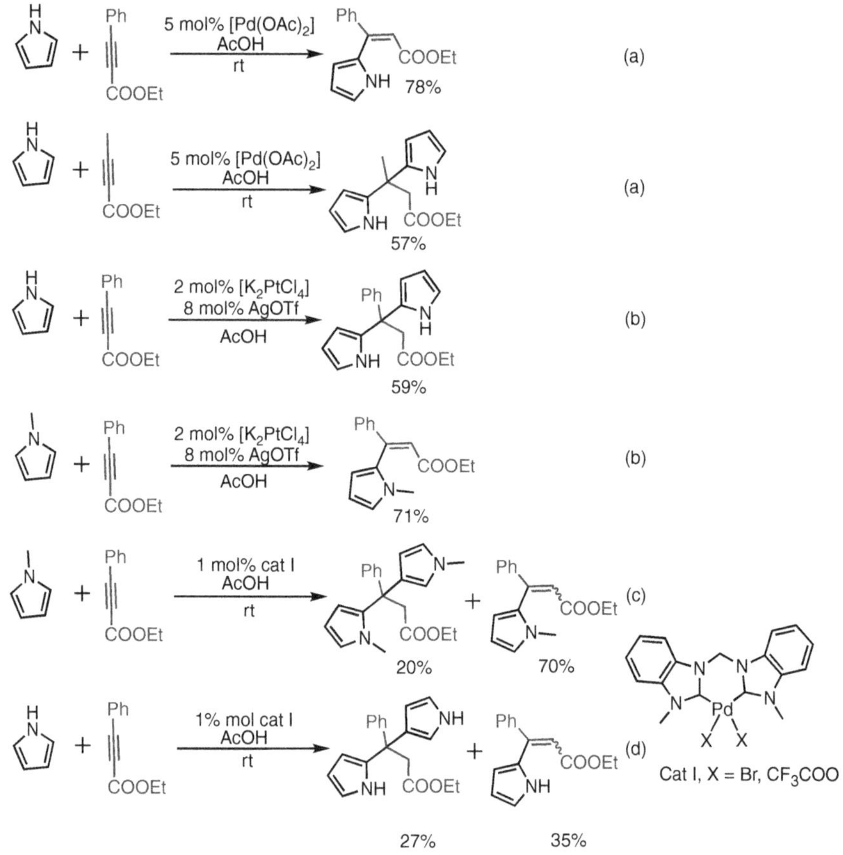

Figure 1.27 Hydroarylation of alkynes at the C2 position of a pyrrole using Pd or Pt catalysts.

Another interesting type of hydroarylation reaction is intramolecular benzannulation. One of the first examples was reported by Fürstner and Mamane [50]. In this case the coupling between a 2-phenylpyrrole and an alkyne catalyzed by $PtCl_2$ allowed the synthesis of 1*H*-benzo[*g*]indoles, while the coupling of *N*-phenylpyrrole gave the pyrrolo[1,2-*a*]quinoline (Figure 1.29a). Other metals such as Ga and In were also used as catalysts that gave very good yields as well [51]. The high electrophilic nature of $GaCl_3$ and $InCl_3$ and their behavior as soft Lewis acids (compared to B and Al) made the catalysts ideal candidates for this kind of chemistry. These heterocycles have shown their importance as anti-tumor agents in medicinal chemistry. The Sames group also proposed the benzannulation of *N*-substituted pyrroles to yield the corresponding dihydroindolizines using an electron-poor alkyne and $PtCl_4$ as catalyst (Figure 1.29b) [47].

Other interesting contributions from Beller and coworkers were the development of a method for the synthesis of pyrroloazepin-4-one via an internal cyclization of the corresponding pyrrolamide using Pt(IV) as catalyst (Figure 1.30a) [69, 70]. Interestingly, the major product was obtained after a rearrangement of the coupling product via a spiro intermediate. The group

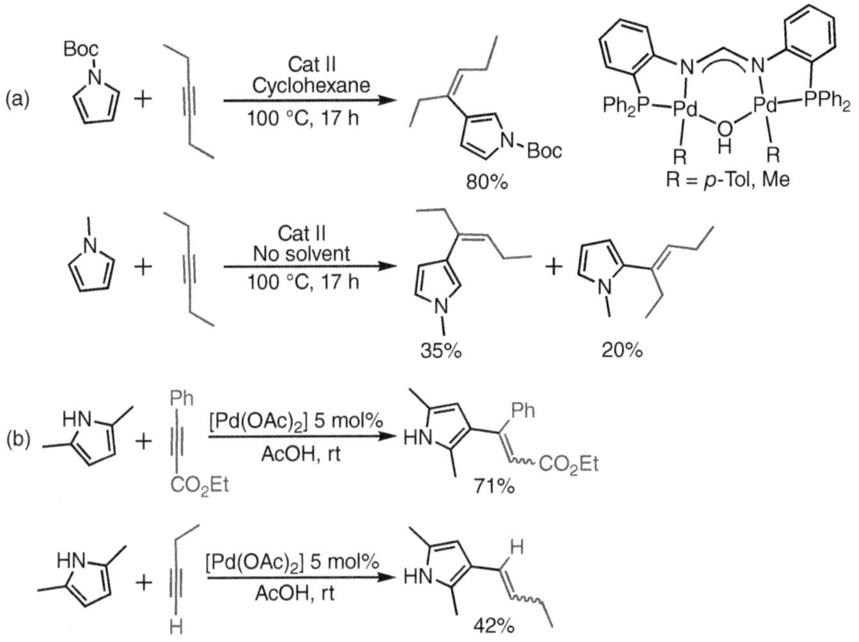

Figure 1.28 Hydroarylation of alkynes at the C3 position of pyrroles.

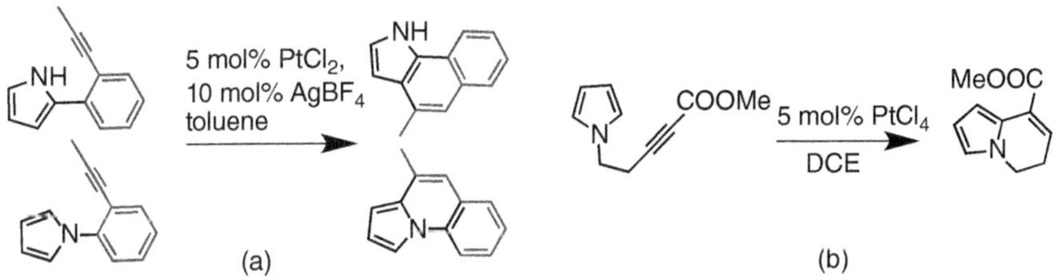

Figure 1.29 Intramolecular benzannulation at the C2 position of pyrroles using tethered alkynes.

of Van der Eycken also successfully attempted the intramolecular synthesis of pyrroloazepinones but by using a different alkyne (Figure 1.30b) [71]. The reaction works using Pt(II) instead of Pt(IV), and the expected product of the coupling was obtained. Finally, Waser and coworker developed a method for the synthesis of substituted indoles from pyrroles via a domino cyclization/alkynylation reaction (Figure 1.30c) [72]. The reaction achieved the 5-substituted indole as the sole product and no traces of other alkynylated positions were observed.

Functionalization of pyridines has been a difficult challenge due to the low reactivity of the pyridine ring using traditional methods such as the Friedel–Crafts alkylation, and therefore examples of this reactivity are scarce. In this context, the hydroarylation of internal alkynes at the C2 position of an *N*-pyridine oxide was developed by Hiyama using a Ni catalyst to obtain mainly the *E*-stereoisomer (Figure 1.31a) [73]. Using a related catalytic system, but with

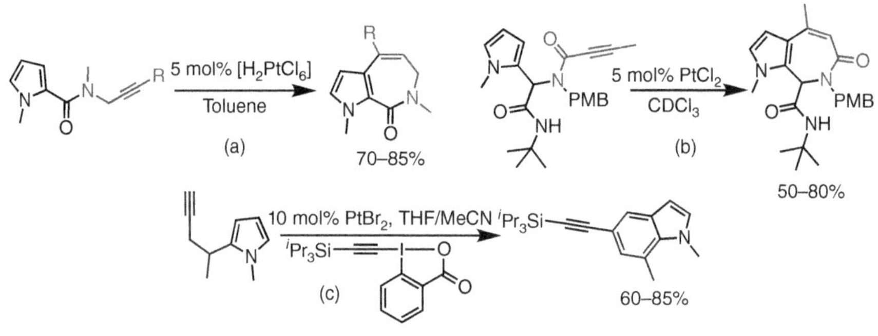

Figure 1.30 Benzannulation at the C3 position of pyrroles.

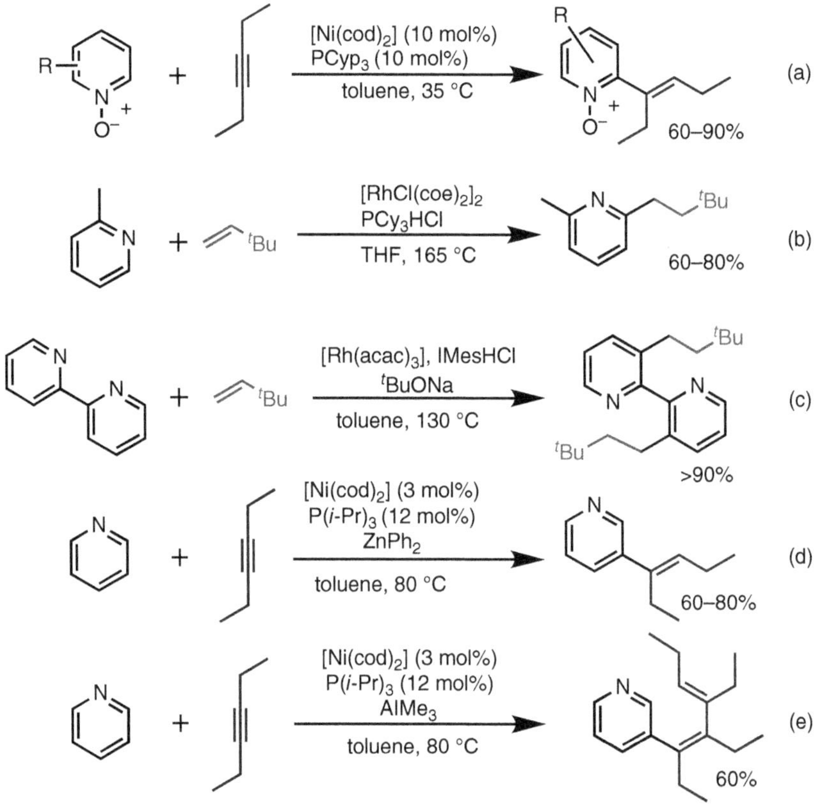

Figure 1.31 Hydroarylation of alkynes and alkenes at C2 and C3 positions of pyridines.

the assistance of $ZnPh_2$ or $AlMe_3$ as Lewis acids, even free pyridines could be successfully hydroarylated (Figures 1.31d and e) [74]. Bergman proposed that the hydroarylation of alkenes using a Rh(I) catalyst and a phosphine was able to functionalize pyridines and quinolines (Figure 1.31b) [75]. Finally, Chang and coworkers described the hydroarylation of alkenes using pyridines and bipyridines with a catalyst of Rh(III) with IMes.HCl as ligand (Figure 1.31c,

IMes = 1,3-bis(2,4,6-trimethylphenyl)imidazol-2-ylidene) [64]. Mono- and bis-inserted compounds were obtained for a wide range of bipyridines and biquinolines with a wide range of alkenes and internal alkynes.

Hiyama and coworkers also developed the hydroarylation of alkynes and alkenes with polifluoropyridines and pyridinones using [Ni(COD)$_2$] as catalyst in combination with PCyp$_3$ (Cyp = cyclopentyl) as co-catalyst [76]. In the case of pyridinones, a Lewis acid such as AlMe$_3$ is needed to obtain optimized results. In the case of fluoropyridines, position 4 is functionalized (Figure 1.32a). However, in the case of pyridinones, position 2 is activated. Montgomery and coworkers studied the reaction mechanism [43]. The catalytic system in this case comprises [Ni(COD)$_2$] and IMes as ligands and highlights the importance of the source of Ni(0) (Figure 1.32b). The use of directing groups such as amides has also been studied. Shi and coworkers performed the hydroarylation of alkynes (mostly diarylacetylenes) using amides as directing products in the *ortho*-position [77]. [RhCp*Cl$_2$]$_2$ (Cp* = C$_5$Me$_5$) as catalyst and Cu(OAc)$_2$ and AgSbF$_6$ as co-catalysts were used and, interestingly, although *E*-isomers are obtained for most pyridines, when picolinamide was used as substrate *Z*-isomers were obtained (Figure 1.32c). Carretero and coworkers used a similar catalytic system to yield phenyltetrasubstituted isoquinolines as a product of a double insertion of diphenylacetylene (Figure 1.32d) [78].

The same catalytic system used by Bergman in the alkylation of pyridines can be applied in the intramolecular cyclization of enol-tethered methylpyridines

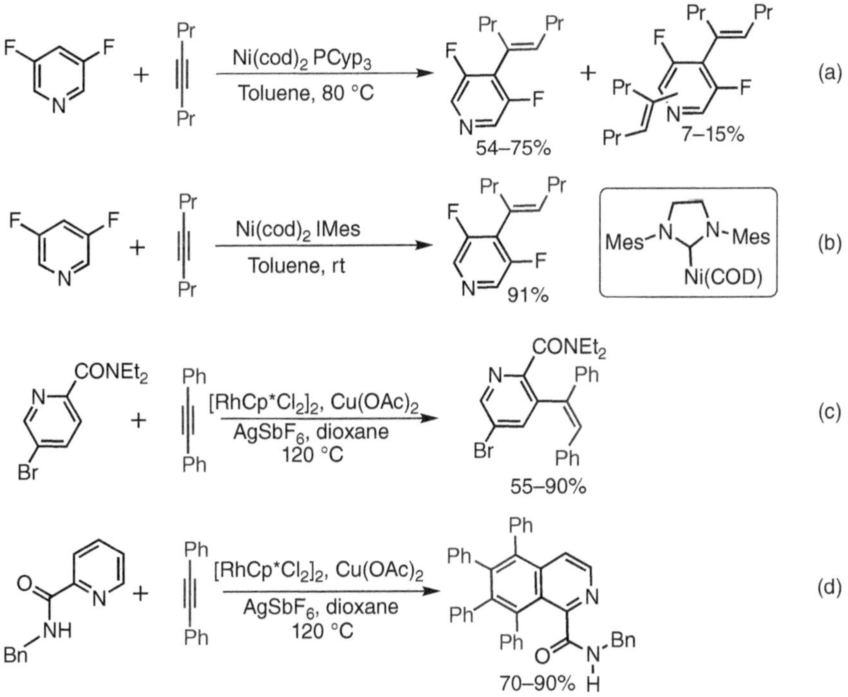

Figure 1.32 Hydroarylation of alkynes and alkenes using pyridines.

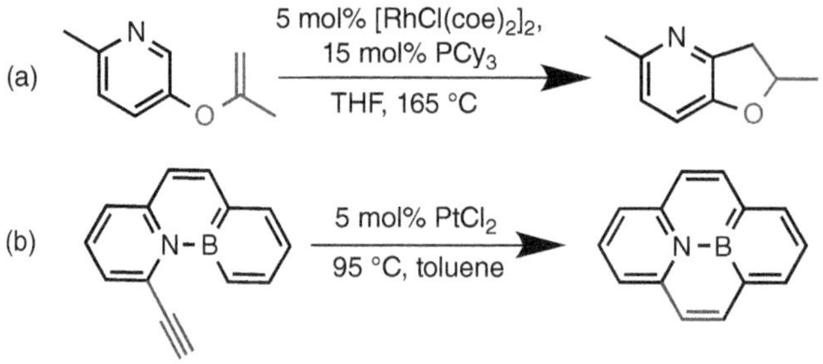

Figure 1.33 Cyclization of pyridines via intramolecular hydroarylation.

to yield the corresponding 2,3-dihydrofuro[3,2-*b*]pyridine (Figure 1.33a) [79]. As explained earlier, $[RhCl(coe)]_2$ and PCy_3 promote the internal cyclization in position 2 of pyridine. This technique was also used with other substrates such as quinolines. In addition, the internal cyclization of an azaboraphenanthrene to yield a 10*a*-aza-10*b*-borapyrene, using $PtCl_2$ as catalyst, was reported (Figure 1.33b). This product displays interesting fluorophore properties with potential applications in sensors and biochemical labeling studies [80].

Indoles are among the most commonly applied *N*-containing heterocyclic substrates in which the hydroarylation reaction has been tested with alkynes as a common coupling partner. One of the first examples was reported by the Hiyama group with a Ni-catalyzed hydroarylation using *N*-methylindoles along with 4-octyne as internal alkyne [42]. The C—H activation occurs at the C2 of the indole (Figure 1.34a). A wide range of indoles and other heterocycles are used under mild conditions to give good yields as it has been described in the preceding paragraphs (see Figure 1.9, conditions a). Yoshikai and coworker proposed the hydroarylation of *N*-pyrimidylindoles functionalizing the C2 position using a $CoBr_2$ catalyst, a phosphine, and a Grignard reagent as a reductant under very mild conditions (Figure 1.34b) [81]. The *syn*-isomer was achieved as reported by the Hiyama and the Yoshikai group. These results were further improved recently by Petit using $Co^0(PMe_3)_4$ as cobalt catalyst and an imine as a directing group in a microwave reactor without the need of a reductant (Figure 1.34c) [82]. The product obtained in this case is the *anti*-isomer.

Another interesting contribution was the work developed by the late K. Fagnou (Figure 1.34d); the directing group used in this case was *N,N*-dimethylcarbamoyl with a Rh catalyst obtaining good yields and selectivities [83]. Yu and coworkers attempted the rare hydroarylation of terminal alkynes with *N*-pyrimidylindoles using a Co(II) catalyst and, as in the case of Yoshikai, avoided the use of Grignard reagents (Figure 1.34e) [65]. The reaction gives regio- and stereoselectively the *E*-isomer.

Functionalization at the C3 position of indoles through hydroarylation of alkynes has also been developed using a Pt catalyst [84]. The group of Cheng developed the hydroarylation of alkynols over the C3 position of indoles using $PtCl_2$ as catalyst (Figure 1.35a). The first step is the cyclization of the alkynol (either by 5-*endo-dig* or 6-*endo-dig* or even 6-*exo-dig* pathways, as shown in Figure 1.35c) promoted by the Pt complex. Next, the most likely mechanism

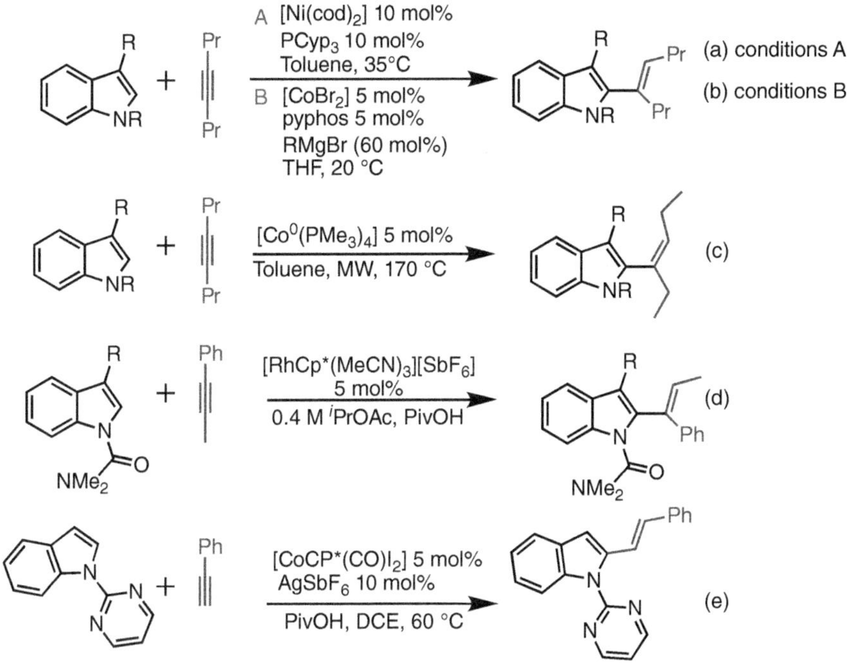

Figure 1.34 Hydroarylation of alkynes at the C2 position of indoles.

Figure 1.35 Hydroarylation of alkynes and allenes at the C3 position of indoles.

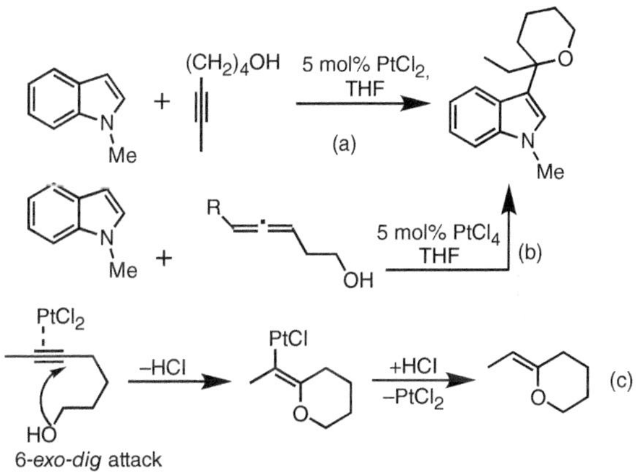

involves the electrophilic metallation of the indol, followed by insertion of the C=C double bond and protodemetallation. This product can also be obtained using allenes as reagents instead of alkynes using $PtCl_4$ as the catalyst precursor. In this case just the pyrane derivative is obtained, as Ma and coworkers demonstrated. In both cases, the selectivity is in the C3 position of the indole (Figure 1.35b) [85, 86].

Hydroarylation of alkenes with indoles has been widely studied. Both C2 and C3 positions of the indoles have been successfully and selectively functionalized using a wide variety of alkenes. There are many examples of functionalization

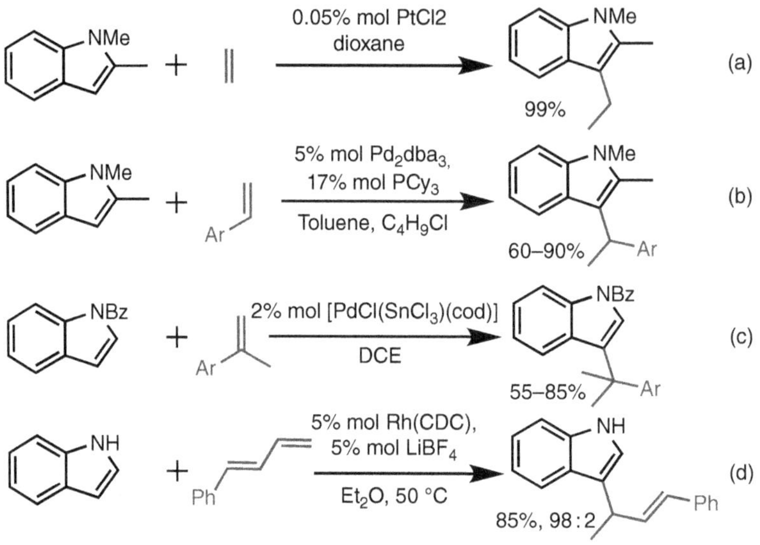

Figure 1.36 Hydroarylation of alkenes at the C3 position of indoles.

of the C3 position of indoles. One of the earliest examples was reported by Widenhoefer and coworkers using different alkenes, from ethylene to styrene, and $PtCl_2$ in dioxane as the catalytic system (Figure 1.36a) [87]. The use of styrene, however, resulted in a mixture that was between the branched and the linear products. A Pd catalyst shows activity for the hydroarylation of vinylphenols with indoles, both on its own and as a heterobimetallic Pd–Sn catalyst. In the first case, $Pd_2(dba)_3$ was used with PCy_3, while alkyl halides were used as a hydride source (Figure 1.36b) [57]. Interestingly, no alkylation of the phenol was observed. In the second case, Roy and coworkers designed a heterobimetallic catalyst to perform the reaction using substituted styrene as starting material [58]. Moderate to good yields were obtained for free and *N*-substituted indoles (Figure 1.36c). More recently, the group of Meek developed a method for the hydroarylation of α,β-unsaturated compounds catalyzed by Rh with CDC ligands (CDC = carbodicarbenes ligands). Good yields and excellent selectivities toward the γ-substituted product in the C3 of the indole were obtained (Figure 1.36d) [88].

There are also a good number of examples of hydroarylation of alkenes at the C2 position of indoles. Hiyama and coworkers studied the hydroarylation of several styrenes using *N*-methylindoles with C3 protected with an electron-withdrawing group (Figure 1.37a) [56]. The catalyst of choice was $[Ni(cod)_2]$ with the carbene IMes to yield the product with excellent yields in most cases (70–90%). Yoshikai and coworker also studied the hydroarylation of styrene using an imine as a directing group, which directed to C2 with $CoBr_2$ as catalyst, Ixyl as ligand (Ixyl = 1,3-bis(2,6-dimethylphenyl)imidazolium chloride), and a Grignard reagent as a reducing agent (Figure 1.37b) [89, 90]. The reaction worked well and just the branched product was obtained for *N*-methylindole (yields around 80–90%). The same author reported the testing of an array of phosphines that

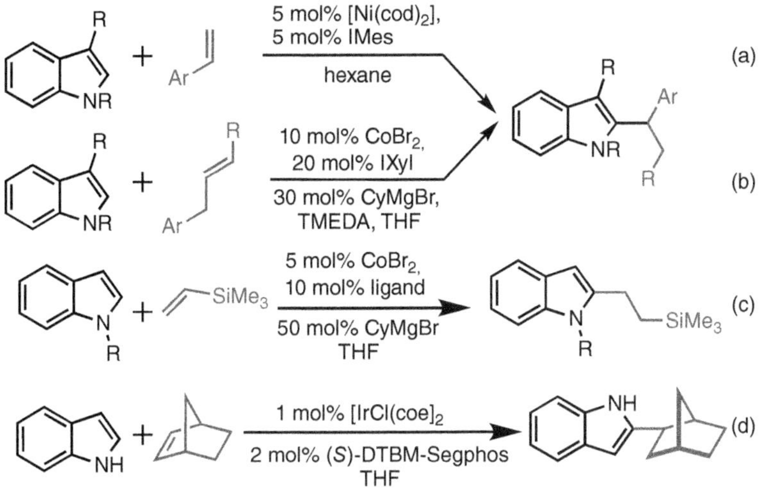

Figure 1.37 Hydroarylation of alkenes at the C2 position of indoles.

led to different results. If the phosphine of choice is triphenylphosphine, the product was predominantly branched with excellent yield (70–90%). However, the use of (dimethoxyphenyl)diphenylphosphine lowers yield (<10%), although overall a change in the selectivity was obtained and the linear product was dominant. More interestingly, use of an additive such as 2-methoxypyridine allowed the synthesis of both products. If the additive was used without a ligand the product was predominantly branched in good yields, and if the additive was used together with the phosphine the product was predominantly linear in good yields. A similar catalytic system was used by the same author but instead of an imine at C3, an *N*-pyrimidylindole was used as starting material with pyrimidyl as the *ortho*-directing group. Different vinylsilanes were hydroarylated. A phenanthroline derivative is the ligand of choice this time allowing the selective synthesis of the linear product (yields around 50–90%; Figure 1.37c) [91]. Finally, Hartwig and coworker developed the enantioselective insertion of indoles to norbornene (Figure 1.37d) [59]. The reaction is performed using an Ir(I) catalyst and a chiral ligand ((*S*)-DTBM-Segphos = (*S*)-(+)-5,5′-bis[di(3,5-di-*tert*-butyl-4-methoxyphenyl)phosphino]-4,4′-bi-1,3-benzodioxole) which allows the functionalization at C2 of the indole with excellent yields (85–95%) and enantioselectivities (99% ee).

One of the first examples of internal cyclization on the C2 of alkenylindoles was developed by Bergman and coworkers [92]. *N*-substituted indoles were cyclized to yield the tricyclic compound using a Rh(I) precursor and a phosphoramidite ligand in toluene (Figure 1.38a). Because the hydroarylation incorporated a substituted alkene, a chiral center was created, and excellent yields and enantioselectivities were obtained. The selectivity to C2 is achieved by using an imine directing group at C3 of the indole. A similar strategy but a completely different catalytic system is used by Yoshikai and coworkers [93]. In this case $CoBr_2$ acted as catalytic precursor, a carbene as ligand and a Grignard as additive was used

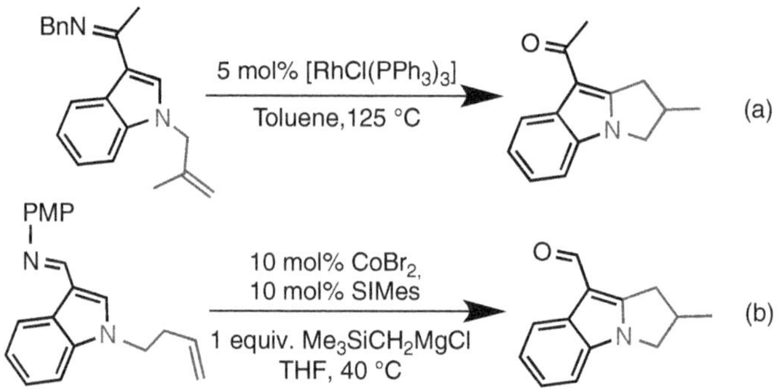

Figure 1.38 Benzannulation at the C2 position of indoles.

at 40 °C to promote cyclization at the C2. Interestingly, the choice of carbene determines the selectivity toward a five- or a six-member ring: SIMes·HCl (SIMes = 1,3-bis(2,4,6-trimethylphenyl)-4,5-dihydroimidazol-2-ylidene) promoted the formation of a five-membered ring (Figure 1.38b) while IPr·HCl promoted the formation of a six-membered ring, this regioselectivity being dependent on the steric hindrance of the NHC ligand.

Intramolecular cyclization at the C3 of indoles has also been developed, although examples are limited. Widenhoefer and coworkers reported the intramolecular cyclization of 2-alkenyl indoles to yield tetrahydrocarbazoles using a Pt(II) complex in dioxane [94–96]. The reaction proceeded in good yields and selectivities, and both substituted and free indoles could be functionalized (Figure 1.39a). The same authors describe an asymmetric variant of this reaction using $PtCl_2$(P-P) (P-P = chiral diphosphine). Several diphosphines were screened ((*S*)-BINAP, (*S*)-tol-BINAP, and others), but the best results were obtained with (*S*)-3,5-*t*-Bu-4-MeOMeOBIPHEP, achieving excellent yields (56–96%) and enantioselectivities (up to 90% ee). Ma and coworkers also achieved the synthesis of carbazoles through internal cyclization (Figure 1.39b) [86]. In this case, a preformed allenol indole was used as starting material and $PtCl_2$ as the catalyst yielding the corresponding carbazole. Many substrates were tested with different

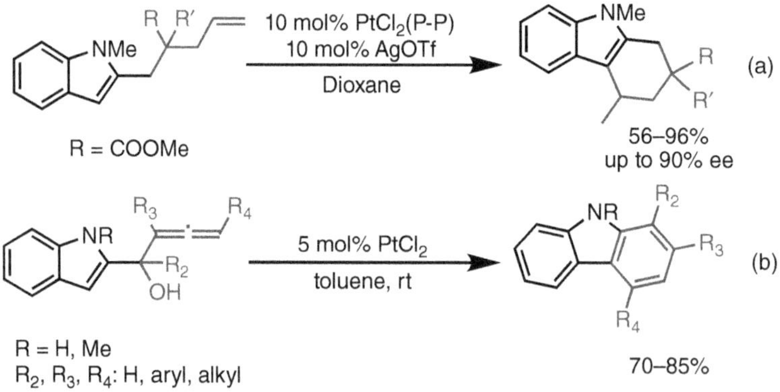

Figure 1.39 Benzannulation at the C3 position of indoles.

substituents in the indole or the allenol moieties obtaining good to excellent yields and demonstrating the versatility of the method.

1.4 Azoles and Other Miscellaneous Heterocycles

Interest in the synthesis of diverse *N*-heterocycles prompted Bergman and coworkers to initiate extensive research in catalytic hydroarylation using a large variety of *N*-heteroaromatic substrates. The first examples of this rich chemistry were related to the intramolecular cyclization of *N*-homoallyl benzimidazoles to give fused carbocycles, as shown in Figure 1.40 [97]. The reaction is fully regioselective, and only the C—H bond in *α* to both N atoms is activated. The reaction scope is quite general with respect to alkene substituents, because terminal and internal alkenes as well as trisubstituted alkenes were successfully cyclized with a high degree of regioselectivity to give new five- or six-membered rings. In general, terminal to internal alkene isomerization is observed before C—C coupling, giving five-membered rings. The authors suggest that either the internal geminal substitution or the allylic α,α-dibranching could be responsible for the formation of the larger six-membered rings. For instance, the internal geminal substitution could lead the formation of the six-membered ring because

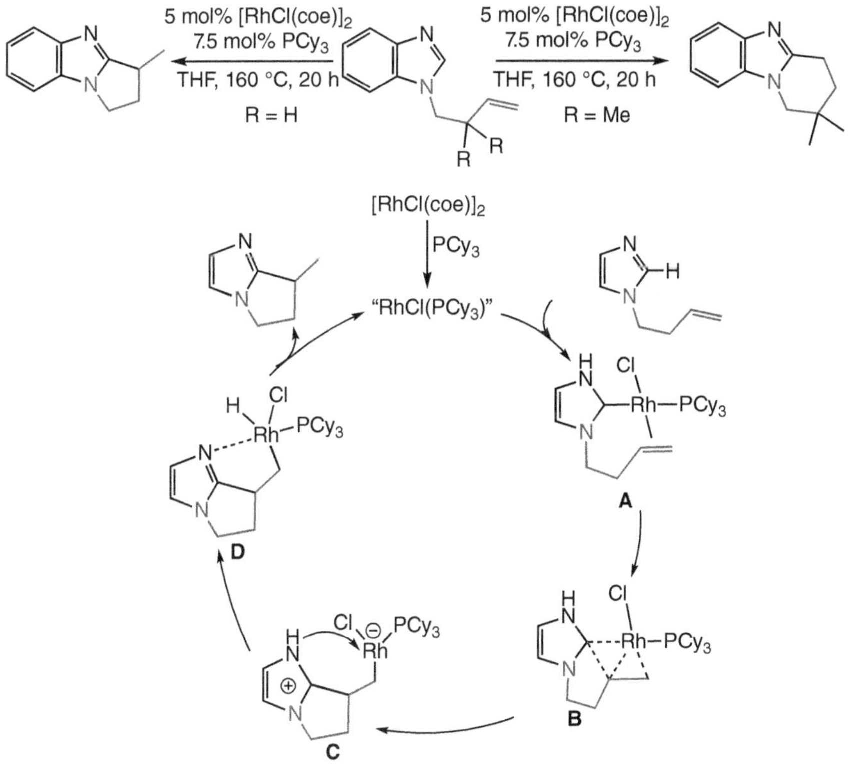

Figure 1.40 Intramolecular hydroarylation of homoallyl-benzimidazoles using a Rh catalyst.

of the difficulty to cyclize forming a quaternary center [8, 97]. In the same way, when the olefin cannot isomerize, the formation of the six-membered ring is favored. Mechanistic study by DFT methods of the reactions shown in Figure 1.40 revealed that the Rh(I) *N*-heterocyclic carbene complex **A** is the likely resting state [98]. This is surprising because a C—H bond activation using electron-rich Rh(I) complexes usually results in a Rh(III)-hydride species [8, 13]. From complex **A**, a transition state leading to insertion of the alkenyl group into the Rh-C(carbene) bond was proposed as the rate-determining step [98]. After alkene insertion the zwitterion **C** is formed, which evolves through intramolecular proton transfer to give hydride **D**. By C—H reductive elimination the final cyclized derivative is obtained and the catalyst is renewed.

Processes shown in Figure 1.40 were notably accelerated in the presence of Brønsted or, even more effectively, Lewis acids. Therefore, additives such as lutidinium chloride, YCl_3, or $MgBr_2$ have a dramatic impact on the reaction rate, although the authors have not provided a rationale for this fact. This acceleration allowed the expansion of this reaction to the first intermolecular hydroarylation of benzimidazole, benzothiazole, benzoxazole, dimethylthiazole, and purine (Figure 1.41) [99]. Optimization of the additive showed that lutidinium chloride provided the highest yields. The scope of the reaction is quite general, both for the heteroaromatics (some of them shown in Figure 1.41) and for the alkenes, activated and unactivated, regardless of being susceptible to possible isomerization. In all studied cases the hydroarylation reactions take place regioselectively to give the linear alkyl derivative. However, no reaction was observed with pyrimidine or indole, showing that the formation of a Rh-NHC intermediate is important. The range of substrates able to be functionalized using this method was notably expanded in a subsequent work by including aryl-substituted benzimidazoles [100]. The reaction conditions were also re-optimized, and the authors found that $[HPCy_3]Cl$, the phosphonium salt of PCy_3, is the best option as phosphine and additive (Lewis acid), which simplifies the experimental setup due to its stability. In this case the authors provide a tentative explanation, based on the formation of a Rh(III) hydride [100]. The reactions presented in

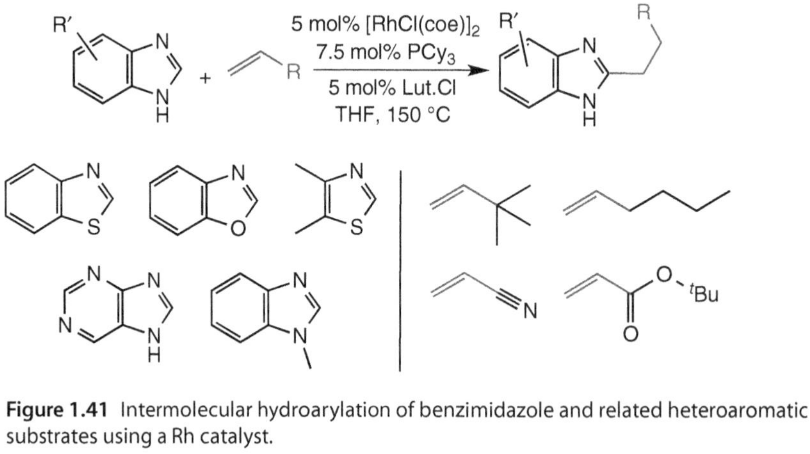

Figure 1.41 Intermolecular hydroarylation of benzimidazole and related heteroaromatic substrates using a Rh catalyst.

Figures 1.40 and 1.41 were reinvestigated using microwave radiation in place of conventional heating [101]. Under optimized conditions the reaction times can be reduced from 20 h to 6–12 min without erosion of the yield.

This powerful and versatile methodology was applied to the synthesis of more complex products. For instance, intramolecular alkene hydroarylation is a key step in the synthesis of a potent kinase inhibitor, more specifically the c-Jun N-Terminal Kinase (JNK3) inhibitor shown in Figure 1.42 [102]. The inclusion of this step in the total synthesis of JNK3 inhibitor reduces the number of synthetic steps from 14 to 11, and increases the total yield from less than 6% to 13%, representing an advance in synthetic methods.

Other heterocyclic substrates able to form *N*-heterocylic carbene (NHC) complexes with Rh have been studied. 4,4-Dimethyl-2-oxazoline reacted with a large variety of alkenes, including substrates with protected aldehyde, acid, alcohol, and amine functional groups, to give linear hydroarylated derivatives with excellent regioselectivity (Figure 1.43 left) [103]. Notably, the reaction temperature in these processes is substantially lower (45 °C, versus 150–180 °C typically used in other reactions). This surprising increase in catalyst activity correlates with observations in other NHC-based catalysts: a higher catalytic efficiency is observed when unsaturated NHC ligands are replaced by their saturated versions in the same complexes [103].

The synthetic possibilities of the dihydroquinazoline skeleton have also been investigated by Bergman and coworkers because of its presence in products with medicinal applications [104]. 3,4-Dihydroquinazoline reacts with alkenes under Rh-catalysis to give the corresponding 2-alkyl-3,4-dihydroquinazolines in good

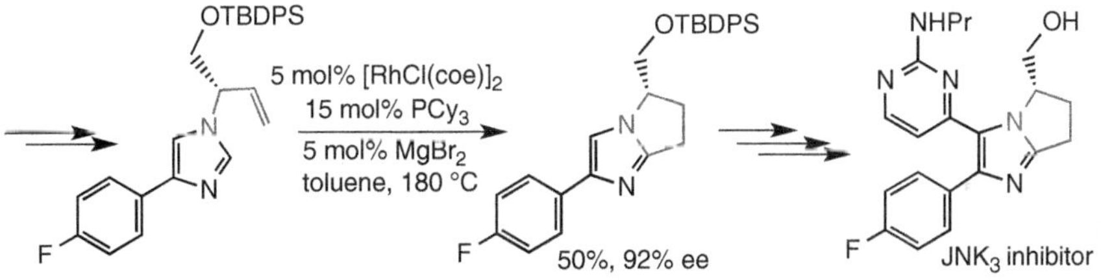

Figure 1.42 Application of intramolecular alkene hydroarylation to the synthesis of high value-added products.

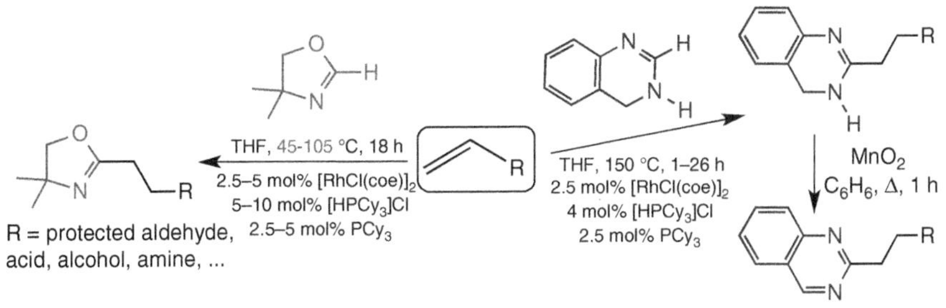

Figure 1.43 Oxazolines and dihydroquinazolines as substrates in hydroarylation reactions catalyzed by Rh complexes.

yields (Figure 1.43 right). Under prolonged heating a mixture with the corresponding aromatized quinazoline is obtained, because a Rh/PCy_3 system can also work as a transfer dehydrogenation catalyst. Another procedure for the production of quinazolines in two steps is described: (i) the controlled Rh-catalyzed hydroarylation of the olefin with the dihydroquinazoline and (ii) the oxidation of the reaction crude with MnO_2 (Figure 1.43) [104]. The intramolecular version of the hydroarylation has been applied to the total synthesis of Vasicoline, a representative of a family of products with biological activity (Figure 1.44) [104].

3-Methyl-3,4-dihydroquinazoline (**A** in Figure 1.45) has also been used as a model for interesting experimental and computational mechanistic studies [104, 105]. These studies were focused on the C—H bond activation of the dihydroquinazoline ligand promoted by $RhCl(PCy_3)_2$ and the concomitant formation of a Rh-NHC species **C**. The results show that complex **B**, where ligand **A** is N1-bonded to the Rh center, is an intermediate in the synthesis of the NHC-carbene derivative **C**. The kinetic analysis of the concentration changes of **A**, **B**, and **C** along the time at 37 °C clearly showed that **B** is a likely intermediate on the pathway to **C**. Based on this, the authors propose an intramolecular H transfer ongoing from **B** to **C** as the most plausible mechanism for C—H bond activation; that is, the H at carbon C2 is transferred to nitrogen N1 without exchange with any other H atom while it is being transferred [105]. All these facts were supported by isotopic labeling experiments.

All examples previously shown in this section shared a common feature, which is the regioselective insertion of the alkene in the C—H bond of the heterocycle to give the linear alkyl derivative. In the following sections the discussion is on

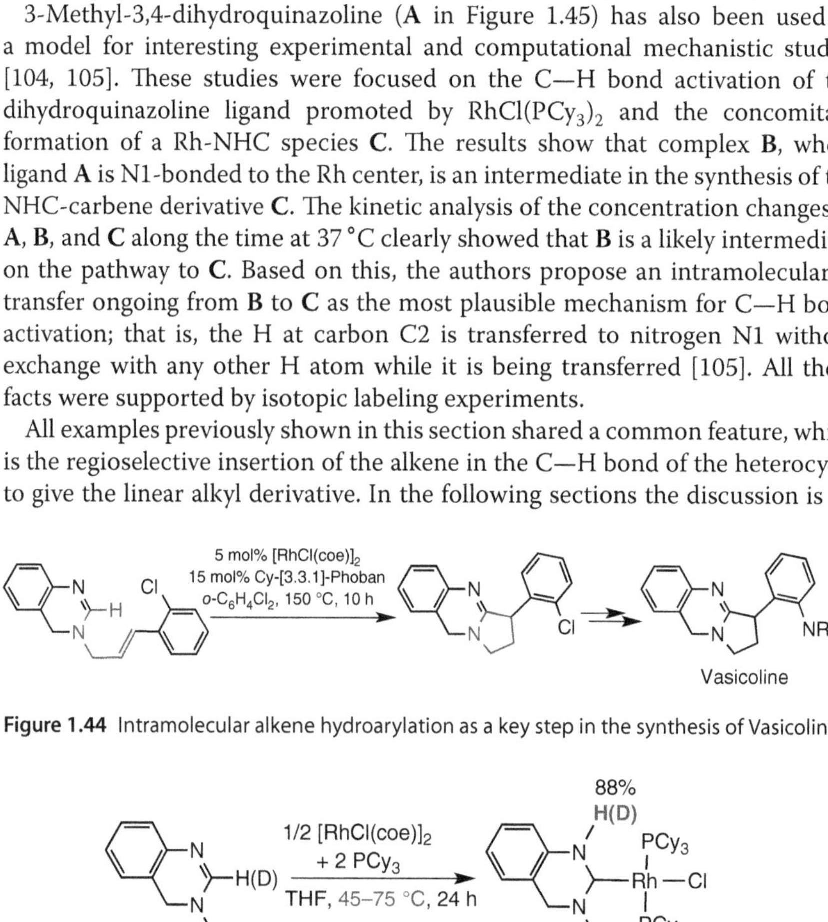

Figure 1.44 Intramolecular alkene hydroarylation as a key step in the synthesis of Vasicoline.

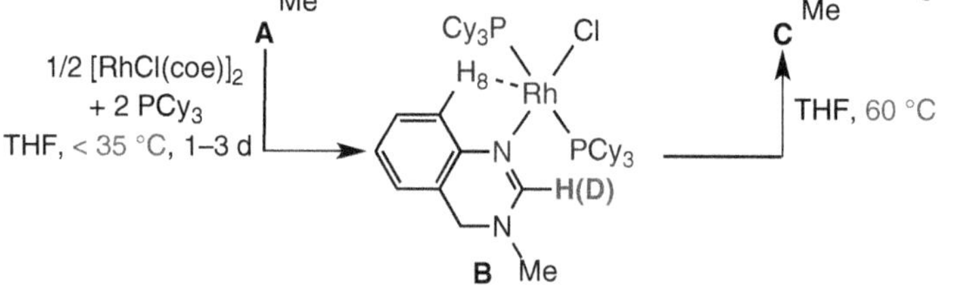

Figure 1.45 Intermediates in the C—H bond activation of 3,4-dihydroquinazolines using Rh catalysts.

Ni catalysts, in combination with Lewis acids, to afford control of the regioselectivity of the alkene insertion, affording either branched (1,1-disubstituted ethane or Markovnikov addition) or linear derivatives (1,2-disubstituted ethane or anti-Markovnikov addition) depending on the reaction conditions [56, 106–108]. Hiyama and coworkers have reported that the hydroarylation of substituted styrenes and other related alkenes with heterocycles, catalyzed by $Ni(cod)_2$ and a very bulky NHC as stabilizing ligand, affords the corresponding 1,1-diarylethanes, that is, the branched derivatives, as it is shown in left part of Figure 1.46 [56]. This method notably expanded the range of substrates able to undergo hydroarylation to give branched products with respect to previous methods. The scope is wide because styrenes with electron-donating and electron-withdrawing groups, and even indene, successfully give the branched 1,1-diaryldeivatives. On the other hand, an aliphatic alkene such as 1-tridecene affords exclusively the linear derivative, probably due to the lack of functionalities with a π-system. In addition, heteroaromatics such as benzimidazole, benzoxazole, oxazole, and benzothiazole, among others, couple with styrene to also give the branched product. Mechanistic studies resulted in the proposal that the reaction starts with oxidative addition of the C—H bond to Ni(0) to give a Ni(II)-H species, followed by coordination of the alkene, hydronickelation by migratory insertion, and reductive elimination to regenerate the Ni(0) catalyst. No explanation about the observed selectivity was provided [56].

Following the seminal work of Hiyama, Ong and coworkers reported control of the regioselectivity of the insertion by choice of additives, as shown in central and left parts of Figure 1.46 [106]. When the reaction between benzimidazole and styrene is performed in the absence of additives the branched product, the 1,1-diarylsubstituted ethane, is formed [106]. The scope of both benzimidazoles and styrenes covers a wide range of substrates with yields of branched products comprised between 50–99%. However, when the reaction between benzimidazole and styrene is performed in the presence of an additional 10% of a Lewis acid, such as $AlMe_3$, the regioselectivity of the process switches completely, and the linear derivative is obtained with full regioselectivity and yields around 90% in almost all studied cases. As for the branched reaction, the scope of alkenes

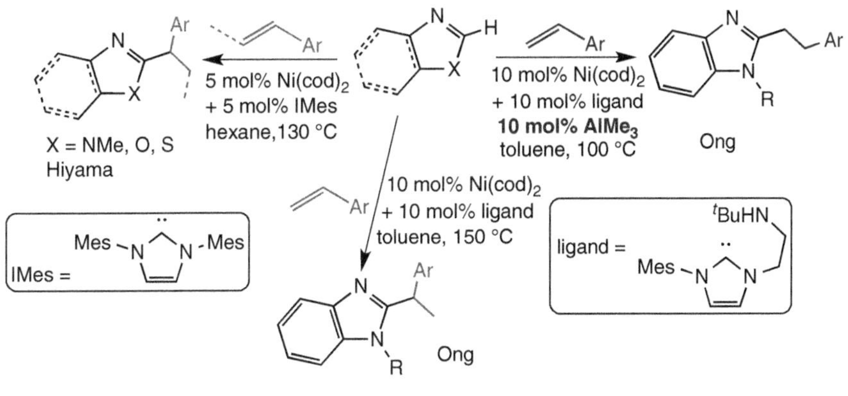

Figure 1.46 Hydroarylation of alkenes using Ni catalysts with control of the regioselectivity.

and azoles able to undergo hydroarylation is quite wide. Interestingly, the reaction conditions in the presence of a Lewis acid are milder (toluene, 100 °C) than in its absence (toluene, 150 °C), showing that the addition of the Lewis acid has an additional accelerating effect.

The switch in the regioselectivity of the hydroarylation processes shown in Figure 1.46 was combined with an isomerization reaction in a subsequent elegant work by Ong and coworkers [107]. As shown in Figure 1.47, branched or linear compounds can be obtained with full selectivity from azoles and allylarenes. In the left part of Figure 1.47, the reaction takes place in toluene at 130 °C to give the corresponding 1,1-diarylpropanes in good to excellent yields and with full regioselectivity. This means that allylbenzene undergoes a previous fast isomerization to *trans*-β-methyl-styrene (process also catalyzed by $Ni(cod)_2$/IMes), which takes place at 80 °C in only 1 h) and further reacts with the azole to give the 1,1-diarylpropane (Markovnikov addition) [107]. On the other hand, the reaction of azoles with allylarenes under the same catalytic conditions but in the presence of $AlMe_3$ gives the linear anti-Markovnikov 1,3-derivative (Figure 1.47 right) [107]. The effect of the addition of $AlMe_3$ as a critical game-changer is multiple: (i) it decreases the rate of the olefin isomerization, making this process slower than the hydroarylation reaction and (ii) switch the regioselectivity towards the linear product [106]. The origin of this outstanding selectivity seems to be related to the subtle balance of the steric versus electronic control of the insertion of the alkene into the Ni—H bond [106, 108].

Recent work of Filloux and Rovis has shown that the enantioselectivity of the hydroarylation can be controlled with careful design of the catalytic system [109]. The method reported the hydroarylation of α-substituted acrylates with different benzoxazoles, a process catalyzed by $[Rh(OAc)(COD)]_2$ in NCMe at 100 °C (Figure 1.48), which affords the corresponding linear C2-alkylated benzoxazoles. The system requires the presence of an enantiomerically pure diphosphine ligand, based on previous reports. Due to this mandatory requirement, other chiral ligands were not tested. The screening of different diphosphines showed that the bulky CTH-(*R*)-xylyl-P-Phos (Figure 1.48) was the best option. Under these conditions high yields (most of them higher than 60%) and enantioselectivities (range

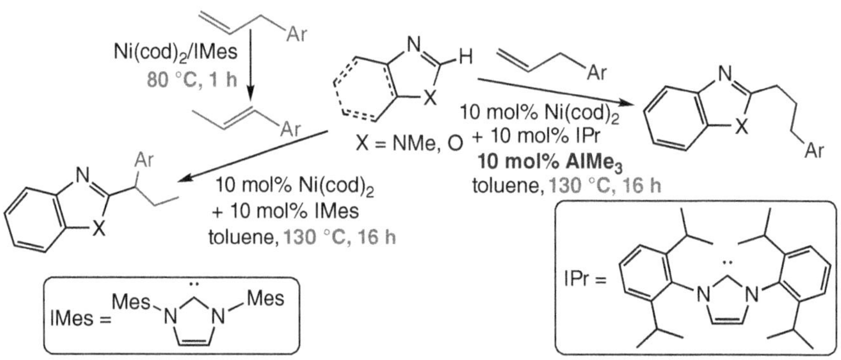

Figure 1.47 Tandem isomerization–hydroarylation using a Ni catalyst with control of regioselectivity.

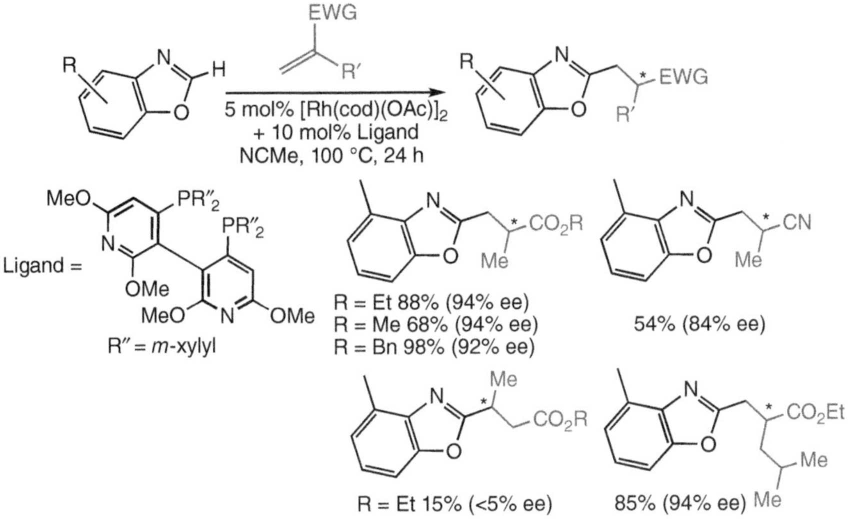

Figure 1.48 Enantioselective hydroarylation of electron-poor alkenes using a Rh catalyst.

68–96%, with most of them higher than 90%) were achieved. Following mechanistic studies that include isotope labeling, it seems that the enantiodeterminant step is the migratory insertion followed by a stereospecific isomerization that insulates the formed stereocenter from epimerization [109].

Cobalt complexes have recently been shown to achieve high activity and selectivity for the hydroarylation of alkynes with a large diversity of heterocycles [110]. Yoshikai and coworkers described chemo-, regio-, and stereoselective addition of substituted azoles to electron-rich internal alkynes catalyzed by $CoBr_2$ to give good yields of the C2-alkenylated derivatives (Figure 1.49) [110]. Under optimized conditions the reaction requires a diphosphine ligand in catalytic amounts (10%) and a Grignard reagent in substoichiometric amounts (50%). The reaction takes place at room temperature selectively with activation of the C2—H bond

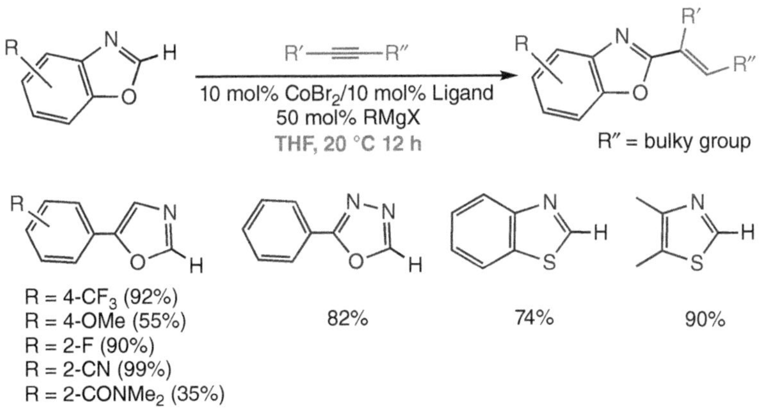

Figure 1.49 Cobalt-catalyzed hydroarylation of alkynes using azoles.

of the heterocycle, in spite of the presence of many other seemingly readily activated C—H bonds in the starting substrates. The reaction occurs with excellent (*E*)-stereoselectivity, since the *E*/*Z* ratio is higher than 99/1 in all cases. Moreover, full regioselectivity was observed in the insertion of asymmetrically substituted alkynes, forming the C—C bond at the less hindered carbon.

Nickel complexes, in combination with different Lewis acids, have shown a high versatility for the hydroarylation of alkenes, as presented in previous paragraphs in this section (Figures 1.46 and 1.47), and this is also true for the hydroarylation of alkynes. Hiyama and coworkers have reported the hydroarylation of alkynes with indoles [42]. In the same contribution a large number of azoles have shown to display the same reactivity, (shown in Figure 1.50a) using $Ni(cod)_2$ and $PCyp_3$ (Cyp = cyclopentyl) as the catalytic system. The reaction occurs at 35 °C and gives excellent levels of stereo- and regioselectivity. The same group has reported the hydroarylation using different imidazoles, represented in Figure 1.50b [111]. When the starting material has substituents at N1 and C4 but not at C2, the functionalization occurs by selective activation of the C2—H bond. The reaction affords in all cases the *E*-isomer, but further isomerization to the *Z*-isomer has been detected in alkynes containing aryl groups. On the other hand, when the starting imidazole contains substituents at N1, C2, and C4, the functionalization takes place selectively at C5 achieving, as expected, excellent

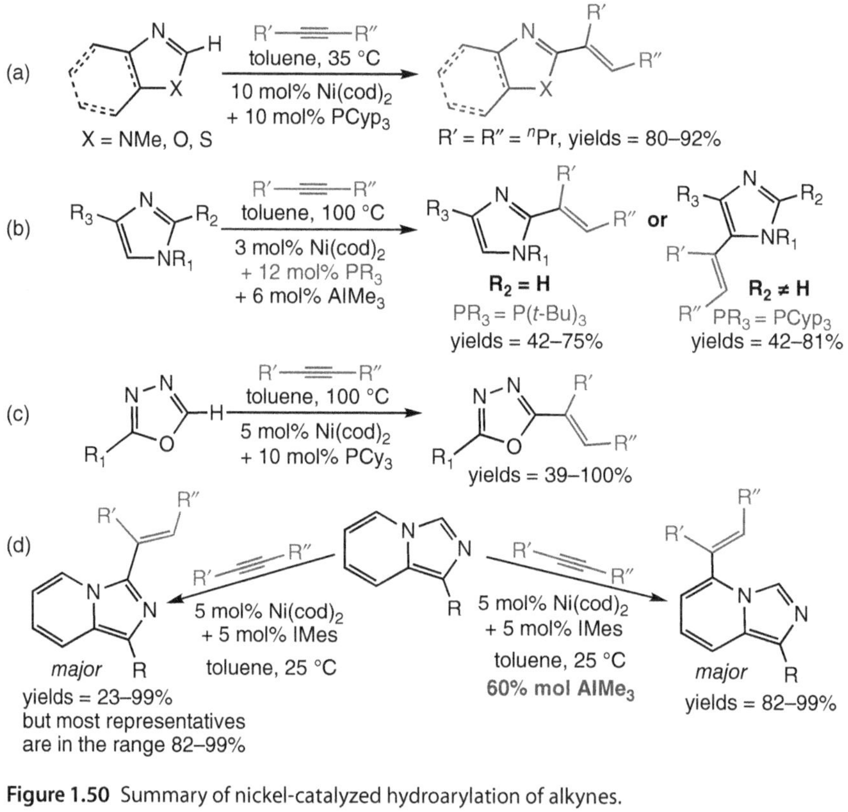

Figure 1.50 Summary of nickel-catalyzed hydroarylation of alkynes.

stereo- and regioselectivities in the final alkenylated derivatives (*E*-isomers with the bulkiest R″ group in *trans* to the imidazole ring) [111].

Miura and coworkers reported an efficient method for the synthesis of alkenylated oxadiazoles, as shown in Figure 1.50c [112]. The oxadiazole core is probably one of the most interesting in the agrochemical and pharmaceutical fields, and the development of efficient methods for their rapid synthesis is of considerable importance in organic synthesis. The reaction is selective at the five position of the oxadiazole ring and stereoselective for formation of the *E*-isomer for electron-rich internal alkynes. Unfortunately, only one example of terminal alkyne was reported, and other terminal alkynes did not show adequate reactivity. Imidazo[1,5-*a*]pyridines are another example of heteroarenes that show adequate control of the reactivity through fine-tuning of the additives. In this respect, recent results of the group of Ong are shown in Figure 1.50d [113]. The starting imidazo[1,5-*a*]pyridine has the C1 atom substituted; therefore, C3—H and C5—H are the bonds most plausible to be activated. When the reaction is performed at 25 °C using $Ni(cod)_2$ as catalyst and IMes as an optimal stabilizing ligand, the incorporation of the alkyne was observed at, mainly, the C3-position (C3/C5 = 5/1 to 7/1, left part) giving the *E*-alkene complete stereoselectivity [113]. The addition of a Lewis acid, such as $AlMe_3$, and its further bonding to the N2 atom, affords the C5-functionalized derivative as the main reaction product (Figure 1.50d, right part). The reaction gives in this case higher C5/C3 ratios (8/1 to 14/1) and complete *E*-selectivity.

From Figure 1.50, it is noteworthy that the same catalytic system, through fine-tuning and undergoing only small changes, is able to display high catalytic activity and remarkable selectivity for different heteroromatic substrates.

1.5 Summary

In summary, the hydroarylation of alkenes and alkynes with heteroarenes is a basic and fundamental tool for the formation of new C—C bonds in metal-catalyzed organic synthesis. The research performed until now has shown that the reaction seems to be unlimited, because it provides the expected coupling products for virtually all types of heteroarenes, both electron-rich (furanes, thiophenes, pyrroles, indoles) as well as electron-deficient (pyridines, quinolines), and also for a large variety of electronically and sterically different alkenes and alkynes. In addition, in most of the cases the coupling can be performed through metal-catalyzed C—H bond activation strategies, avoiding the prefunctionalization of the starting materials, then saving time, energy, and resources, and minimizing the amount of waste. In this chapter we have focused our interest in the transition metals of the 9 and 10 groups, but other metals can also be used. The reaction can be easily tuned by addition of diverse additives, and it is especially accelerated in the presence of Lewis acids. Despite its general character, the reaction can show very high levels of selectivity: stereoselectivity, when the (*Z*/*E*) geometry of the olefin formed by hydroarylation of an alkyne is considered; regioselectivity, with respect to the activated position in the starting heteroarene, or when asymmetric alkenes or alkynes are considered; or even enantioselectivity, when new stereogenic centers are being formed.

Despite these impressive developments, there are still a number of challenges for the future. Further research on electron-deficient rings, clearly less developed than electron-rich ones, generalization of the use of C—H bond activation processes promoted by earth-abundant metals (Fe, Co, Ni), or a better control of the selectivity (mainly the enantioselectivity) are tasks for a close future.

References

1 Meth-Cohn, O. (ed.) (1984) *Comprehensive Heterocyclic Chemistry*, Pergamon, New York, Vol. 1.

2 Bird, C.W. and Cheeseman, G.W.H. (eds) (1984) *Comprehensive Heterocyclic Chemistry*, Pergamon, New York, Vol. 4.

3 Katritzky, A.R. and Pozharskii, A.F. (2000) *Handbook of Heterocyclic Chemistry*, 2nd edn, Pergamon, Oxford.

4 Kakiuchi, F. and Murai, S. (2002) Catalytic C—H/olefin coupling. *Acc. Chem. Res.*, **35**, 826.

5 Kakiuchi, F. and Chatani, N. (2003) Catalytic methods for C—H bond functionalization: application in organic synthesis. *Adv. Synth. Catal.*, **345**, 1077.

6 Nevado, C. and Echavarren, A.M. (2005) Transition metal-catalyzed hydroarylation of alkynes. *Synthesis*, **2**, 167.

7 Kakiuchi, F. and Kochi, T. (2008) Transition-metal-catalyzed carbon—carbon bond formation via carbon—hydrogen bond cleavage. *Synthesis*, 3013.

8 Lewis, J.C., Bergman, R.G., and Ellman, J.A. (2008) Direct functionalization of nitrogen heterocycles via Rh-catalyzed C—H bond activation. *Acc. Chem. Res.*, **41**, 1013.

9 Kitamura, T. (2009) Transition-metal-catalyzed hydroarylation reactions of alkynes through direct functionalization of C—H bonds: a convenient tool for organic synthesis. *Eur. J. Org. Chem.*, 1111.

10 Vasilyev, A.V. (2009) Alkenylation of aromatic compounds. *Russ. J. Org. Chem.*, **45**, 1.

11 Rossi, R., Bellina, F., and Lessi, M. (2010) Alkenylation reactions of heteroarenes by transition metal catalysts. *Synthesis*, 4131.

12 de Mendoza, P. and Echavarren, A.M. (2010) Synthesis of arenes and heteroarenes by hydroarylation reactions catalyzed by electrophilic metal complexes. *Pure Appl. Chem.*, **82**, 801.

13 Colby, D.A., Bergman, R.G., and Ellman, J.A. (2010) Rhodium-catalyzed C—C bond formation via heteroatom-directed C—H bond activation. *Chem. Rev.*, **110**, 624.

14 Cacchi, S. and Fabrizi, G. (2011) Update 1 of: synthesis and functionalization of indoles through palladium-catalyzed reactions. *Chem. Rev.*, **111**, PR215.

15 Andreatta, J.R., McKeown, B.A., and Gunnoe, T.B. (2011) Transition metal catalyzed hydroarylation of olefins using unactivated substrates: recent developments and challenges. *J. Organomet. Chem.*, **696**, 305.

16 Yadav, J.S., Antony, A., Rao, T.S., and Reddy, B.V.S. (2011) Recent progress in transition metal catalysed hydrofunctionalisation of less activated olefins. *J. Organomet. Chem.*, **696**, 16.
17 Nakao, Y. (2011) Hydroarylation of alkynes catalyzed by nickel. *Chem. Rec.*, **11**, 242.
18 Zeng, X. (2013) Recent advances in catalytic sequential reactions involving hydroelement addition to carbon—carbon multiple bonds. *Chem. Rev.*, **113**, 6864.
19 Yamamoto, Y. (2014) Synthesis of heterocycles via transition-metal-catalyzed hydroarylation of alkynes. *Chem. Soc. Rev.*, **43**, 1575.
20 de Haro, T. and Nevado, C. (2014) 5.33 Hydroarylation reactions, in *Comprehensive Organic Synthesis II*, 2nd edn (ed. P. Knochel), Elsevier, Amsterdam, pp. 1621–1659.
21 Gao, K. and Yoshikai, N. (2014) Low-valent cobalt catalysis: new opportunities for C—H functionalization. *Acc. Chem. Res.*, **47**, 1208.
22 Chinchilla, R. and Nájera, C. (2014) Chemicals from alkynes with palladium catalysts. *Chem. Rev.*, **114**, 1783.
23 Abbiati, G., Arcadi, A., Marinelli, F., and Rossi, E. (2014) Sequential addition and cyclization processes of α,β-ynones and α,β-ynoates containing proximate nucleophiles. *Synthesis*, **46**, 687.
24 Yang, L. and Huang, H. (2015) Transition-metal-catalyzed direct addition of unactivated C—H bonds to polar unsaturated bonds. *Chem. Rev.*, **115**, 3468.
25 Khan, M.S., Haque, A., Al-Suti, M., and Raithby, P.R. (2015) Recent advances in the application of group-10 transition metal based catalysts in C—H activation and functionalization. *J. Organomet. Chem.*, **793**, 114.
26 Stephens, D.E. and Larionov, O.V. (2015) Recent advances in the C—H-functionalization of the distal positions in pyridines and quinolines. *Tetrahedron*, **71**, 8683.
27 Boyarskiy, V.P., Ryabukhin, D.S., Bokach, N.A., and Vasilyev, A.V. (2016) Alkenylation of arenes and heteroarenes with alkynes. *Chem. Rev.*, **116**, 5894.
28 Neuhaus, J.D. and Willis, M.C. (2016) Homogeneous rhodium(I)-catalysis in *de novo* heterocycle synthses. *Org. Biomol. Chem.*, **14**, 4986.
29 Hong, P., Yamazaki, H., Sonogashira, K., and Hagihara, N. (1978) Rhodium carbonyl cluster catalyzed addition of arenes to diphenylketene and aryl isocyanates. *Chem. Lett.*, 7, 535.
30 Hong, P. and Yamazaki, H. (1979) Reactions of ethylene and benzenes catalyzed by rhodium carbonyls under carbon monoxide. The formation of styrenes and 3-pentanone. *Chem. Lett.*, 1335.
31 Hong, P., Cho, B.R., and Yamazaki, H. (1979) Reaction of benzenes with acetylenes catalyzed by rhodium carbonyl under carbon monoxide. *Chem. Lett.*, 339.
32 Hong, P., Cho, B.R., and Yamazaki, H. (1980) Addition of 5-membered aromatic heterocyclic compounds to acetylenes catalyzed by tetrarhodium dodecacarbonyl: syntheses of vinyl substituted aromatic heterocyclic compounds. *Chem. Lett.*, 507.
33 Yamazaki, H. and Hong, P. (1983) Organic reactions with rhodium carbonyl cluster catalysis. *J. Mol. Catal.*, **21**, 133.

34 Hong, P. and Yamazaki, H. (1984) Rhodium carbonyl-catalyzed activation of carbon–hydrogen bonds for application in organic synthesis. Phenylation of olefins with benzenes. *J. Mol. Catal.*, **26**, 297.

35 Murai, S., Kakiuchi, F., Sekine, S., Tanaka, Y., Kamatani, A., Sonoda, M., and Chatani, N. (1993) Efficient catalytic addition of aromatic carbon–hydrogen bonds to olefins. *Nature*, **366**, 529.

36 Jia, C., Piao, D., Oyamada, J., Lu, W., Kitamura, T., and Fujiwara, Y. (2000) Efficient activation of aromatic C—H bonds for addition to C—C multiple bonds. *Science*, **287**, 1992.

37 Jia, C., Lu, W., Oyamada, J., Kitamura, T., Matsuda, K., Irie, M., and Fujiwara, Y. (2000) Novel Pd(II)- and Pt(II)-catalyzed regio- and stereoselective *trans*-hydroarylation of alkynes by simple arenes. *J. Am. Chem. Soc.*, **122**, 7252.

38 Lu, W., Jia, C., Kitamura, T., and Fujiwara, Y. (2000) Pd-catalyzed selective addition of heteroaromatic C—H bonds to C—C triple bonds under mild conditions. *Org. Lett.*, **2**, 2927.

39 Keita, M.L., Mizuhara, T., Oyamada, J., and Kitamura, T. (2007) Double thienylation of alkynes in the platinum(II)-catalyzed reaction of thiophenes with propiolic acids. *Chem. Lett.*, **36**, 1150.

40 Oyamada, J. and Kitamura, T. (2009) Pt(II)-catalyzed hydroarylation reaction of alkynes with pyrroles and furans. *Tetrahedron*, **65**, 3842.

41 Buscemi, G., Biffis, A., Tubaro, C., Basato, M., Graiff, C., and Tiripicchio, A. (2010) C—H bond functionalization of aromatic heterocycles with chelating dicarbene palladium(II) and platinum(II) complexes. *Appl. Organomet. Chem.*, **24**, 285.

42 Nakao, Y., Kanyiva, K.S., Oda, S., and Hiyama, T. (2006) Hydroheteroarylation of alkynes under mild nickel catalysis. *J. Am. Chem. Soc.*, **128**, 8146.

43 Nett, A.J., Zhao, W., Zimmerman, P.M., and Montgomery, J. (2015) Highly active nickel catalysts for C—H functionalization identified through analysis of off-cycle intermediates. *J. Am. Chem. Soc.*, **137**, 7636.

44 Lee, P.S., Fujita, T., and Yoshikai, N. (2011) Cobalt-catalyzed, room-temperature addition of aromatic imines to alkynes via directed C—H bond activation. *J. Am. Chem. Soc.*, **133**, 17283.

45 Tsukada, N., Murata, K., and Inoue, Y. (2005) Selective cis-addition of C—H bonds of pyrroles and thiophenes to alkynes catalyzed by a dinuclear palladium complex. *Tetrahedron Lett.*, **46**, 7515.

46 Jia, C., Piao, D., Kitamura, T., and Fujiwara, Y. (2000) New method for preparation of coumarins and quinolinones via Pd-catalyzed intramolecular hydroarylation of C—C triple bonds. *J. Org. Chem.*, **65**, 7516.

47 Pastine, S.J., Youn, S.W., and Sames, D. (2003) Pt(IV)-catalyzed cyclization of arene–alkyne substrates via C—H bond functionalization. *Tetrahedron*, **59**, 8859.

48 Pastine, S.J., Youn, S.W., and Sames, D. (2003) Pt(IV)-catalyzed cyclization of arene–alkyne substrates via intramolecular electrophilic hydroarylation. *Org. Lett.*, **5**, 1055.

49 Kitamura, T. and Otsubo, K. (2012) Palladium-catalyzed intramolecular hydroarylation of 4-benzofuranyl alkynoates. Approach to angelicin derivatives. *J. Org. Chem.*, **77**, 2978.
50 Fürstner, A. and Mamane, V. (2002) Flexible synthesis of phenanthrenes by a $PtCl_2$-catalyzed cycloisomerization reaction. *J. Org. Chem.*, **67**, 6264.
51 Inoue, H., Chatani, N., and Murai, S. (2001) Cycloisomerization of ω-aryl-1-alkynes: 52$GaCl_3$ as a highly electrophilic catalyst for alkyne activation. *J. Org. Chem.*, **67**, 1414.
52 Kang, D., Kim, J., Oh, S., and Lee, P.H. (2012) Synthesis of naphthalenes via platinum-catalyzed hydroarylation of aryl enynes. *Org. Lett.*, **14**, 5636.
53 Martin-Matute, B., Cárdenas, D.J., and Echavarren, A.M. (2001) Pt(II) catalyzed intramolecular reaction of furans with alkynes. *Angew. Chem. Int. Ed.*, **40**, 4754.
54 Martin-Matute, B., Nevado, C., Cárdenas, D.J., and Echavarren, A.M. (2003) Intramolecular reactions of alkynes with furans and electron rich arenes catalyzed by $PtCl_2$: the role of platinum carbenes as intermediates. *J. Am. Chem. Soc.*, **125**, 5757.
55 McKeown, B.A., Foley, N.A., Lee, J.P., and Gunnoe, T.B. (2008) Hydroarylation of unactivated olefins catalyzed by platinum(II) complexes. *Organometallics*, **27**, 4031.
56 Nakao, Y., Kashihara, N., Kanyiva, K.S., and Hiyama, T. (2010) Nickel-catalyzed hydroheteroarylation of vinylarenes. *Angew. Chem. Int. Ed.*, **49**, 4451.
57 Pathak, T.P. and Sigman, M.S. (2011) Palladium-catalyzed hydrofunctionalization of vinyl phenol derivatives with heteroaromatics. *Org. Lett.*, **13**, 2774.
58 Das, D., Pratihar, S., and Roy, S. (2013) Heterobimetallic Pd—Sn catalysis: highly selective intermolecular hydroarylation of α-methyl substituted aryl alkenes. *Tetrahedron Lett.*, **54**, 335.
59 Sevov, C.S. and Hartwig, J.F. (2013) Iridium-catalyzed intermolecular asymmetric hydroheteroarylation of bicycloalkenes. *J. Am. Chem. Soc.*, **135**, 2116.
60 Kruegel, A.C., Rakshit, S., Li, X., and Sames, D. (2015) Constructing *Iboga* alkaloids via C—H bond functionalization: examination of the direct and catalytic union of heteroarenes and isoquinuclidine alkenes. *J. Org. Chem.*, **80**, 2062.
61 Schramm, Y., Takeuchi, M., Semba, K., Nakao, Y., and Hartwig, J.F. (2015) Anti-Markovnikov hydroheteroarylation of unactivated alkenes with indoles, pyrroles, benzofurans, and furans catalyzed by a nickel-*N*-heterocyclic carbene system. *J. Am. Chem. Soc.*, **137**, 12215.
62 Hatano, M., Ebe, Y., Nishimura, T., and Yorimitsu, H. (2016) Asymmetric alkylation of *N*-Sulfonylbenzamides with vinyl ethers via C—H bond activation catalyzed by hydroxoiridium/chiral diene complexes. *J. Am. Chem. Soc.*, **138**, 4010.
63 Gao, K., Lee, P.S., Fujita, T., and Yoshikai, N. (2010) Cobalt-catalyzed hydroarylation of alkynes through chelation-assisted C—H bond activation. *J. Am. Chem. Soc.*, **132**, 12249.

64 Kwak, J., Ohk, Y., Jung, Y., and Chang, S. (2012) Rollover cyclometalation pathway in rhodium catalysis: dramatic NHC effects in the C—H bond functionalization. *J. Am. Chem. Soc.*, **134**, 17778.

65 Wang, S., Hou, J.T., Feng, M.L., Zhang, X.Z., Chen, S.Y., and Yu, X.Q. (2016) Cobalt(III)-catalyzed alkenylation of arenes and 6-arylpurines with terminal alkynes: efficient access to functional dyes. *Chem. Commun.*, **52**, 2709.

66 Ebe, Y. and Nishimura, T. (2015) Iridium-catalyzed branch-selective hydroarylation of vinyl ethers via C—H bond activation. *J. Am. Chem. Soc.*, **137**, 5899.

67 Rubio-Pérez, L., Iglesias, M., Castarlenas, R., Polo, V., Pérez-Torrente, J.J., and Oro, L.A. (2014) Selective C—H bond functionalization of 2-(2-thienyl)pyridine by a rhodium *N*-heterocyclic carbene catalyst. *ChemCatChem*, **6**, 3192.

68 Oyamada, J., Lu, W., Jia, C., Kitamura, T., and Fujiwara, Y. (2002) Direct synthesis of β-alkenylpyrroles by Pd(II)-catalyzed addition of pyrroles to alkynoates. *Chem. Lett.*, **31**, 20.

69 Gruit, M., Michalik, D., Tillack, A., and Beller, M. (2009) Platinum-catalyzed intramolecular cyclizations of alkynes: efficient synthesis of pyrroloazepinone derivatives. *Angew. Chem. Int. Ed.*, **48**, 7212.

70 Gruit, M., Michalik, D., Krüger, K., Spannenberg, A., Tillack, A., Pews-Davtyan, A., and Beller, M. (2010) Synthesis of pyrroloazepinones: platinum- and gold-catalyzed cyclization reactions of alkynes. *Tetrahedron*, **66**, 3341.

71 Modha, S.G., Kumar, A., Vachhani, D.D., Sharma, S.K., Parmar, V.S., and Van der Eycken, E.V. (2012) Gold(I) and platinum(II) switch: a post-Ugi intramolecular hydroarylation to pyrrolopyridinones and pyrroloazepinones. *Chem. Commun.*, **48**, 10916.

72 Li, Y. and Waser, J. (2015) Platinum-catalyzed domino reaction with benziodoxole reagents for accessing benzene-alkynylated indoles. *Angew. Chem. Int. Ed.*, **54**, 5438.

73 Kanyiva, K.S., Nakao, Y., and Hiyama, T. (2007) Nickel-catalyzed addition of pyridine-*N*-oxides across alkynes. *Angew. Chem. Int. Ed.*, **46**, 8872.

74 Nakao, Y., Kanyiva, K.S., and Hiyama, T. (2008) A strategy for C—H activation of pyridines: direct C-2 selective alkenylation of pyridines by nickel/Lewis acid catalysis. *J. Am. Chem. Soc.*, **130**, 2448.

75 Lewis, J.C., Bergman, R.G., and Ellman, J.A. (2007) Rh(I)-catalyzed alkylation of quinolines and pyridines via C—H bond activation. *J. Am. Chem. Soc.*, **129**, 5332.

76 Nakao, Y., Kashihara, N., Kanyiva, K.S., and Hiyama, T. (2008) Nickel-catalyzed alkenylation and alkylation of fluoroarenes via activation of C—H bond over C—F bond. *J. Am. Chem. Soc.*, **130**, 16170.

77 Qian, Z.C., Zhou, J., Li, B., Hu, F., and Shi, B.F. (2014) Rh(III)-catalyzed regioselective hydroarylation of alkynes via directed C—H functionalization of pyridines. *Org. Biomol. Chem.*, **12**, 3594.

78 Martínez, A.M., Echavarren, J., Alonso, I., Rodríguez, N., Gómez-Arrayás, R., and Carretero, J.C. (2015) RhI/RhIII catalyst-controlled divergent aryl/heteroaryl C—H bond functionalization of picolinamides with alkynes. *Chem. Sci.*, **6**, 5802.

79 Yotphan, S., Bergman, R.G., and Ellman, J.A. (2010) Synthesis of multicyclic pyridine and quinoline derivatives via intramolecular C—H bond functionalization. *Org. Lett.*, **12**, 2978.

80 Bosdet, M.J.D., Piers, W.E., Sorensen, T.S., and Parvez, M. (2007) 10*a*-Aza-10*b*-borapyrenes: heterocyclic analogues of pyrene with internalized BN moieties. *Angew. Chem. Int. Ed.*, **46**, 4940.

81 Ding, Z. and Yoshikai, N. (2012) C2-Alkylation of *N*-pyrimidylindole with vinylsilane via cobalt-catalyzed C—H bond activation. *Beilstein J. Org. Chem.*, **8**, 1536.

82 Fallon, B.J., Derat, E., Amatore, M., Aubert, C., Chemla, F., Ferreira, F., Perez-Luna, A., and Petit, M. (2016) C2-Alkylation and alenylation of indoles catalyzed by a low-valent cobalt complex in the absence of reductant. *Org. Lett.*, **18**, 2292.

83 Schipper, D.J., Hutchinson, M., and Fagnou, K. (2010) Rhodium(III)-Catalyzed Intermolecular Hydroarylation of Alkynes. *J. Am. Chem. Soc.*, **132**, 6910.

84 Bhuvaneswari, S., Jeganmohan, M., and Cheng, C.H. (2007) Platinum-catalyzed multistep reactions of indoles with alkynyl alcohols. *Chem. Eur. J.*, **13**, 8285.

85 Kong, W., Cui, J., Yu, Y., Chen, G., Fu, C., and Ma, S. (2009) $PtCl_4$-Catalyzed cyclization reaction of β-allenols in the presence of indoles. *Org. Lett.*, **11**, 1213.

86 Kong, W., Fu, C., and Ma, S. (2009) An efficient synthesis of carbazoles from $PtCl_2$-catalyzed cyclization of 1-(indol-2-yl)-2,3-allenols. *Chem. Commun.*, 4572.

87 Zhang, Z., Wang, X., and Widenhoefer, R.A. (2006) Pt(II)-catalyzed intermolecular hydroarylation of unactivated alkenes with indoles. *Chem. Commun.*, 3717.

88 Roberts, C.C., Matias, D.M., Goldfogel, M.J., and Meek, S.J. (2015) Lewis acid activation of carbodicarbene catalysts for Rh-catalyzed hydroarylation of dienes. *J. Am. Chem. Soc.*, **13**, 6488.

89 Yamakawa, T. and Yoshikai, N. (2014) Alkene isomerization–hydroarylation tandem catalysis: indole C2-alkylation with aryl-substituted alkenes leading to 1,1-diarylalkanes. *Chem. Asian J.*, **9**, 1242.

90 Xu, W. and Yoshikai, N. (2014) Highly linear selective cobalt-catalyzed addition of aryl imines to styrenes: reversing intrinsic regioselectivity by ligand elaboration. *Angew. Chem. Int. Ed.*, **53**, 14166.

91 Gao, K. and Yoshikai, N. (2011) Cobalt–phenanthroline catalysts for the ortho alkylation of aromatic imines under mild reaction conditions. *Angew. Chem. Int. Ed.*, **50**, 6888.

92 Thalji, R.K., Ellman, J.A., and Bergman, R.G. (2004) Highly efficient and enantioselective cyclization of aromatic imines via directed C—H bond activation. *J. Am. Chem. Soc.*, **126**, 7192.

93 Ding, Z. and Yoshikai, N. (2013) Cobalt-catalyzed intramolecular olefin hydroarylation leading to dihydropyrroloindoles and tetrahydropyridoindoles. *Angew. Chem. Int. Ed.*, **52**, 8574.
94 Liu, C., Han, X., Wang, X., and Widenhoefer, R.A. (2003) Platinum-catalyzed intramolecular alkylation of indoles with unactivated olefins. *J. Am. Chem. Soc.*, **126**, 3700.
95 Han, X. and Widenhoefer, R.A. (2006) Platinum-catalyzed intramolecular asymmetric hydroarylation of unactivated alkenes with indoles. *Org. Lett.*, **8**, 3801.
96 Liu, C., Bender, C.F., Han, X., and Widenhoefer, R.A. (2007) Platinum-catalyzed hydrofunctionalization of unactivated alkenes with carbon, nitrogen and oxygen nucleophiles. *Chem. Commun.*, 3607.
97 Tan, K.L., Bergman, R.G., and Ellman, J.A. (2001) Annulation of alkenyl-substituted heterocycles vis rhodium-catalyzed intramolecular C—H activated coupling reactions. *J. Am. Chem. Soc.*, **123**, 2685.
98 Tan, K.L., Bergman, R.G., and Ellman, J.A. (2002) Intermediacy of an *N*-heterocyclic carbene complex in the catalytic C—H activation of a substituted imidazole. *J. Am. Chem. Soc.*, **124**, 3202.
99 Tan, K.L., Bergman, R.G., and Ellman, J.A. (2002) Intermolecular coupling of isomerizable alkenes to heterocycles via rhodium-catalyzed C—H bond activation. *J. Am. Chem. Soc.*, **124**, 13964.
100 Tan, K.L., Park, S., Ellman, J.A., and Bergman, R.G. (2004) Intermolecular coupling of alkenes to heterocycles via C—H bond activation. *J. Org. Chem.*, **69**, 7329.
101 Tan, K.L., Vasudevan, A., Bergman, R.G., Ellman, J.A., and Souers, A.J. (2003) Microwave-assisted C—H bond activation: a rapid entry into functionalized heterocycles. *Org. Lett.*, **5**, 2131.
102 Rech, J.C., Yato, M., Duckett, D., Ember, B., LoGrasso, P.V., Bergman, R.G., and Ellman, J.A. (2007) Synthesis of potent bicyclic bisarylimidazole c-Jun N-terminal kinase inhibitors by catalytic C—H bond activation. *J. Am. Chem. Soc.*, **129**, 490.
103 Wiedemann, S.H., Bergman, R.G., and Ellman, J.A. (2004) Rhodium-catalyzed direct C—H addition of 4,4-dimethyl-2-oxazoline to alkenes. *Org. Lett.*, **6**, 1685.
104 Wiedemann, S.H., Ellmann, J.A., and Bergman, R.G. (2006) Rhodium-catalyzed direct C—H addition of 3,4-dihydroquinazolines to alkenes and their use in the total synthesis of vasicoline. *J. Org. Chem.*, **71**, 1969.
105 Wiedemann, S.H., Lewis, J.C., Ellmann, J.A., and Bergman, R.G. (2006) Experimental and computational studies on the mechanism of *N*-heterocycle C—H activation by Rh(I). *J. Am. Chem. Soc.*, **128**, 2452.
106 Shih, W.C., Chen, W.C., Lai, Y.C., Yu, M.S., Ho, J.J., Yap, G.P.A., and Ong, T.G. (2012) The regioselective switch for amino-NHC mediated C—H activation of benzimidazole via Ni—Al synergistic catalysis. *Org. Lett.*, **14**, 2046.

107 Lee, W.C., Wang, C.H., Lin, Y.H., Shih, W.C., and Ong, T.G. (2013) Tandem isomerization and C—H activation: regioselective hydroheteroarylation of allylarenes. *Org. Lett.*, **15**, 5358.

108 Chen, W.C., Lai, Y.C., Shih, W.C., Yu, M.S., Yap, G.P.A., and Ong, T.G. (2014) Mechanistic study of a switch in the regioselectivity of hydroheteroarylation of styrene catalyzed by bimetallic Ni—Al through C—H activation. *Chem. Eur. J.*, **20**, 8099.

109 Filloux, C.M. and Rovis, T. (2015) Rh(I)-bisphosphine-catalyzed asymmetric, intermolecular hydroheteroarylation of α-substituted acrylate derivatives. *J. Am. Chem. Soc.*, **137**, 508.

110 Ding, Z. and Yoshikai, N. (2010) Cobalt-catalyzed addition of azoles to alkynes. *Org. Lett.*, **12**, 4180.

111 Kanyiva, K.S., Löbermann, F., Nakao, Y., and Hiyama, T. (2009) Regioselective alkenylation of imidazoles by nickel/Lewis acid catalysis. *Tetrahedron Lett.*, **50**, 3463.

112 Mukai, T., Hirano, K., Satoh, T., and Miura, M. (2009) Nickel-catalyzed C—H alkenylation and alkylation of 1,3,4-oxadiazoles with alkynes and styrenes. *J. Org. Chem.*, **74**, 6410.

113 Yu, M.S., Lee, W.C., Chen, C.H., Tsai, F.Y., and Ong, T.G. (2014) Controlled regiodivergent C—H bond activation of imidazo[1,5-*a*]pyridine via synergistic cooperation between aluminum and nickel. *Org. Lett.*, **16**, 4826.

2

Ruthenium Catalysts for the Alkylation of Functionalized Arenes and Heteroaromatic Substrates via Hydroarylation

David J. Burns, Sergei I. Kozhushkov and Lutz Ackermann

Georg-August-Universität, Institut für Organische und Biomolekulare Chemie, Tammannstrasse 2, 37077, Göttingen, Germany

2.1 Introduction

Alkylation of arenes is a central objective of synthesis and has been such since the early beginnings of organic reaction development with examples that include the now famous Friedel–Crafts processes [1]. However, the prevalence of alkenes in biologically relevant structures in addition to themselves functioning as building blocks for the construction of complex molecules makes their use in valorization reactions of utmost importance. Herein, we summarize the development of ruthenium-catalyzed alkylations of aromatics, with a focus on heterofunctionalized arenes and heteroaromatics, using alkene reaction partners via hydroarylation technology. Recently developed C—H alkylation reactions [2] with electrophilic alkyl halides are not discussed within this review [3] as they have been previously summarized [2].

The early reports on the stoichiometric transformation of C—H bonds in the 1960s [4] set the foundation for the development of catalytic methods via chelation-assisted cyclometalation [5]. Such reactions have the power to dramatically shorten synthetic sequences over traditional routes, and thus have pronounced implications toward sustainable and environmentally benign syntheses. In this respect, the rather cheap ruthenium-based catalysts have become a mainstay for the functionalization of C—H bonds [6]. Methods developed include annulation [7], alkenylation [8], and arylation [9] chemistries. Very recently, unique remote activation modes proceeding via chelation-assisted cyclometalation have even allowed access to *meta*-substituted arenes, which represents a powerful tool for the synthesis of complex organic molecules by remote functionalizations [10]. However, one of the earliest and most precedented uses of ruthenium catalysts in C—H bond activation is in the hydroarylation reaction of alkenes, which is highlighted in this chapter. This review covers only those processes that involve chelation-assisted *ortho*-C—H activation, such as C—H cyclometalation, and is split into two sections that are arranged by the distinct reaction pathways operative as a result of complementary modes of C—H bond cleavage (Scheme 2.1). Section 2.2 covers alkylation reactions using

Catalytic Hydroarylation of Carbon-Carbon Multiple Bonds, First Edition.
Edited by Lutz Ackermann, T. Brent Gunnoe and Laurel Goj Habgood.

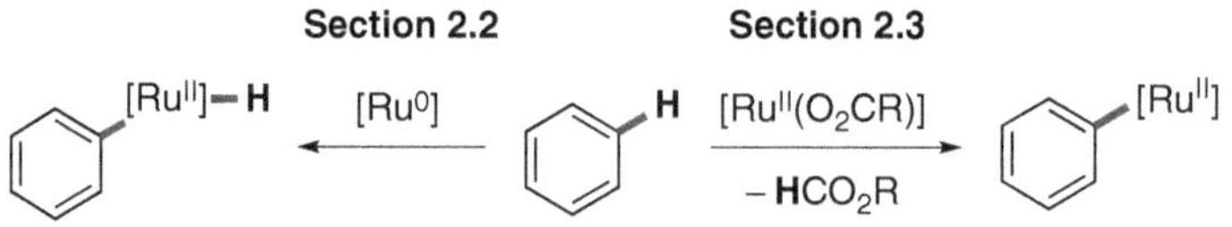

Scheme 2.1 Manifolds of ruthenium-catalyzed C—H alkylations.

ruthenium(0) catalysis that proceed via oxidative-addition C—H activation. Section 2.2.1 highlights methods that are facilitated by ruthenium(II) manifolds via carboxylate-assisted C—H activation.

2.2 Alkylation by Ruthenium(0) Catalysts via Oxidative-Addition C—H Activation

An important breakthrough in the field came in 1986 when Lewis and Smith reported the ruthenium-catalyzed alkylation of an oxy-functionalized arene. Here, they described the *ortho*-alkylation of phenol using a tetraphosphite ruthenium(II) catalyst **1** under a positive pressure of ethylene (Scheme 2.2) [11].

This pioneering work set the stage for the seminal contributions of the Murai group. Their numerous additions to this area of research were initiated in 1993 in a publication reporting one of the first high-yielding catalytic conversion of a C—H bond without the need for a reagent to be present in solvent quantities [12a]. Using aryl ketones **2** with vinyl silanes and terminal alkenes **3** they reported high yields of linear alkylated products **4** employing the $RuH_2(CO)(PPh_3)_3$ catalyst (Scheme 2.3). At the time, this example was the one with the highest turn over number (TON) reported for a catalytic C–H functionalization system. The tolerance of this reaction manifold to heteroaromatics (product **4a**) is notable and is prevalent in the literature.

The selectivity of this process toward the alkylation reaction with no trace of alkenylated reaction products, potentially formed by competitive β-hydride elimination, is accounted for in the reaction pathway. Experimental work [13] within the field as well as DFT calculations [14] has elucidated the following

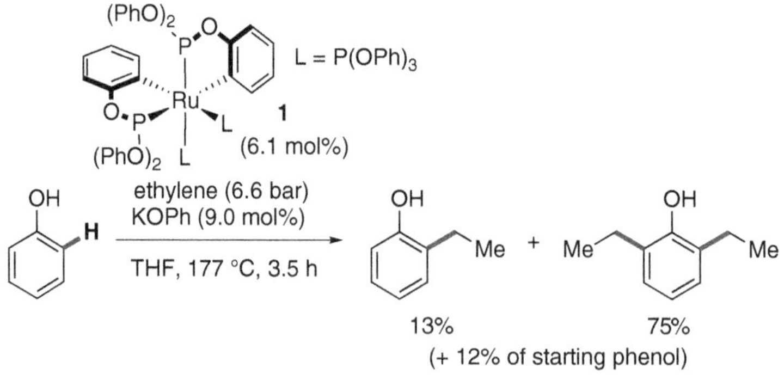

Scheme 2.2 The first catalytic ruthenium-catalyzed C—H alkylation reaction.

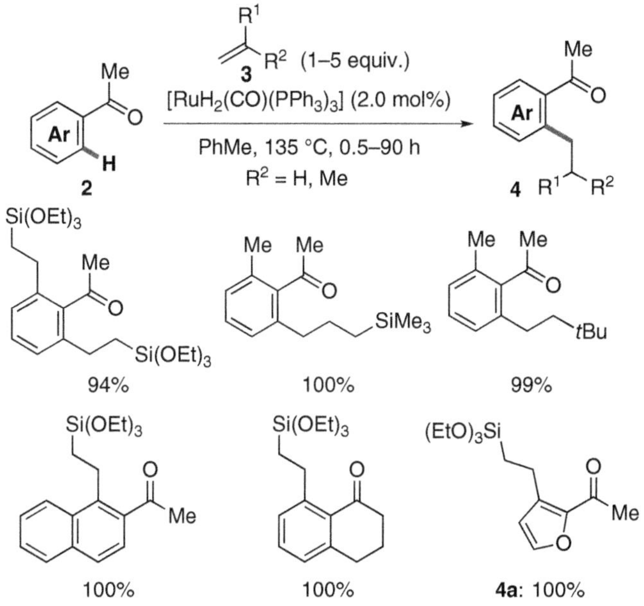

Scheme 2.3 Olefin hydroarylation by the Murai group.

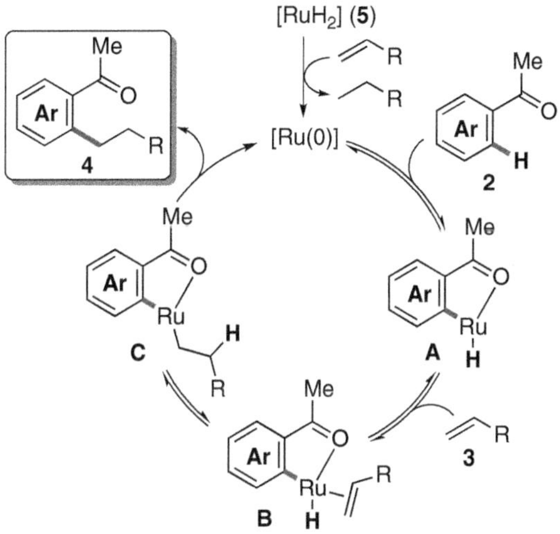

Scheme 2.4 Proposed catalytic cycle via ruthenium(0)-mediated C—H activation through oxidative addition.

reaction manifold (Scheme 2.4). A ruthenium-H_2 precatalyst, such as **5**, is widely accepted to be reduced under the reaction conditions via hydrogenation of the alkene. The exact nature of the ruthenium(0) active catalyst is unclear with the necessity of the carbonyl ligand debated. Nonetheless, it was Chatt and Davidson that initially observed the oxidative addition of ruthenium(0) species into C—H bonds [15a]. In the catalytic cycle, site-selective C—H cleavage by oxidative addition is induced by chelation assistance via the carbonyl's oxygen heteroatom to give a ruthenium(II) hydride intermediate **A**. In studies on *ortho*-selective C—H activation of functionalized benzenes by an iridium complex, which would appear to be a chelate-assisted process, the site-selectivity has been demonstrated to result not from a chelate-assisted process [15b]; instead, the chelation trapped the C—H activation product. In the ruthenium-catalyzed transformation, the C—H activation step is reversible and is therefore not turnover limiting [15a]. Coordination of the alkene reaction partner in **B** then sets up a migratory insertion that proceeds selectively as a hydroruthenation process forming **C**. Deuterium labeling has indicated that every step prior to this point is reversible. The final, and turnover limiting step, is the reductive elimination to form the C—C bond and the final product **4**.

The early driving force in this area of research was that of the Murai group who published contributions based on aryl ketones **2** and elaborated on their original catalytic system (Scheme 2.5) [16]. They demonstrated that vinyl silanes are excellent coupling partners in this reaction. However, at this point, the use of non-activated terminal alkenes in the reaction was poorly selective leading to linear and branched products. Nonetheless, good linear selectivity could be obtained for **4b** when using sterically and electronically biased styrene. At this point, only simple aryl ketones could facilitate the directed *ortho*-C—H alkylation reaction; however, positionally and electronically disparate substituents were well tolerated. Substituents in the *meta*-position resulted in the selective functionalization of the sterically least hindered C—H bond (**4c** and **4d**). Exceptions to this included the *meta*-methoxy substituent that resulted in the opposite regioselectivities through a secondary directing group effect (see, e.g., **4e**). Notably, *ortho*-methoxy acetophenones underwent C—OMe activation. Therefore, these substrates are not tolerated under the optimized reaction conditions. The high reactivity of this catalytic system often led to bis-alkylated products when steric blocking was absent.

The power of this reaction was quickly exploited in challenging applications. Weber and coworkers applied this transformation to the regioselective copolymerization between acetophenone **2** and dienes **6** (Scheme 2.6a) [17]. Subsequently, the same strategy was used for the modification of unsaturated polymers **7** (Scheme 2.6b) [18]. This catalytic modification of polymers had been previously achieved only by using more expensive rhodium catalysts in hydrogenation, hydrosilylation, and hydroformylation reactions [19].

The application of Murai's reaction system to further challenging applications was demonstrated by Woodgate *et al.* They reported on using vinyl silane coupling partners with 2-, 3-, and 4-hydroxyacetophenones **2**, 6-methoxytetralone as well as diterpenoid derivatives **8** (Scheme 2.7) [20]. They thereby highlighted the efficacy of this process on molecules of increasing complexity.

The Trost group was the first to highlight the alkylation of alkenyl C—H bonds under Murai's reaction conditions. They used cyclopentenecarboxylates and α,β-unsaturated esters, which reacted efficiently through β-C–H cleavage to give alkylated products (Scheme 2.8) [21]. The reaction tolerated ketals, thioketals, epoxides, and alkyl bromides on the ester substrate. Galactose diketal **10** with pendant α,β-unsaturated ester underwent effective hydroalkylation followed by

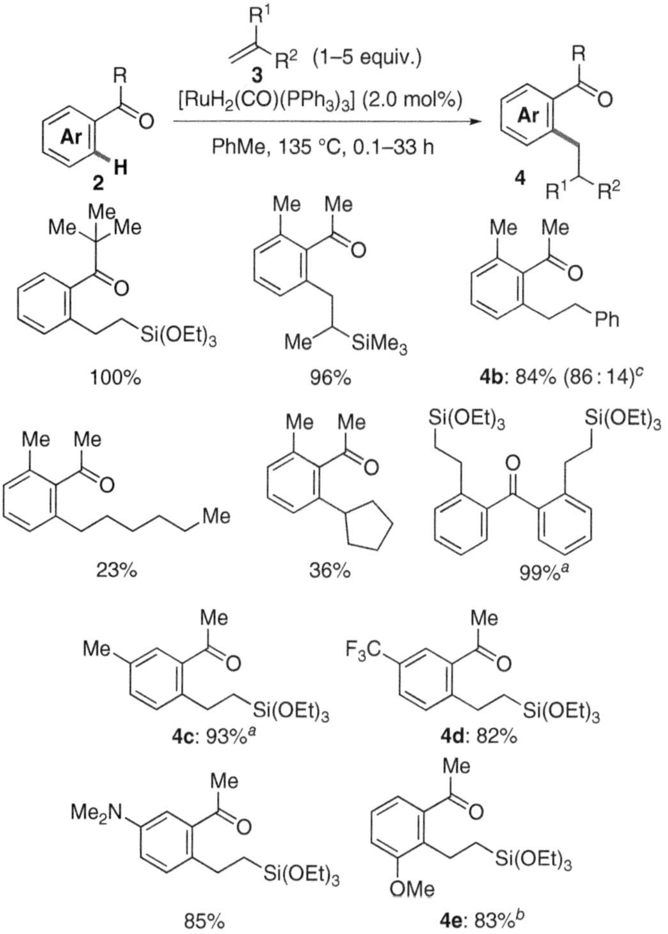

[a]3% of other isomer. [b]10% of other isomer and 7% of bis-alkylated product.
[c]Ratio of linear to branched product is given in parentheses.

Scheme 2.5 Early hydroarylation work using aryl ketones.

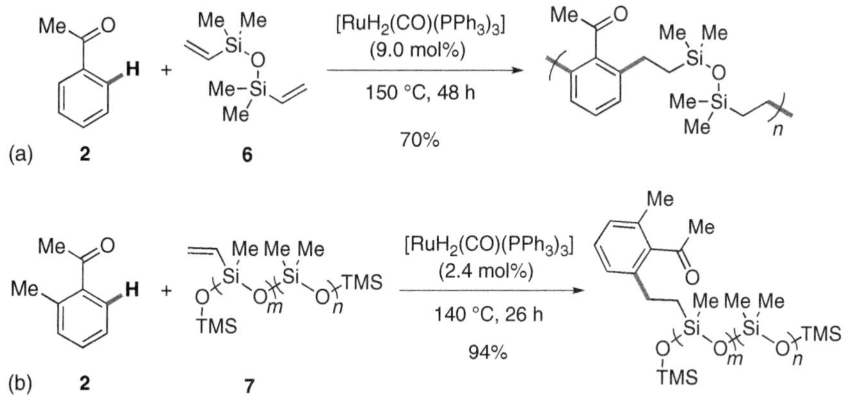

Scheme 2.6 The use of ruthenium-catalyzed C—H alkylations in polymer chemistry.

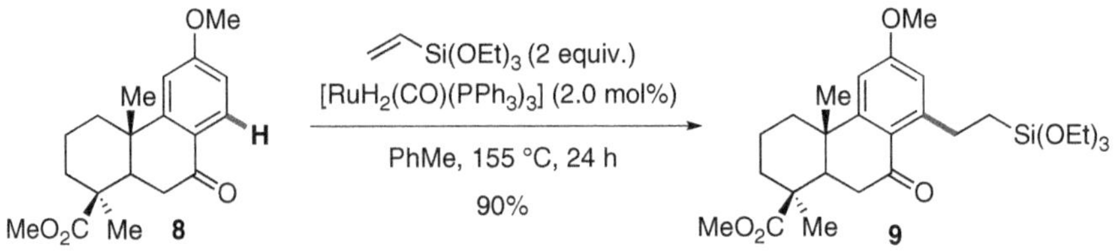

Scheme 2.7 The ruthenium-catalyzed C—H alkylation of the diterpene **8**.

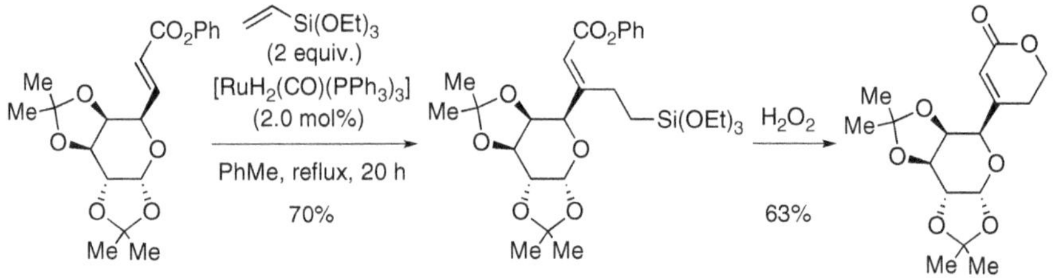

Scheme 2.8 Alkylation of β-C—H bonds in α,β-unsaturated esters by ruthenium(0) catalysis.

oxidative desilylation to form a hydropyrone derivative **11**. It was in this report that investigations began to underpin the nature of the active catalyst species for the first time. NMR studies demonstrated that addition of the alkene to the precatalyst resulted in hydrogenation and the formation of a new catalyst intermediate that is catalytically competent in the reaction process. Strong inhibition by performing the reaction in the presence of CO atmosphere and modest inhibition by performing the reaction in a closed system indicated that loss of CO ligand was also a preactivation step. This led the authors to suggest a $RuL(Ph_3P)_3$ species as the active, on-cycle catalyst. In a subsequent publication, Murai and coworkers also showed the addition of C—H bonds in α,β-unsaturated ketones to alkenes using the same $Ru(H)_2(CO)(PPh_3)_3$ catalyst [22].

Trost's work with acrylate derivatives sparked the use of other carbonyl directing groups in this reaction [21]. In 1996, the Murai group disclosed that aryl esters could produce alkylated products provided the aryl group was activated through an electron-withdrawing substituent (**13a–c**) or when the carbonyl was part of a lactone such as in chromanone substrates **13d** (Scheme 2.9) [23].

Aldehydes **14** had also proven to be a challenging directing group due to the problem of competitive decarbonylation pathways. However, adapting the C—H functionalization partner through increased steric hindrance or electron-donating substituents into the aldehyde carbonyl could disfavor the decarbonylation pathway and allow productive alkylation to occur (Scheme 2.10) [24].

While Murai's original catalyst, $RuH_2(CO)(PPh_3)_3$, has proven to be a robust and widely applicable precatalyst, efforts were made to expand the scope of available precatalysts, although only a few found wide synthetic applications [25]. Notable inclusion to these efforts include the use of $RuH_2(H_2)_2(PCy_3)_2$ catalyst with a labile dihydrogen ligand, which afforded activity even at ambient

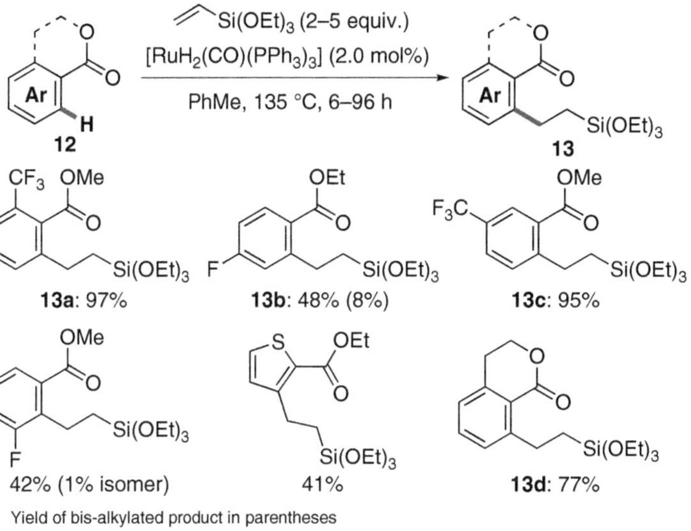

Scheme 2.9 Esters **12** as C–H alkylation substrates.

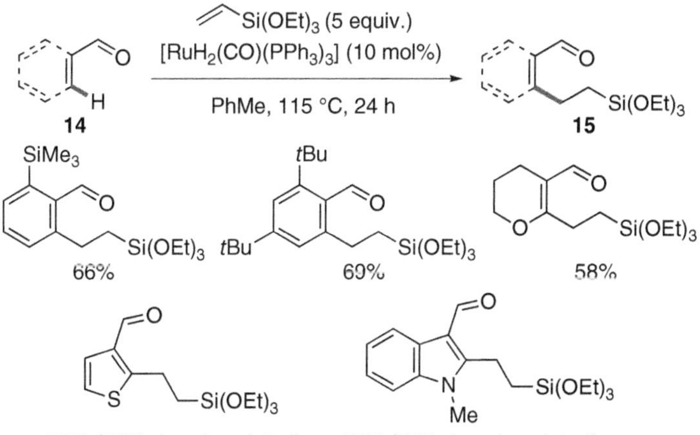

Scheme 2.10 Aldehydes as directing groups.

temperature [26]. More recently, a synthetically significant application was found using a system comprising [RuCl$_2$(*p*-cymene)]$_2$, a phosphine ligand and a formate salt that allowed the formation of an active catalyst species *in situ* (Scheme 2.11) [27]. The advantage of using the commercially available and bench-stable precatalyst [RuCl$_2$(*p*-cymene)]$_2$ over air- and/or moisture-sensitive ruthenium(0) or ruthenium hydride complexes is pronounced. In their initial publication using this system, Darses and Genet showed the excellent reactivity for a range of aryl ketones and even for an acrylamide derivative, which ultimately underwent a double bond migration to give product **17**. Reactivity was

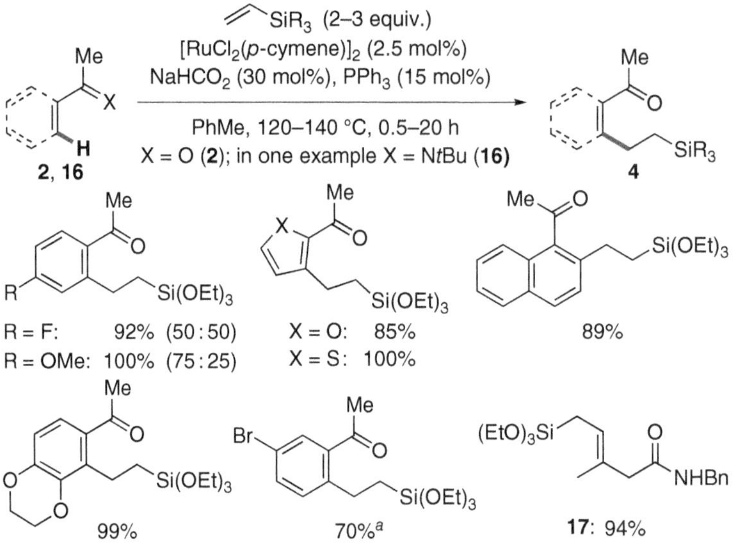

Ratio of obtained mono- to bis-alkylated products is given in parentheses
[a]Isolated as a 85 : 15 mixture of regioisomers, major isomer depicted

Scheme 2.11 The *in situ*-generated catalytic system using $[RuCl_2(p\text{-cymene})]_2$ precatalyst with PPh_3.

shown to decrease as the cone angle of the phosphine ligand increased. Overall, electron-poor, non-hindered phenyl phosphines were best for this reaction.

This $[RuCl_2(p\text{-cymene})]_2$-based system also helped provide a solution to the selectivity issues when using styrene derivatives (Scheme 2.12) [28]. Previous work by Murai had demonstrated the poor selectivity between linear and branched products when using the $RuH_2(CO)(PPh_3)_3$-based catalytic system. However, applying the *in situ* catalyst system discussed allowed access to products **4** using styrene coupling partners with excellent selectivity toward the linear products. The level of selectivity observed was shown to be dependent on the phosphine ligand used. Smaller phosphines gave increased amounts of branched products. For example, PMe_2Ph in a screening reaction gave 83 : 17 branched/linear selectivity. However, using electron-deficient aryl phosphines gave the best conversion and selectivity. As in Murai's results, secondary directing effects for the *meta*-substituted starting materials were shown (**4f** and **4g**).

In a subsequent publication, the Darses group demonstrated the full potential of their *in situ*-generated catalytic system (Scheme 2.13) [29]. In good to excellent yields they outlined the use of aryl ketones and aryl imines with a range of activated and non-activated olefinic reaction partners. They found that a lowered phosphine loading was necessary for the success of aldimine C—H functionalization partners. In addition, substrates without α-substitution gave bis-alkylated products. High regioselectivity was reported for styrenes with aldimines. Along with these synthetic efforts, they carried out a number of mechanistic investigations. Using $[RuCl_2(p\text{-cymene})]_2$ plus thallium formate and triphenylphosphine gave $[Ru(OCOH)_2PPh_3(p\text{-cymene})]$ which was confirmed

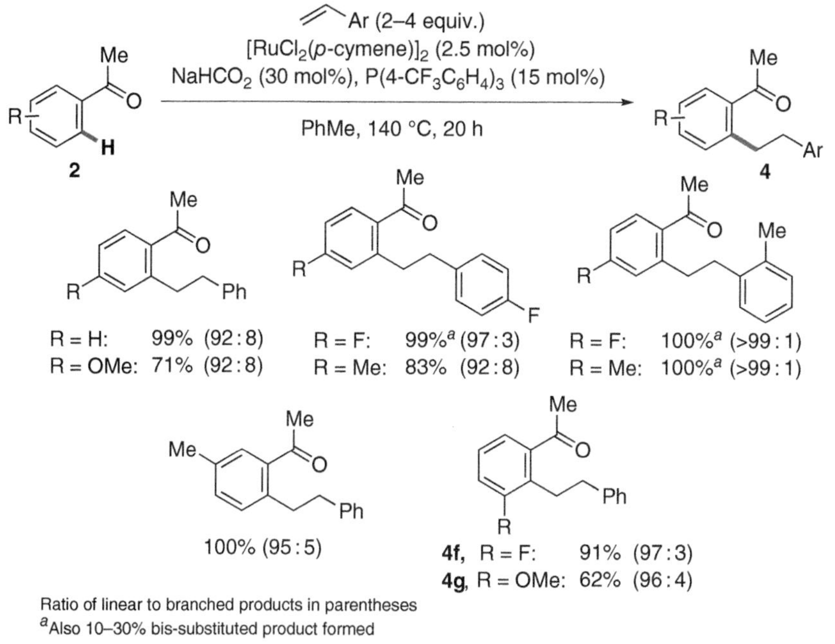

Scheme 2.12 Linear selective C—H alkylation using an *in situ*-generated catalyst.

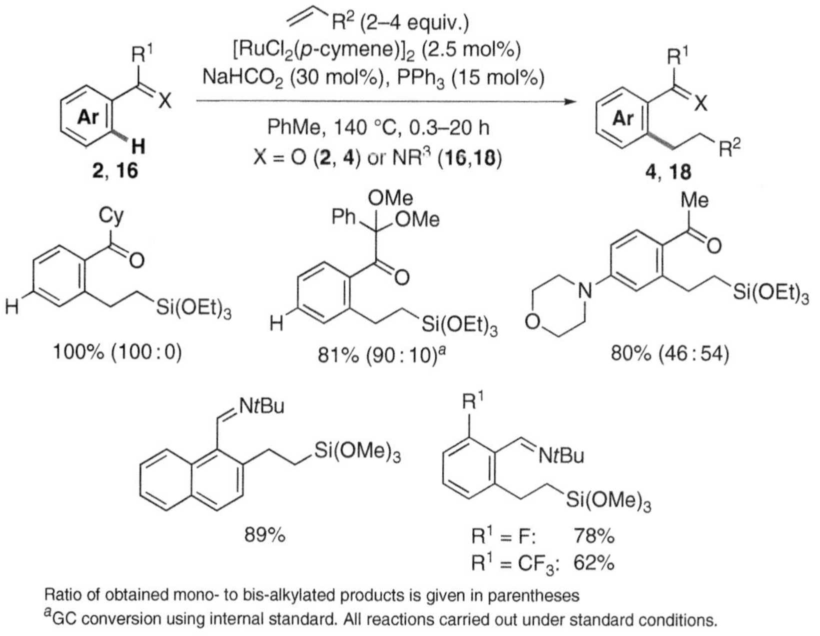

Scheme 2.13 Further development of ruthenium-catalyzed C—H alkylations using an *in situ*-generated catalyst.

by NMR spectroscopy and X-ray crystallography. This same catalyst species was shown to be catalytically competent. In fact, on taking this complex and heating to 50 °C, NMR spectroscopy studies showed arene decomplexation and hydride formation to give [RuH(η^2-OCOH)PPh_3]. At 70 °C a second decarboxylation led to a dihydride complex [$RuH_2(PPh_3)_4$]. Using *i*PrOH as solvent enabled the stable precipitation of the complex [$RuH_2(PPh_3)_4$]. They suggested that this catalyst species undergoes reductive elimination to give the catalytically active ruthenium(0) intermediate.

The advantage of using *i*PrOH in the stabilization of the dihydride complex [$RuH_2(PPh_3)_4$] led Darses and Genet to investigate its use in catalytic reactions as a greener and more catalytically competent solvent system for the hydroarylation reaction (Scheme 2.14) [30]. Indeed, they discovered that *i*PrOH increased the formation of the active catalyst species and demonstrated that acetone acts as a sacrificial hydride acceptor under two alternative sets of catalytic conditions.

During the early development of ruthenium(0)-catalyzed hydroarylation reaction, oxygen-based directing groups have dominated the applications. Nonetheless, nitrogen-containing directing groups have been featured over time. Aldimines **19** are one such moiety and were validated by Murai *et al.* using $Ru_3(CO)_{12}$ as the precatalyst (Scheme 2.15) [31]. In some instances, the reaction did give mixtures of *syn*- and *anti*-imines **20** as well as a tendency toward alkenylation products, presumably formed by a β-hydride elimination pathway. Yields of the alkenylated products varied from 4% to 22% and were more pronounced when using the original catalyst precursor $RuH_2(CO)(PPh_3)_3$.

In an early experiment of catalyst-controlled, site-selective C—H functionalization [32], the group of Murai showed the unique ability to cleave selectively one C—H bond over another through the choice of catalyst (Scheme 2.16) [33]. They found that the $RuH_2(CO)(PPh_3)_3$ complex preferentially provoked a keto-directed C—H bond activation in substrate **21** (Scheme 2.16a), while the reaction with $Ru_3(CO)_{12}$ species proceeded by nitrogen chelation assistance in an aldimine-facilitated process (Scheme 2.16b). This selectivity was robust

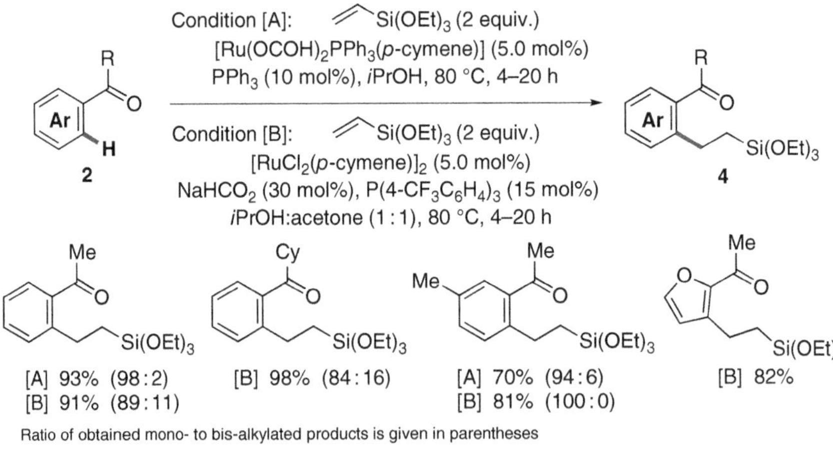

Scheme 2.14 The advantage of using *i*PrOH/acetone as a green solvent medium.

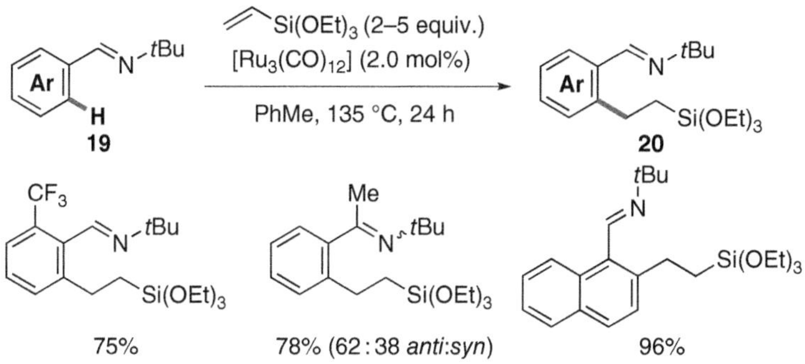

Scheme 2.15 The use of aryl imines in a C—H alkylation reaction.

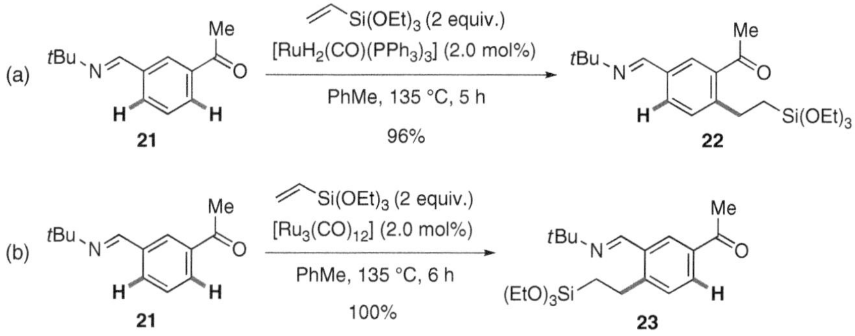

Scheme 2.16 Catalyst-controlled, site-selective C—H alkylations.

enough to work exclusively in the corresponding intermolecular competition experiment.

Using the same $Ru_3(CO)_{12}$-based catalytic system, cyclic imidates **24** and **27** were also shown to be effective as operative directing groups in the alkylation reaction (Scheme 2.17) [34]. However, mixtures of alkylation and alkenylation products **25, 26, 28,** and **29** were again apparent. The selectivity between the two pathways was shown to be dependent on the nature of the directing group and aryl *ortho*-substitution. The authors proposed that a change in pathways between hydro- and carboruthenation of the alkene, based on the stability of the subsequent seven-membered intermediates **A** and **B**, led to a change in chemoselectivity for the migratory insertion step.

In 2001, the Murai group also demonstrated the use of strongly coordinating aryl hydrazones **30** (Scheme 2.18) [35]. This time, using $RuH_2(CO)(PPh_3)_3$ as a catalyst, the key catalytic variation proved a change in solvent to mesitylene. The authors claimed that this alteration gave increased product selectivity for alkylation over alkenylation, which had previously plagued the use of nitrogen-based directing groups. They also saw an increased propensity to form only mono-alkylated over bis-alkylated products in this reaction medium.

It is not just strongly coordinating nitrogen moieties that have been established in this reaction. While a ruthenium catalyst had previously been highlighted to induce the dimerization of acrylonitrile [36], benzonitriles **32** could

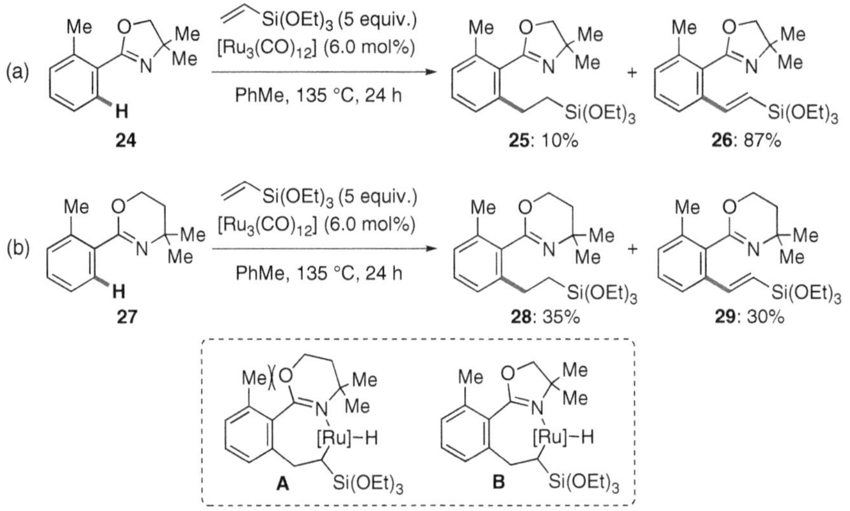

Scheme 2.17 Imidates **24**, **27** and a directing group-dependent change in product selectivity.

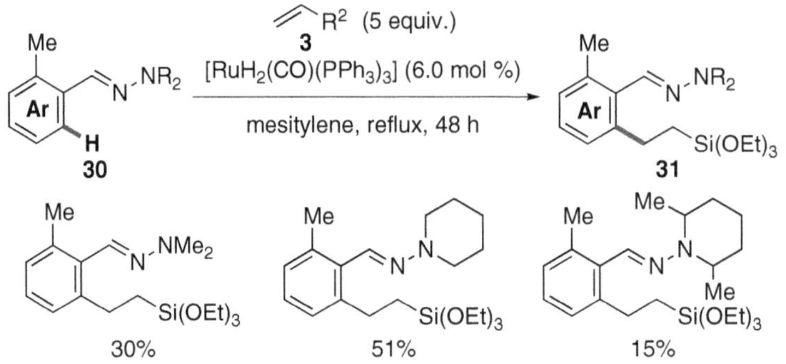

Scheme 2.18 Ruthenium(0)-catalyzed hydroarylation with hydrazone starting materials.

also participate in the alkylation reaction (Scheme 2.19) [37]. Using the $RuH_2(CO)(PPh_3)_3$ complex, the nitrile group was postulated to participate as a directing group via a π-bonded chelation mode by side-on coordination. Yields of the alkylated products **33** were excellent, although the reaction times were rather long for some examples, in many cases resulting in overalkylation.

The alkylation of aniline derivatives has only been disclosed in a solitary example (Scheme 2.20) [38]. Using *N*-methylaniline (**34**), Uchimaru showed an *ortho*-selective branched-type alkylation using solvent quantities of styrene at high temperature to facilitate the alkylation reaction.

Heteroaromatic directing groups have become common in modern C—H bond activation chemistry. However, strongly coordinating directing groups, such as heteroaromatic groups, are not as common in ruthenium(0)-catalyzed C—H alkylation reactions. An initial attempt by the Murai group was centered on an enantioselective reaction using naphthyl-3-methylpyridine, $RuH_2(CO)(PPh_3)_3$, and a chiral ligand [39]. However, both the yield and the level of enantioselectivity

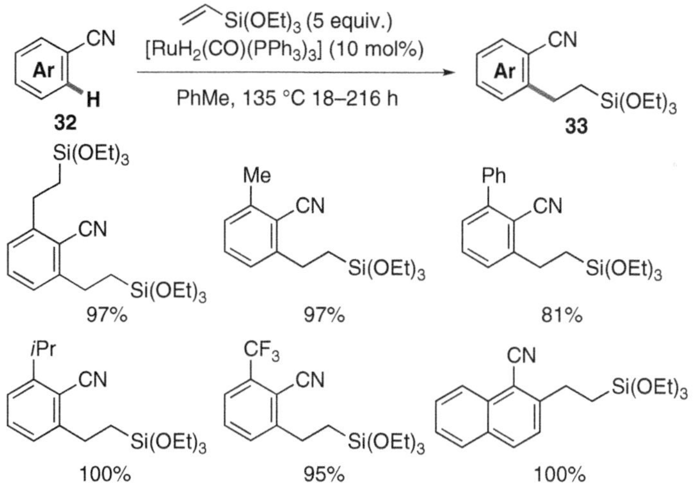

Scheme 2.19 Benzonitriles **32** as substrates in the ruthenium(0)-catalyzed C—H alkylation reaction.

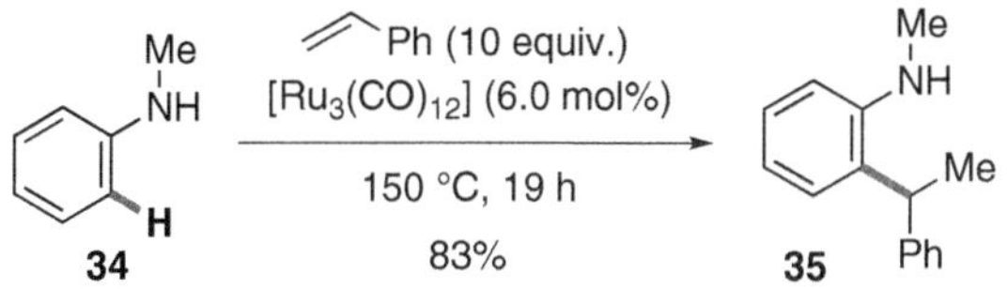

Scheme 2.20 An example of a ruthenium(0)-catalyzed C—H alkylation with aniline **34**.

were disappointing (15% yield, 15% ee). Even the racemic version of a reaction without chiral ligand present suffered from very poor conversion. It was not until the contributions of the Ackermann group that aryl pyridines were successfully utilized in this reaction (Scheme 2.21) [40]. Using $[RuCl_2(COD)]_n$ and the Buchwald-type phosphine ligand **38**, 2-aryl pyridines **36** were successfully added to methylenecyclopropanes **37** to give *cis*-substituted products Cpr-**36**. The compatibility of the strained carbocycle in such a hydroarylation reaction highlights the power of this methodology. The full utility of this method was outlined in a subsequent full paper which also demonstrated aryl oxazoline as compatible C—H functionalization partners [41]. Notably, the Ackermann group reported that acetophenone is not productive under the described reaction conditions.

The kinetically challenging functionalization of $C(sp^3)$—H bonds has also been accomplished in the ruthenium(0)-catalyzed C—H alkylation reaction. Using the $Ru_3(CO)_{12}$ precatalyst, the Murai group reported the alkylation of *N*-pyridyl pyrrolidines to give mono- and bis-substituted products (Scheme 2.22) [42b]. Terminal and internal alkenes were also successful coupling partners. Employing an ethylene atmosphere resulted in ethylated products. However, the ethylation reaction required the additional presence of carbon monoxide.

The addition of an organic acid and employing $(iPr)_2CHOH$ as a solvent allowed for directed $C(sp^3)$—H α-alkylation of piperidines **41** as well as of

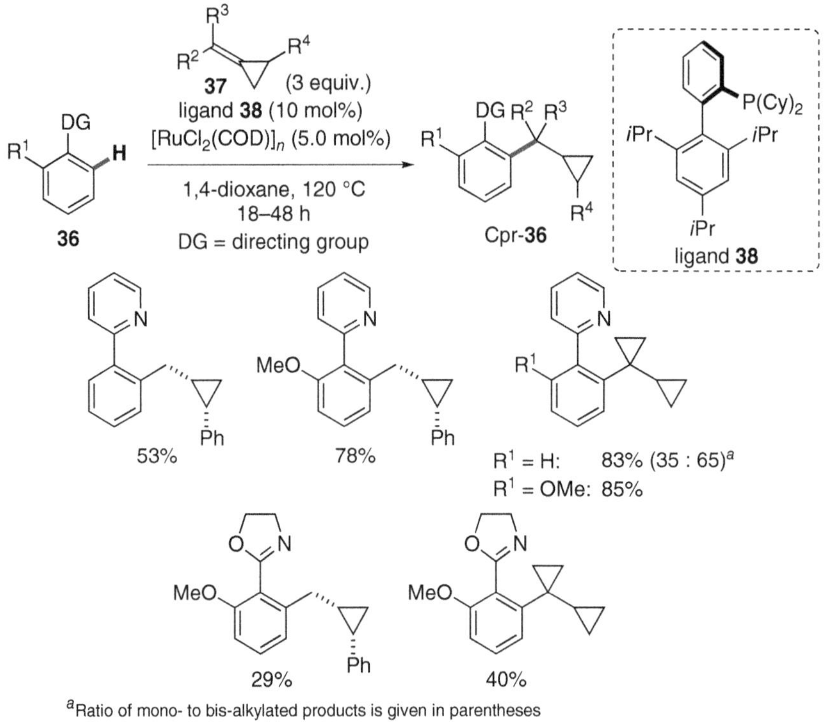

Scheme 2.21 Methylenecyclopropanes **37** as alkylating reagents.

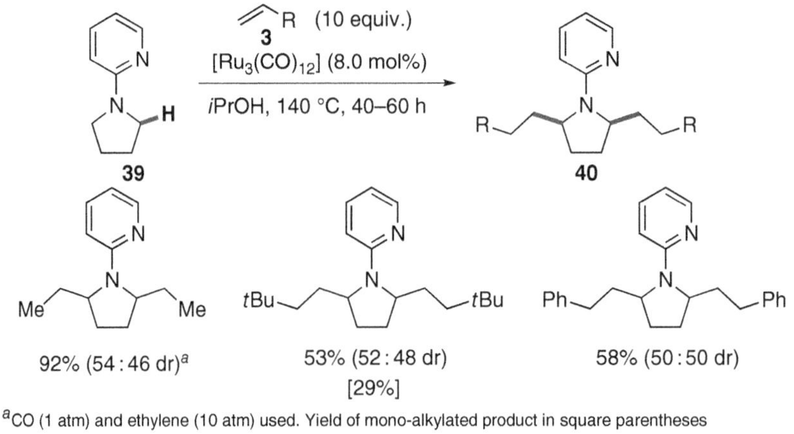

Scheme 2.22 Alkylation of C(sp^3)—H bonds using $Ru_3(CO)_{12}$ catalyst.

selected bicyclic amines using dioxolane-protected alkenones **42**, as was recently reported by Maes *et al.* (Scheme 2.23) [42a]. Notably, 3,4,5-trifluorobenzoic acid (TFBA) appeared to be the additive of choice. However, piperidines **41** with $R^1 = H$ as well as 3-substituted SMs **41** afforded difficulty in separating mixtures of mono- and bis-alkylated products.

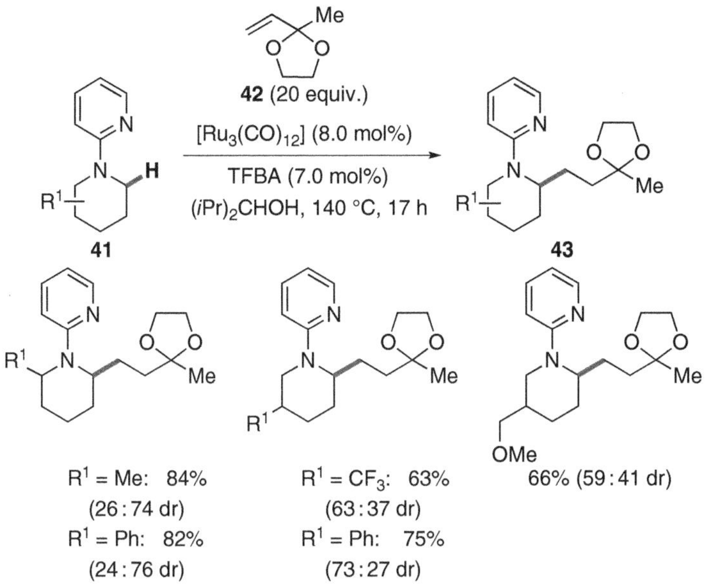

Scheme 2.23 Ruthenium-catalyzed alkylation of C(sp^3)—H bonds with substituted alkenes.

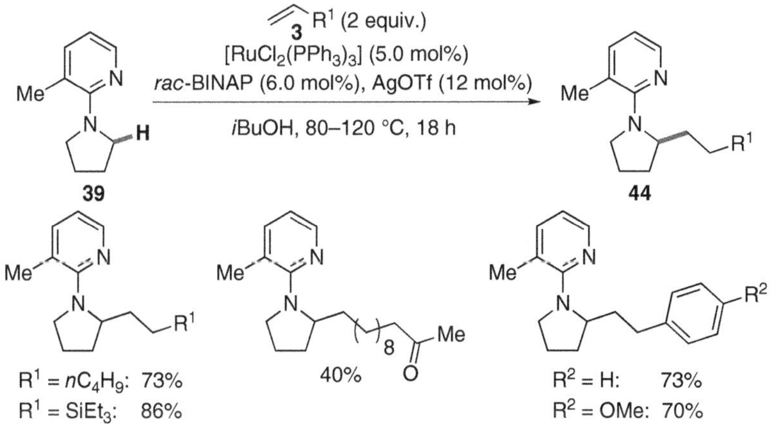

Scheme 2.24 Alkylation of C(sp^3)—H bonds using a $RuCl_2(PPh_3)_3$/*rac*-BINAP catalytic system.

The need for special reaction atmospheres, poor mono- to bis-alkylation selectivity, and the long reaction times led the Ackermann group to demonstrate the use of $RuCl_2(PPh_3)_3$ and a racemic BINAP ligand with (3-methylpyridin-2-yl)pyrrolidines **39** to furnish mono-alkylated products **44** in good to excellent yields (Scheme 2.24) [43].

2.2.1 Alkylation by Ruthenium(II) Catalysts via Carboxylate-Assisted C—H Activation

Over the course of investigations in the field of metal-catalyzed C—H functionalization, two mechanistically distinct reaction pathways became

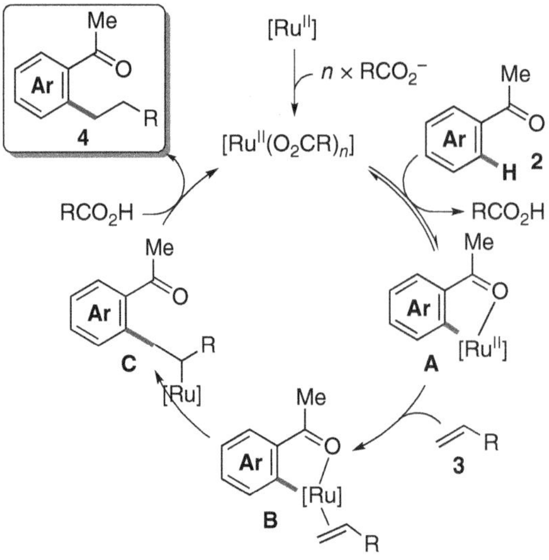

Scheme 2.25 Catalytic cycle via ruthenium(II)-mediated, carboxylate-assisted C—H bond activation.

apparent. The initially applied and *in situ*-generated ruthenium(0) species required the use of air- and moisture-sensitive catalysts and catalytic intermediates. However, a robust and widely applicable mechanistic working mode has emerged within an isohypsic, that is, redox-neutral, ruthenium(II) manifold. Notably, a ruthenium(0) species does not participate in this catalytic cycle (Scheme 2.25). The C—H activation step in **2** proceeds via an isohypsic – that is, redox-neutral carboxylate-assisted pathway [44g]. The formation of intermediate **A** has been shown to be facile, as indicated through experiments with deuterium-labeled substrates [44d, e] and detailed kinetic studies [44b, c]. Therefore, in most cases this elementary C—H activation step does not appear to be turnover limiting. The migratory insertion step in intermediate **B** is also distinct from that previously discussed. A carboruthenation process leads to intermediate **C**. Finally, protodemetalation furnishes the product **4**. This pathway has also been supported by DFT calculations, which indicated the formation of intermediate **A** to proceed via a six-membered transition state through carboxylate assistance [44a].

The Ackermann group was the first to report on such a carboxylate-assisted manifold (Scheme 2.26) [45]. Alkylated phenylpyridines **45a** and **45b**, phenylimidazoles **45c** and **45d**, phenylpyrazoles **45e**, and even biologically relevant indoles **45f** were available through the alkylation reaction in high yields. Both activated and non-activated alkenes worked well under this new carboxylate-assisted system with excellent linear selectivity. Also notable in this new reaction system was the selectivity toward the mono-alkylated products.

This working mode was elegantly extended in 2013 by the Chatani group, who showed that a ruthenium(II) catalyst in the presence of a carboxylate salt could enable the alkylation of benzamides using α,β-unsaturated ketones **47** as coupling partners (Scheme 2.27) [46]. They proposed that the poor substrate scope in the

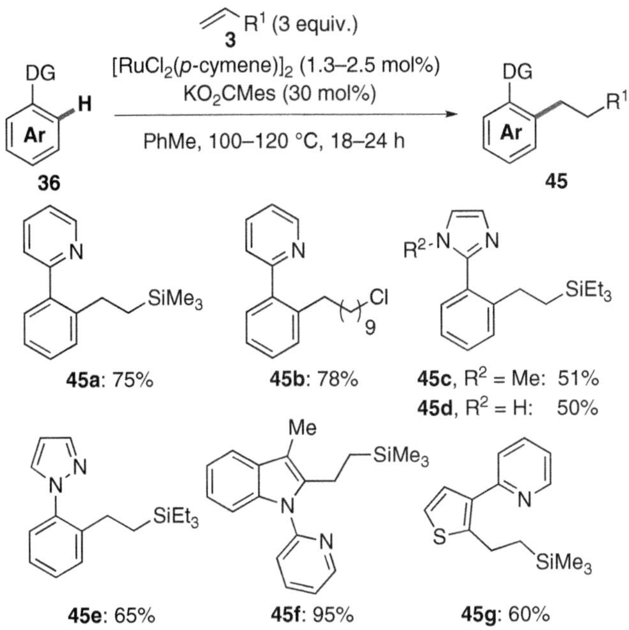

Scheme 2.26 The first report on carboxylate-assisted, ruthenium(II)-catalyzed C—H alkylations via isohypsic C—H activation.

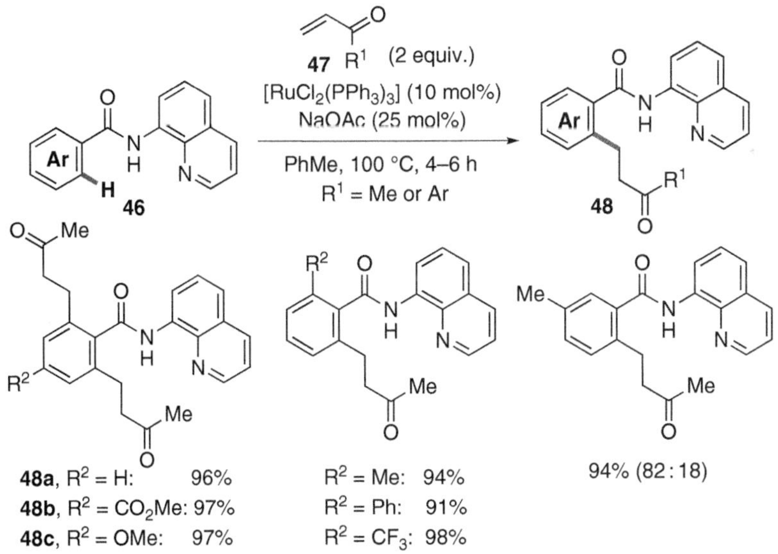

Scheme 2.27 Carboxylate-assisted alkylations using a bidentate directing group.

work of the Darses group was due to the less-effective monodentate directing groups employed. They established that the bidentate 8-aminoquinoline directing group, previously developed by Daugulis *et al.* [47], allowed the alkylation of benzamides **46** [46]. Formation of bis-alkylated products **48a–c** was apparent under these reaction conditions in substrates in the absence of steric blocking.

Utilizing their carboxylate-containing system (*vide supra*), the Ackermann group was able to overcome the need for a bidentate directing group when using benzamides **49** as partners for C—H functionalizations. Employing α,β-unsaturated ketones **47** as coupling partners, they achieved good to excellent yields of alkylated benzamides **50** using water as a reaction medium (Scheme 2.28) [48]. It is noteworthy that these catalytic conditions again favored the selective formation of mono-alkylated products **50**. The site-selectivity mirrored the ruthenium(0) manifold, as sterics dictated the transformation to occur at the least hindered C—H bond in *meta*-substituted benzamides, thereby furnishing compounds **50d** and **50e**. However, secondary directing group effects for certain substituents, such as fluoro or alkoxy, led to the alternative regioisomers **50f** and **50g**.

The Ackermann group also demonstrated the ability of the carboxylate-assisted manifold in the hydroarylations of methylenecyclopropanes **37** (Scheme 2.29) [49]. Two sets of reaction conditions via carboxylate assistance were reported. One set used potassium mesitylenecarboxylate and $[Ru(O_2CMes)_2(p\text{-cymene})]$, while the second incorporated the use of acetate with the $[RuCl_3{\cdot}(H_2O)_n]$

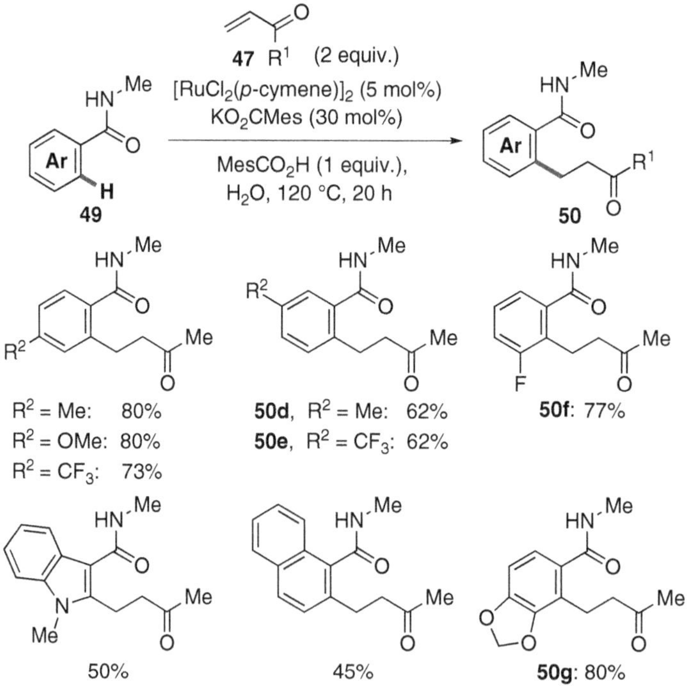

Scheme 2.28 Alkylation of benzamides **49** with monodentate DGs by carboxylate assistance.

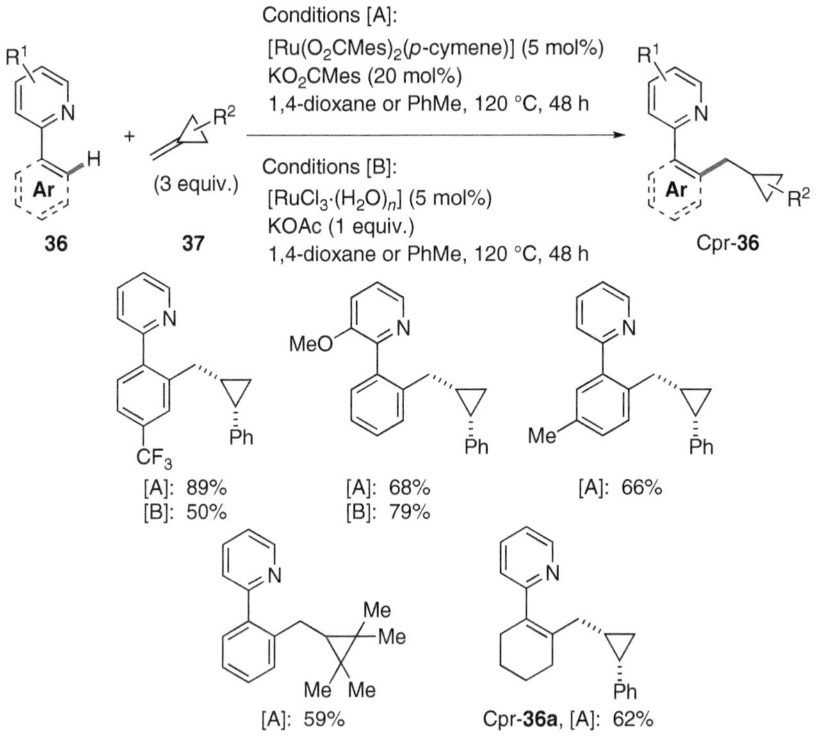

Scheme 2.29 Carboxylate assistance in hydroarylations of methylenecyclopropanes **37**.

precatalyst. Both sets of conditions selectively afforded only *cis*-products Cpr-**36** (cf. Scheme 2.21). The stability of the strained three-membered carbocycle under the reaction conditions is remarkable. For the first time in a carboxylate-assisted manifold the functionalization of a vinylic C—H bond furnishing product Cpr-**36a** was demonstrated.

Aryl phosphine and phosphine oxides are attractive moieties in functional organic materials, ligands, or organocatalysts among others. Despite this utility there has been only a single reported example of such a substrate being used in a ruthenium-catalyzed C—H alkylation reaction using an alkenyl coupling partner. In 2013, the group of Miura and Satoh disclosed an example of an alkylation reaction using triphenylphosphine oxide (**51**) as a C—H functionalization substrate. This was achieved with $[RuCl_2(p\text{-cymene})]_2$ as the catalyst, 1-adamantanecarboxylic acid in the presence of a silver salt and an acrylate (Scheme 2.30) [50]. In this publication, the main body of work focuses on alkenylation processes using alkyne substrates.

As outlined earlier, conditions for the selective branched or linear alkylation of certain types of olefins has been challenging. In 2014, the group of Ramana disclosed two sets of conditions, which gave branched or linear products **55** or **56** with good levels of selectivity using 2-aroylbenzofurans **54** and acrylate reaction partners [51b]. When $RuCl_2(PPh_3)_3$ and potassium carbonate were used, the selective formation of the branched product **55** was observed (Scheme 2.31),

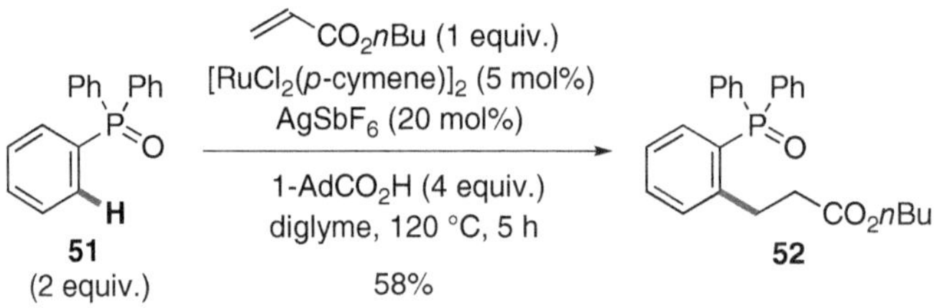

Scheme 2.30 Triphenylphosphine oxide (**51**) in the C—H alkylation reaction.

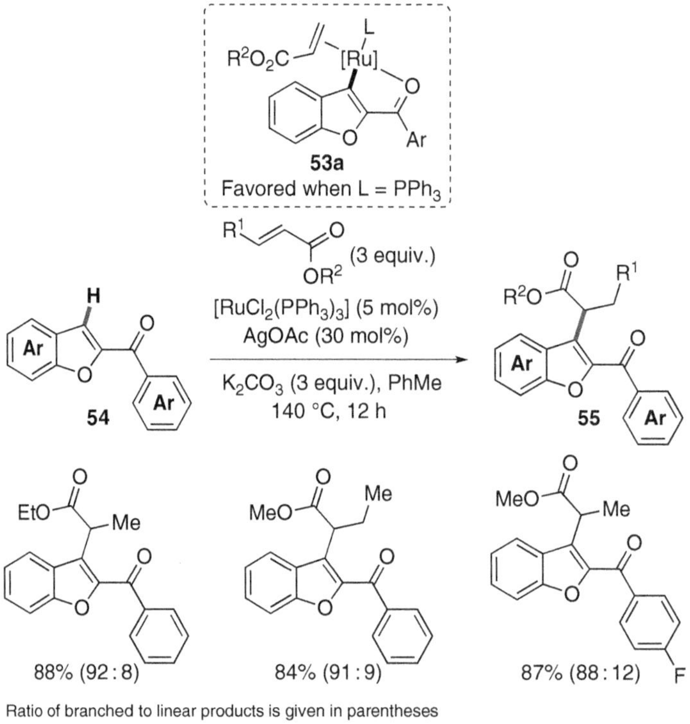

Scheme 2.31 Triphenylphosphine ligand leading to branched product **55** selectivity.

while carbonate-based conditions gave the linear product **56** preferentially (Scheme 2.32). The rationale for changes in selectivity is based on both steric and electronic factors. The authors postulated that the bulkier triphenylphosphine ligand favored coordination of the acrylate such as in **53a**, thus minimizing the steric clash between the acrylate and the ligand (L) of ruthenium. This orientation also placed the more electron-rich ruthenium metal next to the β-carbon of the acrylate ligand leading to the branched product **55**. Conversely, in the conditions using $[RuCl_2(p\text{-cymene})]_2$, the arene ligand in **53b** is smaller and less electron-donating. The Ru—C bond is therefore placed proximal to the β-carbon of the acrylate ligand ultimately giving the linear product **56**.

However, such explanations continue to be somewhat speculative and the mechanisms await further elucidation. For example, similar starting materials,

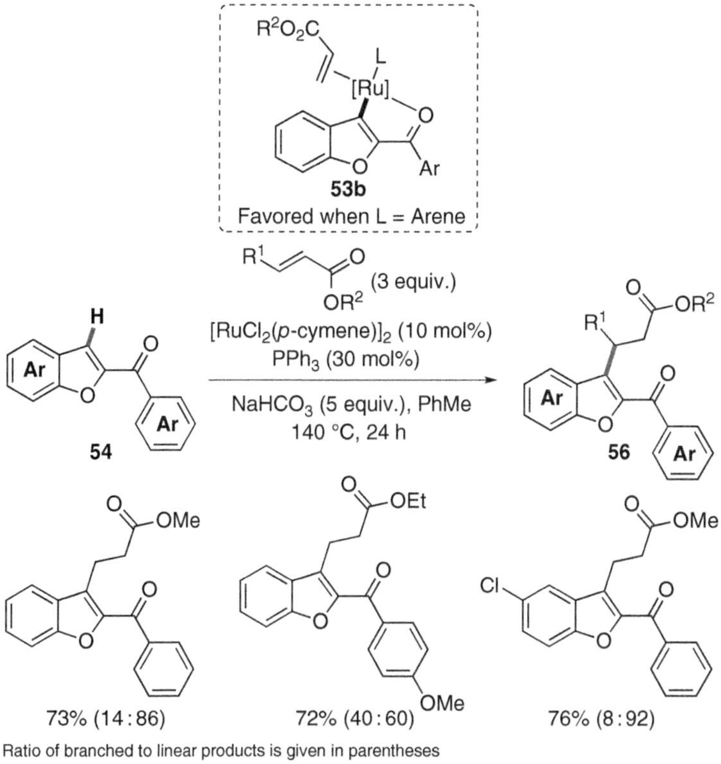

Scheme 2.32 Arene ligand leading to linear products.

such as **54**, but with pyridin-2-yl as the directing group, underwent only alkenylation under the reaction conditions shown in Scheme 2.31. Surprisingly, without potassium carbonate the reaction can be switched from alkenylation to linear alkylation [51a].

In 2015, the group of Prabhu also demonstrated keto-directed alkylation in the carboxylate-assisted reaction manifold (Scheme 2.33) [52]. Acetophenone derivatives **2** were shown to react with maleimides **57** in the presence of the $[RuCl_2(p\text{-cymene})]_2$ catalyst, a silver salt, and superstoichiometric amounts of copper acetate. Detailed DFT calculations supported the mechanism that had been previously proposed (*vide supra*).

In both reaction manifolds described earlier the hydroarylated partner has remained structurally and electronically simple. In 2015, the Ackermann group described a first example using an allene **60** in a carboxylate-assisted, ruthenium-catalyzed alkylation reaction (Scheme 2.34) [53]. This reaction proceeded with excellent regioselectivity toward the alkene giving the linear allylated product **61**. Detailed mechanistic experimental studies, employing intermolecular competition experiments as well as isotopically labeled compounds, were suggestive of a kinetically relevant C—H metalation step, which occurs by the assistance of a carboxylate moiety. This hydroarylation also set the stage for unprecedented C—H allylations with removable directing groups

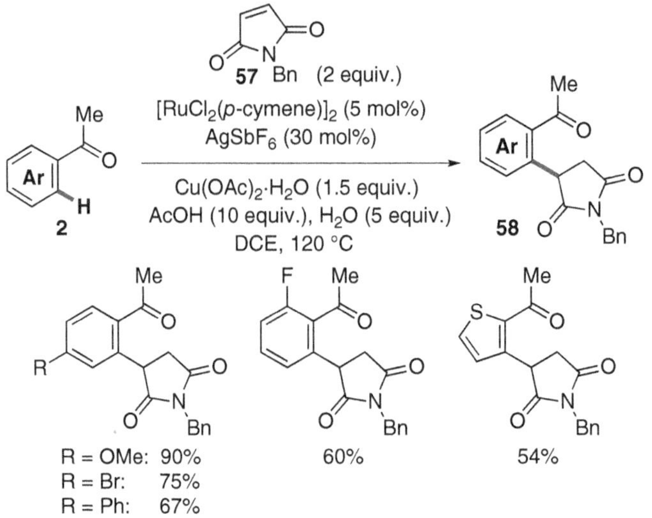

Scheme 2.33 The C—H alkylation of acetophenones **2** using maleimide **57**.

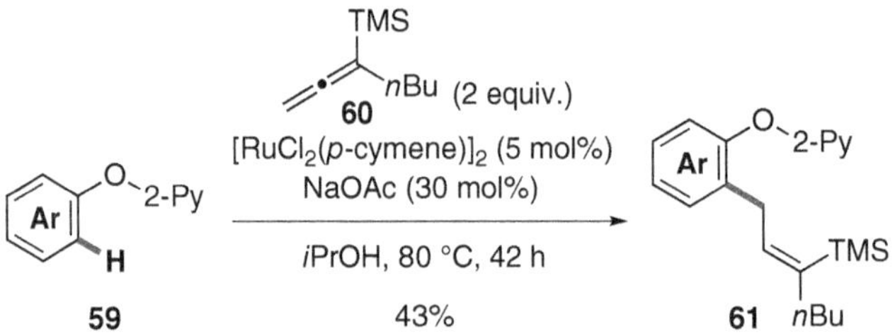

Scheme 2.34 Allene **60** in the ruthenium(II)-catalyzed carboxylate-assisted C—H alkylation reaction.

[54] in a step-economical fashion. The use of such a complex coupling partner as **60** points to the promising future of ruthenium(II)-catalyzed alkylations with alkenes and employing more complex reaction systems.

2.3 Summary and Conclusions

During the past decade, a number of valuable protocols for the direct alkylation of C(sp^2)—H and C(sp^3)—H bonds via the hydroarylation of alkenes have been developed. Among them, ruthenium-catalyzed C—H functionalizations have occupied a prominent position due to the versatility of ruthenium catalysts. Direct alkylation permits efficient introduction of alkyl groups with various functionalities in an atom- and step-economical fashion. While early work featured air- and/or moisture-sensitive ruthenium(0) catalysts, nowadays this chemistry operates with advantageous commercially available and bench-stable ruthenium(II) precatalysts, primarily based on [$RuCl_2$(arene)]$_2$. The excellent chemo- and positional-selectivities of the carboxylate- assisted ruthenium(II) catalysis strategy were, for instance, reflected by robust hydroarylations of non-activated alkenes as well as C—H functionalizations with methylenecyclopropanes via conservation of the sensitive cyclopropane motifs. This platform is broadly applicable and enables C—H functionalization by weakly coordinating and removable directing groups with excellent levels of positional selectivity. Considering the practical importance of C—H bond functionalization for the practitioner in chemical industries, material sciences, and medicinal chemistry, significant further progress is expected in this rapidly evolving research area. Development of methods for asymmetric C—H bond functionalizations under mild reaction conditions as well as further strategies for C(sp^3)—H transformations are expected to have a major impetus on this research area.

Abbreviations

1-Ad	1-Adamantyl
Ar	Aryl
BINAP	[2,2′-Bis(diphenylphosphino)-1,1′–binaphthyl]
Bn	Benzyl
Cy	Cyclohexyl
DCE	Dichloroethane
DG	Directing group
Equiv.	Equivalents
*i*BuOH	*iso*-Butanol
*i*PrOH	*iso*-Propanol
Mes	2,4,6-trimethylphenyl
NMR	Nuclear magnetic resonance
Ph	Phenyl
PhMe	Toluene
Py	Pyridine
[Ru]	Ruthenium catalyst
SM	Starting material
*t*Bu	*tert*-Butyl
TFBA	3,4,5-Trifluorobenzoic acid
THF	Tetrahydrofuran
TMS	Trimethylsilyl
TON	Turn over number

References

1 Friedel, C. and Crafts, J.M. (1877) Sur une nouvelle méthode générale de synthèse d'hydrocarbures, d'acétones, etc. *Compt. Rend.*, **84**, 1392–1450.
2 Ackermann, L. (2010) Metal-catalyzed direct alkylations of (hetero)arenes via C—H bond cleavages with unactivated alkyl halides. *Chem. Commun.*, **46**, 4866–4877.

3 For selected examples of C—H alkylation using alkyl halides see: (a) Ackermann, L., Hofmann, N., and Vicente, R. (2011) Carboxylate-assisted ruthenium-catalyzed direct alkylations of ketimines. *Org. Lett.*, **13**, 1875–1877; (b) Ackermann, L., Novák, P., Vicente, R., and Hofmann, N. (2009) Ruthenium-catalyzed regioselective direct alkylation of arenes with unactivated alkyl halides through C—H bond cleavage. *Angew. Chem. Int. Ed.*, **48**, 6045–6048, and references cited therein; recent review: (c) Yi, C.S. (2016) Metal-catalyzed C—H alkylation using RX compounds. In: *Catalytic Transformations via C—H Activation*, Vol. 1 (Yu, J.-Q., Ed.). *Sci. Synth.*, **4** (3), 229–270.

4 For early reviews on cyclometalated compounds see: (a) Rothwell, I.P. (1985) Carbon hydrogen bond activation in early transition-metal systems. *Polyhedron*, **4**, 177–200; (b) Omae, I. (1980) Organometallic intramolecular-coordination compounds containing a phosphorus donor ligand. *Coord. Chem. Rev.*, **32**, 235–271; (c) Bruce, M.I. (1977) Cyclometalation reactions. *Angew. Chem. Int. Ed.*, **16**, 73–86; (d) Dehand, J. and Pfeffer, M. (1976) Cyclometallated compounds. *Coord. Chem. Rev.*, **18**, 327–352.

5 For selected reviews on chelation-assisted C—H functionalization see (a) Moselage, M., Li, J., and Ackermann, L. (2016) Cobalt-catalyzed C—H activation. *ACS Catal.*, **6**, 498–525; (b) Borie, C., Ackermann, L., and Nechab, M. (2016) Enantioselective syntheses of indanes: from organocatalysis to C—H functionalization. *Chem. Soc. Rev.*, **45**, 1368–1386; (c) Ye, B. and Cramer, N. (2015) Chiral cyclopentadienyls: enabling ligands for asymmetric Rh(III)-catalyzed C—H functionalizations. *Acc. Chem. Res.*, **48**, 1308–1318; (d) Daugulis, O., Roane, J., and Tran, L.D. (2015) Bidentate, monoanionic auxiliary-directed functionalization of carbon–hydrogen bonds. *Acc. Chem. Res.*, **48**, 1053–1064; (e) Kuhl, N., Schröder, N., and Glorius, F. (2014) Formal S_N-type reactions in rhodium(III)-catalyzed C—H bond activation. *Adv. Synth. Catal.*, **356**, 1443–1460; (f) Ackermann, L. (2014) Carboxylate-assisted ruthenium-catalyzed alkyne annulations by C—H/Het—H bond functionalizations. *Acc. Chem. Res.*, **47**, 281–295; (g) Rouquet, G. and Chatani, N. (2013) Catalytic functionalization of C(sp^2)—H and C(sp^3)—H bonds by using bidentate directing groups. *Angew. Chem. Int. Ed.*, **52**, 11726–11743; (h) Satoh, T. and Miura, M. (2010) Oxidative coupling of aromatic substrates with alkynes and alkenes under rhodium catalysis. *Chem. Eur. J.*, **16**, 11212–11222; (i) Giri, R., Shi, B.-F., Engle, K.M., Maugel, N., and Yu, J.-Q. (2009) Transition metal-catalyzed C—H activation reactions: diastereoselectivity and enantioselectivity. *Chem. Soc. Rev.*, **38**, 3242–3272; (j) Chen, X., Engle, K.M., Wang, D.-H., and Yu, J.-Q. (2009) Palladium(II)-catalyzed C—H activation/C—C cross-coupling reactions: versatility and practicality. *Angew. Chem. Int. Ed.*, **48**, 5094–5115; (k) Ackermann, L., Vicente, R., and Kapdi, A.R. (2009) Transition-metal-catalyzed direct arylation of (hetero)arenes by C—H bond cleavage. *Angew. Chem. Int. Ed.*, **48**, 9792–9826; (l) Bergman, R.G. (2007) Organometallic chemistry: C—H activation. *Nature*, **446**, 391–393, and references cited therein.

6 For selected recent reviews on ruthenium-catalyzed C—H bond functionalization, see (a) Dana, S., Yadav, M.R., and Sahoo, A.K. (2016) Ruthenium-catalyzed C—N and C—O bond-forming processes from C—H bond functionalization. *Top. Organomet. Chem.*, **55**, 189–215; (b) Molnar, A. and Papp, A. (2015) Ruthenium-catalyzed C—H activation and coupling reactions in organic synthesis. *Curr. Org. Chem.*, **20**, 381–458; (c) Bruneau, C. (2014) sp^3 C—H bond functionalization with ruthenium catalysts. *Top. Organomet. Chem.*, **48**, 195–236; (d) De Sarkar, S., Liu, W., Kozhushkov, S.I., and Ackermann, L. (2014) Weakly-coordinating directing groups for ruthenium(II)-catalyzed C—H activation. *Adv. Synth. Cat.*, **356**, 1461–1479; (e) Thirunavukkarasu, V.S., Kozhushkov, S.I., and Ackermann, L. (2014) C—H nitrogenation and oxygenation by ruthenium catalysis. *Chem. Commun.*, **50**, 29–39; (f) Li, B. and Dixneuf, P.H. (2013) sp^2 C—H bond activation in water and catalytic cross-coupling reactions. *Chem. Soc. Rev.*, **42**, 5744–5767; (g) Arockiam, P.B., Bruneau, C., and Dixneuf, P.H. (2012) Ruthenium(II)-catalyzed C—H bond activation and functionalization. *Chem. Rev.*, **112**, 5879–5918; (h) Ackermann, L. (2010) Transition metal-catalyzed direct arylations via C—H bond cleavages. *Pure Appl. Chem.*, **82**, 1403–1413; (i) Ackermann, L. and Vicente, R. (2010) Ruthenium-catalyzed direct arylations through C—H bond cleavages. *Top. Curr. Chem.*, **292**, 211–229.

7 For selected examples of ruthenium-catalyzed annulations via C—H functionalization see: (a) Warratz, S., Kornhaaß, C., Cajaraville, A., Niepötter, B., Stalke, D., and Ackermann, L. (2015) Ruthenium(II)-catalyzed C—H activation/alkyne annulation by weak coordination with O_2 as the sole oxidant. *Angew. Chem. Int. Ed.*, **54**, 5513–5517; (b) Banerjee, A., Santra, S.K., Mohanta, P.R., and Patel, B.K. (2015) Ruthenium(II) catalyzed regiospecific C—H/O—H annulations of directing arenes via weak coordination. *Org. Lett.*, **17**, 5678–5681; (c) Li, R., Hu, Y., Liu, R., Hu, R., Li, B., and Wang, B. (2015) Ruthenium(II)-catalyzed oxidative annulation reactions of arylimidazolium salts via *N*-heterocyclic carbene-directed C—H activation. *Adv. Synth. Catal.*, **357**, 3885–3892; (d) Li, J. and Ackermann, L. (2014) Ruthenium-catalyzed oxidative alkyne annulation by C—H activation on ketimines. *Tetrahedron*, **70**, 3342–3348; (e) Kornhaaß, C., Kuper, C., and Ackermann, L. (2014) Ferrocenylalkynes for ruthenium-catalyzed isohypsic C—H/N—O bond functionalizations. *Adv. Synth. Catal.*, **356**, 1619–1624; (f) Nakanowatari, S. and Ackermann, L. (2014) Ruthenium(II)-catalyzed synthesis of isochromenes by C—H activation with weakly coordinating aliphatic hydroxyl groups. *Chem. Eur. J.*, **20**, 5409–5413; (g) Li, J., John, M., and Ackermann, L. (2014) Amidines for versatile ruthenium(II)-catalyzed oxidative C—H activations with internal alkynes and acrylates. *Chem. Eur. J.*, **20**, 5403–5408; (h) Yang, F. and Ackermann, L. (2014) Dehydrative C—H/N—OH functionalizations in H_2O by ruthenium(II) catalysis: subtle effect of carboxylate ligands and mechanistic insight. *J. Org. Chem.*, **79**, 12070–12082; (i) Deponti, M., Kozhushkov, S.I., Yufit, D.S., and Ackermann, L. (2013) Ruthenium-catalyzed C—H/O—H and C—H/N—H bond functionalizations: oxidative annulations of cyclopropyl-substituted alkynes. *Org. Biomol. Chem.*, **11**, 142–148; (j) Li, B., Wang, N., Liang, Y., Xu, S., and Wang, B. (2013)

Ruthenium-catalyzed pyrrole synthesis via oxidative annulation of enamides and alkynes. *Org. Lett.*, **15**, 136–139; (k) Kornhaass, C., Li, J., and Ackermann, L. (2012) Cationic ruthenium catalysts for alkyne annulations with oximes by C—H/N—O functionalizations. *J. Org. Chem.*, **77**, 9190–9198; (l) Thirunavukkarasu, V.S., Donati, M., and Ackermann, L. (2012) Hydroxyl-directed ruthenium-catalyzed C—H bond functionalization: versatile access to fluorescent pyrans. *Org. Lett.*, **14**, 3416–3419; (m) Ma, W.B., Graczyk, K., and Ackermann, L. (2012) Ruthenium-catalyzed alkyne annulations with substituted 1*H*-pyrazoles by C—H/N—H bond functionalizations. *Org. Lett.*, **14**, 6318–6321; (n) Ackermann, L., Pospech, J., Graczyk, K., and Rauch, K. (2012) Versatile synthesis of isocoumarins and alpha-pyrones by ruthenium-catalyzed oxidative C—H/O—H bond cleavages. *Org. Lett.*, **14**, 930–933; (o) Ackermann, L., Wang, L., and Lygin, A.V. (2012) Ruthenium-catalyzed aerobic oxidative coupling of alkynes with 2-aryl-substituted pyrroles. *Chem. Sci.*, **3**, 177–180; (p) Ackermann, L. and Lygin, A.V. (2012) Cationic ruthenium(II) catalysts for oxidative C—H/N—H bond functionalizations of anilines with removable directing group: synthesis of indoles in water. *Org. Lett.*, **14**, 764–767; (q) Reddy Chidipudi, S., Khan, I., and Lam, H.W. (2012) Functionalization of C(sp^3)-H and C(sp^2)-H bonds: synthesis of spiroindenes by enolate-directed ruthenium-catalyzed oxidative annulation of alkynes with 2-aryl-1,3-dicarbonyl compounds. *Angew. Chem. Int. Ed.*, **51**, 12115–12119; (r) Ackermann, L., Lygin, A.V., and Hofmann, N. (2011) Ruthenium-catalyzed oxidative annulation by cleavage of C—H/N—H bonds. *Angew. Chem. Int. Ed.*, **50**, 6379–6382; (s) Ackermann, L. and Fenner, S. (2011) Ruthenium-catalyzed C—H/N—O bond functionalization: green isoquinolone syntheses in water. *Org. Lett.*, **13**, 6548–6551; (t) Ackermann, L., Lygin, A.V., and Hofmann, N. (2011) Ruthenium-catalyzed oxidative synthesis of 2-pyridones through C—H/N—H bond functionalizations. *Org. Lett.*, **13**, 3278–3281; and references cited therein.

8 For selected examples of ruthenium-catalyzed alkenylations via C—H functionalization see: (a) Bechtoldt, A., Tirler, C., Raghuvanshi, K., Warratz, S., Kornhaaß, C., and Ackermann, L. (2016) Ruthenium oxidase catalysis for site-selective C—H alkenylations with ambient O_2 as the sole oxidant. *Angew. Chem. Int. Ed.*, **55**, 264–267; (b) Tirler, C. and Ackermann, L. (2015) Ruthenium(II)-catalyzed cross-dehydrogenative C—H alkenylations by triazole assistance. *Tetrahedron*, **71**, 4543–4551; (c) Ma, W., Mei, R., Tenti, G., and Ackermann, L. (2014) Ruthenium(II)-catalyzed oxidative C—H alkenylations of sulfonic acids, sulfonyl chlorides and sulfonamides. *Chem. Eur. J.*, **20**, 15248–15251; (d) Ma, W. and Ackermann, L. (2013) Ruthenium(II)-catalyzed C—H alkenylations of phenols with removable directing groups. *Chem. Eur. J.*, **19**, 13925–13928; (e) Reddy Chidipudi, S., Wieczysty, M.D., Khan, I., and Lam, H.W. (2013) Synthesis of benzopyrans by Pd(II)- or Ru(II)-catalyzed C—H alkenylation of 2-aryl-3-hydroxy-2-cyclohexenones. *Org. Lett.*, **15**, 570–573; (f) Kozhushkov, S.I. and Ackermann, L. (2013) Ruthenium-catalyzed direct oxidative alkenylation of arenes through twofold C—H bond functionalization. *Chem. Sci.*, **4**, 886–896; (g) Li, J., Kornhaaß, C., and

Ackermann, L. (2012) Ruthenium-catalyzed oxidative C—H alkenylation of aryl carbamates. *Chem. Commun.*, **48**, 11343–11345; (h) Graczyk, K., Ma, W.B., and Ackermann, L. (2012) Oxidative alkenylation of aromatic esters by ruthenium-catalyzed twofold C—H bond cleavages. *Org. Lett.*, **14**, 4110–4113; (i) Ackermann, L., Wang, L., Wolfram, R., and Lygin, A.V. (2012) Ruthenium-catalyzed oxidative C—H alkenylations of anilides and benzamides in water. *Org. Lett.*, **14**, 728–731; (j) Singh, K.S. and Dixneuf, P.H. (2012) Ruthenium(II)-catalyzed alkenylation of ferrocenyl ketones via C—H bond activation. *Organometallics*, **31**, 7320–7323; (k) Ackermann, L. and Pospech, J. (2011) Ruthenium-catalyzed oxidative C—H bond alkenylations in water: expedient synthesis of annulated lactones. *Org. Lett.*, **13**, 4153–4155; and references cited therein.

9 For selected examples of ruthenium-catalyzed arylations via C—H functionalization see: (a) Hubrich, J., Himmler, T., Rodefeld, L., and Ackermann, L. (2015) Ruthenium(II)-catalyzed C—H arylation of anilides with boronic acids, borinic acids and potassium trifluoroborates. *Adv. Synth. Catal.*, **357**, 474–480; (b) Hubrich, J., Himmler, T., Rodefeld, L., and Ackermann, L. (2015) Ruthenium(II)-catalyzed C—H arylation of azoarenes by carboxylate assistance. *ACS Catal.*, **5**, 4089–4093; (c) Ackermann, L. (2015) Robust ruthenium(II)-catalyzed C—H arylations: carboxylate assistance for the efficient synthesis of angiotensin-II-receptor blockers. *Org. Process Res. Dev.*, **19**, 260–269; (d) Al-Mamari, H.H., Diers, E., and Ackermann, L. (2014) Triazole-assisted ruthenium-catalyzed C—H arylation of aromatic amides. *Chem. Eur. J.*, **20**, 9739–9743; (e) Kumar, N.Y.P., Jeyachandran, R., and Ackermann, L. (2013) $C(sp^3)$—H bond arylations catalyzed by well-defined $[Ru(O_2CMes)_2(p\text{-cymene})]$. *J. Org. Chem.*, **78**, 4145–4152; (f) Ackermann, L., Diers, E., and Manvar, A. (2012) Ruthenium-catalyzed C—H bond arylations of arenes bearing removable directing groups via six-membered ruthenacycles. *Org. Lett.*, **14**, 1154–1157; (g) Ackermann, L. and Lygin, A.V. (2011) Ruthenium-catalyzed direct C—H bond arylations of heteroarenes. *Org. Lett.*, **13**, 3332–3335; (h) Arockiam, P.B., Fischmeister, C., Bruneau, C., and Dixneuf, P.H. (2010) C–H bond functionalization in water catalyzed by carboxylato ruthenium(II) systems. *Angew. Chem. Int. Ed.*, **49**, 6629–6632; (i) Ackermann, L., Born, R., and Vicente, R. (2009) Ruthenium-catalyzed direct arylations of *N*-aryl 1,2,3-triazoles with aryl chlorides as electrophiles. *ChemSusChem*, **2**, 546–549; (j) Ackermann, L., Vicente, R., and Althammer, A. (2008) Assisted ruthenium-catalyzed C–H bond activation: carboxylic acids as cocatalysts for generally applicable direct arylations in apolar solvents. *Org. Lett.*, **10**, 2299–2302; (k) Ackermann, L., Althammer, A., and Born, R. (2006) Catalytic arylation reactions by C—H bond activation with aryl tosylates. *Angew. Chem. Int. Ed.*, **45**, 2619–2622; (l) Ackermann, L. (2005) Phosphine oxides as preligands in ruthenium-catalyzed arylations via C—H bond functionalization using aryl chlorides. *Org. Lett.*, **7**, 3123–3125; (m) Oi, S., Fukita, S., Hirata, N., Watanuki, N., Miyano, S., and Inoue, Y. (2001) Ruthenium complex-catalyzed direct ortho arylation and alkenylation of 2-arylpyridines with organic halides. *Org. Lett.*, **3**, 2579–2581; and references cited therein.

10 (a) Teskey, C.J., Lui, A.Y.W., and Greaney, M.F. (2015) Ruthenium-catalyzed *meta*-selective C—H bromination. *Angew. Chem. Int. Ed.*, **54**, 11677–11680; (b) Yu, Q., Hu, L.A., Wang, Y., Zheng, S., and Huang, J. (2015) Directed *meta*-selective bromination of arenes with ruthenium catalysts. *Angew. Chem. Int. Ed.*, **54**, 15284–15288; (c) Paterson, A.J., St John-Campbell, S., Mahon, M.F., Press, N.J., and Frost, C.G. (2015) Catalytic *meta*-selective C—H functionalization to construct quaternary carbon centres. *Chem. Commun.*, **51**, 12807–12810; (d) Li, J., Warratz, S., Zell, D., De Sarkar, S., Ishikawa, E.E., and Ackermann, L. (2015) *N*-Acyl amino acid ligands for ruthenium(II)-catalyzed *meta*-C—H *tert*-alkylation with removable auxiliaries. *J. Am. Chem. Soc.*, **137**, 13894–13901; (e) Hofmann, N. and Ackermann, L. (2013) *meta*-Selective C—H Bond Alkylation with Secondary Alkyl Halides. *J. Am. Chem. Soc.*,**135**, 5877–5884; (f) Reynolds, W.R., Liu, P.M., Kociok-Köhn, G., and Frost, C.G. (2013) Sequential chelation-assisted aromatic C—H functionalization via catalytic *meta* sulfonation. *Synlett*, **24**, 2687–2690; (g) Saidi, O., Marafie, J., Ledger, A.E.W., Liu, P.M., Mahon, M.F., Kociok-Köhn, G., Whittlesey, M.K., and Frost, C.G. (2011) Ruthenium-catalyzed *meta*-sulfonation of 2-phenylpyridines. *J. Am. Chem. Soc.*, **133**, 19298–19301; (h) Ackermann, L., Hofmann, N., and Vicente, R. (2011) Carboxylate-assisted ruthenium-catalyzed direct alkylations of ketimines. *Org. Lett.*, **13**, 1875–1877; Reviews:(i) Li, J., De Sarkar, S., and Ackermann, L. (2016) *meta-* and *para*-selective C—H functionalization by C—H activation. *Top. Organomet. Chem.*, **55**, 217–257; (j) Yang, J. (2015) Transition metal catalyzed *meta*-C—H functionalization of aromatic compounds. *Org. Biomol. Chem*, **13**, 1930–1941.

11 Lewis, L.N. and Smith, J.F. (1986) Catalytic carbon-carbon bond formation *via ortho*-metalated complexes. *J. Am. Chem. Soc.*, **108**, 2728–2735.

12 (a) Murai, S., Kakiuchi, F., Sekine, S., Tanaka, Y., Kamatani, A., Sonoda, M., and Chatani, N. (1993) Efficient catalytic addition of aromatic carbon-hydrogen bonds to olefins. *Nature*, **366**, 529–531; For selected reviews on this topic, see: (b) Andreatta, J.R., McKeown, B.A., and Gunnoe, T.B. (2011) Transition metal catalyzed hydroarylation of olefins using unactivated substrates: recent developments and challenges. *J. Organomet. Chem.*, **696**, 305–315; (c) Foley, N.A., Lee, J.P., Ke, Z., Gunnoe, T.B., and Cundari, T.R. (2009) Ru(II) catalysts supported by hydridotris(pyrazolyl)borate for the hydroarylation of olefins: reaction scope, mechanistic studies, and guides for the development of improved catalysts. *Acc. Chem. Res.*, **42**, 585–597; (d) Kakiuchi, F. (2007) Catalytic addition of C—H bonds to C—C multiple bonds. *Top. Organomet. Chem.*, **24**, 1–33; (e) Kakiuchi, F. and Chatani, N. (2003) Catalytic methods for C—H bond functionalization: application in organic synthesis. *Adv. Synth. Catal.*, **345**, 1077–1101.

13 (a) Kakiuchi, F., Ohtaki, H., Sonoda, M., Chatani, N., and Murai, S. (2001) Mechanistic study of the $Ru(H)_2(CO)(PPh_3)_3$-catalyzed addition of C—H bonds in aromatic esters to olefins. *Chem. Lett.*, **30**, 918–919; (b) Hiraki, K., Ishimoto, T., and Kawano, H. (2000) An NMR study on the ruthenium complex-catalyzed C—H/olefin coupling reaction. *Bull. Chem. Soc. Jpn.*, **73**, 2099–2108.

14 (a) Helmstedt, U. and Clot, E. (2012) Hydride ligands make the difference: density functional study of the mechanism of the murai reaction catalyzed by $[Ru(H)_2(H_2)_2(PR_3)_2]$ (R=cyclohexyl). *Chem. Eur. J.*, **18**, 11449–11458; (b) Matsubara, T., Koga, N., Musaev, D.G., and Morokuma, K. (2000) Density functional study on highly *ortho*-selective addition of an aromatic C—H bond to olefins catalyzed by a $Ru(H)_2(CO)(PR_3)_3$ complex. *Organometallics*, **19**, 2318–2329; (c) Matsubara, T., Koga, N., Musaev, D.G., and Morokuma, K. (1998) Density functional study on activation of *ortho*-C—H bond in aromatic ketone by Ru complex. Role of unusual five-coordinated d6 metallacycle intermediate with agostic interaction. *J. Am. Chem. Soc.*, **120**, 12692–12693.
15 (a) Chatt, J. and Davidson, J.M. (1965) Tautomerism of arene and ditertiary phosphine complexes of ruthenium(0) and preparation of new types of hydridocomplexes of ruthenium(II). *J. Chem. Soc.*, 843–855; (b) Zhang, X., Kanzelberger, M., Emge, T.J., and Goldman, A.S. (2004) Selective addition to iridium of aryl C–H bonds ortho to coordinating groups. Not chelation-assisted. *J. Am. Chem. Soc.*, **126**, 13192–13193.
16 (a) Fujii, N., Kakiuchi, F., Yamada, A., Chatani, N., and Murai, S. (1998) Transition metal-catalyzed intramolecular cyclization of 1,5- and 1,6-dienes via direct cleavage and addition of the carbon–hydrogen bond. *Bull. Chem. Soc. Jpn.*, **71**, 285–298; (b) Murai, S., Chatani, N., and Kakiuchi, F. (1997) Catalytic addition of C—H bonds to multiple bonds. *Pure Appl. Chem.*, **69**, 589–594; (c) Sonoda, M., Kakiuchi, F., Chatani, N., and Murai, S. (1997) Ruthenium-catalyzed addition of carbon–hydrogen bonds in aromatic ketones to olefins. The effect of various substituents at the aromatic ring. *Bull. Chem. Soc. Jpn.*, **70**, 3117–3128; (d) Sonoda, M., Kakiuchi, F., Chatani, N., and Murai, S. (1995) Directing effect of functional groups in ruthenium-catalyzed addition of substituted acetophonones to an olefin. *J. Organomet. Chem.*, **504**, 151–152; (e) Kakiuchi, F., Sekine, S., Tanaka, Y., Kamatani, A., Sonoda, M., Chatani, N., and Murai, S. (1995) Catalytic addition of aromatic carbon–hydrogen bonds to olefins with the aid of ruthenium complexes. *Bull. Chem. Soc. Jpn.*, **68**, 62–83; (f) Murai, S., Kakiuchi, F., Sekine, S., Tanaka, Y., Kamatani, A., Sonoda, M., and Chatani, N. (1994) Catalytic C—H olefin coupling. *Pure Appl. Chem.*, **66**, 1527–1534.
17 (a) Guo, H. and Weber, W.P. (1995) Ruthenium catalyzed regioselective step-growth copolymerization of 4′-methoxyacetophenone, or 4′-phenoxyacetophenone and α,ω-dienes. *Polym. Bull.*, **35**, 259–264; (b) Guo, H.J. and Weber, W.P. (1994) Ruthenium-catalyzed regioselective copolymerization of aceotphenone and alpha, omega-dienes – addition of aromatic *ortho* C—H bonds across C—C double-bonds of alpha, omega-dienes. *Polym. Bull.*, **32**, 525–528.
18 (a) Gupta, S.K. and Weber, W.P. (2002) Ruthenium-catalyzed chemical modification of poly(vinylmethylsiloxane) with 9-acetylphenanthrene. *Macromolecules*, **35**, 3369–3373; (b) Guo, H.J., Tapsak, M.A., and Weber, W.P. (1994) Ruthenium-catalyzed chemical modification of unsaturated polymers. *Polym. Bull.*, **33**, 417–423.
19 (a) Ohmori, S., Ito, S., and Yamamoto, M. (1990) Excimer formation and energy trapping in Langmuir–Blodgett films of poly(vinyl octal) containing

pyrene chromophore. *Macromolecules*, **23**, 4047–4053; (b) Mohammadi, N.A. and Rempel, G.L. (1988) *Chemical Reactions on Polymers Vol. 364 ACS Symposium Series Ch. 28*, American Chemical Society, pp. 393–408.
20 (a) Harris, P.W.R. and Woodgate, P.D. (1997) Ruthenium-catalysed *ortho*-alkylation of hydroxyacetophenones; the functionalization of ring C aromatic diterpenoids. *J. Organomet. Chem.*, **530**, 211–223; (b) Harris, P.W.R. and Woodgate, P.D. (1996) Ruthenium-catalysed *ortho*-alkylation of hydroxyacetophenones; the functionalization of ring C aromatic diterpenoids. *J. Organomet. Chem.*, **506**, 339–341.
21 Trost, B.M., Imi, K., and Davies, I.W. (1995) Elaboration of conjugated alkenes initiated by insertion into a vinylic C—H bond. *J. Am. Chem. Soc.*, **117**, 5371–5372.
22 Kakiuchi, F., Tanaka, Y., Sato, T., Chatani, N., and Murai, S. (1995) Catalytic addition of olefinic C—H bonds to olefins. *Chem. Lett.*, **24**, 679–680.
23 Sonoda, M., Kakiuchi, F., Kamatani, A., Chatani, N., and Murai, S. (1996) Ruthenium-catalyzed addition of aromatic esters at the *ortho*-C—H bonds to olefins. *Chem. Lett.*, **25**, 109–110.
24 Kakiuchi, F., Sato, T., Igi, K., Chatani, N., and Murai, S. (2001) The ruthenium-catalyzed addition of *ortho*-C—H bonds in aldehydes to olefins. *Chem. Lett.*, **30**, 386–387.
25 (a) Kakiuchi, F., Kochi, T., Mizushima, E., and Murai, S. (2010) Room-temperature regioselective C–H/olefin coupling of aromatic ketones using an activated ruthenium catalyst with a carbonyl ligand and structural elucidation of key intermediates. *J. Am. Chem. Soc.*, **132**, 17741–17750; (b) Drouin, S.D., Amoroso, D., Yap, G.P.A., and Fogg, D.E. (2002) Multifunctional ruthenium catalysts: a novel borohydride-stabilized polyhydride complex containing the basic, chelating diphosphine 1,4-bis(dicyclohexylphosphino)butane and its application to hydrogenation and Murai catalysis. *Organometallics*, **21**, 1042–1049; (c) Jazzar, R.F.R., Mahon, M.F., and Whittlesey, M.K. (2001) Synthesis and X-ray structural characterization of $Ru(PPh_3)_3(CO)(C_2H_4)$ and $RuH(o\text{-}C_6H_4C(O)CH_3)(PPh_3)2L$ (L = PPh_3, CO, DMSO): ruthenium complexes with relevance to the Murai reaction. *Organometallics*, **20**, 3745–3751.
26 (a) Simon, M.-O., Genet, J.-P., and Darses, S. (2010) Ruthenium chloride as an efficient catalytic precursor for hydroarylation reactions via C–H bond activation. *Org. Lett.*, **12**, 3038–3041; (b) Guari, Y., Castellanos, A., Sabo-Etienne, S., and Chaudret, B. (2004) $RuH_2(H_2)_2(PCy_3)_2$: a room temperature catalyst for the Murai reaction. *J. Mol. Catal. A: Chem.*, **212**, 77–82; (c) Busch, S. and Leitner, W. (2001) Ruthenium-catalysed Murai-type couplings at room temperature. *Adv. Synth. Catal.*, **343**, 192–195; (d) Guari, Y., Sabo-Etienne, S., and Chaudret, B. (1998) Exchange couplings between a hydride and a stretched dihydrogen ligand in ruthenium complexes. *J. Am. Chem. Soc.*, **120**, 4228–4229.
27 Martinez, R., Chevalier, R., Darses, S., and Genet, J.-P. (2006) A versatile ruthenium catalyst for C—C bond formation by C—H bond activation. *Angew. Chem. Int. Ed.*, **45**, 8232–8235.

28 Martinez, R., Genet, J.-P., and Darses, S. (2008) Anti-Markovnikov hydroarylation of styrenes catalyzed by an *in situ* generated ruthenium complex. *Chem. Commun.*, 3855–3857.

29 Martinez, R., Simon, M.O., Chevalier, R., Pautigny, C., Genet, J.P., and Darses, S. (2009) C—C bond formation via C—H bond activation using an *in situ*-generated ruthenium catalyst. *J. Am. Chem. Soc.*, **131**, 7887–7895.

30 Simon, M.O., Genet, R., and Darses, S. (2010) C—C bond formation via C—H bond activation under protic conditions: on the role of phosphane ligand and cosolvent. *J. Org. Chem.*, **75**, 208–210.

31 Kakiuchi, F., Yamauchi, M., Chatani, N., and Murai, S. (1996) Ruthenium-catalyzed addition of aromatic imines at the *ortho*-C—H bonds to olefins. *Chem. Lett.*, **25**, 111–112.

32 For a review on catalyst controlled site selective synthesis see: (a) Mahatthananchai, J., Dumas, A.M., and Bode, J.W. (2012) Catalytic selective synthesis. *Angew. Chem. Int. Ed.*, **51**, 10954–10990; For examples of catalyst-controlled, site-selective C—H functionalization see: (b) Dooley, J.D., Chidipudi, S.R., and Lam, H.W. (2013) Catalyst-controlled divergent C—H functionalization of unsymmetrical 2-aryl cyclic 1,3-dicarbonyl compounds with alkynes and alkenes. *J. Am. Chem. Soc.*, **135**, 10829–10836; (c) Campbell, A.N., Meyer, E.B., and Stahl, S.S. (2011) Regiocontrolled aerobic oxidative coupling of indoles and benzene using Pd catalysts with 4,5-diazafluorene ligands. *Chem. Commun.*, **47**, 10257–10259; (d) Ueda, K., Yanagisawa, S., Yamaguchi, J., and Itami, K. (2010) A general catalyst for the β-selective C—H bond arylation of thiophenes with iodoarenes. *Angew. Chem. Int. Ed.*, **49**, 8946–8949.

33 Kakiuchi, F., Sato, T., Tsujimoto, T., Yamauchi, M., Chatani, N., and Murai, S. (1998) New protocol for the siteselective alkylation and vinylation of aromatic compounds. Catalyst-specific reactions. *Chem. Lett.*, **27**, 1053–1054.

34 Kakiuchi, F., Sato, T., Yamauchi, M., Chatani, N., and Murai, S. (1999) Ruthenium-catalyzed coupling of aromatic carbon–hydrogen bonds in aromatic imidates with olefins. *Chem. Lett.*, **28**, 19–20.

35 Kakiuchi, F., Tsujimoto, T., Sonoda, M., Chatani, N., and Murai, S. (2001) The transition metal-catalyzed addition of C—H bonds in aromatic hydrazones to olefins. *Synlett*, **2001**, 948–951.

36 Kashiwagi, K., Sugise, R., Shimakawa, T., Matuura, T., Shirai, M., Kakiuchi, F., and Murai, S. (1997) Catalytic dimerization of acrylonitrile. *Organometallics*, **16**, 2233–2235.

37 Kakiuchi, F., Sonoda, M., Tsujimoto, T., Chatani, N., and Murai, S. (1999) The ruthenium-catalyzed addition of C—H bonds in aromatic nitriles to olefins. *Chem. Lett.*, **28**, 1083–1084.

38 Uchimaru, Y. (1999) N—H activation vs. C—H activation: ruthenium-catalysed regioselective hydroamination of alkynes and hydroarylation of an alkene with *N*-methylaniline. *Chem. Commun.*, 1133–1134.

39 Kakiuchi, F., Le Gendre, P., Yamada, A., Ohtaki, H., and Murai, S. (2000) Atropselective alkylation of biaryl compounds by means of transition metal-catalyzed C—H/olefin coupling. *Tetrahedron: Asymmetry*, **11**, 2647–2651.

40 Kozhushkov, S.I., Yufit, D.S., and Ackermann, L. (2008) Ruthenium-catalyzed hydroarylations of methylenecyclopropanes: mild C—H bond functionalizations with conservation of cyclopropane rings. *Org. Lett.*, **10**, 3409–3412.

41 Ackermann, L., Kozhushkov, S.I., and Yufit, D.S. (2012) Ruthenium-catalyzed hydroarylation of methylenecyclopropanes through C—H bond cleavage: scope and mechanism. *Chem. Eur. J.*, **18**, 12068–12077.

42 (a) Kulago, A.A., Van Steijvoort, B.F., Mitchell, E.A., Meerpoel, L., and Maes, B.U.W. (2014) Directed ruthenium-catalyzed C(sp^3)—H α-alkylation of cyclic amines using dioxolane-protected alkenones. *Adv. Synth. Cat.*, **356**, 1610–1618; (b) Chatani, N., Asaumi, T., Yorimitsu, S., Ikeda, T., Kakiuchi, F., and Murai, S. (2001) $Ru_3(CO)_{12}$-catalyzed coupling reaction of sp^3 C—H bonds adjacent to a nitrogen atom in alkylamines with alkenes. *J. Am. Chem. Soc.*, **123**, 10935–10941.

43 Schinkel, M., Wang, L.H., Bielefeld, K., and Ackermann, L. (2014) Ruthenium(II)-catalyzed C(sp^3)-H α-alkylation of pyrrolidines. *Org. Lett.*, **16**, 1876–1879.

44 (a) Gray, A., Tsybizova, A., and Roithova, J. (2015) Carboxylate-assisted C—H activation of phenylpyridines with copper, palladium and ruthenium: a mass spectrometry and DFT study. *Chem. Sci.*, **6**, 5544–5553; (b) Fabre, I., von Wolff, N., Le Duc, G., Flegeau, E.F., Bruneau, C., Dixneuf, P.H., and Jutand, A. (2013) Autocatalytic intermolecular versus intramolecular deprotonation in C—H bond activation of functionalized arenes by ruthenium(II) or palladium(II) complexes. *Chem. Eur. J.*, **19**, 7595–7604; (c) Flegeau, E.F., Bruneau, C., Dixneuf, P.H., and Jutand, A. (2011) Autocatalysis for C—H bond activation by ruthenium(II) complexes in catalytic arylation of functional arenes. *J. Am. Chem. Soc.*, **133**, 10161–10170; (d) Ackermann, L., Vicente, R., Potukuchi, H.K., and Pirovano, V. (2010) Mechanistic insight into direct arylations with ruthenium(II) carboxylate catalysts. *Org. Lett.*, **12**, 5032–5035; (e) Prades, A., Poyatos, M., and Peris, E. (2010) (η^6-Arene)ruthenium(*N*-heterocyclic carbene) complexes for the chelation-assisted arylation and deuteration of arylpyridines: catalytic studies and mechanistic insights. *Adv. Synth. Cat.*, **352**, 1155–1162; (f) Ackermann, L., Vicente, R., and Althammer, A. (2008) Assisted ruthenium-catalyzed C–H bond activation: carboxylic acids as cocatalysts for generally applicable direct arylations in apolar solvents. *Org. Lett.*, **10**, 2299–2302; Review: (g) Ackermann, L. (2011) Carboxylate-assisted transition-metal-catalyzed C–H bond functionalizations: mechanism and scope. *Chem. Rev.*, **111**, 1315–1345.

45 Schinkel, M., Marek, I., and Ackermann, L. (2013) Carboxylate-assisted ruthenium(II)-catalyzed hydroarylations of unactivated alkenes through C—H cleavage. *Angew. Chem. Int. Ed.*, **52**, 3977–3980.

46 Rouquet, G. and Chatani, N. (2013) Ruthenium-catalyzed *ortho*-C—H bond alkylation of aromatic amides with alpha,beta-unsaturated ketones via bidentate-chelation assistance. *Chem. Sci.*, **4**, 2201–2208.

47 (a) Tran, L.D. and Daugulis, O. (2012) Nonnatural amino acid synthesis by using carbon–hydrogen bond functionalization methodology. *Angew. Chem. Int. Ed.*, **51**, 5188–5191; (b) Tran, L.D., Popov, I., and Daugulis, O.

(2012) Copper-promoted sulfenylation of sp(2) C—H bonds. *J. Am. Chem. Soc.*, **134**, 18237–18240; (c) Shabashov, D. and Daugulis, O. (2010) Auxiliary-assisted palladium-catalyzed arylation and alkylation of sp(2) and sp(3) carbon–hydrogen bonds. *J. Am. Chem. Soc.*, **132**, 3965–3972; (d) Zaitsev, V.G., Shabashov, D., and Daugulis, O. (2005) Highly regioselective arylation of sp(3) C—H bonds catalyzed by palladium acetate. *J. Am. Chem. Soc.*, **127**, 13154–13155.

48 Li, J. and Ackermann, L. (2015) Carboxylate-assisted ruthenium(II)-catalyzed C—H activations of monodentate amides with conjugated alkenes. *Org. Chem. Front.*, **2**, 1035–1039.

49 Schinkel, M., Wallbaum, J., Kozhushkov, S.I., Marek, I., and Ackermann, L. (2013) Carboxylate assistance for catalyzed hydroarylations of methylenecyclopropanes. *Org. Lett.*, **15**, 4482–4484.

50 Itoh, M., Hashimoto, Y., Hirano, K., Satoh, T., and Miura, M. (2013) Ruthenium-catalyzed *ortho*-alkenylation of phenylphosphine oxides through regio- and stereoselective alkyne insertion into C—H bonds. *J. Org. Chem.*, **78**, 8098–8104.

51 (a) Kommagalla, Y., Mullapudi, V.B., Fredi Francis, F., and Ramana, C.V. (2015) Ruthenium(II)-catalyzed switchable C3-alkylation versus alkenylation with acrylates of 2-pyridylbenzofurans via C—H bond activation. *Catal. Sci. Technol.*, **5**, 114–117; (b) Kommagalla, Y., Srinivas, K., and Ramana, C.V. (2014) Ru-catalyzed branched versus linear selective C3-alkylation of 2-aroylbenzofurans with acrylates via C—H activation. *Chem. Eur. J.*, **20**, 7884–7889.

52 Bettadapur, K.R., Lanke, V., and Prabhu, K.R. (2015) Ru(II)-catalyzed C—H activation: ketone-directed novel 1,4-addition of *ortho*-C—H bond to maleimides. *Org. Lett.*, **17**, 4658–4661.

53 Nakanowatari, S. and Ackermann, L. (2015) Ruthenium(II)-catalyzed C—H functionalizations with allenes: versatile allenylations and allylations. *Chem. Eur. J.*, **21**, 16246–16251.

54 For selected reviews on removable directing groups, see: (a) Zhang, F. and Spring, D.R. (2014) Arene C—H functionalization using a removable/modifiable or a traceless directing group strategy. *Chem. Soc. Rev.*, **43**, 6894–6905; (b) Wang, C. and Huang, Y. (2013) Expanding structural diversity; removable and manipulable directing groups for C—H activation. *Synlett*, **24**, 145–149.

3

Alkylation of Arenes Without Chelation Assistance: Transition Metal Catalysts with d^6 Electron Configurations

Bradley A. McKeown[1], *Laurel Goj Habgood*[2], *Thomas R. Cundari*[3] and *T. Brent Gunnoe*[1]

[1] *University of Virginia, Department of Chemistry, McCormick Road, Charlottesville, VA 22904-4319, USA*
[2] *Rollins College, Department of Chemistry, 1000 Holt Avenue, Winter Park, FL 32789, USA*
[3] *University of North Texas, Center for Advanced Scientific Computing and Modeling (CASCaM), Department of Chemistry, 1155 Union Circle, #305070, Denton, TX 76203-5017, USA*

3.1 Transition Metal-Mediated Arene Alkylation: Overview

Alkyl arenes are an important class of commodity chemicals that are used as fuel additives and provide the basis for the preparation of detergents, thermoplastics, and elastomers. The synthesis of alkyl arenes from arenes and olefins has traditionally been accomplished using Friedel–Crafts catalysis, which utilizes a strong Lewis acid in combination with a Brønsted acid (e.g., $AlCl_3$ and HF) [1, 2]. Unfortunately, the Friedel–Crafts processes suffer from several deficiencies including limited benzene conversion, polyalkylation, and the inability to recycle the catalyst [2]. Most of the issues surrounding Friedel–Crafts alkylation of arenes result from the acid-catalyzed mechanism, which involves electrophilic addition to the olefin, generating a carbocation intermediate, followed by electrophilic aromatic substitution (Scheme 3.1). As a result, mono-alkylation of the parent substrate results in a more electron-rich arene that is often more reactive than the starting arene, which promotes the formation of undesired polyalkylated products. For example, the synthesis of ethylbenzene using $AlCl_3$ as the catalyst produces substantial amounts of polyethylbenzenes, and a second transalkylation process is required to optimize the yield of ethylbenzene [2]. Electron-deficient arenes are often unreactive with the carbocation and do not undergo alkylation. Furthermore, when using Friedel–Crafts catalysis, the reactivity of substituted olefins is highly or exclusively selective for the Markovnikov addition product, and alternative regioselectivity for substituted olefins is difficult to achieve [3].

Alternative industrial processes that use solid acid catalysts for the alkylation of simple arenes with light olefins have largely obviated the use of corrosive $AlCl_3$/HX catalyst systems [4–8]. The most common class of solid acid catalyst implemented for industrial applications is zeolites, which are crystalline microporous aluminosilicates [6, 9–11]. Zeolites are attractive candidates as they are

Catalytic Hydroarylation of Carbon-Carbon Multiple Bonds, First Edition.
Edited by Lutz Ackermann, T. Brent Gunnoe and Laurel Goj Habgood.

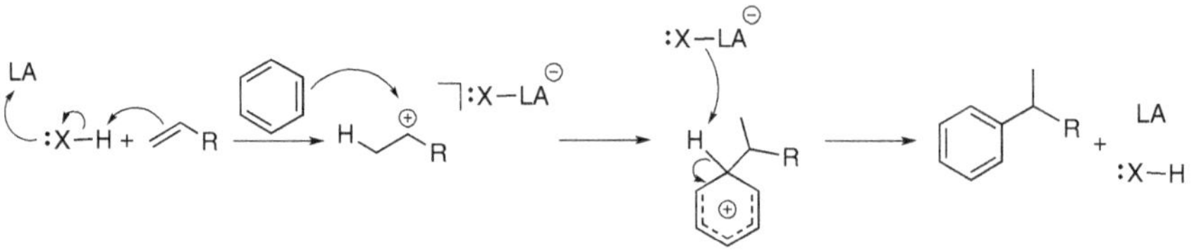

Scheme 3.1 Olefin hydroarylation using Friedel–Crafts catalysis (LA = Lewis acid, such as $AlCl_3$; HX = Brønsted acid, such as HF or HCl).

readily isolated from the product mixture and can be easily recycled. However, their acidity is often reduced as compared to the traditional Friedel–Crafts catalysts using a Lewis and Brønsted acid; thus, elevated temperatures are often required to achieve the same level of activity [12].

An alternative to acid-mediated arene alkylation involves a catalyst that mediates olefin insertion into a metal–aryl bond followed by arene C—H activation (Scheme 3.2). This class of catalyst offers several opportunities to improve upon current acid-based methodologies, including: (i) reduced polyalkylation by designing catalysts that are more reactive with the starting unsubstituted arene than the resulting alkylated arene; (ii) control over the regioselectivity for dialkylation reactions (e.g., ortho, meta, or para substitution); (iii) direct conversion of an arene and an olefin to a vinyl arene (i.e., oxidative vinylation) through a β-hydride elimination reaction; (iv) complementary selectivity (i.e., anti-Markovnikov addition) for hydroarylation of substituted olefins; (v) arene alkylation reactions with electron-deficient arenes. There is significant precedent using transition metals for the two key steps for the desired overall process, olefin insertion into a metal–aryl bond and metal-mediated arene C—H activation [13–19]. However, avoiding side reactions, which are likely to have similar

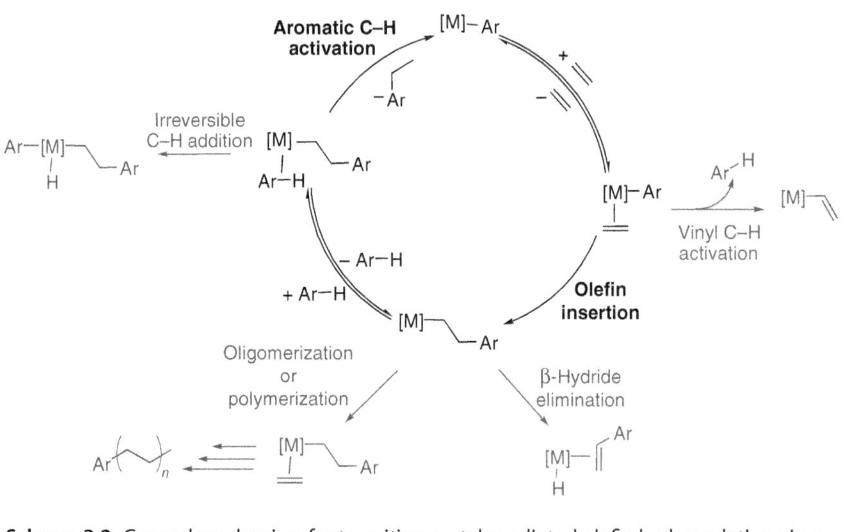

Scheme 3.2 General mechanism for transition metal-mediated olefin hydroarylation via a non-acid pathway with possible side reactions.

activation barriers compared to desired "on cycle" reactions, is a substantial challenge (Scheme 3.2). For example, the formation of stable metal–olefin complexes might result in large activation barriers for olefin insertion, which could result in undesirable olefinic C—H activation. Thus, the catalyst must be able to activate arene sp^2 hybridized C—H bonds but not vinyl sp^2 hybridized C—H bonds. Further, olefin insertion cannot be too facile, as low activation energies for olefin coordination and insertion into M—C bonds could promote olefin oligomerization or polymerization. Following olefin insertion, the newly formed M–alkyl moiety can be susceptible to β-hydride elimination, resulting in the formation of a coordinated vinyl arene and a metal hydride. Catalysts based on rhodium yield vinyl arenes in the presence of an oxidant [20–23]. While vinyl arenes are desirable in some cases, catalysts that selectively form saturated alkyl arenes must favor arene C—H activation over β-hydride elimination and vinyl arene dissociation (regardless of the exact mechanism of vinyl arene dissociation). Thus, understanding the interplay between catalyst properties and the propensity to undergo β-hydride elimination/vinyl arene dissociation versus arene C—H activation is of central importance in the design of catalysts for the synthesis of alkyl arenes. A common strategy for transition metal-mediated C—H activation is to use electron-rich metal centers that can insert into C—H bonds to give oxidative addition products [24]. However, if C—H oxidative addition is highly exothermic, slow reversal (i.e., reductive elimination) to generate free alkyl arenes can inhibit catalysis. Despite these challenges, significant achievements in the area of transition metal-mediated olefin hydroarylation with unactivated hydrocarbon substrates (benzene, toluene, xylenes, ethylene, propylene, etc.) have been reported.

3.2 Octahedral d^6 Transition Metal Catalysts for Olefin Hydroarylation: Scorpionate Supported Ru(II) Catalysts

In 2003, the Gunnoe group reported a low-valent Ru(II) complex, TpRu(CO)(NCMe)Ph (**Ru-CO**; Tp = hydridotris(pyrazolyl)borate), that catalyzes olefin hydroarylation using unfunctionalized substrates (e.g., benzene and ethylene to yield ethylbenzene) [25]. The facial coordination of the scorpionate ligand is well suited for the olefin hydroarylation reaction as it imposes a cis-orientation between the aryl ligand and coordinated olefin, which is required for olefin insertion. Catalytic olefin hydroarylation using **Ru-CO** has been studied using combined experimental data and theoretical calculations to facilitate mechanistic understanding and to define the limits of selectivity and yields. For catalysis using **Ru-CO**, the ancillary CO is a spectator ligand. Thus, the base architecture of the ruthenium catalyst, TpRu(L)(NCMe)Ph (where L represents various neutral two-electron donor ligands), is amenable for probing catalytic reactivity and product distribution as a function of steric and electronic properties of L [26]. More recent efforts have examined the effect of replacing the formally anionic Tp with neutral poly(pyrazolyl)alkanes, which provide cationic catalyst precursors [27].

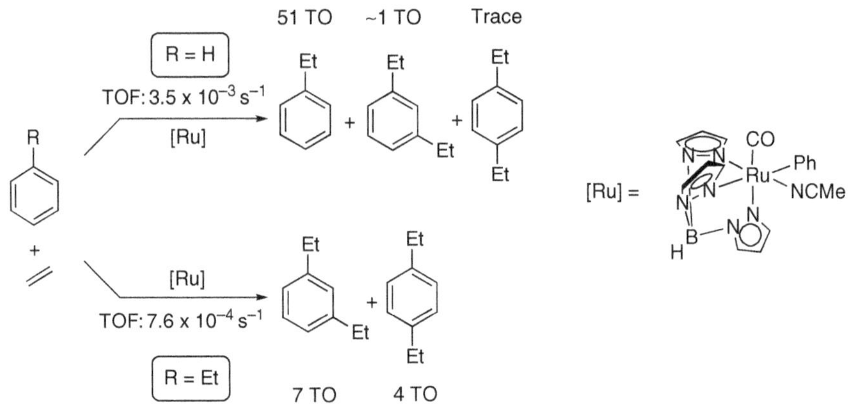

Scheme 3.3 Hydroarylation of ethylene (0.2 MPa) using 0.1 mol% TpRu(CO)(NCMe)Ph (**Ru-CO**) in benzene or ethylbenzene after 4 h at 90 °C. TOF = turnover frequency ([product]/[**Ru-CO**] × time).

The initially reported complex **Ru-CO** catalyzes ethylene hydrophenylation under mild conditions [25, 28]. At 90 °C and 0.2 MPa of ethylene, **Ru-CO** (0.1 mol% relative to benzene) catalyzes 51 and 77 turnovers (TOs) of ethylbenzene after 4 and 24 h, respectively, with only trace quantities of 1,3- and 1,4-diethylbenzenes detected (Scheme 3.3). Although conversions of benzene are low under these conditions, it was demonstrated that formation of ethylbenzene from benzene and ethylene is about five times faster than the formation of diethylbenzenes from ethylbenzene and ethylene using **Ru-CO** (Scheme 3.3) [28]. The comparative rates of ethylation of benzene versus ethylbenzene is proof of concept that catalysts that operate via olefin insertion and metal-mediated arene C—H activation can be more reactive with the unfunctionalized substrate, and therefore offer the opportunity to achieve better selectivity for mono-alkylated arenes than Friedel–Crafts catalysts. In addition, the selectivity for 1,3- and 1,4-diethylbenzenes (with no observation of 1,2-diethylbenzene) demonstrates selectivity that is unique compared to acid-based catalysis, which generally produces all three products with ortho/para favored over meta regioselectivity. In a recent comparative study it was reported that **Ru-CO** gives greater than 400 total TO of ethylbenzene at lower catalyst loadings (0.025 mol% relative to benzene; Table 3.1) [29]. Greater catalyst longevity with lower catalyst loadings suggests that the mechanism of catalyst deactivation is likely second-order in **Ru-CO**.

The hydrophenylation of propylene and 1-hexene by **Ru-CO** provides propylbenzenes and hexylbenzenes with linear to branched ratios of 1.6 : 1 and 1.7 : 1, respectively. For the reaction with 1-hexene, the formation of the 3-phenyl isomer is not observed, suggesting that either olefin isomerization is not competitive with catalytic olefin hydrophenylation or that hydrophenylation of 2-hexene does not occur. A thorough analysis of the various factors that control the regioselectivity for hydrophenylation of α-olefins has not been undertaken, and the origin of the linear/branched ratio has not been understood in detail to date [30]. The overall selectivity could be controlled either by the selectivity for

Table 3.1 Comparison of data for catalytic hydrophenylation of ethylene using TpRu(L)(NCMe)Ph complexes.

L	$E_{Ru(III/II)}$ [a)] (V)	k_{obs} [b)] (10^{-5} s^{-1})	$\Delta G^{‡}_{insertion}$ (kcal/mol)	TON[c)]
:C≡O:	1.03	0.462(3)	17.7	415
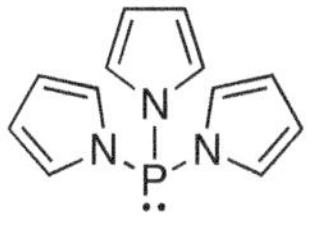	0.82	N/A	23.2	0
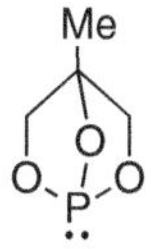	0.69	0.72(5)	19.4	90
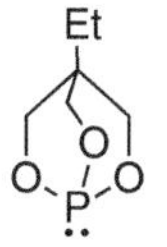	0.55	1.20(2)	20.1	20
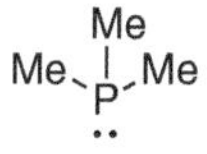	0.29	1.36(4)	23.4	0

a) Referenced versus NHE.
b) Reactions performed in C_6D_6 with 0.065 mmol of added NCMe at 60 °C.
c) Turnover number (TON) of ethylbenzene for ethylene hydrophenylation with 0.025 mol% Ru, relative to C_6H_6, and 0.1 MPa of ethylene at 90 °C.

the olefin insertion into the Ru—Ph bond or by the rate of subsequent benzene C—H activation (i.e., Curtin–Hammett conditions). One proposed explanation for selective formation of the anti-Markovnikov products is that the ancillary ligands display a preferred orientation of the coordinated α-olefin that minimizes steric interactions with the pyrazolyl rings of Tp and the β-hydrogens of the phenyl ligand prior to insertion [31]. Catalytic activity could not be extended to 1,1-disubstituted olefins as heating the combination of 0.1 mol% **Ru-CO** in benzene with isobutylene fails to yield alkylated benzenes [28].

Several results point to a non-acid based reaction for olefin hydroarylation using **Ru-CO**. Most importantly, the hydrophenylation of propylene and 1-hexene selectively produces the straight-chain *n*-alkyl benzene products, which are not observed for acid-catalyzed pathways. Additionally, the reduced formation of 1,2-diethylbenzene in the reaction of ethylbenzene with ethylene supports the proposed non-acid-catalyzed pathway. The conversion of ethylbenzene and ethylene to diethylbenzenes forms an approximate 2 : 1 ratio of 1,3-diethylbenzene and 1,4-diethylbenzene under the same conditions as the hydrophenylation of ethylene. Finally, attempted transalkylation to produce ethylbenzene in a 1 : 1 mixture of 1,4-diethylbenzene and benzene at 90 °C in the presence of **Ru-CO** does not occur, as is often observed for Friedel–Crafts catalysts.

The incorporation of functionality into mono-substituted olefin substrates is incompatible with catalysis using **Ru-CO** [32–34]. With olefins containing electron-withdrawing substituents (e.g., styrene, methyl methacrylate, or acrylonitrile), attempted catalytic hydrophenylation results in radical polymerization of the olefin [32]. The reaction of **Ru-CO** with olefins that possess electron-donating substituents, such as ethylvinyl sulfide and 2,3-dihydrofuran, results in C—O or C—S bond cleavage to produce either ring-opened products or μ-SEt bridged Ru dimers with free styrene (Scheme 3.4) [34].

For the catalyst precursor **Ru-CO**, the use of aromatic heterocycles, such as furan and thiophene, results in catalytic TO with regioselective alkylation at the α-position, although the catalytic efficiency is decreased compared to that with ethylene hydrophenylation [35]. For example, heating 0.1 mol% of **Ru-CO** with ethylene and excess furan or thiophene provides 17 and 3 TO of 2-ethylfuran and 2-ethylthiophene, respectively, after 24 h (Scheme 3.5). Pyrrole is catalytically inactive as it undergoes metal-promoted C—C bond formation with the acetonitrile ligand to provide a stable κ^2-metallacycle [36].

Focusing on the hydrophenylation of ethylene, mechanistic studies are consistent with the proposed catalytic cycle in Scheme 3.6 [25, 26, 28, 29]. The catalytic cycle is initiated by acetonitrile dissociation followed by ethylene coordination

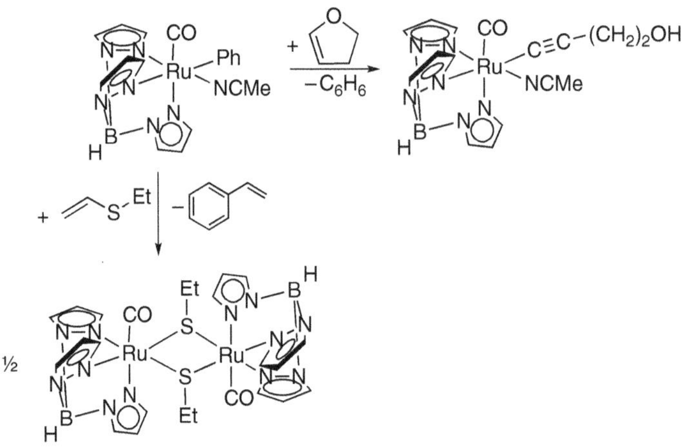

Scheme 3.4 Attempted catalytic hydrophenylation of olefins possessing electron-donating substituents results in C—O or C—S bond cleavage.

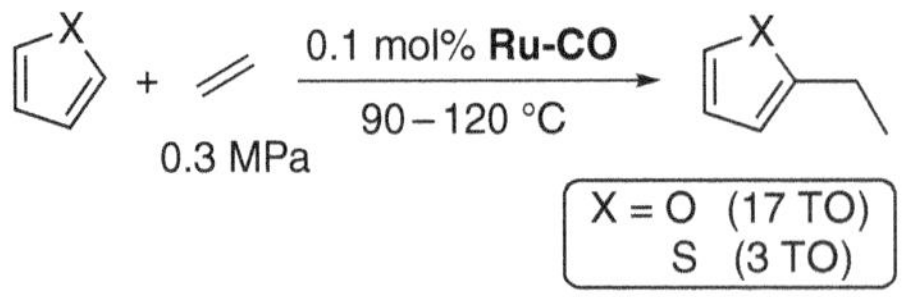

Scheme 3.5 Catalytic ethylene hydroarylation with heteroaromatic substrates.

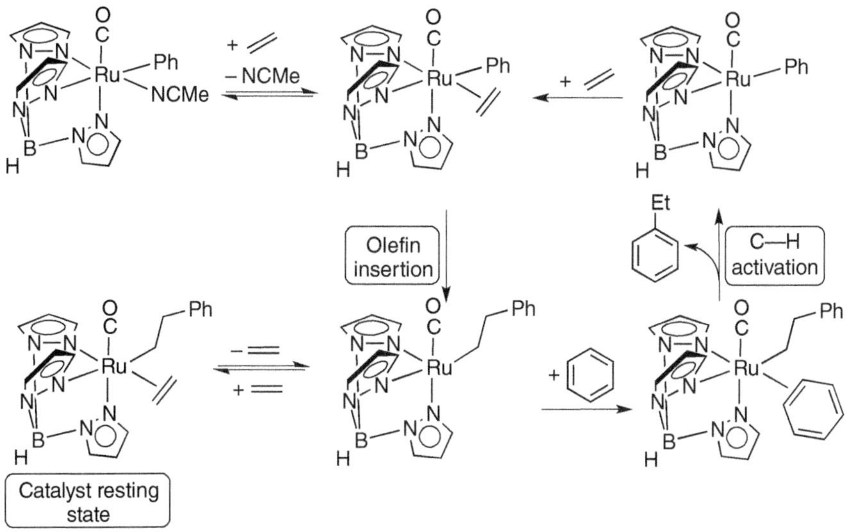

Scheme 3.6 Proposed catalytic cycle for the hydrophenylation of ethylene using **Ru-CO**.

and insertion into the Ru—Ph bond. Benzene coordination and subsequent C—H activation result in ethylbenzene formation. Ethylbenzene dissociation and coordination of ethylene complete the catalytic cycle. Using both experimental and computational data, it has been proposed that the benzene C—H activation step is rate-limiting [26]. For example, comparison of the rate of hydrophenylation of ethylene in a 1 : 1 molar mixture of C_6H_6 and C_6D_6 revealed a kinetic isotope effect (KIE) with $k_H/k_D = 2.1(1)$ [28]. This value is statistically identical to the KIE for stoichiometric benzene activation by TpRu(CO)(NCMe)Ph (C—D activation of C_6D_6) and TpRu(CO)(NCMe)Ph-d_5 (C—H activation of C_6H_6), which was determined to be $k_H/k_D = 2.5(5)$. However, one set of density functional theory (DFT) calculations is consistent with ethylene insertion into the M—Ph bond as the rate-limiting step for both **Ru-CO** and an Ir(III) catalyst (see Section 3.4) [30, 37]. The reaction of **Ru-CO** and ethylene in acetonitrile produces TpRu(CO)(CH_2CH_2Ph)(NCMe), demonstrating that the 16-electron TpRu(CO)Ph fragment can indeed coordinate and insert ethylene into the Ru—Ph bond. Thus, ethylene insertion into the Ru—Ph bond of **Ru-CO** is a viable step in the catalytic process. TpRu(CO)(CH_2CH_2Ph)(η^2-C_2H_4) has been observed while monitoring catalysis by ^1H NMR spectroscopy and has been proposed as the catalyst resting state. Consistent with the proposed catalyst resting state, which is TpRu(CO)(η^2-C_2H_4)(CH_2CH_2Ph), catalytic ethylene hydrophenylation using **Ru-CO** reveals an inverse dependence on ethylene concentration. This suggests that ethylene effectively competes with benzene for coordination to the intermediate TpRu(CO)(CH_2CH_2Ph). Coordination of ethylene to this intermediate removes Ru from the catalytic cycle via formation of TpRu(CO)(CH_2CH_2Ph)(η^2-C_2H_4), slowing the rate of ethylbenzene formation.

Since benzene C—H activation is proposed to be the rate-limiting step for catalysis using **Ru-CO**, the pathway for cleavage of the benzene C—H bond is of interest for the design of improved catalysts. The mechanism of benzene C—H activation by **Ru-CO** has been investigated by DFT calculations and has been suggested to occur through a concerted process featuring a four-membered transition state, similar to classic σ-bond metathesis (see Section 3.4 and Scheme 3.14) [26, 28, 31, 38–40]. Preceding the C—H bond cleavage, initial coordination of the C—H bond to the Ru center increases the acidity of the C—H bond, similar to related acidity increases for coordination of dihydrogen by transition metal complexes [41]. In the transition state for C—H activation, Ru donates electron density to the activated hydrogen atom, forming a transient Ru—H bond. This Ru—H interaction has been used to differentiate a single-step concerted C—H activation by some d^6 complexes compared to related concerted σ-bond metathesis transition states by d^0 transition metal complexes. The Ru—H interaction in the calculated transition state (calculated Ru—H bond distance in the transition state is 1.65 Å) suggests oxidative character for Ru (i.e., donation of Ru-centered electrons to the activated hydrogen atom increases the positive charge on Ru) [26, 28, 37]. This process for d^6 complexes has been termed oxidative hydrogen migration (OHM) [31, 37, 42, 43]. Oxidative addition to provide a seven-coordinate Ru(IV) intermediate was determined by calculations to be less favorable than the proposed OHM C—H activation. Aromatic C—H activation by TpRu(CO)(NCMe)Ph has also been extended to sp^3 hybridized C—H bonds of other substrates [44].

3.2.1 Structure–Activity Relationships with TpRu(L)(NCMe)Ph: Examination of Elementary Steps and Catalytic Hydrophenylation of Ethylene as a Function of Ligand L

The carbonyl ligand of **Ru-CO** can be formally substituted by a variety of phosphine-based ligands, which provides a means to study structure/activity relationships for catalytic olefin hydroarylation using TpRu(L)(NCMe)Ph catalysts. A series of Ru(II) complexes TpRu(L)(NCMe)Ph {L = PMe_3 (**Ru-PMe$_3$**), P(*N*-pyrrolyl)$_3$ (**Ru-P(pyr)$_3$**), $P(OCH_2)_3CEt$ (**Ru-BPhos**), or 2,6,7-trioxa-1-phosphabicyclo[1,2,2]heptane (**Ru-SPhos**)} were prepared and studied to understand the impact of the electronic and steric properties of the Ru catalyst precursor on catalytic performance within a consistent structural motif [25, 26, 28, 29, 35, 45, 46]. The reversible Ru(III/II) potentials for the complexes TpRu(L)(NCMe)Ph, determined by cyclic voltammetry, are a suitable proxy to determine the relative electron density at the Ru(II) center. Using the reversible Ru(III/II) potentials, the relative electron densities of the Ru complexes were found to be **Ru-PMe$_3$** > **Ru-BPhos** > **Ru-SPhos** > **Ru-P(pyr)$_3$** > **Ru-CO** (Table 3.1). [26, 29]

Due to the proposed role of Ru in the calculated benzene C—H activation transition state, in which Ru donates electron density to the activated H atom, it was speculated that increasing the electron density of the Ru center would result in a lower overall activation barrier to benzene C—H activation. The stoichiometric reactions of the complexes TpRu(L)(NCMe)Ph with benzene-d_6 to yield TpRu(L)(NCMe)Ph-d_5 and benzene-d_1 were studied (Table 3.1). Indeed, replacement of CO with more electron-donating phosphine ligands results in more rapid overall benzene C—D bond activation. For example, **Ru-PMe$_3$**

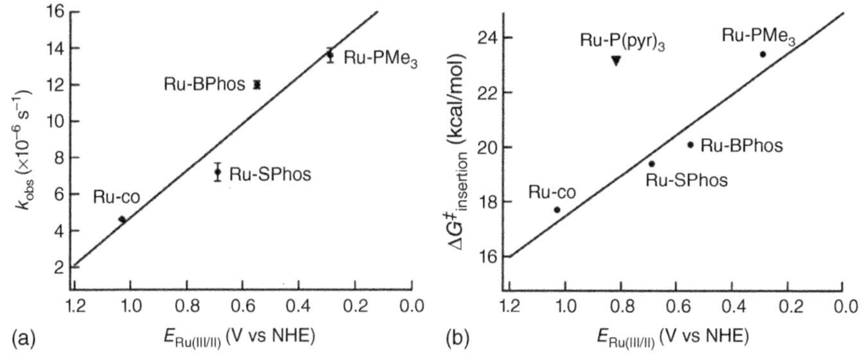

Figure 3.1 (a) Plot of k_{obs} for stoichiometric C—D activation of C_6D_6 by TpRu(L)(NCMe)Ph (L = CO, BPhos, SPhos, or PMe_3) at 60 °C with 0.065 mmol of added NCMe versus reversible Ru(III/II) potentials (vs NHE). (b) Plot of calculated $\Delta G^{\ddagger}_{insertion}$ for ethylene insertion into the Ru—Ph bond from TpRu(L)(η^2-C_2H_4)Ph complexes. Data using **Ru-P(pyr)$_3$** is not included in the linear regression since this complex decomposes in competition with C_6D_6 activation.

undergoes C—D activation of benzene-d_6 approximately three times faster than does **Ru-CO**. A plot of k_{obs} for benzene C—D activation versus the reversible Ru(III/II) potentials for TpRu(L)(NCMe)Ph complexes reveals a correlation (Figure 3.1a). The P(pyr)$_3$ ligand of **Ru-P(pyr)$_3$** was found to be sterically cumbersome, inhibiting benzene coordination and producing benzene-d_1 in less than 70% yield [47]. Thus, data for **Ru-P(pyr)$_3$** are not included in Figure 3.1a.

Computational modeling suggests that increased electron-density at the Ru center of TpRu(L)(NCMe)Ph complexes has an opposite effect (compared to the rates of benzene C—H activation) on the calculated activation barriers for ethylene insertion into the Ru—Ph bond from TpRu(L)(η^2-C_2H_4)Ph complexes. As shown in Figure 3.1b, the calculated free energy activation barriers for ethylene insertion into the Ru—Ph bonds (calculated from TpRu(L)(η^2-C_2H_4)Ph intermediates) increases linearly with decreasing Ru(III/II) potential. This effect is attributed to increased Ru-to-ethylene π-back-donation, which stabilizes the Ru-(η^2-C_2H_4) adduct. The **Ru-P(pyr)$_3$** complex deviates from the observed trend with a calculated $\Delta G^{\ddagger}_{insertion}$ comparable to that calculated for **Ru-PMe$_3$**. This effect is possible due to steric repulsions between the P(pyr)$_3$ ligand and ethylene, which hinders ethylene coordination to the Ru center.

Although the rate of stoichiometric benzene C—H/D activation is accelerated by more strongly donating ancillary ligands, the increased metal electron-density results in a dramatic decrease in catalyst longevity for ethylene hydrophenylation for TpRu(L)(NCMe)Ph. The most electron-rich complex, **Ru-PMe$_3$**, was found to be inactive for ethylene hydrophenylation under most conditions investigated (e.g., 60–180 °C; 0.2–6 MPa of ethylene) [45]. It was determined that the more electron-rich Ru center of **Ru-PMe$_3$** (relative to **Ru-CO**) performs ethylene C—H activation more rapidly than ethylene insertion into the Ru—Ph bond. The initial ethylene C—H activation produces TpRu(PMe$_3$)(C_2H_3)(η^2-C_2H_4), which ultimately leads to the formation of the stable η^3-allyl complex TpRu(PMe$_3$)(η^3-C_4H_7) (Scheme 3.7) [26, 29, 45, 46].

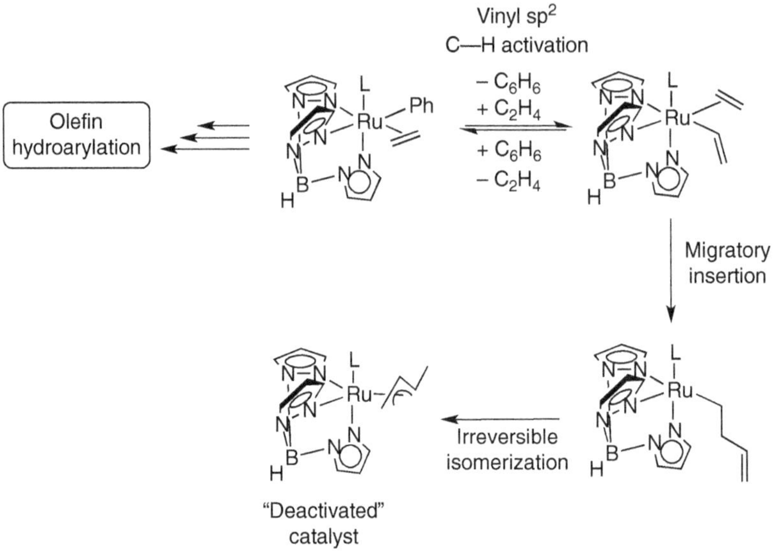

Scheme 3.7 Pathway for the formation of TpRu(L)(η^3-C_4H_7) (L = PMe_3, BPhos, and SPhos), the primary mechanism of catalyst deactivation for TpRu(II)-catalyzed ethylene hydrophenylation.

The bicyclic phosphites of the complexes **Ru-BPhos** and **Ru-SPhos** are less donating than PMe_3, likely because of reduced σ-donation and increased π-acidity. These complexes were predicted to favor ethylene insertion into Ru—Ph bonds and yield catalytic TO for ethylene hydrophenylation. Catalytic studies revealed that **Ru-BPhos** is a catalyst for ethylene hydrophenylation, but catalytic activity ceases after approximately 20 TO with 0.025 mol% Ru and 0.1 MPa of ethylene at 90 °C [46]. Similar to **Ru-PMe$_3$**, catalyst deactivation is a result of competitive ethylene C—H activation, which leads to the formation of the η^3-allyl complex TpRu(BPhos)(η^3-C_4H_7). The ligand SPhos is less electron donating than BPhos, consistent with reversible Ru(III/II) potentials of **Ru-BPhos** and **Ru-SPhos** (Table 3.1), and greater catalytic TO are observed for **Ru-SPhos** compared to **Ru-BPhos** under identical conditions. Catalysis with 0.025 mol% of **Ru-SPhos** and 0.1 MPa of ethylene at 90 °C provides 90 TO of ethylbenzene before catalysis is halted, again as a result of competitive ethylene C—H activation to yield TpRu(SPhos)(η^3-C_4H_7). The P(pyr)$_3$ complex **Ru-P(pyr)$_3$** is not a catalyst for the hydrophenylation of ethylene [47].

Despite the catalytic rate-determining step being benzene C—H activation for ethylene hydrophenylation for **Ru-CO**, the influence of the ancillary ligand L of TpRu(L)(NCMe)Ph catalyst precursors on the activation barrier for ethylene insertion into Ru—Ph bonds was shown to be critical for catalyst longevity. For instance, the PMe_3, BPhos, and SPhos ligands increase the rate of stoichiometric benzene C—D activation by TpRu(L)(NCMe)Ph complexes relative to **Ru-CO**, but the relative catalyst longevities follows the opposite trend [29, 45, 46]. This systematic structure/activity relationship addressing both ancillary ligand electronics and sterics (i.e., the P(pyr)$_3$ ligand is too large) illustrates the delicate balance between olefin insertion and arene C—H activation for efficient catalysis.

In a computational modeling study, a general trend for octahedral d^6 complexes was predicted whereby changes in electron density at the metal center would have opposite effects on the activation barriers for olefin insertion into M–aryl bonds and arene C—H activation [37]. Studies of TpRu(L)(NCMe)Ph are consistent with this prediction.

3.2.2 Ethylene Hydrophenylation Catalyzed by Cationic Ru(II) Complexes Ligated by Poly(pyrazolyl)alkanes

As discussed earlier, studies examining ethylene hydrophenylation catalyzed by TpRu(L)(NCMe)Ph complexes have indicated a clear trend in catalyst longevity. As the electron density of the Ru catalyst precursor is increased via modulation of the ligand L, fewer catalytic TO are observed due to more facile formation η^3-allyl complexes as a result of ethylene C—H activation [26, 29]. This suggests that more electron-deficient Ru(II) complexes (compared to **Ru-CO**) should increase catalyst longevity and turnover number (TON) for ethylene hydrophenylation. An alternative strategy (compared to changing the ligand L of TpRu(L)(NCMe)Ph complexes) to tune the electron density is to replace the Tp ligand with an isostructural but less strongly donating and charge neutral poly(pyrazolyl)alkane ligand, which would result in a complex with a net cationic charge. To this end, the Ru(II) complex [(Tpm^5)Ru(BPhos)(NCMe)Ph][BAr$'_4$] (**Tpm^{5}Ru-BPhos**; Tpm^5 = tris(5-methylpyrazolyl)methane; Ar$'$ = 3,5-bis(CF_3)C_6H_3) was prepared to explore catalytic ethylene hydrophenylation [27]. Installation of the methyl substituent in the 5-position of the pyrazolyl rings protects against intramolecular C—H activation, which has been observed for the related complex [(C(pz)$_4$)Ru(BPhos)(Me)(NCMe)][BAr$'_4$] (C(pz)$_4$ = tetrakis(pyrazolyl)methane) [48].

The cyclic voltammagram of **Tpm^{5}Ru-BPhos** displays a reversible Ru(III/II) couple at 0.83 V (vs NHE), which is a +0.29 V shift in redox potential compared to the corresponding TpRu complex **Ru-BPhos** (0.54 V vs NHE) [46]. **Tpm^{5}Ru-BPhos** is an active catalyst precursor for ethylbenzene production from benzene and ethylene and possesses greater thermal stability than its Tp counterparts [27]. At 150 °C and 0.1 MPa of ethylene, a benzene solution of **Tpm^{5}Ru-BPhos** (0.025 mol%) selectively yields approximately 400 TO of ethylbenzene before catalyst deactivation. However, a direct comparison of the apparent TOF for **Tpm^{5}Ru-BPhos** and **Ru-BPhos** reveals that the Tp catalyst precursor **Ru-BPhos** is slightly more active (by a factor of ~1.3). The increased stability of **Tpm^{5}Ru-BPhos** permits catalysis at elevated temperatures (compared to TpRu(L)(NCMe)Ph catalysts) and renders **Tpm^{5}Ru-BPhos** among the most rapid molecular catalysts for ethylene hydrophenylation with an apparent TOF of approximately $2 \times 10^{-2}\ s^{-1}$ at 150 °C. An Ir(III) catalyst demonstrates a TOF of approximately $4 \times 10^{-2}\ s^{-1}$ (see Section 3.3), but the data for the Ir(III) catalyst are obtained from a reaction at 180 °C indicating that **Tpm^{5}Ru-BPhos** likely has a similar or better activity than the Ir(III) catalyst [49]. In contrast, nearly all TpRu(L)(NCMe)Ph complexes undergo rapid decomposition at temperatures greater than 90 °C. Upon lowering the reaction temperature to 90 °C, **Tpm^{5}Ru-BPhos** catalyzes over 550 TO of ethylbenzene, corresponding

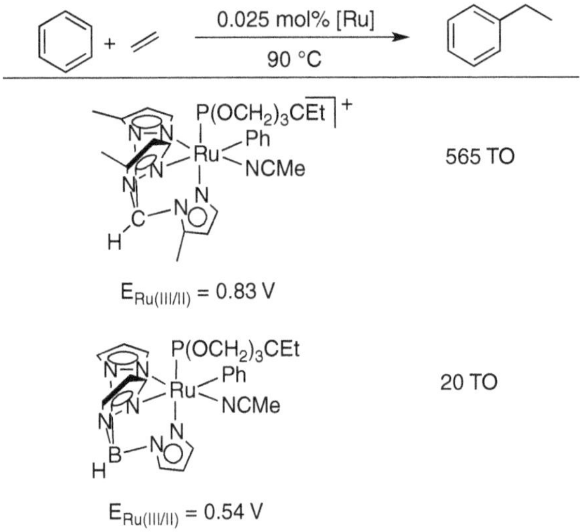

Scheme 3.8 Direct comparison of TO for ethylbenzene using **Tpm^{5}Ru-BPhos** and **Ru-BPhos** with 0.1 MPa of ethylene and 0.025 mol% Ru (relative to benzene) at 90 °C.

to a 95% yield based on ethylene, which is the limiting reagent. In contrast, under these same conditions, the TpRu analog **Ru-BPhos** only achieves 20 TO of ethylbenzene before catalyst deactivation (Scheme 3.8). As observed with the TpRu(II) complexes, **Tpm^{5}Ru-BPhos** eventually undergoes deactivation by formation of the η^3-allyl complex [Tpm^{5}Ru(BPhos)(η^3-C_4H_7)][BAr$'_4$]. Comparison of the TON obtained for each catalyst precursor versus their corresponding $E_{Ru(III/II)}$ reveals a distinct trend that could aid future catalyst design (Figure 3.2). Increasing the electron density at the Ru center (i.e., more negative $E_{Ru(III/II)}$) results in decreased catalytic TO for ethylene hydrophenylation, likely due to increased rates of catalyst deactivation via olefin C—H activation and subsequent Ru(η^3-allyl) formation. Thus, more electron-deficient catalyst precursors

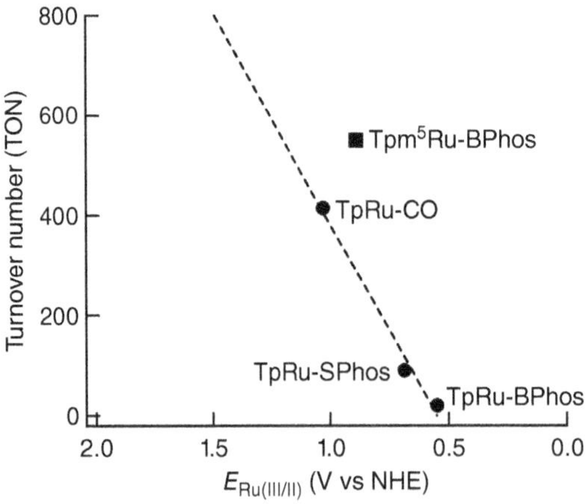

Figure 3.2 Turnover number as a function of $E_{Ru(III/II)}$ for ethylene hydrophenylation using TpRu(L)(NCMe)Ph catalysts and **Tpm^{5}Ru-BPhos** using 0.025 mol% Ru (relative to C_6H_6) and 0.1 MPa of ethylene at 90 °C.

($E_{Ru(III/II)} \geq 0.9$ V) are predicted to sufficiently minimize deactivation and provide even greater catalyst longevity.

3.3 Olefin Hydroarylation Catalyzed by Octahedral d^6 Ir(III) Supported by the Acetylacetonate Ligand

Periana, Goddard and coworkers have developed Ir(III) complexes that serve as catalyst precursors for olefin hydroarylation [38, 49–53]. Initial reports showed that the dimer [Ir(μ-acac-*O,O,C*^{3})-(acac-*O,O*)-(acac-*C*^{3})]$_2$ (acac = acetylacetonate) (**Ir-acac**) catalyzes the formation of ethylbenzene at 180 °C from ethylene and benzene with a TOF = 4.2×10^{-2} s^{-1} (Scheme 3.9) [49]. Diethylbenzenes are produced in a relatively small quantity compared to that of ethylbenzene (~31 : 1 ratio). The active catalyst is proposed to be monomeric *cis*-(*O,O*-acac)$_2$Ir(Ph)(η^2-ethylene), formed *in situ* from benzene C—H activation and ethylene coordination (see Section 3.3.1).

The alkylation of benzene with α-olefins (e.g., propylene, 1-hexene, isobutylene, and styrene) using **Ir-acac** was examined (Scheme 3.10) [49, 50]. In each reaction, both the linear and branched isomers are formed in varied

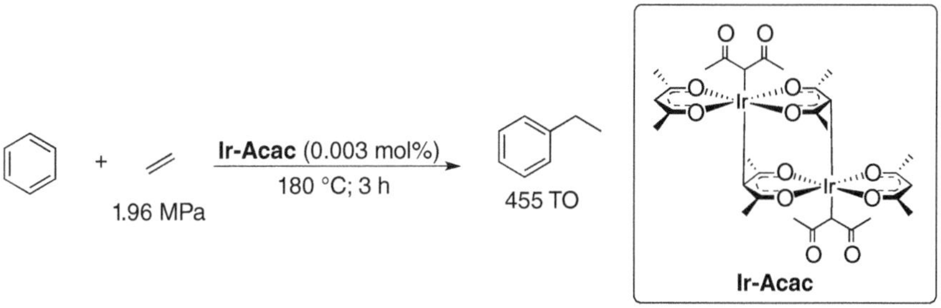

Scheme 3.9 Hydrophenylation of ethylene using [Ir(μ-acac-*O,O,C*^{3})-(acac-*O,O*)-(acac-*C*^{3})]$_2$ (**Ir-acac**) to provide ethylbenzene in a H_2O-saturated benzene solution (mol% of **Ir-acac** relative to benzene).

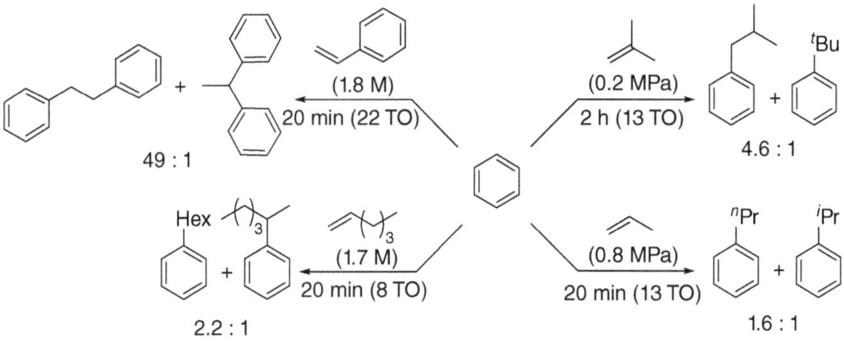

Scheme 3.10 Hydrophenylation of substituted olefins using **Ir-acac** (0.003 mol% relative to benzene) at 180 °C.

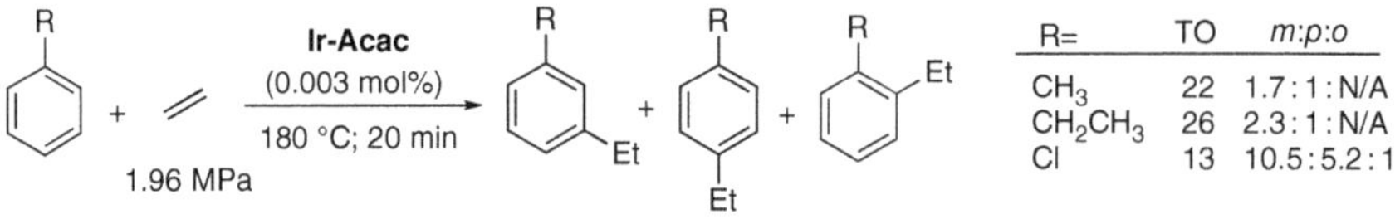

Scheme 3.11 Hydroarylation of ethylene using **Ir-acac**.

ratios. However, the regioselectivity is predisposed to the formation of the linear (i.e., anti-Markovnikov) product, with styrene achieving 98% selectivity for anti-Markovnikov addition. To probe the selectivity of the reaction with respect to the arene, several arenes were substituted for benzene with products exhibiting preference for meta-substitution (Scheme 3.11). Alkylation of naphthalene with ethylene is selective for the β-position, which greatly contrasts the selectivity for alkylation (71% α-substitution and 29% β-substitution) when a Brønsted acid catalyst was used.

3.3.1 Mechanism of Catalytic Olefin Hydrophenylation using Ir(III) Supported by the Acetylacetonate Ligand

Formation of a monomeric Ir complex as the active catalyst using **Ir-acac** as the catalyst precursor is supported by the synthesis of Ir(acac)$_2$Ph(L) complexes (L = acac, H_2O or pyridine) (Figure 3.3). The complexes Ir(acac)$_2$Ph(L) exhibit activity similar to that of the binuclear Ir catalyst precursor **Ir-acac**, and the slight differences in reactivity are proposed to be a result of ground state effects [51].

Experimental studies and DFT calculations have been used to probe the mechanism of Ir(III)-catalyzed ethylene hydrophenylation, and the proposed pathway is shown in Scheme 3.12 [38]. Following the dissociation of the labile ligand, L, *trans*-(κ^2-*O,O*-acac)$_2$Ir(Ph) undergoes *trans*/*cis* isomerization of the acac ligands. Next, coordination of ethylene and migratory insertion into the Ir—Ph bond occurs to produce the five-coordinate intermediate *cis*-(*O,O*-acac)$_2$Ir(CH_2CH_2Ph). The ethylene insertion step is proposed to be the rate-determining step with a calculated activation enthalpy ($\Delta H_{0\,K}$) of 25.2 kcal/mol. The experimentally determined activation energy for ethylene hydrophenylation using **Ir-acac** as the catalyst precursor is 28.7 kcal/mol [50]. Coordination of benzene and subsequent C—H activation liberates ethylbenzene and regenerates the Ir—Ph bond, closing the catalytic cycle. Several mechanisms for C—H bond activation were considered, and computational modeling studies support an OHM process with a formally Ir(V) transition state (see Section 3.4 and Scheme 3.14) [38]. The formation of a discrete Ir(V) intermediate from

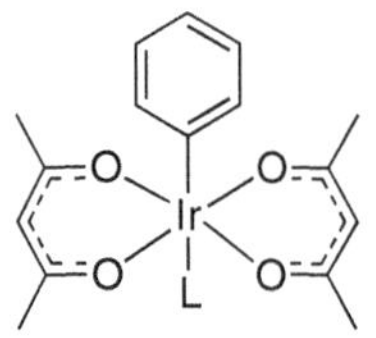

L = acac, H_2O, or pyridine

Figure 3.3 Mononuclear Ir(III) complexes used as catalyst precursors for olefin hydroarylation (L = acac refers to C-bound κ^1 ligand).

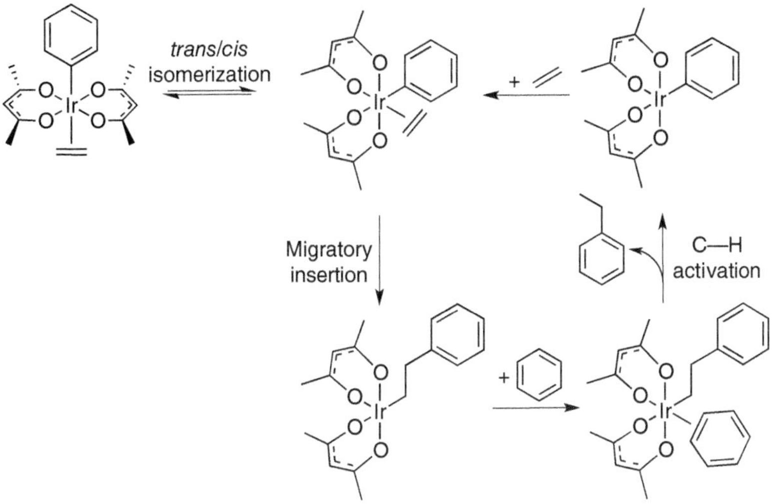

Scheme 3.12 Proposed mechanism for ethylene hydrophenylation catalyzed by *cis*-(κ^2-*O,O*-acac)$_2$Ir(η^2-C_2H_4)Ph.

C—H oxidation addition could not be located computationally, suggesting that it is not accessible under catalytic conditions.

The regioselectivity favoring the linear versus branched isomers when α-olefins are utilized is attributed to a composite of steric and electronic factors [38]. Using the simplest α-olefin as an example, the calculated transition state for the 2,1-insertion ($TS_{2,1}$) of propylene (i.e., the pathway to yield *n*-propylbenzene) is more favorable than the transition state for 1,2-insertion ($TS_{1,2}$), which would ultimately yield cumene, by 0.4 kcal/mol ($\Delta\Delta H^{\ddagger}_{0\,K}$; Scheme 3.13). This is in remarkably good agreement with the $\Delta\Delta G^{\ddagger}$ calculated from the experimental results of 0.4 kcal/mol (180 °C or 453 K). Examination of geometry optimized

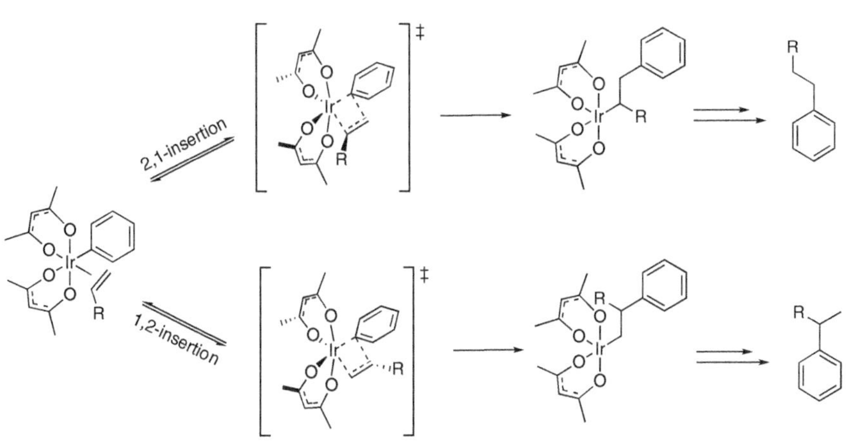

Scheme 3.13 Product selectivity from the hydrophenylation of substituted olefins is ultimately dictated by the mode of insertion into the Ir—Ph bond.

structures of $TS_{2,1}$ and $TS_{1,2}$ reveals that this small yet significant preference for $TS_{2,1}$ is proposed to arise due to the minimization of steric interactions between the olefin substituent and the phenyl group by orienting the substituent into an area of empty space above one of the acac ligands. It can be deduced that a steric effect cannot be the only factor that determines product selectivity because use of isobutylene gave a lower ratio of linear to branched products (4.6 : 1) compared to styrene (49 : 1) (Scheme 4.9). The electronic contribution is related to the electron donating/withdrawing attributes of the olefin substituent. Electron-withdrawing groups localize more positive charge on the substituted C—H unit of the olefin. Therefore, it is more favorable to generate an Ir—C bond through $TS_{2,1}$, as carbon is more electron-deficient than iridium [38].

3.3.2 Other d^6 Ir(III) Catalysts

Ancillary ligands other than acac were explored for Ir(III)-catalyzed olefin hydroarylation, including bipyrimidine and cyclopentadienyl, but these complexes were found to be inactive [50]. Studies of O-donor ligands were expanded through substitution of the O,O-donor bidentate ligand from acac to trop (trop = κ^2-*O,O*-tropolonato, Figure 3.4) [52]. The tropolone-based ligand is expected to be less electron-donating than acac due to increased delocalization of electrons into the aromatic backbone, and the smaller bite angle of the trop ligand (five-membered chelate compared to the six-membered for acac) should provide a different steric environment compared to acac. The complex $(trop)_2Ir(py)Ph$ (**Ir-trop**, py = pyridine) was found to mediate aromatic C—H activation approximately 50 times faster than $(acac)_2Ir(py)Ph$ [54]. However, upon evaluation of catalytic activity for olefin hydrophenylation, **Ir-trop** displayed a comparable or slightly decreased catalytic activity compared to **Ir-acac**. For example, the catalytic hydrophenylation of propylene under identical conditions resulted in 8 and 9 TO after 30 min for **Ir-trop** and **Ir-acac**, respectively. An identical distribution of *n*-propylbenzene and cumene in a 1.6 : 1 ratio was obtained for both complexes. For the hydrophenylation of ethylene, **Ir-trop** appears to be approximately three times less active than **Ir-acac**. It should be noted that the catalytic ethylene hydrophenylation reactions were not conducted under identical conditions, and therefore an exact comparison is not possible [52]. Consistent with experimental results, the calculated activation enthalpy for benzene C—H activation is significantly lower for **Ir-trop** compared to **Ir-acac** with a calculated $\Delta\Delta H^{\ddagger}$ of 3.5 kcal/mol. However, the activation barrier for olefin insertion (the proposed rate-determining step for both catalysts) is increased by 1.5 kcal/mol for **Ir-trop**, relative to **Ir-acac**, with activation enthalpies of 26.7 and 25.2 kcal/mol, respectively, relative to the *cis*-Ir(η^2-C_2H_4)Ph intermediate [52].

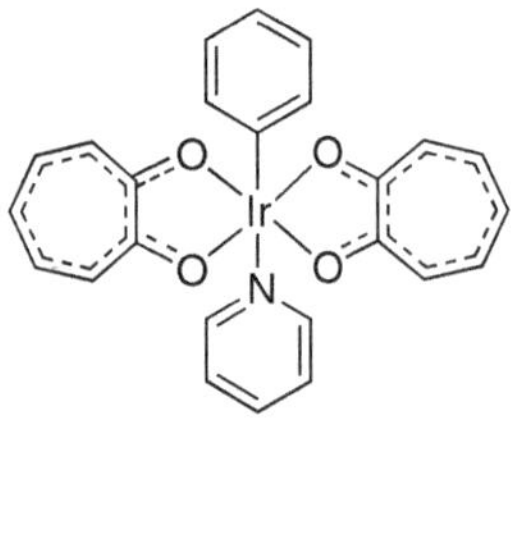

Figure 3.4 The Ir(III) complex $(trop)_2Ir(py)Ph$ (trop = κ^2-*O,O*-tropolonato; **Ir-trop**).

3.4 Summary: Comparison of Ru(II) and Ir(III) Catalysts for Olefin Hydroarylation

Computational studies by Periana, Goddard, and coworkers have directly compared catalysis by the two closely related d^6 catalysts **Ir-acac** and **Ru-CO** and proposed that they produce alkyl arenes by similar reaction pathways [37]. Based upon their computations, the rate-determining step for Ir(III)- and Ru(II)-catalyzed ethylene hydrophenylation is proposed to be ethylene insertion into metal-phenyl bonds (Scheme 3.14). Relative to the η^2-ethylene intermediates, {[M](η^2-C_2H_4)Ph ([M] = *cis*-(*O*,*O*-acac)$_2$Ir or TpRu(CO)}, the $\Delta H^{\ddagger}$ for ethylene insertion is calculated to be lower for TpRu(CO)(η^2-C_2H_4)Ph (calculated $\Delta H^{\ddagger} = 21.1$ kcal/mol) compared to *cis*-(*O*,*O*-acac)$_2$Ir(η^2-C_2H_4)Ph (calculated $\Delta H^{\ddagger} = 24.5$ kcal/mol). For both Ru and Ir, benzene C—H activation has a lower calculated activation enthalpy relative to ethylene insertion into the M—Ph bond. Benzene C—H activation from [M](CH_2CH_2Ph)(η^2-C_6H_6) is evaluated to be more favorable for the Ru complex (calculated $\Delta H^{\ddagger} = 11.8$ kcal/mol) than the Ir complex (calculated $\Delta H^{\ddagger} = 18.0$ kcal/mol) system by 6.2 kcal/mol. The increased activation barrier to olefin insertion and overall decreased activity of the *cis*-(*O*,*O*-acac)$_2$IrPh catalyst relative to TpRu(CO)(NCMe)Ph has been primarily attributed to ligand effects rather than the metal identity and oxidation state of the metal center [37]. For example, the calculated $\Delta H^{\ddagger}$ values for ethylene insertion for isoelectronic [TpIr(CO)(η^2-C_2H_4)(Ph)]$^+$ and *cis*-(*O*,*O*-acac)$_2$Ru(η^2-C_2H_4)(Ph)]$^-$, are 17.0 and 28.3 kcal/mol, respectively. For

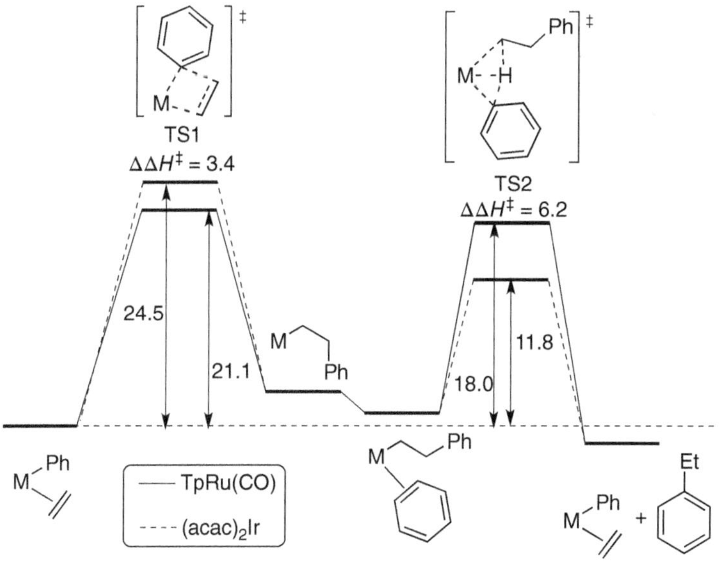

Scheme 3.14 Comparison of calculated enthalpies (kcal/mol) of activation for ethylene insertion and benzene C—H activation during ethylene hydrophenylation catalyzed by *cis*-(*O*,*O*-acac)$_2$Ir and TpRu(CO) [37].

Ru(II), replacing the ancillary ligand set results in a 7.2 kcal/mol increase in the activation enthalpy of the predicted rate-limiting step.

Both catalyst precursors, **Ru-CO** and **Ir-acac**, are predicted to activate aromatic C—H bonds by a mechanism that might be considered a hybridization of classic concerted σ-bond metathesis and oxidative addition. The calculated pathway for C—H activation predicts a concerted process that involves four-membered transition states in which the metal center directly interacts with the hydrogen that is transferred from the arene to the alkyl ligand (TS2 in Scheme 3.14) [26, 28, 31, 38]. As the transition state is oxidative in character with respect to the metal (i.e., back donation of metal-centered electrons to the activated hydrogen), the C—H activation processes could be argued to possess transient Ru(IV) and Ir(V) *formal* oxidation states. Thus, C—H activation processes that go through these transition states have been called OHM [26, 28, 31, 38]. For TpRu(CO)R and *cis*-(*O*,*O*-acac)$_2$Ir(R), oxidative addition of benzene C—H bonds to M^n–R to form M^{n+2}(H)(R)(Ph) intermediates is predicted to exhibit higher energy barriers (compared to the OHM transition states) possibly due to the formation of seven-coordinate intermediates. Examples of seven-coordinate Ir(V) complexes as a result of oxidative addition reactions are known [55–57], but the increased steric profile and reduced basicity of the Ir(III) center of *cis*-(*O*,*O*-acac)$_2$Ir(R) [compared to, for example, $[Cp^*(L)IrMe(X)]^{n+}$ ($Cp^* = \eta^5\text{-}C_5Me_5$; L = PMe_3 or $P(OMe)_3$; X = OTf, $n = 0$ or X = CH_2Cl_2, $n = 1$) due to the acac ligand framework result in aromatic C—H oxidative addition exhibiting higher energy than concerted OHM [38].

An important point to be taken from the experimental and computational studies of the **Ru-CO** and **Ir-acac** catalysts for olefin hydroarylation is the apparent inverse relationship between activation barriers for olefin insertion and aromatic C—H activation. Tuning the ancillary ligand set to decrease the activation barrier for one process along the catalytic cycle (e.g., olefin insertion) results in an increase in the activation barrier for its companion transformation (e.g., aromatic C—H activation) [26, 37]. Indeed, this could be a general feature for catalytic olefin hydroarylation using six-coordinate d^6 catalyst precursors that operate by the pathways discussed herein. In such a case, there is a narrow window for optimization as continual enhancement in catalyst activity is not possible because eventually the rate-limiting step would change and the activation barrier would increase.

3.5 Future Outlook: Extension of Olefin Hydroarylation using Hydrocarbons to Earth Abundant Metals

One of the challenges for the design of new catalysts is to extend transformations that are mediated by precious metals to earth-abundant first row transition metals. Iron is increasingly being explored in catalytic chemistry and has displayed reactivity toward C—C and C—E (E = B, N, O, P, or S) bond forming reactions as well as hydrogenation reactions [58–64]. However, the development of iron catalysts for olefin hydroarylation is challenging as examples of d^6 Fe^{II}-mediated aromatic C—H activation are rare and are typically focused on the more activated C–H bonds of heteroaromatic substrates [65–67].

Despite the challenges, at least two examples of Fe(II)-mediated benzene C–H activation by a non-radical pathway have been reported [65–67]. Using a coordinatively unsaturated iron-methyl complex, Cp*Fe(Me-NHC)Me (Me-NHC = 1,3,4,5-tetramethyl-imidazol-2-ylidene), Tatsumi and coworkers reported a rare example of Fe(II)-mediated benzene C—H activation in addition to the C—H activation of other heteroaromatic substrates (Scheme 3.15) [66]. In contrast to the quantitative (determined by ^{1}H NMR spectroscopy) C—H activation reactions using furan and thiophene, the reaction with benzene required seven days at 80 °C to obtain Cp*Fe(Me-NHC)Ph in 40% yield. Another Cp*Fe(II) complex, Cp*Fe(CO)(NCMe)Ph (**Cp*FeCO**), was similarly found to activate the C—H(D) bonds of benzene, furan, thiophene, and 2-methylfuran (Scheme 3.16) [67]. Heating a solution of **Cp*FeCO** in benzene-d_6 at 50 °C revealed the formation of Cp*Fe(CO)(NCMe)(Ph-d_5) in approximately 80% yield (determined by ^{1}H NMR spectroscopy). For heteroaromatic substrates, including 2-methylfuran, regioselective C—H activation of the α-position of

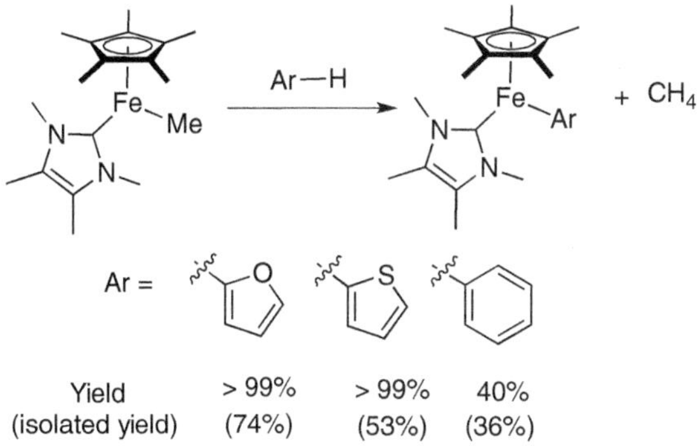

Scheme 3.15 Examples of aromatic C—H activation mediated by Cp*Fe(Me)(Me-NHC) (Me-NHC = 1,3,4,5-tetramethyl-imidazol-2-yl).

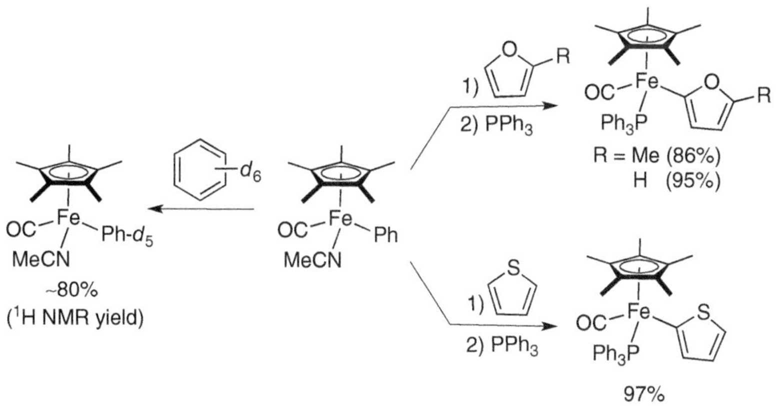

Scheme 3.16 Examples of aromatic C—H activation mediated by Cp*Fe(CO)(NCMe)Ph.

the aromatic ring is observed for both Cp*Fe(II) complexes. As **Cp*FeCO** has been demonstrated to efficiently mediate aromatic C—H activation and olefin insertion into Fe—C_{aryl} bonds is known for Fe(II) half-sandwich complexes [68], the stage has been set to extend olefin hydroarylation from precious metal systems to Fe-based catalysts.

References

1 Olah, G.A. and Molnár, Á. (2003) *Hydrocarbon Chemistry*, 2nd edn, New York, Wiley-Interscience.
2 Weissermel, K. and Arpe, H.-J. (1997) *Industrial Organic Chemistry*, VCH Publishers, New York.
3 Carey, F.A. (2000) Alcohols and alkyl halides, in *Organic Chemistry*, McGraw-Hill, New York, pp. 126–167.
4 Perego, C. and Ingallina, P. (2002) Recent advances in the industrial alkylation of aromatics: new catalysts and new processes. *Catal. Today*, **73**, 3–22.
5 Perego, C. and Ingallina, P. (2004) Combining alkylation and transalkylation for alkylaromatic production. *Green Chem.*, **6**, 274–279.
6 Vermeiren, W. and Gilson, J.P. (2009) Impact of zeolites on the petroleum and petrochemical industry. *Top. Catal.*, **52**, 1131–1161.
7 Macquarrie, D.J. (2009) Industrial Friedel–Crafts chemistry, in *Catalytic Asymmetric Friedel–Crafts Alkylations* (eds M. Bandini and A. Umani-Ronchi), Wiley-VCH, Weinheim, pp. 271–288.
8 Perego, C. and Pollesel, P. (2010) Chapter 2 – Advances in aromatics processing using zeolite catalysts, in *Advances in Nanoporous Materials* (ed. E. Stefan), Elsevier, Vol. 1, pp. 97–149.
9 Jan, D.Y. and Barger, P.T. (2010) Processes on industrial C—C bond formation, in *Zeolites in Industrial Separation and Catalysis* (ed. S. Kulprathipanja), John Wiley & Sons, pp. 505–533.
10 Yilmaz, B. and Müller, U. (2009) Catalytic applications of zeolites in chemical industry. *Top. Catal.*, **52**, 888–895.
11 Corma, A. and Martínez, A. (2005) Zeolites in refining and petrochemistry, in *Studies in Surface Science and Catalysis* (eds J. Čejka and H. van Bekkum), Elsevier Vol. 157, pp. 337–366.
12 Marcilly, C. (2006) Main acids, superacids and bases of interest in catalysis, in *Acido-Basic Catalysis*, Editions Technip, France Vol. 1, pp. 57–130.
13 Guari, Y., Sabo-Etienne, S., and Chaudret, B. (1999) Catalytic formation of carbon-carbon bonds by activation of carbon-hydrogen bonds. *Eur. J. Inorg. Chem.*, **1999**, 1047–1055.
14 Trost, B.M., Toste, F.D., and Pinkerton, A.B. (2001) Non-metathesis ruthenium-catalyzed C—C bond formation. *Chem. Rev.*, **101**, 2067–2096.
15 Ritleng, V., Sirlin, C., and Pfeffer, M. (2002) Ru-, Rh-, and Pd-catalyzed C—C bond formation involving C—H activation and addition on unsaturated substrates: reactions and mechanistic aspects. *Chem. Rev.*, **102**, 1731–1769.
16 Jia, C., Kitamura, T., and Fujiwara, Y. (2001) Catalytic functionalization of arenes and alkanes via C—H bond activation. *Acc. Chem. Res.*, **34**, 633–639.

17 McKeown, B.A., Gonzalez, H.E., Friedfeld, M.R., Gunnoe, T.B., Cundari, T.R., and Sabat, M. (2011) Mechanistic studies of ethylene hydrophenylation catalyzed by bipyridyl Pt(II) complexes. *J. Am. Chem. Soc.*, **133**, 19131–19152.
18 Goj, L.A. and Gunnoe, T.B. (2005) Developments in catalytic aromatic C—H transformations: promising tools for organic synthesis. *Curr. Org. Chem.*, **9**, 671–685.
19 MacDonald, M.G., Kostelansky, C.N., White, P.S., and Templeton, J.L. (2006) Ethylene insertion in Tp′Pt(Ph)(C_2H_4) and C—H activation of ethylbenzene to form a platinum(IV) *ortho*-metalated phenethyl complex. *Organometallics*, **25**, 4560–4570.
20 Matsumoto, T. and Yoshida, H. (2000) Oxidative arylation of ethylene with benzene to produce styrene. *Chem. Lett.*, **29**, 1064–1065.
21 Hong, P. and Yamazaki, H. (1979) Reactions of ethylene and benzenes catalyzed by rhodium carbonyls under carbon monoxide. the formation of styrenes and 3-pentanone. *Chem. Lett.*, **8**, 1335–1336.
22 Matsumoto, T., Periana, R.A., Taube, D.J., and Yoshida, H. (2002) Direct synthesis of styrene by rhodium-catalyzed oxidative arylation of ethylene with benzene. *J. Catal.*, **206**, 272–280.
23 Vaughan, B.A., Webster-Gardiner, M.S., Cundari, T.R., and Gunnoe, T.B. (2015) A rhodium catalyst for single-step styrene production from benzene and ethylene. *Science*, **348**, 421–424.
24 Labinger, J.A. and Bercaw, J.E. (2002) Understanding and exploiting C—H bond activation. *Nature*, **417**, 507–514.
25 Lail, M., Arrowood, B.N., and Gunnoe, T.B. (2003) Addition of arenes to ethylene and propene catalyzed by ruthenium. *J. Am. Chem. Soc.*, **125**, 7506–7507.
26 Foley, N.A., Lee, J.P., Ke, Z., Gunnoe, T.B., and Cundari, T.R. (2009) Ru(II) catalysts supported by hydridotris(pyrazolyl)borate for the hydroarylation of olefins: reaction scope, mechanistic studies, and guides for the development of improved catalysts. *Acc. Chem. Res.*, **42**, 585–597.
27 Burgess, S.A., Joslin, E.E., Gunnoe, T.B., Cundari, T.R., Sabat, M., and Myers, W.H. (2014) Hydrophenylation of ethylene using a cationic Ru(II) catalyst: comparison to a neutral Ru(II) catalyst. *Chem. Sci.*, **5**, 4355–4366.
28 Lail, M., Bell, C.M., Conner, D., Cundari, T.R., Gunnoe, T.B., and Petersen, J.L. (2004) Experimental and computational studies of Ru(II) catalyzed addition of arene C—H bonds to olefins. *Organometallics*, **23**, 5007–5020.
29 Joslin, E.E., McMullin, C.L., Gunnoe, T.B., Cundari, T.R., Sabat, M., and Myers, W.H. (2012) Catalytic hydroarylation of ethylene using TpRu(L)(NCMe)Ph (L = 2,6,7-trioxa-1-phosphabicyclo[2,2,1]heptane): comparison to TpRu(L′)(NCMe)Ph systems (L′ = CO, PMe_3, $P(pyr)_3$, or $P(OCH_2)_3CEt$). *Organometallics*, **31**, 6851–6860.
30 Andreatta, J.R., McKeown, B.A., and Gunnoe, T.B. (2011) Transition metal catalyzed hydroarylation of olefins using unactivated substrates: recent developments and challenges. *J. Organomet. Chem.*, **696**, 305–315.
31 Oxgaard, J. and Goddard, W.A. III (2004) Mechanism of Ru(II)-catalyzed olefin insertion and C—H activation from quantum chemical studies. *J. Am. Chem. Soc.*, **126**, 442–443.

32 Arrowood, B.N., Lail, M., Gunnoe, T.B., and Boyle, P.D. (2003) Radical polymerization of styrene and methyl methacrylate with ruthenium(II) complexes. *Organometallics*, **22**, 4692–4698.

33 Lail, M., Gunnoe, T.B., Barakat, K.A., and Cundari, T.R. (2005) Conversions of ruthenium(III) alkyl complexes to ruthenium(II) through Ru-C_{alkyl} bond homolysis. *Organometallics*, **24**, 1301–1305.

34 Goj, L.A., Lail, M., Pittard, K.A., Riley, K.C., Gunnoe, T.B., and Petersen, J.L. (2006) Reactions of TpRu(CO)(NCMe)(Ph) with electron-rich olefins: examples of stoichiometric C—S, C—O and C—H bond cleavage. *Chem. Commun.*, 982–984.

35 Pittard, K.A., Lee, J.P., Cundari, T.R., Gunnoe, T.B., and Petersen, J.L. (2004) Reactions of TpRu(CO)(NCMe)(Me) (Tp = hydridotris(pyrazolyl)borate) with heteroaromatic substrates: stoichiometric and catalytic C—H activation. *Organometallics*, **23**, 5514–5523.

36 Pittard, K.A., Cundari, T.R., Gunnoe, T.B., Day, C.S., and Petersen, J.L. (2005) Ruthenium(II)-mediated carbon–carbon bond formation between acetonitrile and pyrrole: combined experimental and computational study. *Organometallics*, **24**, 5015–5024.

37 Oxgaard, J., Periana, R.A., and Goddard, W.A. III (2004) Mechanistic analysis of hydroarylation catalysts. *J. Am. Chem. Soc.*, **126**, 11658–11665.

38 Oxgaard, J., Muller, R.P., Goddard, W.A. III, and Periana, R.A. (2004) Mechanism of homogeneous Ir(III) catalyzed regioselective arylation of olefins. *J. Am. Chem. Soc.*, **126**, 352–363.

39 Waterman, R. (2013) σ-Bond metathesis: a 30-year retrospective. *Organometallics*, **32**, 7249–7263.

40 DeYonker, N.J., Foley, N.A., Cundari, T.R., Gunnoe, T.B., and Petersen, J.L. (2007) Combined experimental and computational studies on the nature of aromatic C—H activation by octahedral ruthenium(II) complexes: evidence for sigma-bond metathesis from Hammett studies. *Organometallics*, **26**, 6604–6611.

41 Kubas, G.J. (2001) *Metal Dihydrogen and σ-Bond Complexes*, Springer, New York, NY.

42 Balcells, D., Clot, E., and Eisenstein, O. (2010) C—H bond activation in transition metal species from a computational perspective. *Chem. Rev.*, **110**, 749–823.

43 Lam, W.H., Jia, G., Lin, Z., Lau, C.P., and Eisenstein, O. (2003) Theoretical studies on the metathesis processes, $[Tp(PH_3)MR(\eta^2\text{-H-CH}_3)] \rightarrow [Tp(PH_3)M(CH_3)(\eta^2\text{-H-R})]$ (M = Fe, Ru, and Os; R = H and CH_3). *Chem. Eur. J.*, **9**, 2775–2782.

44 Foley, N.A., Gunnoe, T.B., Cundari, T.R., Boyle, P.D., and Petersen, J.L. (2008) Activation of sp^3 carbon–hydrogen bonds by a ruthenium(II) complex and subsequent metal-mediated C—C and C—N bond formation. *Angew. Chem. Int. Ed.*, **47**, 726–730.

45 Foley, N.A., Lail, M., Lee, J.P., Gunnoe, T.B., Cundari, T.R., and Petersen, J.L. (2007) Comparative reactivity of TpRu(L)(NCMe)Ph (L = CO or PMe_3):

impact of ancillary ligand L on activation of carbon–hydrogen bonds including catalytic hydroarylation and hydrovinylation/oligomerization of ethylene. *J. Am. Chem. Soc.*, **129**, 6765–6781.

46 Foley, N.A., Ke, Z., Gunnoe, T.B., Cundari, T.R., and Petersen, J.L. (2008) Aromatic C—H activation and catalytic hydrophenylation of ethylene by TpRu{P(OCH$_2$)$_3$CEt}(NCMe)Ph. *Organometallics*, **27**, 3007–3017.

47 Foley, N.A., Lail, M., Gunnoe, T.B., Cundari, T.R., Boyle, P.D., and Petersen, J.L. (2007) Combined experimental and computational study of TpRu{P(pyr)$_3$}(NCMe)Me (pyr = *N*-pyrrolyl): inter- and intramolecular activation of C—H bonds and the impact of sterics on catalytic hydroarylation of olefins. *Organometallics*, **26**, 5507–5516.

48 Joslin, E.E., Quillian, B., Gunnoe, T.B., Cundari, T.R., Sabat, M., and Myers, W.H. (2014) C—H activation of pyrazolyl ligands by Ru(II). *Inorg. Chem.*, **53**, 6270–6279.

49 Matsumoto, T., Taube, D.J., Periana, R.A., Taube, H., and Yoshida, H. (2000) Anti-markovnikov olefin arylation catalyzed by an iridium complex. *J. Am. Chem. Soc.*, **122**, 7414–7415.

50 Matsumoto, T., Periana, R.A., Taube, D.J., and Yoshida, H. (2002) Regioselective hydrophenylation of olefins catalyzed by an Ir(III) complex. *J. Mol. Catal. A: Chem.*, **180**, 1–18.

51 Periana, R.A., Liu, X.Y., and Bhalla, G. (2002) Novel bis-acac-*O,O*-Ir(III) catalyst for anti-Markovnikov hydroarylation of olefins by arene ch activation. *Chem. Commun.*, **24**, 3000–3001.

52 Bhalla, G., Oxgaard, J., Goddard, W.A. III, and Periana, R.A. (2005) Anti-Markovnikov hydroarylation of unactivated olefins catalyzed by a bis-tropolonato iridium(III) organometallic complex. *Organometallics*, **24**, 3229–3232.

53 Bhalla, G., Bischof, S.M., Ganesh, S.K., Liu, X.Y., Jones, C.J., Borzenko, A., Tenn, W.J. III, Ess, D.H., Hashiguchi, B.G., Lokare, K.S., Leung, C.H., Oxgaard, J., Goddard, W.A. III, and Periana, R.A. (2011) Mechanism of efficient anti-Markovnikov olefin hydroarylation catalyzed by homogeneous Ir(III) complexes. *Green Chem.*, **13**, 69–81.

54 Bhalla, G. and Periana, R.A. (2005) C—H activation of alkanes and arenes catalyzed by an O-donor bis(tropolonato)iridium(III) complex. *Angew. Chem. Int. Ed.*, **44**, 1540–1543.

55 Alaimo, P.J. and Bergman, R.G. (1999) Modeling the proposed intermediate in alkane carbon–hydrogen bond activation by Cp*(PMe$_3$)Ir(Me)OTf: synthesis and stability of novel organometallic iridium(V) complexes. *Organometallics*, **18**, 2707–2717.

56 Klei, S.R., Tilley, T.D., and Bergman, R.G. (2000) The mechanism of silicon–hydrogen and carbon–hydrogen bond activation by iridium(III): production of a silylene complex and the first direct observation of Ir(III)/Ir(V) C—H bond oxidative addition and reductive elimination. *J. Am. Chem. Soc.*, **122**, 1816–1817.

57 Tellers, D.M., Yung, C.M., Arndtsen, B.A., Adamson, D.R., and Bergman, R.G. (2002) Electronic and medium effects on the rate of arene C—H bond activation by cationic Ir(III) complexes. *J. Am. Chem. Soc.*, **124**, 1400–1410.

58 Bauer, E.B. (2008) Recent advances in iron catalysis in organic synthesis. *Curr. Org. Chem.*, **12**, 1341–1369.
59 Plietker, B. (ed.) (2008) *Iron Catalysis in Organic Chemistry: Reactions and Applications*, John Wiley & Sons.
60 Xiaoli, S., Jilai, L., Xuri, H., and Chiachung, S. (2012) Recent advances in iron-catalyzed C—H bond activation reactions. *Curr. Inorg. Chem.*, **2**, 64–85.
61 Tondreau, A.M., Atienza, C.C.H., Weller, K.J., Nye, S.A., Lewis, K.M., Delis, J.G.P., and Chirik, P.J. (2012) Iron catalysts for selective anti-Markovnikov alkene hydrosilylation using tertiary silanes. *Science*, **335**, 567–570.
62 Nesbit, M.A., Suess, D.L.M., and Peters, J.C. (2015) E—H bond activations and hydrosilylation catalysis with iron and cobalt metalloboranes. *Organometallics*, **34**, 4741–4752.
63 Bauer, I. and Knölker, H.-J. (2015) Iron catalysis in organic synthesis. *Chem. Rev.*, **115**, 3170–3387.
64 Noda, D., Tahara, A., Sunada, Y., and Nagashima, H. (2016) Non-precious-metal catalytic systems involving iron or cobalt carboxylates and alkyl isocyanides for hydrosilylation of alkenes with hydrosiloxanes. *J. Am. Chem. Soc.*, **138**, 2480–2483.
65 Ohki, Y., Hatanaka, T., and Tatsumi, K. (2008) C—H bond activation of heteroarenes mediated by a half-sandwich iron complex of *N*-heterocyclic carbene. *J. Am. Chem. Soc.*, **130**, 17174–17186.
66 Hatanaka, T., Ohki, Y., and Tatsumi, K. (2010) C—H bond activation/borylation of furans and thiophenes catalyzed by a half-sandwich iron *N*-heterocyclic carbene complex. *Chem. Asian J.*, **5**, 1657–1666.
67 Kalman, S.E., Petit, A., Gunnoe, T.B., Ess, D.H., Cundari, T.R., and Sabat, M. (2013) Facile and regioselective C—H bond activation of aromatic substrates by an Fe(II) complex involving a spin-forbidden pathway. *Organometallics*, **32**, 1797–1806.
68 Yasuda, S., Yorimitsu, H., and Oshima, K. (2010) Arylation of styrenes with aryliron complexes [$CpFe(CO)_2Ar$]. *Organometallics*, **29**, 2634–2636.

4

Hydroarylation of Olefins with Complexes Bearing d^8 Metal Centers

Benjamin A. Suslick[1,2] *and T. Don Tilley*[1,2]

[1] *University of California, Berkeley, Department of Chemistry, Berkeley, CA 94720, USA*
[2] *Lawrence Berkeley National Laboratory, Chemical Sciences Division, 1 Cyclotron Road, Berkeley, CA 94720, USA*

4.1 Introduction

Alkylated arenes are essential intermediates in the manufacture of fine chemicals and widely used plastics such as polystyrene. Polystyrene is produced globally on a massive scale and consumed at an annual rate of 1.2×10^7 tons/year [1–3]. Styrene is consumed on a comparable scale during the production of polystyrene, at a rate of 2.6×10^7 tons/year [2, 3]. A majority of the styrene used in the chemical industry is generated from the dehydrogenation of ethylbenzene (about 85%) [1]. Synthesis of simple alkyl arenes such as ethylbenzene, therefore, has become a large and profitable industry. While traditional routes for the synthesis of ethylbenzene employ Lewis acid catalysts (e.g., $AlCl_3$/HF) to couple olefins with an arene, current industrial methods typically employ heterogeneous catalysts (such as zeolites) to generate ethylbenzene from ethylene and benzene via a Friedel–Crafts mechanism [1, 4]. Such alkylation systems, however, have drawbacks. The resulting alkyl arene products are often more reactive than the starting materials, resulting in undesired polyalkylation. Additionally, catalysis with α-alkene substrates can result in rapid isomerization to an internal olefin, inhibiting generation of the desired anti-Markovnikov, linear alkyl arene.

In academia, synthetic chemists have extensively investigated the process of generating C—C bonds from cheap and abundant feedstock. While synthetic methods exist to transform simple arenes into more complicated structures (such as the Stille or Suzuki cross-coupling reactions), these routes often require aryl substituents such as SnR_3, BR_2, or ZnX [5]. Installation of these groups can be complex and typically involves additional synthetic steps. Since these substituents are not incorporated into the final product, cross-coupling processes are not atom-economical. Moreover, potentially dangerous or toxic waste (e.g., Sn) is generated as a byproduct. As a result, there is a need to directly couple arene C—H bonds with simple alkyl groups (such as olefins) in a selective and efficient manner, using highly active and selective catalysts.

Catalytic Hydroarylation of Carbon-Carbon Multiple Bonds, First Edition.
Edited by Lutz Ackermann, T. Brent Gunnoe and Laurel Goj Habgood.

In contrast to traditional acid-mediated alkylation catalysis or cross-coupling reactions, transition-metal-catalyzed hydroarylation of C—C multiple bonds offers potential advantages.

Recent investigations into homogeneous catalytic hydroarylations of olefins have focused heavily on catalyst precursors bearing a d^8 electronic configuration. This chapter discusses such cases in the context of olefin hydroarylations with benzene and substituted arenes. More specifically, three types of olefin hydroarylation by d^8 catalysts will be discussed: those involving unfunctionalized arenes (e.g., benzene and alkyl arenes), those assisted (and directed) by *ortho*-arene donor groups, and finally palladium-catalyzed hydroarylations that involve reaction sequences of a Heck-type coupling followed by reduction of the C=C double bond with an exogenous hydrogen source. The first two reaction types involve arene C—H activation and olefin insertion as key mechanistic steps. The d^8 metal centers so far identified with hydroarylation catalysis are platinum, rhodium, iridium, palladium, and nickel. In these studies, the identity of the metal center influences the operative mechanism as well as the substrate scope.

The simplest (and perhaps most challenging) catalytic hydroarylations with d^8 metals (e.g., complexes of Pt^{II}) involve unfunctionalized arenes. Research in this area has focused on the coupling of small olefins (i.e., ethylene and propylene) with benzene to generate the corresponding alkyl arene while avoiding over-alkylation (e.g., diethylbenzene formation). For α-olefins such as propylene, Pt^{II} catalysts tend to favor branched, Markovnikov alkyl arene products, but new ligand scaffolds exhibit promise for controlling the regioselectivity for production of linear products. With group 9 metals (Rh^I and Ir^I), directed *ortho*-hydroarylation is possible with arenes bearing coordinating functionalities (imines, ketones, aldehydes, etc.). With these substrates, an initial binding of the aryl-directing group orients the *ortho*-aryl hydrogen into a particular position for C—H activation. Alkylation, therefore, only occurs *ortho* to the directing group. Surprisingly, complexes of Rh^I and Ir^I appear to inherently impart different regioselectivities for hydroarylation. For Rh^I complexes, anti-Markovnikov, linear products are favored, whereas branched products are primarily formed with Ir^I catalysts.

In contrast to Pt^{II}-, Rh^I-, and Ir^I-catalyzed hydroarylations, those mediated by Pd^{II} have not been reported to promote hydroarylations with unactivated arenes. This difference can perhaps be partially explained by the greater reluctance of palladium to engage in C—H activation processes. However, several studies point to the ability of Pd^{II} to mediate hydroarylations with different substrates, namely aryl halides, in combination with an olefin. Such reactions resemble Heck couplings, but require an external hydrogen source to accomplish the net addition of aryl and hydrogen groups across the olefin. Finally, recently reported hydroarylations with nickel complexes indicate that this metal may be promising for the development of new catalyst systems.

4.2 Pt^{II} Catalyzed Hydroarylation

Since hydroarylation requires both aryl C—H activation and olefin-insertion steps, Pt has proven to be the focus in the early developments of hydroarylation catalysis. Significantly, Pt^{II} complexes are known to activate aryl and alkyl C—H bonds [6]. In addition, Pt–olefin complexes are able to undergo migratory insertion reactions [7] that functionalize the olefin through an inner sphere pathway. Unlike other transition metal hydroarylation catalysts that require use of a directing aryl substituent or a strained olefin (e.g., norbornene), Pt^{II} complexes allow the coupling of simple olefins (e.g., ethylene, propylene) with unfunctionalized arenes (benzene, toluene) to generate alkyl arenes (e.g., ethylbenzene). Several types of supporting ligand systems have been investigated, and, in particular, complexes bearing neutral and anionic bidentate NN ligands have been described as competent catalysts. Additionally, several other types of non-nitrogen-based ligands have been briefly explored. A brief discussion of these Pt^{II}-based systems is given below.

4.2.1 Pt^{II} Hydroarylation Catalysts Bearing Anionic Bidentate (NN) Ligands

Studies of Pt-based catalysts for hydroarylation have employed chelating, nitrogen-donor ligands of the XL type. The first such report, by Tilley and coworkers, involved pyridyl-indolate (PyInd) ligands [8–10]. These ligands were synthesized by condensation of the appropriate phenyl hydrazine compounds with acetylpyridine, followed by cyclization of the resulting aryl-hydrazone via the Fisher-indole reaction to afford the desired PyInd ligand precursors **1** and **2** (Scheme 4.1).

PyInd-based potassium salts were used to obtain various Pt^{II} complexes **3–6** (Figure 4.1) by treatment with $[(C_2H_4)PtCl(\mu\text{-}Cl)]_2$ (Zeise's dimer), $[(\eta^3\text{-}CH_2CMeCH_2)Pt(\mu\text{-}Cl)]_2$, or $[Me_2Pt(\mu\text{-}SMe_2)]_2$, respectively [8, 9]. Protonation of the methyl ligand on the anionic complex **6** with $[^iPr_2EtNH][BPh_4]$ in MeCN generated the neutral complex **7**.

The hydroarylation ability of these complexes was examined using norbornene and benzene as substrates (Table 4.1). Complexes **3** and **6** did not exhibit catalytic activity but catalysis was observed with the addition of activating, chloride abstraction reagents such as AgOTf or $AgBF_4$ (for **3**), or the methyl abstraction

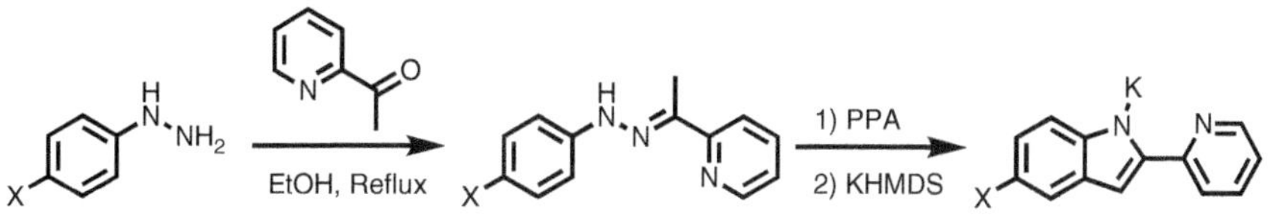

1: X = H
2: X = F

Scheme 4.1 General synthesis of PyInd ligands.

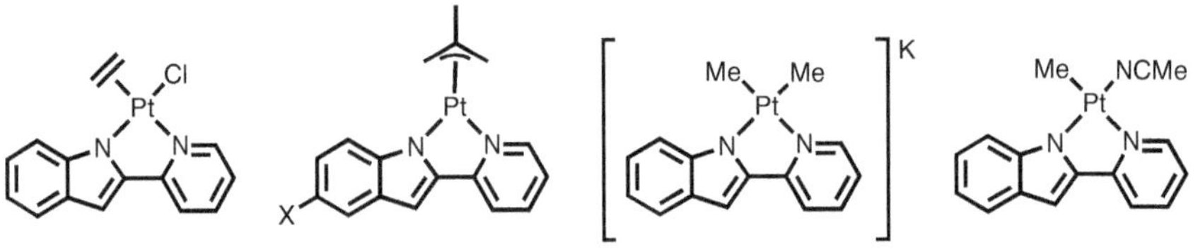

3 **4**: X = H **5**: X = F **6** **7**

Figure 4.1 PyInd-ligated Pt complexes for olefin hydroarylation.

Table 4.1 Hydroarylation of norbornene with benzene using (PyInd)Pt^{II} complexes [8, 9].

Catalyst	Solvent	Ar—H	Temperature (°C)	Time (h)	Yield (%)[a]
3[b]	o-$C_6Cl_2H_4$	C_6H_6	115	20	79
3[c]	o-$C_6Cl_2H_4$	C_6H_6	92	2	92
4	C_6H_6	C_6H_6	140	16	38
4	C_6MeH_5	C_6MeH_5	140	16	28
4	C_6ClH_5	C_6ClH_5	140	16	58
4	C_6FH_5	C_6FH_5	140	16	0
4	p-$C_6Me_2H_4$	p-$C_6Me_2H_4$	140	16	0
5	C_6H_6	C_6H_6	140	16	41
6[d]	C_6H_6	C_6H_6	140	16	27
7	C_6H_6	C_6H_6	140	16	26

a) Percentage (%) yield refers to the yield of the hydroarylation product.
b) Reaction performed with added AgOTf.
c) Reaction performed with added $AgBF_4$.
d) Reaction performed with added $B(C_6F_6)_3$.

reagent $B(C_6F_5)_3$ (for complex **6**). Thus, an empty coordination site is presumably required for substrate activation and functionalization.

Several key differences in the activities were observed among these various catalysts. For complexes **4–7**, significant quantities of a norbornene dimer byproduct were identified (yields of ~30–45%), which suggests that vinylic C—H activation is possible with some Pt species. However, this product was not observed with complex **3**. Additionally, high temperatures (140 vs 110 °C) and long reaction times (16 vs 5 h) were required for catalysts **4–7** [8, 9]. Minor differences were observed between the activities of complexes bearing PyInd ligand **1** and those involving the fluorinated PyInd ligand **2**. The reduced activity for complexes **6** and **7** versus the ethylene-bound complex **3** or the allylic complexes **4** and **5** is possibly due to the preactivation necessary to replace a methyl ligand for a phenyl ligand.

Complex **6** requires the addition of a borane reagent to abstract a methyl ligand to generate an open coordination site for C—H activation to occur; in contrast, complex **7** undergoes a C—H activation step to eliminate methane and generate the corresponding Pt—Ph complex, which presumably is the catalytically active species [8, 9].

The scope of hydroarylation with PyInd complexes was probed with additional substrates. It was found that other cyclic olefins such as cyclohexene and cyclopentene were coupled to benzene at 80 °C providing yields of 65% and 62%, respectively, after 2 h using 5 mol% of Zeise's dimer and 10 mol% of $AgBF_4$ to generate an active catalyst. Simple olefins such as propylene underwent hydroarylation in high yield (79%) under the same conditions as described (see page 110–111). With catalyst **4**, hydroarylation of norbornene was observed using benzene, toluene, or chlorobenzene. A mixture of regioisomers was observed for both toluene and chlorobenzene, with an *o:m:p* ratio for toluene of 0.4 : 1.0 : 0.6. Interestingly, *p*-xylenes and fluorobenzene were not suitable arene substrates and only norbornene dimers were observed as products.

A pyridyl-pyrrole (PyPyr) ligand set published by Goldberg and colleagues [11–13] illustrates the ability of organoplatinum complexes to promote both C—H activation and olefin coordination. In an initial report [11], five-coordinate [(PyPyr)$Pt^{IV}Me_3$] complexes **8** and **9** were observed to reductively eliminate ethane in the presence of benzene and ethylene to generate a transient Pt^{II} species (Scheme 4.2). This species undergoes a rapid intramolecular C—H activation reaction under ethylene to generate the corresponding $Pt^{II}(C_2H_4)$ complexes **10** or **11**. Based on these observations, this system was studied in the context of hydroarylation catalysis.

Further investigations [12] with PyPyr complexes indicated that hydroarylation was indeed possible with either Pt^{II} or Pt^{IV} precatalysts (Figure 4.2). Complexes bearing a dimethyl-substituted pyridyl-pyrrole ligand (PyPyr-Me_2) with labile SMe_2 and a "pre-activated" phenyl ligand were generated with either a d^6 Pt^{IV} (**12**) or a d^8 Pt^{II} (**13**) metal center. Upon exposure of **13** to ethylene in benzene-d_6, olefin insertion and subsequent ethylene coordination was observed to afford [(PyPyr-Me_2)$Pt^{II}(C_2H_4)(CH_2CH_2Ph)$] (**14**). Thermoylsis of complex **14** at 100 °C in benzene-d_6 generated the hydroarylation product, $C_6D_5CH_2CH_2D$. A similar complex (**15**) bearing an unsubstituted PyPyr ligand was also prepared.

The hydroarylation activity [12, 13] of complexes **12**, **13**, and **15** is summarized in Table 4.2. Several key differences between these catalysts were observed, and a

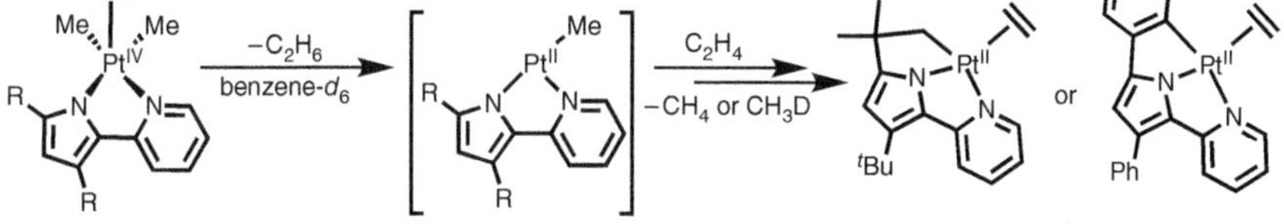

8: R = ${}^tBu$
9: R = Ph

10 **11**

Scheme 4.2 Ethane elimination and intermolecular C—H activation with (PyPyr)Pt^{IV} complexes at 85–100 °C for 1 day.

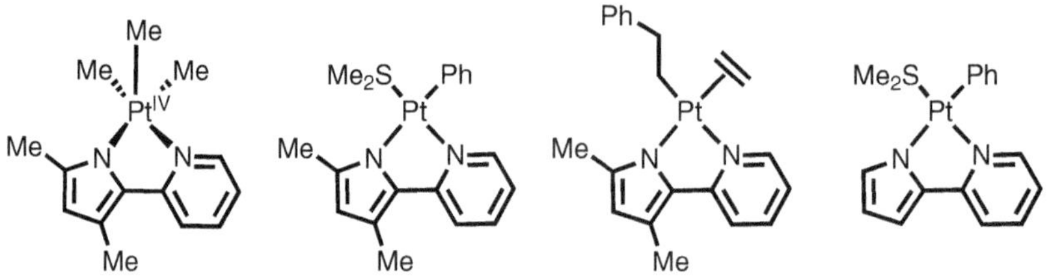

12 13 14 15

Figure 4.2 Several PyPyr ligated complexes for olefin hydroarylation.

potential key observation is an increase in reactivity for Pt^{II} versus Pt^{IV} precatalysts. With benzene and ethylene, complex **12** gave 26 turnover numbers (TONs) while **13** gave 36 TONs after a similar period of time. More striking, however, is the difference in regioselectivities during propylene hydroarylation; complex **13** afforded mainly the Markovnikov (branched) product in a 86 : 14 ratio whereas catalyst **15** generated nearly equivalent amounts of both regioisomers. Finally, it should be noted that substituents on either the aryl or olefin substrate tend to

Table 4.2 Hydroarylation substrate scope for (PyPyr)Pt^{II} complexes [12, 13].

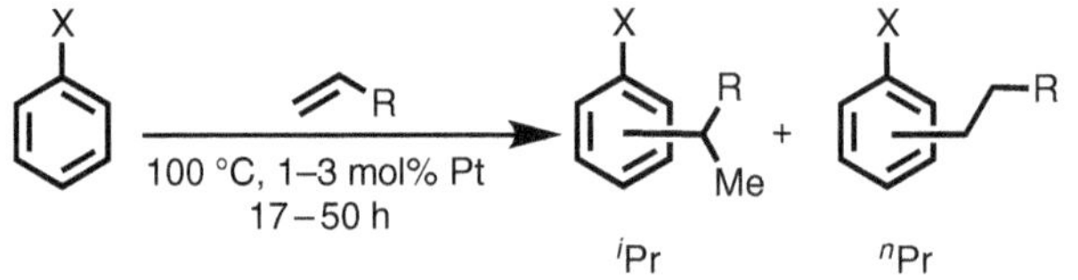

Catalyst	X	R	TON	*o:m:p*	Branched:linear
12	H	H	26	—	—
		Me	8	—	86 : 14
	Me	H	4	7 : 93a)	—
		Me	2	10 : 63 : 27	85 : 15
	CF_3	H	2	6 : 62 : 32	—
13	H	H	36	—	—
		Me	18	—	85 : 15
		^{n}Bu	12	—	83 : 16
	Me	H	12	6 : 94a)	—
		Me	3	9 : 66 : 25	84 : 16
15	H	Me	16	—	51 : 49
		^{n}Bu	12	—	43 : 57

Reactions were performed at either 100 or 120 °C with catalyst loadings between 1 and 3 mol%, for 17–50 h.

a) *meta-* and *para-*Isomers were not separated and are instead listed as the sum of the two isomers.

reduce the overall activity of the catalyst, noted by a reduction in TON. While more distal C—H bonds (i.e., *meta* or *para*) were easier to activate, a mixture of all three isomers was obtained.

Based on these results, a mechanism was proposed for the hydroarylation of ethylene with benzene (Figure 4.3) [12]. In this mechanism, a [(PyPyr)Pt^{II}(C_2H_4)Ph] complex (**20**) (likely generated by a ligand exchange of SMe_2 for C_2H_4) undergoes ethylene insertion and subsequent ethylene coordination to generate complex **14**. An intramolecular aryl C—H activation produces the Pt^{IV}–hydride intermediate **16**, which can undergo reductive elimination and then coordination of benzene (**17**). A reversible oxidative addition of a benzene C—H bond provides a second Pt^{IV}–hydride intermediate (**18**), which again can undergo reductive elimination to form the ethylbenzene complex **19**. Ligand substitution with ethylene regenerates **20** and free ethylbenzene. It is possible, however, that the mechanism can proceed via a single C—H activation pathway. In this alternative mechanism, dissociation of C_2H_4 from **14** can occur and is followed by C—H activation of the solvent to generate the intermediate complex [(PyPyr)Pt^{IV}(H)(CH_2CH_2Ph)Ph]. Reductive elimination from this species generates complex **19**.

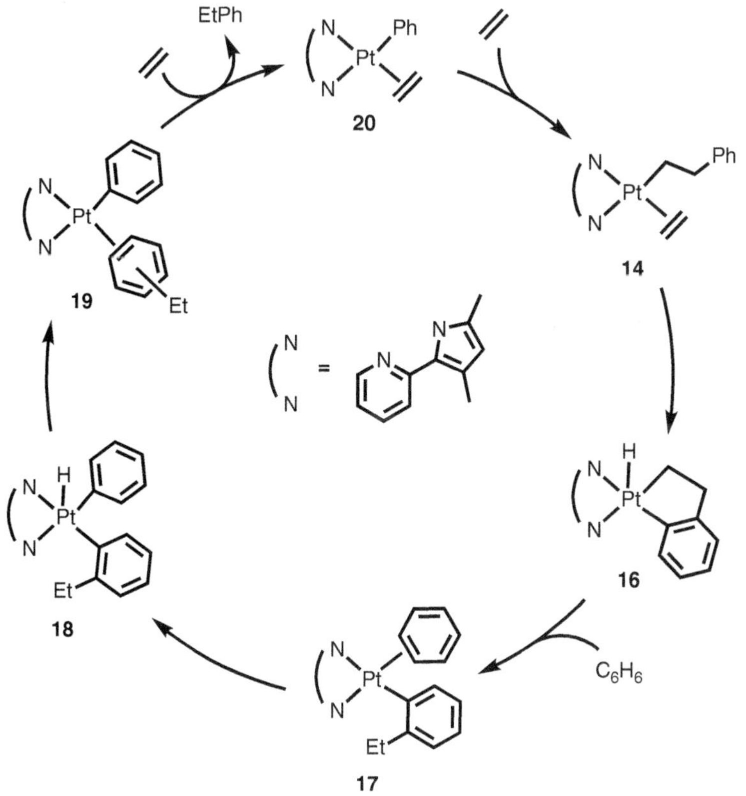

Figure 4.3 Proposed mechanism for (PyPyr)Pt^{II}- catalyzed olefin hydroarylation.

4.2.2 Pt^{II} Hydroarylation Catalysts Bearing Neutral Bidentate (NN) Ligands

Initial reports by Tilset and coworkers [14, 15] demonstrated that cationic $[L_2Pt^{II}Me(OH)_2]^+$ complexes supported by simple diimine ligands activate the C—H bonds of benzene and other substituted arenes, such as toluene or *p*-xylene, to generate the corresponding Pt—Ar complexes. Similarly, Tilley and coworkers [16] showed that a dicationic $[(bpy)Pt^{II}(solvent)_2][NTf_2]_2$ complex ($NTf_2 = {}^-N(SO_2CF_3)_2$, bpy = bipyridine) activates allylic C—H bonds. Given their ability to cleave unactivated C—H bonds, cationic Pt complexes have been thoroughly investigated as potential hydroarylation catalysts. An initial investigation by Gunnoe *et al.* [17] demonstrated that the neutral $[(bpy)Pt^{II}Ph_2]$ complex catalyzes the hydroarylation of ethylene and benzene at elevated temperatures, upon addition of a non-coordinating acid such as $H[BAr'_4]$ ($Ar' = 2,5\text{-}(CF_3)_2C_6H_3$) as an activator. While catalysis with this platinum-based system gave low yields of product (3.7 and 11.9 TONs after 4 and 16 h, respectively), it was hypothesized that a cationic complex of the type $[(bpy)Pt^{II}PhL]^+$ might be more active since preactivation would not be required.

Several cationic complexes were therefore synthesized and isolated (Scheme 4.3) [17, 18]. A *tert*-butyl-substituted ligand (^{t}bpy) was installed on Pt^{II} to afford the neutral complex $[(^tbpy)Pt^{II}PhCl]$ (**21**). Chloride removal with $Na[BAr'_4]$ in THF yielded the cationic complex **22** with $[BAr'_4]^-$ as the counter ion. This complex, however, proved stable only in the presence of an excess of THF. In the presence of other coordinating solvents such as acetonitrile or perfluoropyridine, ligand exchange occurred to give the corresponding $[(^tbpy)Pt^{II}PhL][BAr'_4]$ complexes (**23** and **24**). Substitution of THF with other ligands, however, was demonstrated to reduce the overall hydroarylation activity. Complexes **22**, **23**, and **24**, exhibited 15.7, 8.1, and 7.4 TONs of ethylbenzene, respectively, after 4 h. A labile ancillary L ligand, therefore, promotes increased reactivity since ligand loss is required for catalysis to occur.

Optimization of initial reaction conditions with precatalyst **22** [17] revealed the effects of temperature and ethylene pressure (using 0.1 mol% catalyst loadings, relative to benzene). It was determined that as the temperature increases from 90 to 140 °C (at 15 psi of ethylene) the overall TON for ethylbenzene after 4 h increased from 7.2 to 35.8. However, elevated temperatures also increased the amount of dialkylated products. A mixture of *ortho*-, *meta*-, and

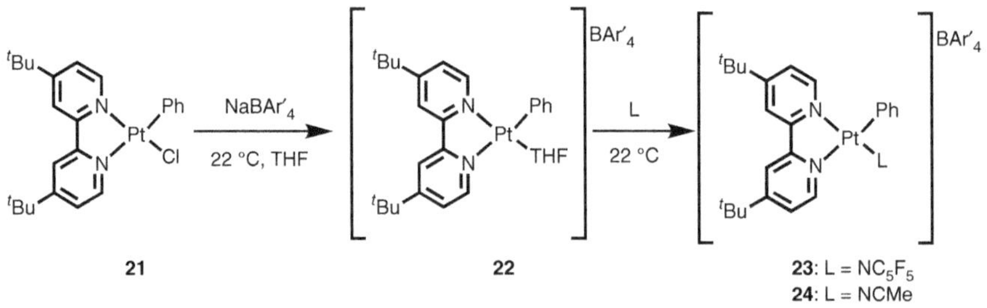

Scheme 4.3 Synthesis of cationic $[(^tbpy)Pt^{II}PhL]^+$ complexes bearing labile ancillary L-type ligands.

para-diethylbenzene side products were observed with 0.3, 7.3, and 3.3 TON, respectively, after 4 h at 140 °C. Interestingly, increasing the ethylene pressure reduced the overall activity of the system: an increase in ethylene pressure from 15 to 60 psi (at 100 °C) resulted in a decrease in TON by a factor of almost 50. While the overall activity dropped at higher ethylene pressures, the product ratio of ethylbenzene to the mixture of diethylbenzenes remained relatively constant at 2.3. Hydroarylations of several other substrates (e.g., ethylene with furan or propylene with benzene) were observed with **22** as catalyst.

The operative mechanism for these cationic, Pt-catalyzed hydroarylations was investigated using both experimental evidence and DFT calculations [18]. Attempts to determine a rate law for the hydroarylation of ethylene, with benzene and precatalyst **22**, revealed a first-order dependence on the concentration of **22** whereas benzene displayed saturation kinetics. The kinetics indicates a complex dependence on ethylene, however, as increases in the ethylene pressure result in rapid reduction of the product yield.

DFT calculations were used to explore the free energy preferences of various intermediates and transition states for multiple possible pathways. The energetics of two different ethylene-insertion pathways was investigated, starting from an analog of **22** without ^{t}Bu groups. The initial ligand substitution of THF with ethylene can occur either through an associative or dissociative pathway, and DFT calculations indicate that these two routes are energetically similar, with activation energies $\Delta G^{\ddagger}$ of 19.0 and 19.7 kcal/mol, respectively. Aryl C—H activation can occur through two different pathways: a concerted, one-step σ-bond metathesis route or two-step oxidative addition/reductive elimination route (Figure 4.4). The calculated relative energies (given in Figure 4.4) reveal a 2.3 kcal/mol energy preference for oxidative addition over σ-bond metathesis.

As a further probe to the mechanism, an isotope experiment with catalyst **22** was conducted to compare the TON for ethylbenzene, using benzene or benzene-d_6 as the arene substrate, to give a TO_H/TO_D (TO − turnovers) ratio of 1.8(4). Given these results from experiment and DFT, two possible mechanisms were proposed (Figure 4.5). In the first mechanism, a $[L_2Pt^{II}(C_2H_4)Ph]$ complex (**28**) undergoes ethylene insertion to afford a quasi-three-coordinate Pt^{II} species (**29**), likely with either an agostic phenethyl C—H interaction or an arene π-coordination to the Pt^{II} center. Upon rapid coordination of ethylene (**30**) and reversible exchange with benzene (**31**), C—H activation occurs either as a one- or two-step process as discussed earlier. Ligand exchange with ethylene regenerates **28** and liberates an equivalent of ethylbenzene. The second proposed mechanism has many similar steps and intermediates, with the exception that complex **30** lies off-cycle in equilibrium with **29**. Given the experimental and DFT evidence, however, both mechanisms are possible and likely indistinguishable.

The selectivities associated with longer chain α-olefins (e.g., propylene) have also been investigated with several (bpy)Pt^{II} complexes (Table 4.3) [19]. By modifying the substitution on the ligand at the 4 and 4′ position, effects on the overall activity and selectivity for anti-Markovnikov products were observed. By changing the substituents from donating groups such as OMe (**33**) or ^{t}Bu (**22**) to electron-withdrawing groups such as NO_2 (**37**), the amount of anti-Markovnikov product increased. Moreover, a decrease in the overall activity for hydroarylation

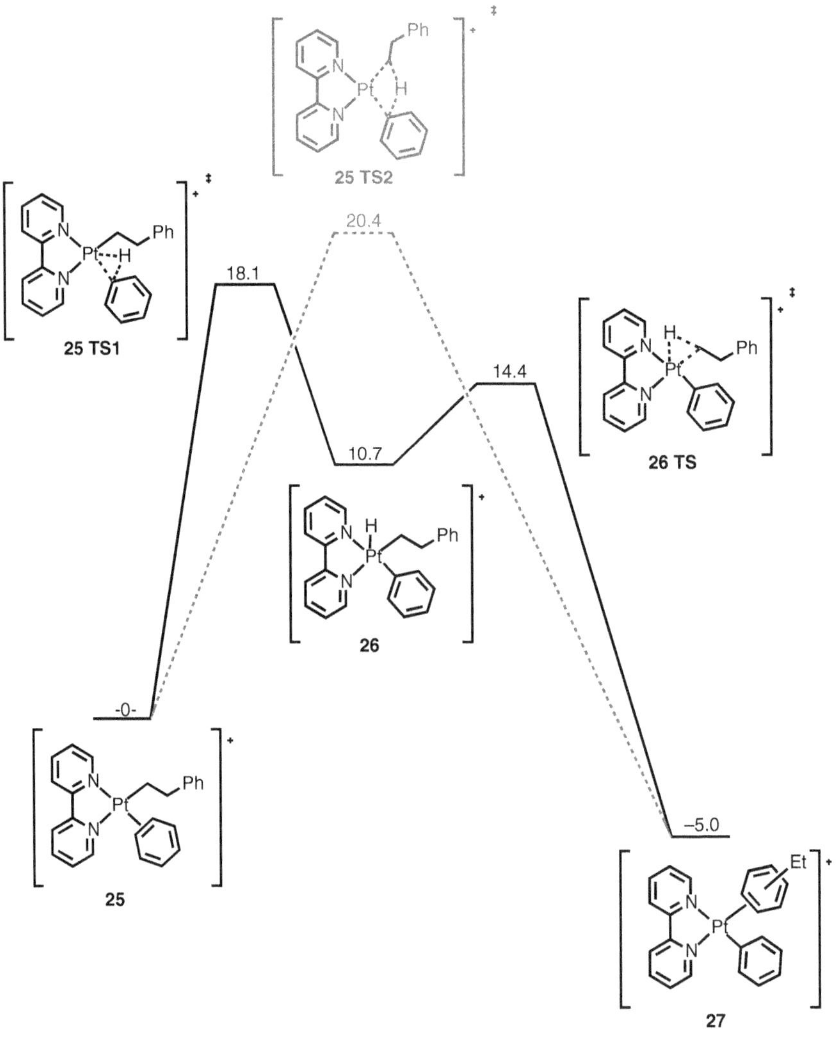

Figure 4.4 Calculated barriers for C—H activation with cationic Pt complexes via either a two-step oxidative addition/reductive elimination route (black, solid) or a single-step σ-bond metathesis (gray, dashed) pathway. Energies are given in kcal/mol.

was noted for most substituents, especially with the Br-substituted complex **35**, and **37**. A slight trend toward branch-selective catalysis was observed with more electron-withdrawing substituents. A Hammett plot was generated to better illustrate the magnitude of this trend, and a linear fit was observed using the σ_p value and the ratio of the two products (i.e., ^{i}PrPh/^{n}PrPh). A small slope was observed ($\rho = 0.2$), indicating only a slight substituent effect on the overall selectivity.

Ligand modifications resulted in slight changes in the product selectivity (ethylbenzene vs styrene) for the hydrophenylation of ethylene (Table 4.4) [20]. The ratio of ethylbenzene to styrene for electron-withdrawing substituents (**35–37**)

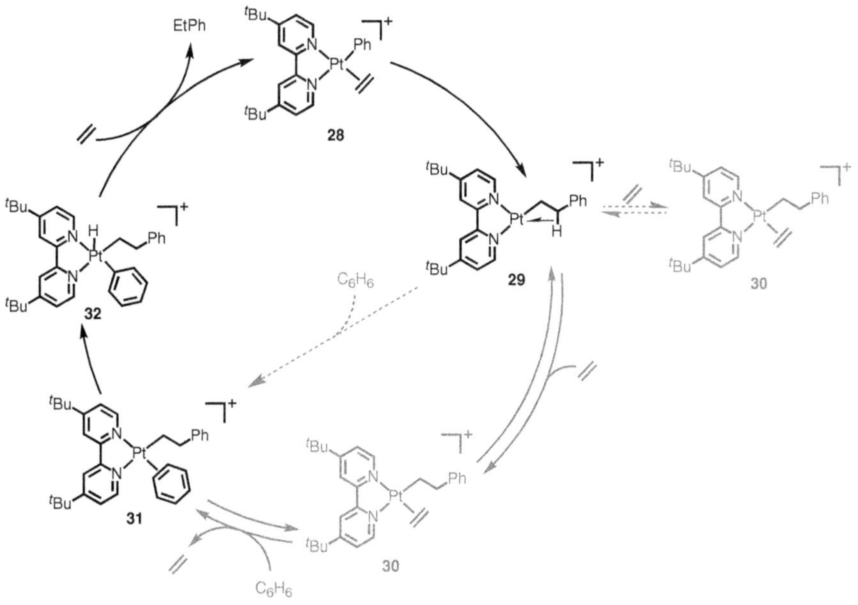

Figure 4.5 Possible mechanisms for cationic Pt^{II}-catalyzed hydrophenylation of ethylene. Complex **30** is on the catalytic cycle for mechanism 1 (gray, solid), whereas it is an off-cycle resting state in mechanism 2 (gray, dashed).

Table 4.3 Catalytic hydrophenylation of propylene with $[(bpy)Pt^{II}Ph(THF)]^+$ derivatives [19].

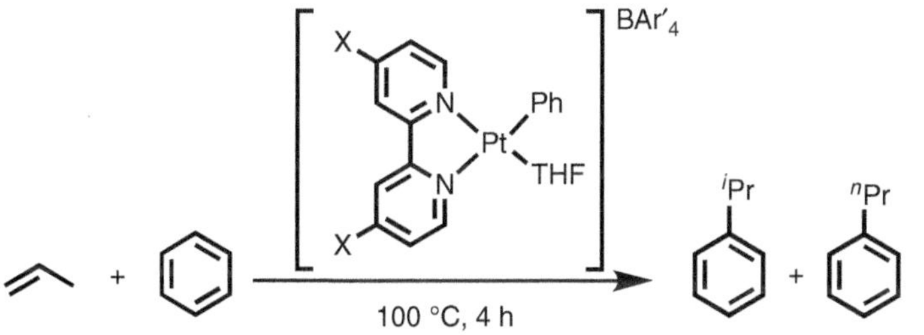

X	σ_p	TON ^{i}PrPh	TON ^{n}PrPh	^{i}PrPh + ^{n}PrPh	^{i}PrPh/^{n}PrPh
OMe (**33**)	−0.27	10.6	3.7	14.3	2.9
^{t}Bu (**22**)	−0.2	25.0	8.5	33.5	2.9
H (**34**)	0.0	25.8	8.0	33.8	3.2
Br (**35**)	0.23	2.5	0.7	3.2	3.8
CO_2Et (**36**)	0.45	12.1	3.3	15.4	3.7
NO_2 (**37**)	0.78	4.1	0.9	5.0	4.6

Conditions: 0.01 mol% catalyst loading relative to benzene, 0.1 MPa propylene, 100 °C for 4 h. σ_p represents the Hammett parameter for *para*-substitutions with negative values representing electron-donating groups and positive values denoting electron-withdrawing ability.

Table 4.4 Catalytic hydrophenylation of ethylene with $[(bpy)Pt^{II}Ph(THF)]^+$ derivatives [20].

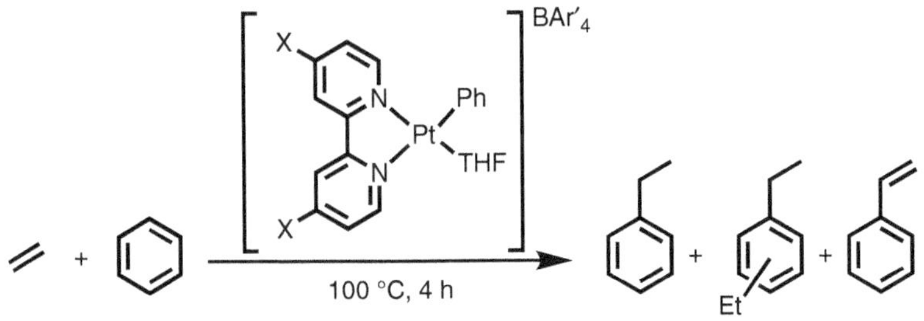

X	TON (EtPh)	TON ($Et_2C_6H_4$)	TON (styrene)	EtPh/styrene
OMe (**33**)	3.3	0.7	0.4	8.3
^{t}Bu (**22**)	4.0	0.9	0.5	8.0
H (**34**)	5.5	1.4	0.7	7.9
Br (**35**)	0.2	0	1.3	0.2
CO_2Et (**36**)	1.9	0.5	1.2	1.6
NO_2 (**37**)	0	0	1.0	NA

Conditions: 0.01 mol% catalyst loading relative to benzene, 0.3 MPa ethylene, 100 °C for 4 h.

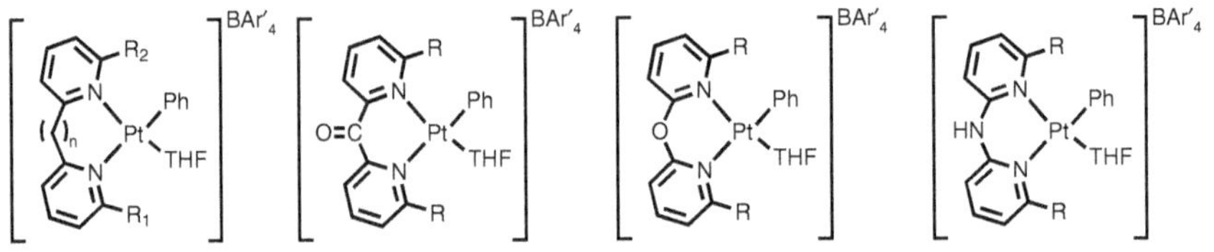

38: $R_1 = R_2 = H$, $n = 1$
39: $R_1 = H$, $R_2 = Me$, $n = 1$
40: $R_1 = R_2 = Me$, $n = 1$
41: $R_1 = R_2 = Me$, $n = 0$
42: $R_1 = R_2 = H$, $n = 2$

43 **44** **45**

Figure 4.6 Alternative ligand designs using a linked bpy motif.

was smaller than that for electron-donating substituents (**22**, **33**, and **34**), both at 4 and 16 h time points. It should be noted that the observed TONs for styrene were fairly constant for complexes **22** and **33–37**. The change in ratio, therefore, was attributed to a reduction in overall hydrophenylation activity for complexes bearing ligands with electron-withdrawing substituents.

Alternative bis(pyridyl)-based, chelating ligands have been employed recently by Gunnoe and coworkers [21, 22]. The pyridyl fragments are connected through a linker (e.g., CH_2), and were used to elucidate the effect of larger ligand bite angles (Figure 4.6). Heteroatom linkers were also used to determine their electronic influence on the catalysis. As with the bpy systems described earlier, catalysts were generated by reaction of $[L_2Pt^{II}Ph_2]$ complexes with $[H(Et_2O)_2][BAr'_4]$ in THF at −70 °C, to afford the corresponding $[L_2Pt^{II}Ph(THF)]^+$ species.

Table 4.5 Catalytic activity in the hydrophenylation of ethylene using linked bis(pyridyl) complexes [22].

Catalyst	TON (EtC_6H_5)	TON ($Et_2C_6H_4$)	TON (styrene)
38	55.3	10.6	0.4
39	31.2	1.3	2.7
40	19.5	0.0	1.7
41	0.0	0.0	0.7
42	0.7	0.0	0.4
43	3.3	0.0	1.9
44	1.6	0.0	2.9
45	18.6	4.0	2.5

Also given are the TONs for the mixture of *ortho-*, *meta-*, *para*-diethylbenzene ($Et_2C_6H_4$) and styrene. Conditions: 0.01 mol% catalyst, at 100 °C with 0.1 MPa of ethylene for 4 h.

The differences in catalytic activities for complexes **38–45** in the hydrophenylation of ethylene are summarized in Table 4.5 [22]. Increasing the linker length from $n=0$ (**22**) to $n=1$ (**38**) increased the amount of product generated, with a 55.3 TON observed for ethylbenzene after 4 h. Unfortunately, a large amount of diethylbenzenes also formed with **38**. By increasing the steric bulk at the metal center with either 6- or 6′-Me substituents on the pyridyl groups (**39–41**), the generation of diethylbenzenes was dramatically reduced while only moderately reducing the overall activity. Complexes bearing longer linkers in the supporting ligand (**42**) or with oxygen-containing linkers (**43** and **44**), however, demonstrated very little catalytic activity. The nitrogen-linked complex **45** demonstrated comparable activity (18.6 ton) toward hydroarylation. The selectivity for branched versus linear products with **38–45** was examined during the hydrophenylation of propylene. These catalysts, however, were observed to give mostly branched products (>4.0 ratio of branched/linear products).

4.2.3 Pt^{II} Hydroarylation Catalysts Supported by Nonnitrogen-based Ligands

Simple Pt^{II} complexes with trifluoromethylsulfonate (OTf) or cyclooctadiene (cod) ligands have been observed to generate alkyl arenes. In a report by Tilley and coworkers [23], [(cod)Pt^{II}(OTf)$_2$] (**46**) catalyzed the coupling of cyclic olefins with non-functionalized arenes such as mesitylene and benzene in an unreactive aromatic solvent. Initial studies with norbornene and benzene demonstrated complete consumption of the olefin after 5 h using 10 mol% **46** in *ortho*-dichlorobenzene at elevated temperatures, though only 41% of the hydroarylation product **47** was formed (Eq. (4.1)). Interestingly, products containing a dichlorophenyl fragment were not observed, which suggests that electron-deficient arenes are not suitable substrates for this type of hydroarylation. To examine the side reactions generated, the reaction was probed at lower temperatures (25 °C) by ^1H NMR spectroscopy. It was observed that **46** promoted the isomerization of norbornene to the highly strained nortricyclene **48** without generating **47**. This suggested that this rearrangement was likely competent at elevated temperatures and therefore accounts for the low yields for the desired alkyl arene **47**.

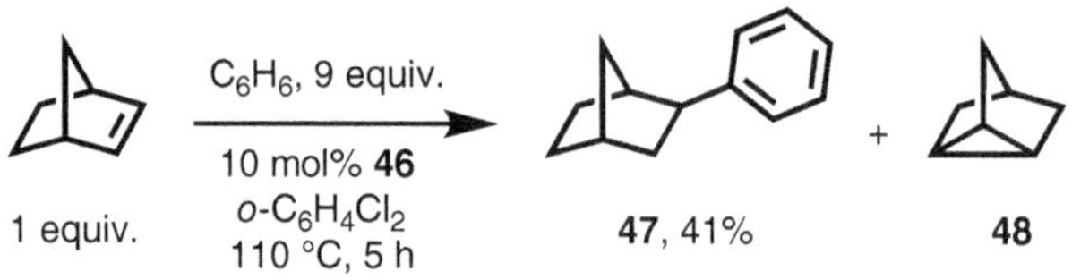

(4.1)

Under conditions similar to those described earlier, cyclohexene reacts with benzene and mesitylene to generate aryl-substituted cyclohexane products (Eq. (4.2)). Dialkylation occurred with benzene; a yield of 36% for the mono-alkylated arene product (**49**) and 21% for the dialkylated species (**50**) was observed after 17 h. In contrast, bulkier arenes such as mesitylene afforded only the mono-alkylated compound (**51**) in quantitative yield after 4 h.

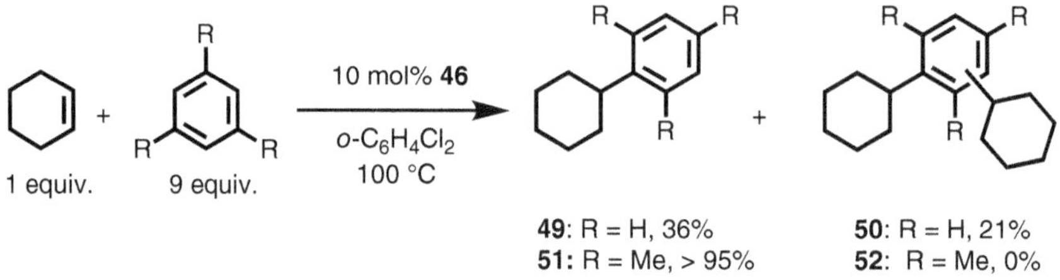

(4.2)

To probe the catalytic mechanisms associated with **46**, the influence of a hindered base (2,6-di-*tert*-butyl-4-methylpyridine) on hydroarylation was examined. Notably, the addition of this non-coordinating base greatly inhibited the reaction of cyclohexene with mesitylene, and complete inhibition was observed in the presence of 20 mol% of the base. In a similar control experiment, **46** was replaced with 10 mol% of HOTf. This resulted in quantitative formation of the hydroarylation product **51** suggesting the possibility of an acid-catalyzed pathway. Thus, the proposed mechanism of catalysis with **46** involves an initial coupling of two cyclohexene molecules to produce a platinum-allyl derivative and an equivalent of HOTf. This type of metal-mediated olefin coupling with ethylene and cyclopentene was first identified with electrophilic platinum complexes by Sharp and coworker [24]. The acid generated in this coupling chemistry then initiates a Friedel–Crafts-type hydroarylation via protonation of cyclohexene to give a carbocation, which then reacts with the arene (Figure 4.7).

The observation of a metal-initiated acid-catalyzed hydroarylation (with **46**) is reminiscent of the platinum-mediated proton-transfer mechanism proposed for hydroaminations of olefins [25], and raises broader questions regarding the possible role of Brønsted acids in reported, metal-catalyzed hydroarylations. In a study designed to differentiate metal-centered C—H activation catalyzed hydroarylation from adventitious acid-mediated hydroarylation, Tilley and

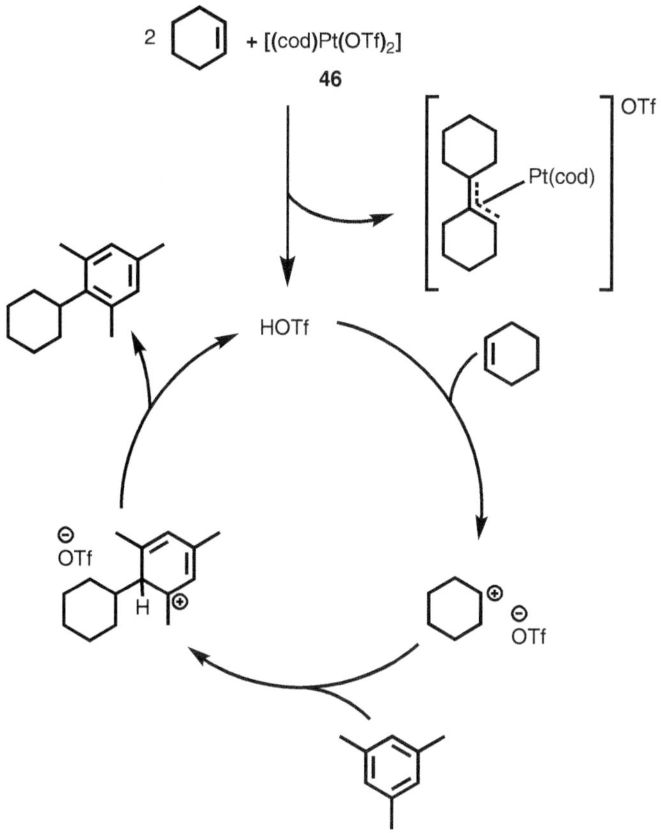

Figure 4.7 Proposed catalytic cycle for the hydroarylation of cyclohexene with mesitylene with complex **46**.

coworkers [26] described the reactivity of [(bpy)Pt^{II}Ph(NTf_2)] (**53**) with the bicyclic olefin substrate *spiro*-[bicyclo[2.2.1]hept-2-ene-7,1′-cyclopropane] (Figure 4.8). It was expected that a mechanism involving direct olefin insertion into a Pt—Ph bond would favor formation of the expected type of hydroarylation product. Alternatively, if an acid-catalyzed mechanism was operative, an acid-mediated Wagner–Meerwein carbocation rearrangement would occur more rapidly than hydroarylation. The resulting rearrangement product could then undergo an acid-catalyzed hydroarylation event, yielding a rearranged hydroarylation product. The result of this experiment, however, yielded a surprising result, in that hydroarylation did not occur. Instead a platinum-mediated C—C bond rearrangement reaction induced by a C—H activation step occurred to generate 1,2,4,7,7*a*-pentahydroindene as the only product.

Several other ligands for cationic Pt^{II} hydroarylation catalysts have also been described, mostly based on chelating bisphosphine donors (**54–61**) [27] (Figure 4.9). Unfortunately, these complexes exhibit poor activities as hydroarylation catalysts. Thus, less than 2 equiv. of ethylbenzene was observed for each of these complexes after 4 h at 100 °C. Interestingly, styrene formation was

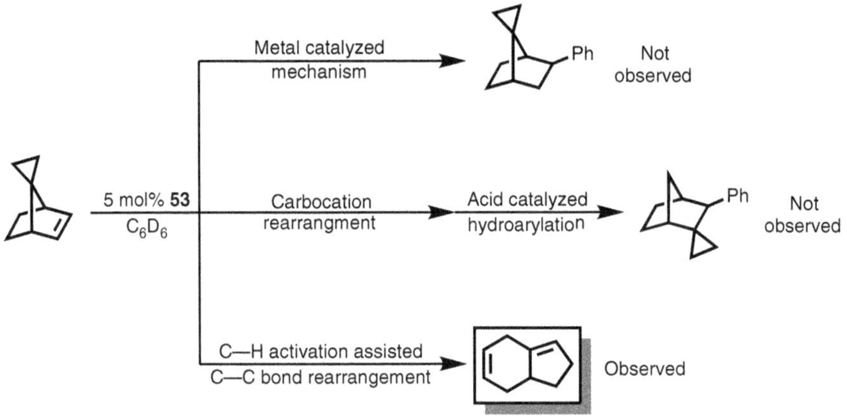

Figure 4.8 Possible reaction pathways for the reaction of a bicyclic olefin with complex **53** in benzene-d_6. No hydroarylation reactions occurred; only the rearrangement product was observed.

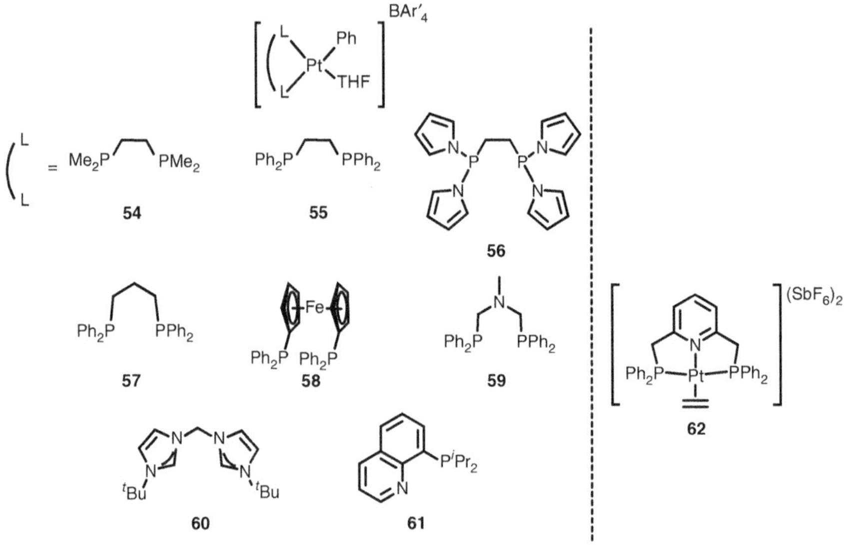

Figure 4.9 Complexes of the type $[L_2Pt^{II}(THF)Ph][BAr'_4]$ and $[L_3Pt^{II}(C_2H_4)]$ bearing phosphine, mixed phosphine, or bis-*N*-heterocyclic carbene-based ligands.

observed for complexes **54–60**, giving 1–2 turnovers after 4 h. Additionally, complex **61** bearing a phosphino-quinolyl PN ligand proved to be catalytically inactive. DFT calculations were performed to elucidate the cause for the switch in product formation, and it was determined that a low energy pathway involving β-hydride elimination (8.8 kcal/mol barrier) followed by benzene C—H activation led to the generation of styrene, along with an equivalent of H_2. The tridentate PNP-ligated (PNP = 2,6-bis(diphenylphosphinomethyl)pyridine), dicationic complex $[(PNP)Pt^{II}(C_2H_4)][SbF_6]_2$ (**62**) has also been employed in the hydroarylation of ethylene [28] with only electron-rich, methoxy-substituted benzene derivatives at elevated temperatures. In this system, however, high yields for the dialkylated product were observed using 3-methylanisole as the arene source.

4.2.4 Summary of Pt^{II} Catalyzed Hydroarylations

Pt^{II} complexes have been employed as catalysts for the hydroarylation of ethylene, using both cationic and neutral complexes. The supporting ligands have most often been chelating, nitrogen-based supporting ligands; bipyridine [17–22] and other nitrogen-containing heterocycles (i.e., indole [8–10] or pyrrole [11–13]) are the most commonly employed ligand motifs. Ligand substituent effects on the few systems investigated so far are relatively minor. While other types of ligands with non-heteroaromatic donors (e.g., phosphines, *N*-heterocycliccarbenes, diimines) have been studied, such complexes exhibit reduced activity toward hydroarylation, and are often either completely inactive or generate primarily styrene. This information would seem to suggest that platinum-based catalysts are more effective with hard donors as supporting ligands, but further confirmation of this hypothesis is required.

For Pt^{II} species, several related mechanisms [10, 12, 18] have been proposed. Two key steps are required during catalysis: arene C—H cleavage and olefin insertion into a Pt—Ph bond. Species of the type $[L_2Pt^{II}(C_2H_4)Ph]^+$ have been implicated as intermediates during catalysis. Rapid olefin insertion and subsequent olefin coordination result in a $[L_2Pt^{II}(C_2H_4)(CH_2CH_2Ph)]$ species, which is presumed to be the catalytic resting state on the basis of DFT and experimental evidence [18]. Aryl C—H activation can then occur to generate a Pt(ethylbenzene) π-complex, either through a single-step σ-bond metathesis pathway or through a two-step oxidative addition/reductive elimination route (via a Pt^{IV}–hydride intermediate). Upon substitution by ethylene, an equivalent of ethylbenzene is liberated along with the regenerated active catalyst. It has also been shown that $[(cod)Pt^{II}(OTf)_2]$ initiates an acid-mediated catalytic cycle [23] with several substrates.

The electrophilic nature of the metal center may play an important role in determining whether a Pt- or acid-based mechanism prevails. Catalyst precursors that can serve as a source of dicationic Pt centers (i.e., the most electrophilic complexes) produce protons likely via electrophilic displacement of a proton from the hydrocarbon (e.g., olefin). These species, therefore, catalyze hydroarylation through a Friedel–Crafts mechanism. In contrast, catalyst precursors that initially possess a Pt—aryl bond appear to proceed via a pathway that features a metal-centered insertion mechanism. Such catalysts are typically monocationic, and therefore less electrophilic. Current data suggests that catalysts with bidentate NN ligands operate via a metal-centered insertion mechanism, though further studies are required to rigorously test this hypothesis.

Pt catalysts, nonetheless, do present several unique advantages. Unlike other metal centers employed in hydroarylation, Pt^{II} complexes may be used with undirected (unactivated) arene substrates; in contrast, other metal centers (see Sections 4.3–4.6) are often only able to promote hydroarylation *ortho-* to a directing aryl substrate (e.g., imine, ketone). Additionally, the stability of Pt^{II}

complexes permits the isolation and characterization of a number of proposed intermediates, providing further insight into the operative hydroarylation mechanism and potentially aiding in the development of more active species.

Though Pt-based catalysts seem to offer potential for hydroarylation with unfunctionalized substrates (i.e., ethylene and benzene), their productivity is quite low. In many cases, only tens to hundreds of TONs are observed. When compared to other metal centers, Pt^{II} catalysts often require neat substrates and exhibit lower yields to achieve hydroarylation catalysis. Moreover, these systems all seem to suffer from competing catalyst deactivation caused either by decomposition to metallic platinum or through unproductive catalysis to generate undesired side products. To make homogeneous Pt^{II} catalysis industrially applicable and commercially useful, dramatic improvements are required to improve their activities, selectivities, and stabilities.

4.3 Rh^I-Catalyzed Hydroarylation

Hydroarylations catalyzed by rhodium have been thoroughly investigated, and reports on Rh^I catalysts have provided the most information on d^8 systems for this catalysis. This may relate to the well-known ability of Rh^I complexes to undergo C—H activation chemistry with aryl substrates [29, 30]. In 1989, Ghosh and Graham [31] demonstrated that a trispyrazolylborate ($HBPz_3$) complex of Rh^I binds ethylene and activates a C—H bond in benzene at room temperature under ultraviolet irradiation. The resulting Rh–hydride complex undergoes rapid ethylene insertion, to give the complex $[(HBPz_3)Rh^{III}(CO)(C_2H_5)Ph]$ with both alkyl and aryl ligands bound to the metal center. Upon exposure to CO, insertion into the Rh–alkyl bond occurs to generate the acyl complex $[(HBPz_3)Rh^{III}(CO)(COC_2H_5)Ph]$. Treatment of this complex with $ZnBr_2$ afforded propiophenone as the final organic product. This work demonstrated that Rh^I species can mediate both olefin insertions and C—H activations. Subsequent studies confirmed that some Rh^I complexes (e.g., Wilkinson's catalyst, $[(PPh_3)_3RhCl]$) can in fact catalyze olefin hydroarylation [32, 33]. Several reviews have been published on Rh^I-mediated hydroarylation [34–36], which emphasize the versatility of this metal center in this type of catalysis. Three classes of arene substrates have been employed in Rh^I-catalyzed hydroarylation: unfunctionalized arenes, arenes with directing groups, and perfluorinated aryl reagents. Catalysis using each of these substrate types is discussed in the following text.

4.3.1 Reactions of Unfunctionalized Arenes with Rh^I Complexes Proceeding via Hydroarylation-Like Mechanisms

A recent report by Gunnoe and coworkers [37] highlighted a new method to selectively generate styrene from ethylene and benzene in a one-pot system. Initial synthesis and reactivity studies of a diimine Rh^I catalyst indicated promising activity toward C—H activation [38]. Complexes **63** and **64** with auxiliary cyclooctene (coe) [38] and ethylene [37] ligands, respectively, were synthesized from the corresponding $[L_2Rh^I(\mu\text{-TFA})]_2$ dimer (Scheme 4.4; TFA = trifluoroacetate).

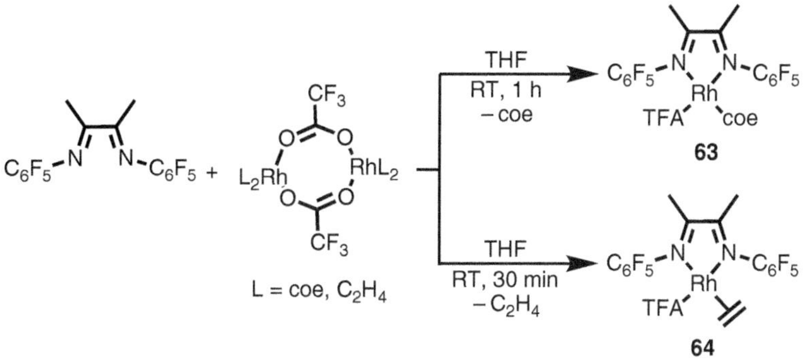

Scheme 4.4 Synthesis of Rh^I complexes of the type [L_2Rh(olefin)(TFA)].

In the presence of trifluoroacetate-*d* (TFA-*d*), 1.6 mol% of complex **63** catalyzes arene H(D) exchange at 150 °C [38]. After 2 h, complex **63** converted benzene to a mixture of $C_6D_nH_{6-n}$ species with 456 TONs. Using TFA and toluene-d_8, **63** catalyzed H(D) exchange to give a mixture of *ortho-*, *meta-*, and *para-* H(D) exchanged toluene species with a ratio of 6.9 : 1.0 : 6.4, respectively. In a subsequent report, complex **64** also exhibited activity toward arene H(D) exchange [39]. Given this relatively fast rate of C—H(D) activation and exchange, it was hypothesized that [L_2Rh^I(TFA)(olefin)] complexes **63** and **64** would be competent catalysts for hydroarylation and hydrovinylation.

Using complex **64** and a Cu^{II} salt as an external oxidant, the catalytic coupling of ethylene and benzene to generate styrene was achieved [37] at very low catalyst loadings (0.001 mol% relative to neat benzene). After 24 h at 150 °C, styrene (60 TONs, which is quantitative yield based on Cu^{II} as limiting reagent) was produced using 120 equiv. of $Cu(OAc)_2$ relative to **64**. Increasing the amount of $Cu(OAc)_2$ to 240 or 2400 equiv. resulted in a corresponding increase in product formation, to 115 and 835 TONs, respectively. Moreover, the use of **64** and $Cu(OAc)_2$ selectively generated styrene; hydroarylation to form ethylbenzene was not observed. Other Cu^{II} oxidants such as $Cu(OHex)_2$ or $Cu(TFA)_2$ were examined, but $Cu(OAc)_2$ resulted in the most stable system.

Kinetics experiments probed the effect of ethylene pressure on the rates of reaction (using TOFs). The reaction rate was observed to have a first-order dependence on ethylene, which is in contrast to related systems that generally exhibit an inverse dependence on olefin concentration (e.g., Pt^{II}-based hydroarylation catalysts; see Section 4.2). A competition experiment with equimolar amounts of benzene and benzene-d_6 generated a mixture of styrene and styrene-d_5 in a 3.1(2) : 1.0 ratio, and products containing more than five deuterons were not observed. This rate difference appears to correspond to an isotope effect that is consistent with other C—H activation reactions. On the basis of this data, the mechanism of Figure 4.10 was proposed. The ethylene complex undergoes benzene C—H activation to generate a [$L_2Rh^IPh(C_2H_4)$] complex, which then proceeds to a $Rh(CH_2CH_2Ph)$ derivative via olefin insertion. In other catalytic

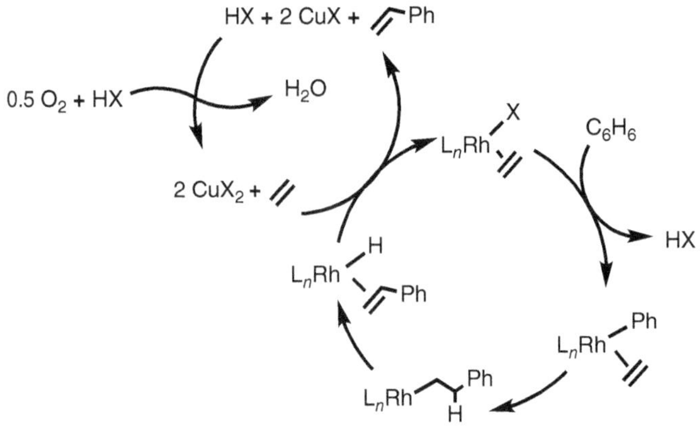

Figure 4.10 Mechanism for styrene formation with a Rh^I hydroarylation-like system.

hydroarylation systems, β-hydride elimination often leads to catalyst decomposition and a reduction in activity. In this system, however, β-hydride elimination appears to be a key step in the formation of styrene. A secondary catalytic cycle involving a Cu^{II} species liberates styrene and regenerates the starting Rh^I complex. The resulting reduced Cu^I species can then be reoxidized by O_2 and an equivalent of HX.

An informative mechanistic investigation of this system has been recently reported by Gunnoe and coworkers [40]. Kinetic data reflect a complex dependence on the concentration of catalyst, which can vary from first- to half-order depending on the reaction temperature and ethylene concentration. The catalysis exhibits a first-order dependence on the ethylene concentration (saturation kinetics at sufficiently high concentrations) and zero-order dependence on the concentration of Cu^{II} oxidant. Rate constants were measured in independent reactions with benzene and benzene-d_6, which afforded a k_H/k_D of 3.3(2). A similar experiment but with equimolar amounts of benzene and benzene-d_6 gave the product ratio styrene:styrene-d_5 of 3.0(1). These kinetic isotope effects (KIEs) are consistent with a rate-limiting C—H activation step. Computational studies were performed to elucidate activation parameters for individual steps in the proposed catalytic mechanism.

4.3.2 Directed *ortho*-Hydroarylation Catalyzed by Rh^I Complexes

Most Rh^I hydroarylation systems have been developed for use with arenes bearing directing or coordinating substituents [32, 41–66]. Commercially available catalysts such as Wilkinson's catalyst ([$(PPh_3)_3Rh^ICl$], **65**) are typically employed. For this type of hydroarylation, olefin insertion occurs only at aryl C—H bonds that are *ortho*- to the directing group. While slight variations in the operative mechanism have been suggested [32, 41–66], the generalized mechanism of Figure 4.11 seems to be operative for all of the systems discussed herein. Initially, pre-coordination of the directing aryl substituent likely occurs, to place the *ortho*-proton proximal to the metal center. Subsequently, C—H activation, assisted by a chelation effect, occurs while also generating a Rh–hydride bond.

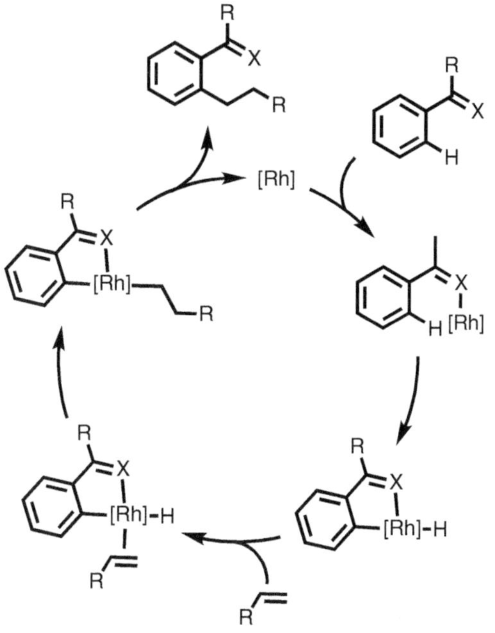

Figure 4.11 Generalized mechanism for Rh^{I}-catalyzed *ortho*-hydroyarlation.

Olefin coordination and insertion into this Rh–hydride bond generates the corresponding Rh(alkyl)(aryl) complex. Upon reductive elimination, the desired *ortho*-hydroarylated product is generated. Displacement of product by another arene completes the cycle.

Related chemistry reported by Jun and coworkers involves the catalytic hydroacylation of ketones [41] and aldimines [42] using **65** and 2-amino-3-picoline as a chelation cocatalyst. For ketone substrates, an initial condensation with the picoline additive generates an aldimine. Transimination occurs instead with aldimine substrates to replace a phenyl group with a pyridine. The newly installed pyridine fragment is capable of binding to the Rh^{I}, anchoring the substrate to the metal center. Chelation-assisted C—H activation of the aldimine hydrogen then occurs, followed by olefin coordination. Migratory insertion of the olefin followed by C—C reductive elimination produces the hydroacylated product. Condensation of this hydroacylated imine with water regenerates 2-amino-3-picoline and forms the desired ketone.

With a variation of this hydroacylation system, Jun and coworkers [32] observed that **65** catalyzes both the hydroacylation and hydroarylation of *tert*-butylethylene with the aryl aldimine **66** (Eq. (4.3)), in the presence of 2-amino-3-picoline at elevated temperatures. It is worth noting that the initial transimination of **66** with 2-amino-3-picoline affords a new aldimine with two positions available for C—H activation: an imine hydrogen and an *ortho*-aryl hydrogen. With this substrate, therefore, both hydroarylation and hydroacylation occur. Upon acid-mediated hydrolysis, the ketone **67** was generated in high yield. In addition, a minor hydroacylation product (**68**) was observed. It was noted that in the absence of 2-amino-3-picoline, product formation was not observed with this substrate.

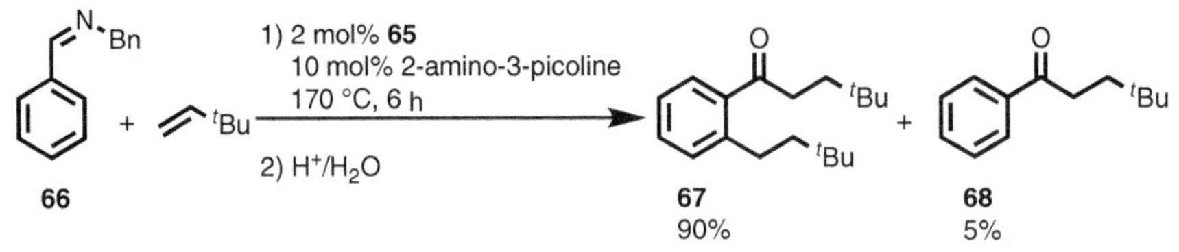

(4.3)

When a ketamine was used in place of an aldimine, only hydroarylation products were obtained. Moreover, the addition of picoline as a reagent was no longer required for catalysis. To further investigate the scope of the system, a wide variety of arene and olefin substrates were tested [32, 43], and the results are summarized in Table 4.6. Three different substituents were varied: the aryl

Table 4.6 Representative *ortho*-hydroarylation substrate scope using **65** [32, 43].

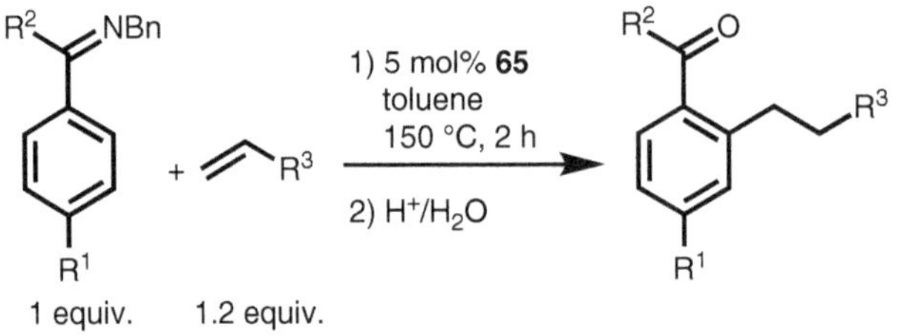

R^1	R^2	R^3	Isolated yield (%)
H	Me	^{t}Bu	97
		^{n}Bu	94
		^{n}Hex	71
		Cy	65
		$Si(Me)_3$	92
		C_6F_5	91
		CO_2Me	94
		CO_2Et	93
		CONMe	81
		SO_2Ph	43
		CN	32
	Et	$Si(Me)_3$	93
	^{n}Pent	$Si(Me)_3$	73
CF_3	Me	^{t}Bu	76[a)]
		CO_2Me	95
OMe	Me	^{t}Bu	42[a)]
		CO_2Me	90

a) Catalyst loading reduced to 1 mol% of **65** and the reaction was run for 30 min at 130 °C.

group (R^1), the imine (R^2), and the olefin (R^3). Hydroarylation was achieved with both electron-withdrawing ($R^1 = CF_3$) and electron-donating ($R^1 = OMe$) aryl substituents with minimal differences in the isolated yields. Changing the length of the ketamine substituent had a minimal effect, with a change in yield observed only with longer R^2 groups (i.e., pentyl). Finally, olefins with bulky aliphatic groups ($R^3 = {}^t$Bu, ^{n}Bu, ^{n}Hex, Cy, $SiMe_3$) tended to give high (often quantitative) yields. Ester- and amide-containing olefin substituents afforded hydroarylated products in high yield. Electron-withdrawing groups such as SO_2Ph and CN resulted in a large reduction in overall activity, to give yields of 43% and 32%, respectively.

A similar experiment with Wilkinson's complex **65**, using various ratios of 1-hexene (**69**) and *N,N*-dimethylacrylamide (**70**) with ketamine **71** afforded a mixture of the corresponding hydroarylated products **72** and **73**, respectively (Eq. (4.4)). With an equimolar amount of **69** and **70**, a product ratio (**72** : **73**) of 10 : 90 was observed. By increasing the relative ratio of **69** to **70**, a corresponding increase in the product ratio was observed. With an olefin ratio (**69** : **70**) of 5.0 : 1.2, the product ratio increased to 38 : 62. Even with a substantial excess of **69**, larger amounts of the products derived from the activated olefin (**73**) were observed, suggesting that functionalized olefins are preferred substrates. Solvent-free conditions were also reported for the hydroarylation with **65** [44]. Using a microwave reactor, the hydroarylation of various substrates was achieved with high temperatures and short reaction times. Using this method, yields similar to those obtained with high-boiling solvents were achieved after 15 min of microwave heating.

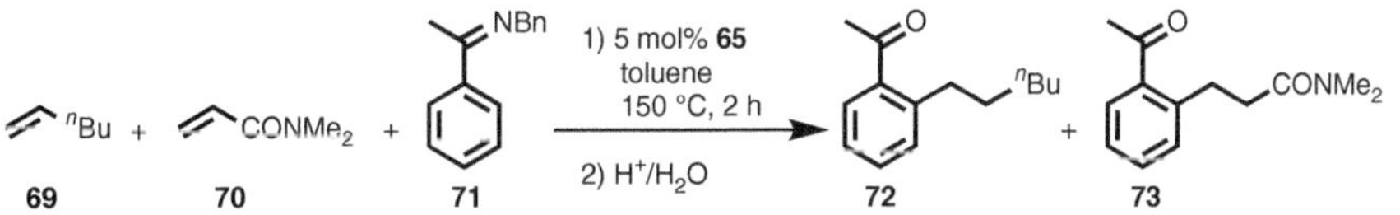

(4.4)

In an effort to develop a recyclable system, Jun and coworkers [45] generated phosphine-containing, extended supramolecular structures to use as ligands. The extended assembly was synthesized from two small molecules, a barbiturate derivative and 2,4,6-triaminopyrimidine, which self-assembled into a larger aggregate (Figure 4.12). $[(coe)_2Rh^I(\mu\text{-}Cl)]_2$ (**74**) was then added to the barbiturate/pyrimidine mixture. The resulting complex readily formed an extended hydrogen-bonding network (depicted in following page) at room temperature to afford a solid-supported precatalyst. At elevated temperatures, however, this hydrogen-bonded network cannot form and liberates the metal complex as a soluble small molecule; the reaction was catalyzed by a homogenous species. Cooling and the addition of *n*-pentane reform the extended network resulting in the precipitation of the solid-supported precatalyst. This solid was then isolated by centrifugation and filtration after hydroarylation. The recovered catalyst was then recycled, and it exhibited only a small reduction in overall yield over eight catalytic runs with several substrates.

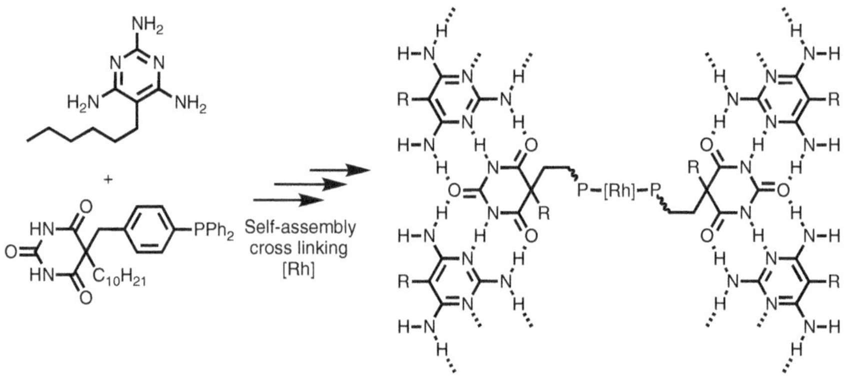

Figure 4.12 Self-assembly of a barbiturate derivative and 2,4,6-triaminopyrimidine with Rh^I, generating a solid-supported hydroarylation catalyst.

Bergman, Ellman, and coworkers [46] demonstrated that an intramolecular *ortho*-hydroarylation reaction was achieved for imine-substituted arenes with a pendent olefin. For intramolecular hydroarylation, cyclized products were generated. Upon hydrolysis with acid, acyl-substituted aromatic bicyclic products were isolated in reasonable yields. While the operative mechanism is similar to the general mechanism discussed previously (*vide supra*), a key difference is proposed [47] to selectively generate the cyclized product over intermolecular coupling (Figure 4.13). While the initial pre-coordination and oxidative addition steps are identical, a rapid coordination of the pendent olefin occurs preferentially over intermolecular olefin binding. This coordination step anchors the substrate through a tridentate-binding motif. Olefin insertion into

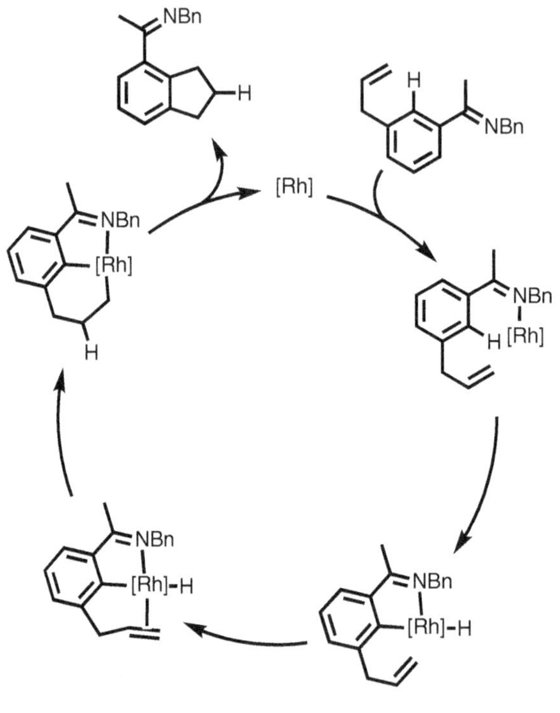

Figure 4.13 Mechanism for intramolecular *ortho*-hydroarylation with Rh^I.

the Rh–hydride bond followed by subsequent reductive elimination affords the cyclized hydroarylation product (in this case, a substituted indane).

To demonstrate the versatility of this intramolecular *ortho*-hydroarylation system, a wide variety of olefin-containing imine-functionalized arenes were cyclized [46, 47] using conditions similar to those described earlier and the substrate scope is summarized in Table 4.7. While some substrates can potentially isomerize to generate an internal olefin that may inhibit cyclization, cyclic products were typically isolated in high yields. Substrates with substituted olefins (entries 3–7) tended to provide higher yields, potentially due to reduced isomerization. Five-membered cyclized products were observed. Some substrates could potentially generate multiple bicyclic products due to olefin insertion occurring either in a Markovnikov or anti-Markovnikov fashion. The new heterocycles generated were either five- or six-membered. With a longer chain linker (entry 2), for example, both bicyclic products were observed in a 1 : 1 ratio. Heteroatom-tethered olefins (entries 8–12) in general exhibited higher activity toward ring-closing (*vs.* their aliphatic analogs), producing substituted dihydrobenzofuran and indoline compounds. Overall, cyclization with **65** afforded products with a wide variety of substrates.

While Wilkinson's catalyst **65** has been successfully employed in hydroarylation chemistry, the PPh_3 ancillary ligands appear to limit its versatility. Since more complex phosphine ligands have exhibited useful properties in other catalytic reactions (e.g., enantioselectivity), additional Rh^I precatalysts have been explored. Lim and coworkers [48, 49] discovered that upon using the phosphine-free Rh^I precursor **74** ($[(coe)_2Rh^I(\mu\text{-Cl})]_2$) along with added PCy_3 ligands, the *ortho*-hydroarylation of aldimines was achieved. In contrast, complex **65** catalyzed the hydroarylation of aldimines only upon addition of a picoline cocatalyst [32].

Using this system, a series of substituted aldimine and ketimine substrates were alkylated (Table 4.8) to generate the corresponding monohydroarylated (**75**) and dihydroarylated (**76**) products. Moreover, most aldimines ($R^1 = H$) exhibited higher yields for **76**, with selectivities generally greater than 95% for dihydroarylated products from *para*-substituted aldimines. With highly electron-withdrawing substituents a switch in selectivity was observed, with *p*-NO_2, *m*-F, and *m*-Cl aryl substituents giving reduced selectivities for **76**. Similarly, *meta*- and *ortho*-substituents generally afforded **75** in higher yields. Reducing the steric bulk of the olefin using ^{n}Bu rather than ^{t}Bu substituents decreased the amount of **76** formed. However, this was achieved at the expense of reduced overall reactivity. A similar trend toward producing monohydroarylated products was noted for ketimine substrates ($R^1 = Me$ or Et). These conditions were also used in the heterohydroarylation of imine-substituted furan, thiophene, and naphthalene substrates [49].

In a subsequent publication by Lim and coworkers [50], compound **77** was used as a probe to determine chemoselectivity in *ortho*-hydroarylations. Since each aryl C—H bond in **77** is *ortho* to an imine-directing group, all four positions can potentially undergo hydroarylation. Using an excess of *tert*-butylethylene with **77** under conditions similar to those described above, multiple hydroarylation products were observed (Eq. (4.5)). While the monohydroarylation product **78** was observed, dihydroarylation led to the major product, **79**. With several other olefin substrates (1-butene, 1-pentene, 1-hexene, 3-methyl-1-butene) **80**

Table 4.7 Substrate scope for intermolecular *ortho*-hydroarylation with **65** [46, 47].

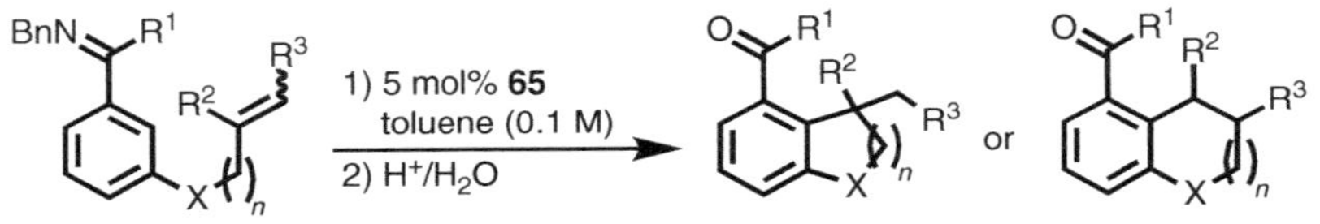

$n = 0,1$ X = CH_2, O, NR

Entry	Substrate	Temperature (°C)	Time (h)	Product	Yield (%)[a]
1	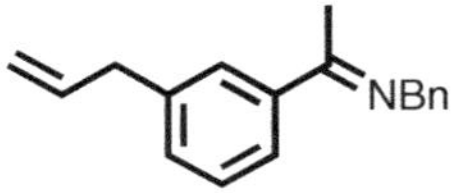	125	1	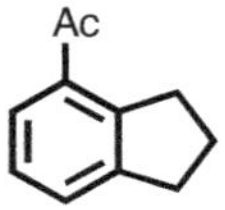	52
2	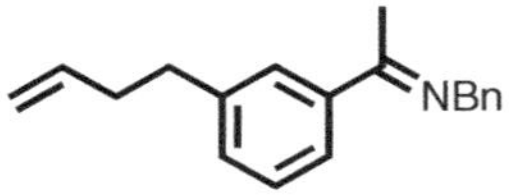	150	48	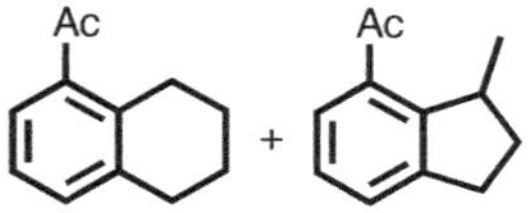	50 (1 : 1)
3	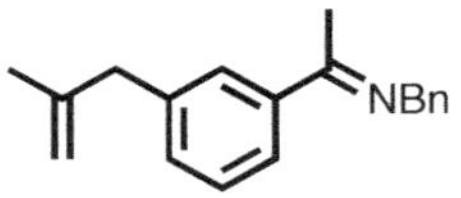	125	4	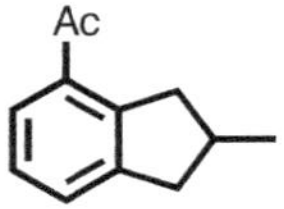	71
4	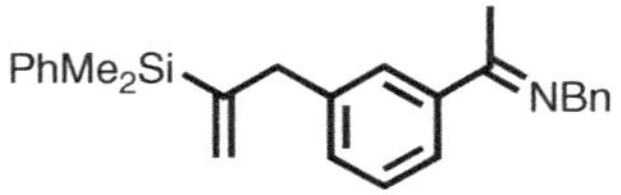	150	36	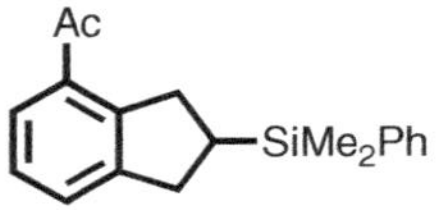	58
5	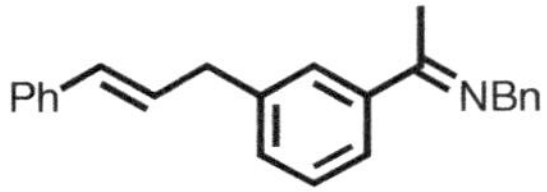	150	16	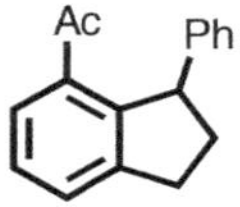	68
6	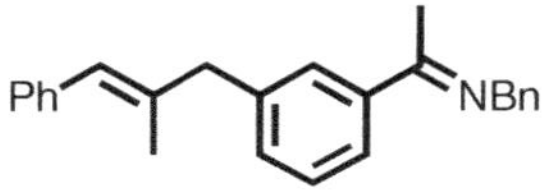	150	36	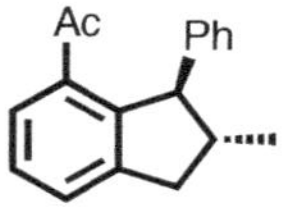	50 (7 : 1 *cis:trans*)

7	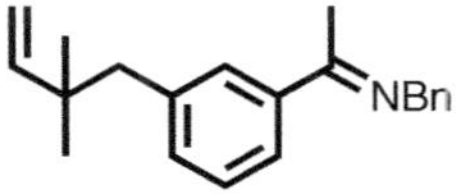	125	2	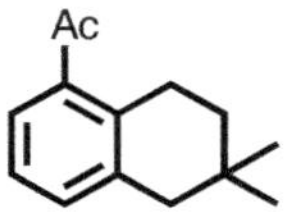	85
8	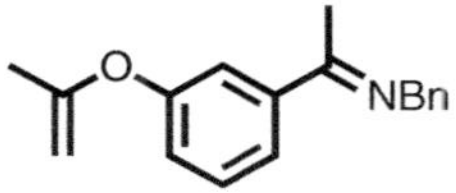	175	0.75	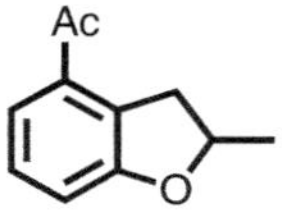	90
9	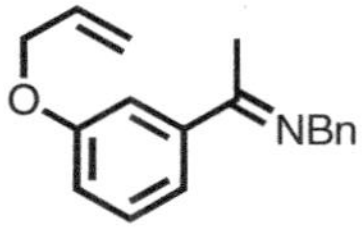	150	16	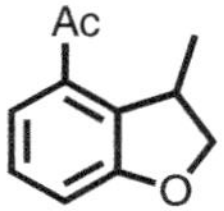	59
10	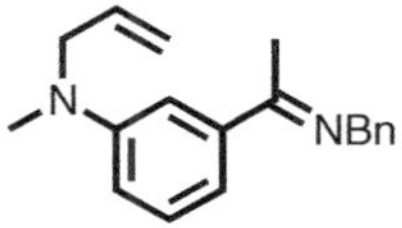	150	3	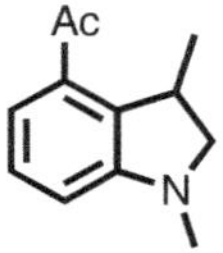	50
11	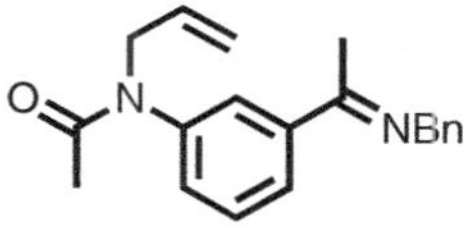	125	12	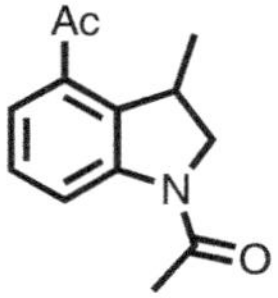	53
12	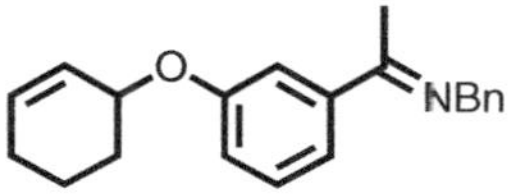	150	72	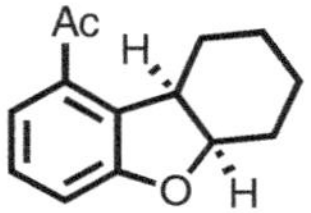	41
13	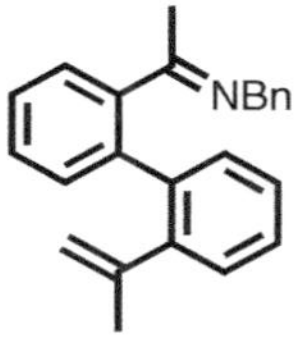	150	6	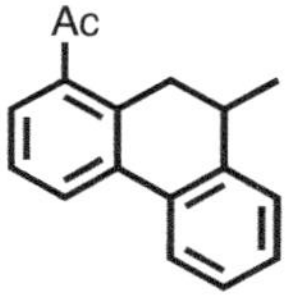	81

a) Measured as isolated yields.

Table 4.8 Substrate scope with **74** and PCy_3 [48, 49].

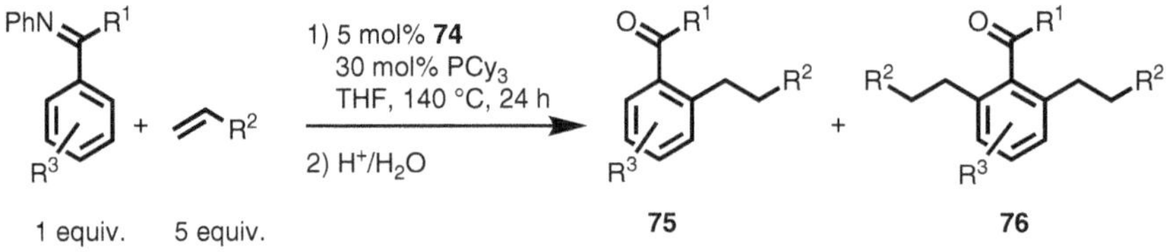

R^1	R^2	R^3	75 : 76	Isolated yield (%)
H	^{t}Bu	H	11 : 89	93
		p-OMe	1 : 99	90
		p-Me	5 : 95	87
		p-F	1 : 99	84
		p-Cl	3 : 97	88
		p-CF_3	2 : 98	90
		p-NO_2	88 : 22	6
		m-OMe	97 : 3	50
		m-F	26 : 74	62
		m-Cl	77 : 23	63
		o-Me	100 : 0	80
		o-Cl	—	0
		o-NO_2	—	0
	^{n}Bu	H	34 : 66	19
Me	^{t}Bu	H	28 : 72	99
Et	^{t}Bu	H	97 : 3	86
	^{n}Pr	H	100 : 0	65

was formed as a minor product (typically < 3%) and **81** was not observed by ^1H NMR spectroscopy. It was suggested, therefore, that the pathways to form **80** and **81** proceed via intermediates with sterically hindered Rh^I centers, for which a second C—H activation at positions *ortho-* or *para-* to the first alkylation is difficult. In contrast, C—H activation *meta-* to the first alkylation generates a less sterically crowded intermediate and compound **79** as the major product.

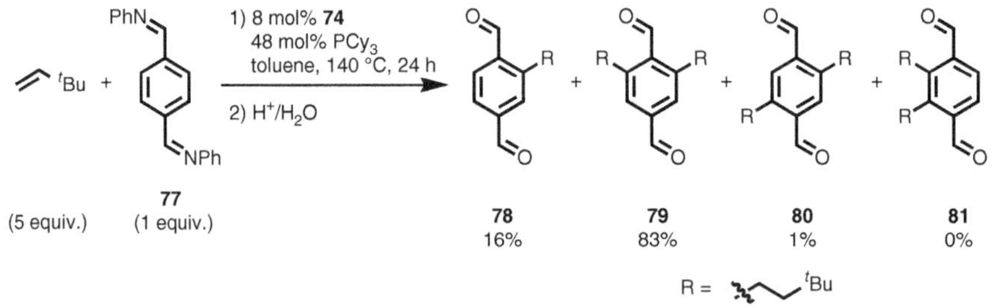

(4.5)

While the **74**/PCy_3 catalyst system has been extensively used, other phosphorous-containing ligands have been investigated [47]. In particular, ferrocene-containing phosphine ligands ($FpPCy_2$) demonstrated higher yields and lower reaction times than the corresponding alkyl (PCy_3) or aryl (PPh_3) analogs. Chiral ligands have also been studied by Ellman, Bergman, and colleagues [51]. In these investigations, intramolecular *ortho*-hydroarylation was observed using a series of chiral phosphorous ligands (**82–98**, depicted in Figure 4.14).

An excess of chiral ligand (**82–97**) relative to **74** was used to determine the yields and enantioselectivities for the intramolecular *ortho*-hydroarylation of **99** to the cyclized product **100** (Table 4.9). Ligands containing P—N bonds (**82**, **83**, and **85**) provided poor yields, likely due to ligand chelation that inhibits substrate binding to the Rh^{I} center. In contrast, ligands containing P—O bonds (**84**, **86**, and **88–97**) exhibited enhanced yields with nearly quantitative conversions in many cases. Phosphoramidite ligands **93–97** demonstrated the highest enantioselectivities, producing the *S*-stereoisomer in greater than 50% enantiomeric excess (ee). The (*S*)-binol phosphoramidite ligands (**95–97**) exhibited the highest conversions and enantioselectivities while also requiring the shortest reaction time. Note that both *R*- and *S*-products could be obtained depending on the ligand choice. In a subsequent publication [52], the optimized conditions for enantioselective intramolecular *ortho*-hydroarylation were described for several other ketimine substrates using a similar set of chiral P—O containing ligands. It is clear that there is much room for improvement in the enantioselectivities of these reactions.

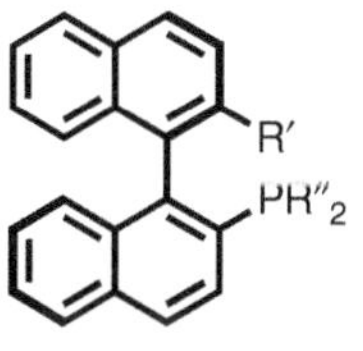

82: R′ = NMe_2, R″ = Ph
83: R′ = NMe_2, R″ = Cy
84: R′ = OMe, R″ = Ph

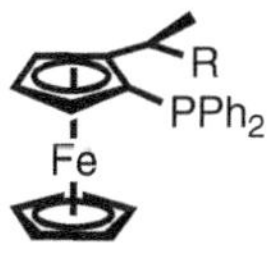

85: R = NMe_2
86: R = OMe

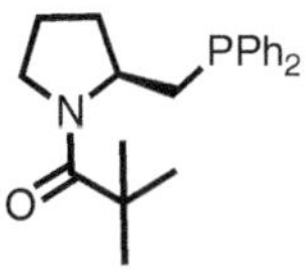

87

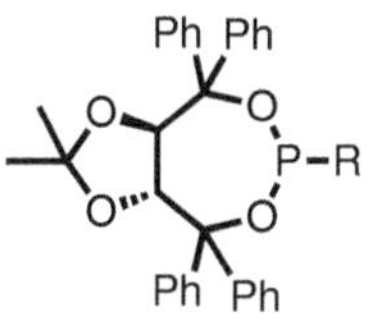

88: R = OMe
89: R = OPh
90: R = OMenthyl
91: R = Ph

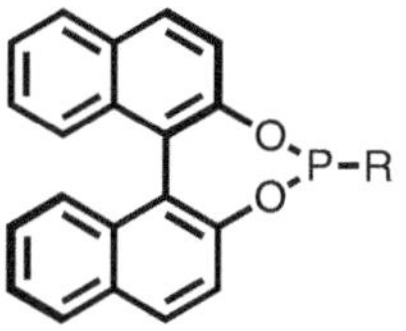

92: R = OPh
93: R = NMe_2
94: R = NBn_2
95: R = $N(^{i}Pr)_2$
96: R = (*S*,*R*,*R*)-$N(CHCH_3Ph)_2$
97: R = (*S*,*S*,*S*)-$N(CHCH_3Ph)_2$

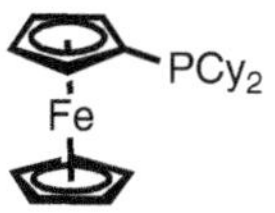

98

Figure 4.14 Asymmetric, chiral phosphorous-containing ligands used for enantioselective hydroarylation.

Table 4.9 Asymmetric intramolecular *ortho*-hydroarylation using chiral phosphorous ligands [51, 52].

Ligand	Temperature (°C)	Time (h)	Yield (%)	ee (%)
82	125	20	5	NA
83	125	20	Trace	NA
84	125	20	48	8 (*S*)
85	125	20	9	NA
86	75	6	99	35 (*R*)
87	100	6	56	23 (*S*)
88	75	20	93	17 (*R*)
89	75	20	94	9 (*R*)
90	75	20	91	38 (*R*)
91	125	20	34	0
92	125	20	6	NA
93	125	2.5	15	19 (*S*)
94	125	2.5	52	58 (*S*)
95	125	<2	100	83 (*S*)
96	125	<2	100	88 (*S*)
97	125	<2	99	87 (*S*)

Benzyl and phenyl imines have been extensively employed as hydroarylation substrates [32–52] due to their simple synthesis. More recently, Bergman, Ellman, and coworkers [53] have demonstrated that chiral aminoindane-substituted aldimines promote enantioselectivity for hydroarylation reactions, even though the aminoindane functionality is ultimately removed by hydrolysis. A series of aminoindane substrates were cyclized using **74** and the phosphine ligand **98** ($FpPCy_2$, Figure 4.14). While **98** is chiral, previous studies [51, 52] demonstrated that similar ferrocene-based phosphine ligands (**85** and **86**) exhibited poor enantioselectivities in cyclizations. In conjunction with these chiral aminoindane-substituted imines, however, **98** exhibited enhanced enantioselectivity (Table 4.10). For all substrates tested, yields greater than 60% were observed with high ee. Temperatures at or below 75 °C resulted in cyclized products using aminoindane-substituted imines, whereas benzyl imine substrates required elevated temperatures for cyclization.

While ketimines and aldimines have been extensively used to bind the substrate to the Rh^{I} center, other coordinating groups have also been investigated. Murai and coworkers [54] demonstrated that the pyridine-substituted

Table 4.10 Substrate scope for enantioselective cyclization using chiral substrates [53].

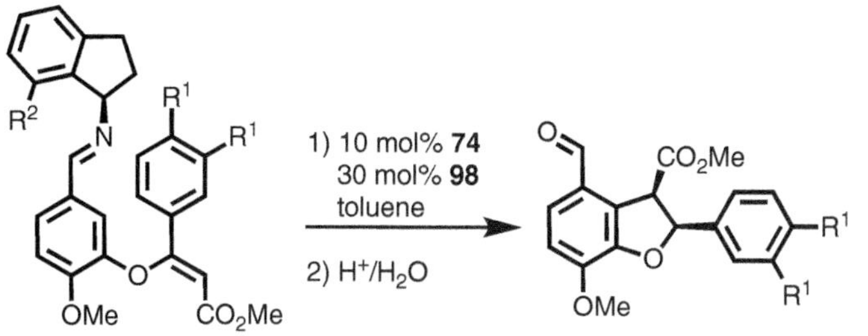

R^1	R^2	Temperature (°C)	Time (h)	Yield (%)	ee (%)
OMe	H	75	16	76	76
	Me	75	16	62	80
	F	60	36	70	90
	Ph	75	16	90	83
H	H	75	16	80	65
	F	75	16	81	70

naphthalene compound **101** could undergo *ortho*-hydroarylation (Eq. (4.6)) using trimethylvinylsilane or ethylene with **74** and added chiral phosphine ligand to generate the corresponding products **102** and **103**, respectively. The yields using a pyridyl-directing moiety, however, are generally lower than those derived from imine-based groups.

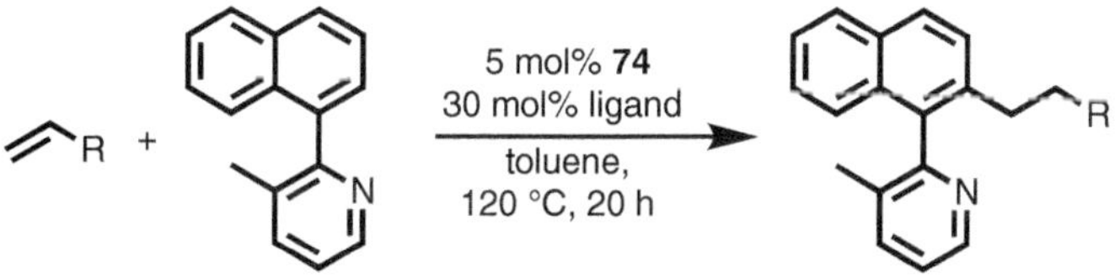

R = SiMe$_3$ or H **101** **102**: R = SiMe$_3$, 55% yield
103: R = H, 37% yield (4.6)

A recent series of investigations utilized a bidentate, pre-coordinating scaffold for the *ortho*-hydroarylation of α,β-unsaturated esters [55] and substituted styrenes [56]. Specifically, an 8-quinolylamide group directed the hydroarylation to selectively generate *ortho*-hydroarylation products in the presence of [(cod)Rh^{I}(μ-Cl)]$_2$ (**104**) and added acetate ligand. Moreover, only anti-Markovnikov products were observed with a variety of substrates, as expected when using Michael Acceptors. The results of the substrate scope using the quinolylamide motif are summarized in Table 4.11. For nearly all substrates, products were isolated in high yields. Interestingly, both α,β-unsaturated esters and substituted styrenes were observed to undergo high conversion, with a minimal effect on yields for a variety of olefins. Similarly, there were minimal effects of arene substituents on the overall yields.

Table 4.11 Substrate scope of *ortho*-hydroarylation using quinolylamide-directing substrates [55, 56].

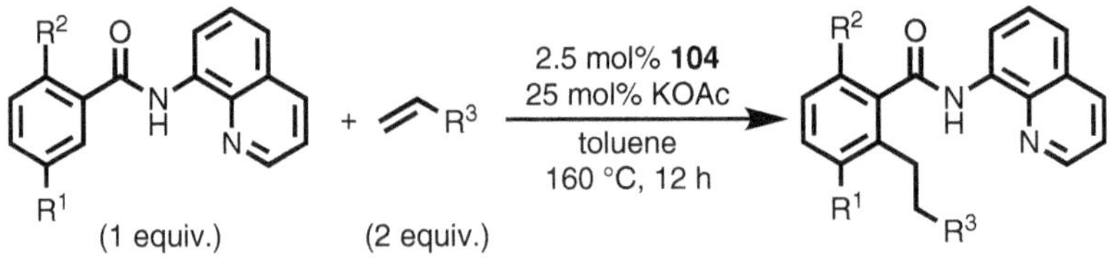

R^1	R^2	R^3	Isolated yield (%)
H	OMe	CO_2Me	77
	Ph	CO_2Me	85
	CF_3	CO_2Me	84
	F	CO_2Me	86
	Me	CO_2Me	86
		$CO_2{}^tBu$	71
		CO_2Bn	81
		$CONMe_2$	68
		SO_2Ph	48
		Ph	86
		p-OMePh	88
		p-^{t}BuPh	79
		p-tolyl	84
		p-FPh	85
Me	Me	CO_2Me	91
F	Me	CO_2Me	90
Br	Me	CO_2Me	68

Using deuterium labeling, several key features of the mechanism were elucidated by Chatani and coworker [55, 56]. Deuterated aryl substrate **105** and a naphthalene-containing α,β-unsaturated ester (**106**) were treated with **104** and added acetate (Eq. (4.7)). After heating at 160 °C for 10 min, the products were identified and the location of the deuterium atoms was determined. Three separate species were identified after reaction: H(D) scrambled **105**, unreacted **106**, and the hydroarylation product **107**. It was noted that **107** contained 2.87 hydrogen atoms, indicating that on average, nearly one deuterium was incorporated into the product, as expected. The location of this deuterium atom, however, was scrambled across both methylene positions, indicating that a reversible C—H(D) activation step occurs. Moreover, scrambling of deuterium atoms into the olefin was not observed.

To probe the H(D) exchange further, **105** was treated with only **104**. Interestingly, H(D) scrambling across the *ortho*-aryl C—D bond of **105** was observed either with or without added acetate, although faster scrambling was noted in the presence of the acetate (0.41 H incorporation after 15 min with added acetate vs 0.19 H incorporation after 12 h without added acetate). This suggests that the role of acetate is to transform **104** into the active catalytic species, likely with a bound acetate. To test this hypothesis, a similar experiment was performed using the acetate-ligated Rh species, $[(cod)Rh^I(OAc)]_2$. Using this catalyst, H(D) scrambling was observed with **105** as a substrate, to incorporate the same number of protons either with or without added acetate (about 0.4 H incorporation after 15 min). A similar deuterium-labeling experiment using styrene-d_8 and **105**-d_0 [56], revealed similar H(D) scrambling.

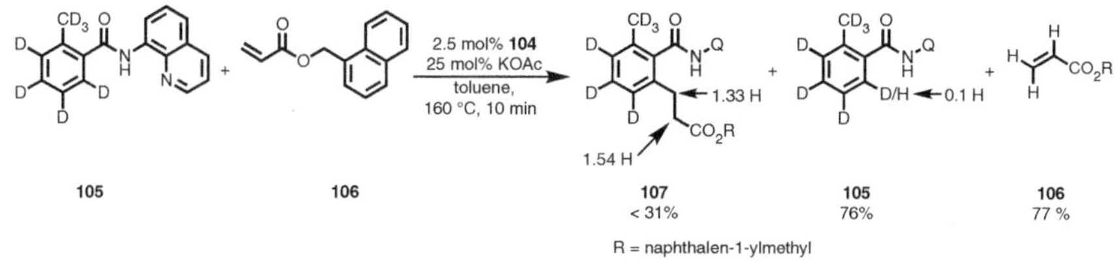

(4.7)

A mechanism for this hydroarylation reaction was proposed (Figure 4.15) to account for both scrambling and the presence of only anti-Markovnikov products [55]. Initially, ligand exchange of a chloride for an acetate ligand occurs. The quinolylamide substrate can bind to this $Rh^I(OAc)$ species, anchoring it in a bidentate fashion to the Rh center. An *ortho*-aryl C—H activation step can then occur to afford a Rh^{III}–hydride species. Coordination and insertion of an olefin yield the resulting Rh–alkyl fragment. This can occur, however, in either a Markovnikov or an anti-Markovnikov fashion. According to the authors, steric congestion proximal to the metal center results in a reductive elimination that can only occur with the linear Rh–alkyl fragment thereby generating the anti-Markovnikov products selectively. Therefore, only one regioisomer is expected during the reaction. Upon reductive elimination of the alkyl and aryl substituents on the Rh^{III} center, acetate displaces the product to regenerate the active Rh^I species.

To demonstrate the practicality of intramolecular *ortho*-hydroarylation, Bergman, Ellman, and coworkers reported the total syntheses of (+)-lithospermic acid [53, 57] and several mescaline derivatives [47, 58] using **74** with added phosphorus-based ligands. In these total syntheses, hydroarylation was used as a cyclization step, significantly reducing the number of steps required to generate each natural product, in decent yields. (+)-Lithospermic acid (**108**) was synthesized with a 5.9% yield over 10 steps (Scheme 4.5); similarly, the

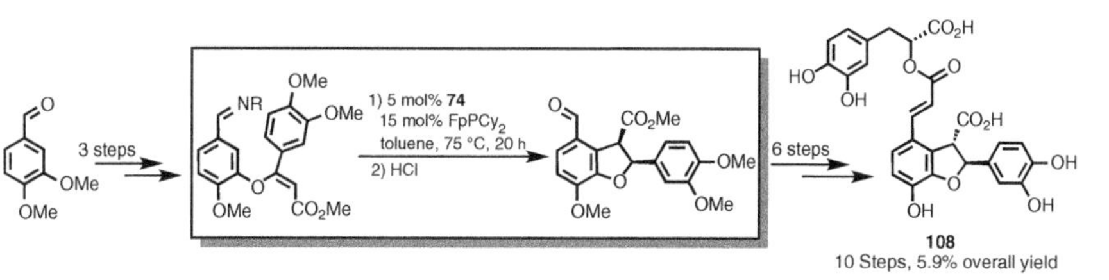

Scheme 4.5 Synthesis of (+)-lithospermic acid (**108**) via Rh catalyzed hydroarylation.

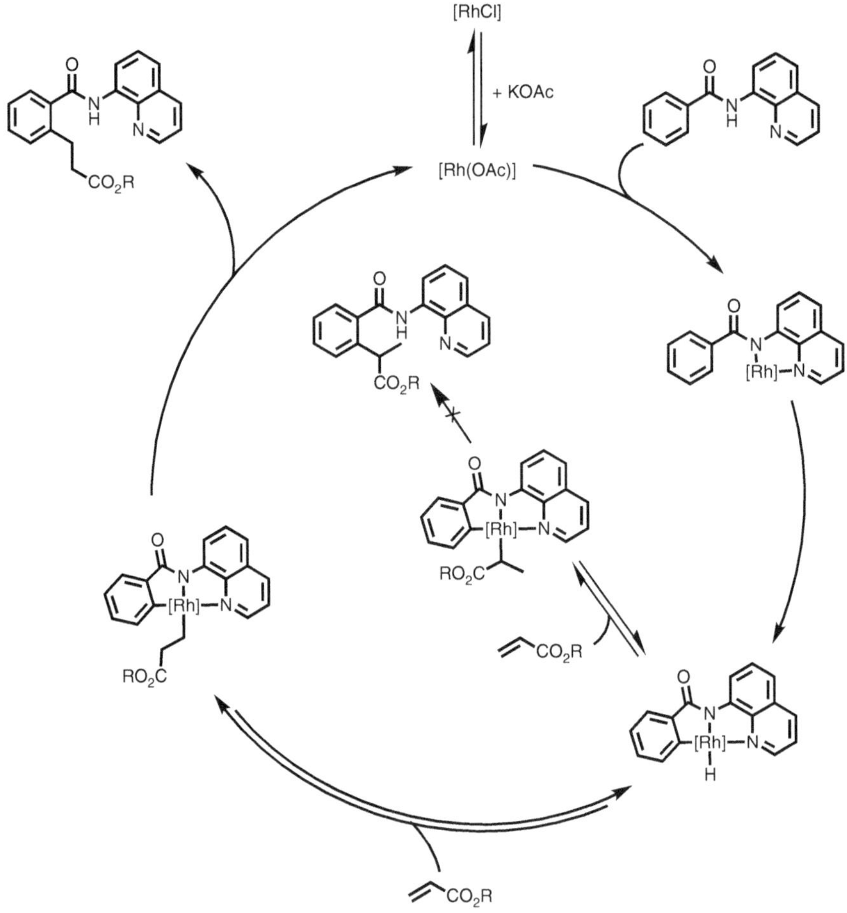

Figure 4.15 Mechanism for quinolylamide-directed *ortho*-hydroarylation.

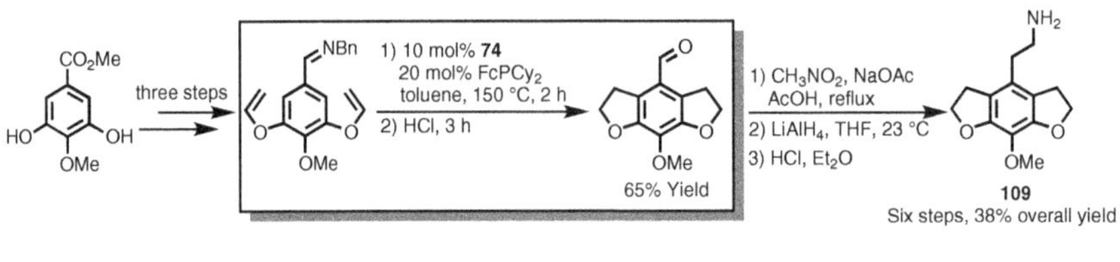

Scheme 4.6 Total synthesis of mescaline derivative **109** via Rh catalyzed hydroarylation.

depicted tetrahydrobis(benzofuran) functionalized mescaline derivative (**109**) was isolated in 38% over 6 steps (Scheme 4.6).

Ortho-hydroarylation has also been successfully employed toward strained olefin substrates such as norbornene. In one such system [59], several Rh^I precatalysts were used to catalyze the hydroarylation of norbornene using the biaryl-substituted ketone compound **111** (Eq. (4.8)). Using 10 mol% of $[(C_2H_4)_2Rh^I(\mu\text{-}Cl)]_2$ (**110**), two products were observed in the attempted hydroarylation of norbornene with **111**: the desired hydroarylated product (**112**) and a C—C activated product (**113**). This reaction was observed with several different norbornene and biaryl ketone substrates, generating both types of products.

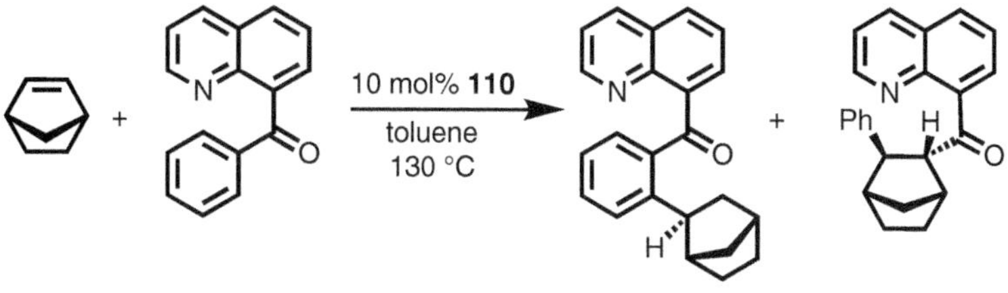

111 **112** **113**

(4.8)

A similar *ortho*-hydroarylation of norbornene with substituted anilines has been reported by Brunet and coworkers [60, 61] using $[(PEt_3)_2Rh^I(\mu\text{-}Cl)]_2$ (**114**). With this complex, both hydroamination and hydroarylation were observed (Eq. (4.9)). Using a mixture of aniline and the corresponding lithiated aniline salt, the resulting hydroamination (**115**) and hydroarylation (**116**) products were observed, with **116** as the major product. While the yields using aniline were relatively low, the use of *p*-toluidine increased the overall yield to 90% giving a 10 : 90 ratio for hydroamination and hydroarylation products.

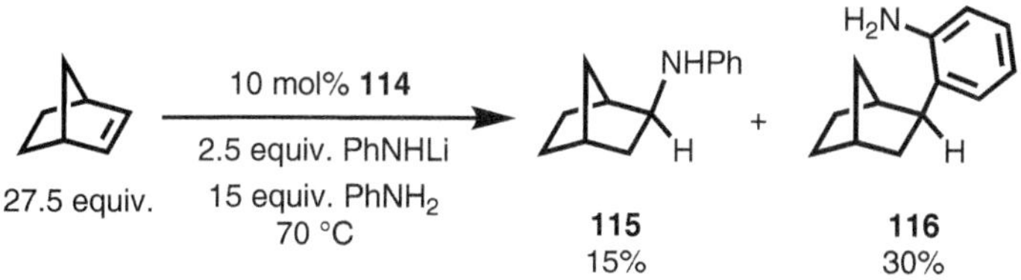

27.5 equiv. **115** 15% **116** 30% (4.9)

While Rh^I complexes with phosphorous ligands have been extensively used in *ortho*-hydroarylation, other ligand sets have proven useful. In a report by the Brookhart group [62], the pentamethylcyclopentadienyl (Cp*) Rh^I complex $[(Cp^*)Rh^I(CH_2CHSiMe_3)_2]$ (**117**) was used as a hydroarylation precatalyst (Eq. (4.10)). Using trimethylvinylsilane and benzophenone (**118**) substrates, both the monohydroarylated (**119**) and dihydroarylated (**120**) products were observed. Note that only anti-Markovnikov products were formed. Several other olefin substrates were probed with **117**, affording yields of greater than 60% in most cases after 10 h.

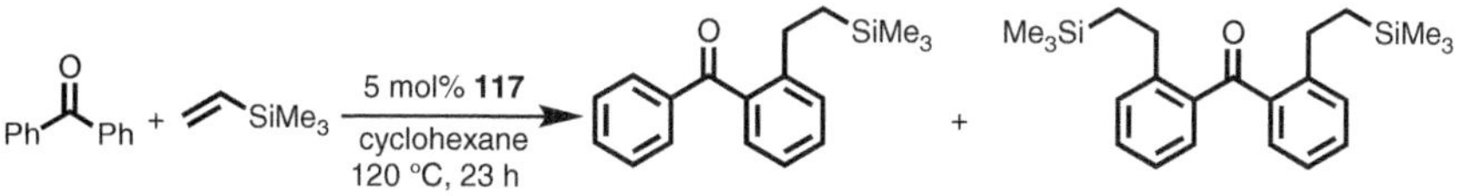

118 **119** 88% **120** 11%

(4.10)

4.3.3 Rh^I-Catalyzed Hydroarylation with Fluorinated Arenes

Hydroarylation with a Rh^I hydroxide was described in a recent report by Zhao and coworkers [63]. It was previously reported by Bercaw *et al.* [64] as well as by Goldberg and coworkers [65] that Rh^I hydroxides undergo stoichiometric aryl C—H bond activations to generate Rh–aryl species. For example, [(cod)Rh^I(μ-OH)]$_2$ (**121**) promotes the C—H activation of indane, to give a [(cod)Rh^I(η^3-indenyl)] complex (**122**) along with an equivalent of water [64]. Moreover, Lautens and coworkers [66] demonstrated that **121** catalyzes the hydroarylation of allyl amines with arylboronic acids in high yields.

Similar C—H bond cleavage reactivity was observed with a Rh^I complex possessing a 2,6-bis[(di-*tert*-butylphosphino)methyl]pyridine (PNP) pincer ligand, [(PNP)Rh^I(OH)] (**123**) [65]. Using **123**, benzene-d_6 or benzene was activated to generate the corresponding Rh–phenyl complexes [(PNP)Rh^I(C_6D_5)] and [(PNP)Rh^{I}Ph], respectively (Scheme 4.7). In a control experiment, H(D) scrambling into the methylene bridge of the ligand backbone occurred in benzene upon exposure of [(PNP)Rh^{I}Ph] to 10 equiv. of D_2O.

On the basis of these observations, complex **121** with added 1,2-bis(diphenylphosphino)benzene (dppbz) ligand was investigated in hydroarylation [63]. Electron-deficient arenes were examined as coupling partners with α,β-unsaturated carbonyl compounds to generate the desired hydroarylation product (Table 4.12). In this reaction, both the direct C—C coupled (**124**) and the hydroarylation products (**125**) were observed. For ester and ketone substrates, high yields were observed, whereas amide-containing substrates resulted in lower yields. All substrates exhibited high selectivities for generation of **125** over **124**, with nearly quantitative conversion to **125** in some cases. Moreover, only anti-Markovnikov products were observed for all substrates tested.

To rationalize the generation of both products, a mechanism was proposed as shown in Figure 4.16. The initial [Rh^I(OH)] complex undergoes hydroxide-assisted C—H activation of the fluorinated arene to generate a Rh–aryl complex along with an equivalent of water. Upon coordination and insertion of olefin, a Rh–alkyl fragment is generated. This species can then either undergo β-hydride elimination to produce **124**, or abstract a proton from water to form **125** and the regenerated Rh^I complex **121**.

4.3.4 Summary of Rh^I-Catalyzed Hydroarylation

Many Rh^I-based systems have been used as catalysts for olefin hydroarylation. Three major types of arenes have been employed in these systems:

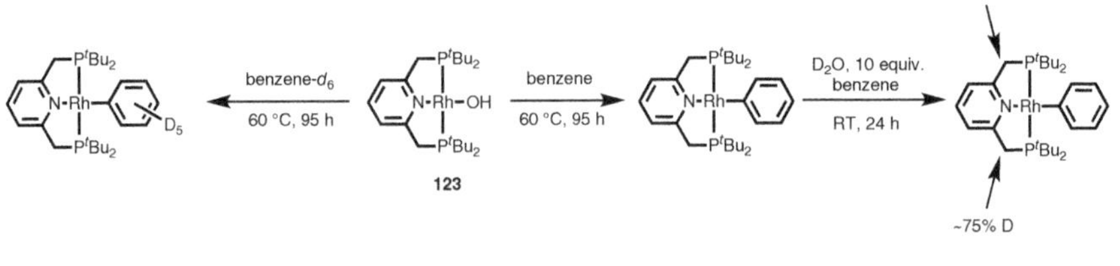

Scheme 4.7 Arene C—H(D) activations and H(D) scrambling with Rh^I complex **123**.

Table 4.12 Hydroarylation of pentafluorobenzene with α,β-unsaturated esters, amines, and ketones [63].

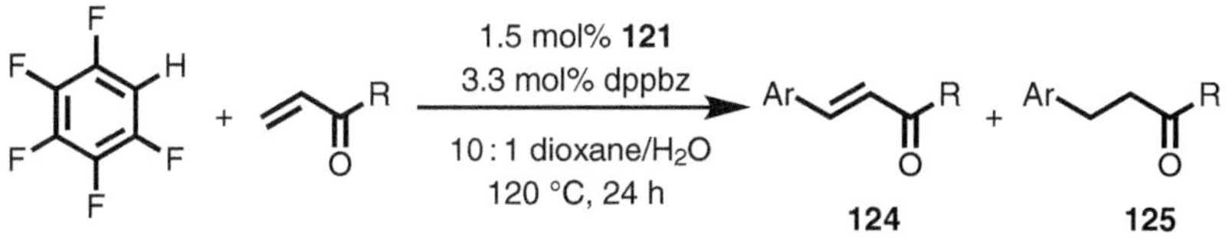

R	124 : 125	Overall Yield (%)
O^nBu	1 : 17	88
O^tBu	1 : 40	87
O^iBu	1 : 20	91
OEt	1 : 12	80
NH	1 : 50	52
NMe	1 : 11	43
Me	1 : 50	78
Et	1 : 50	85

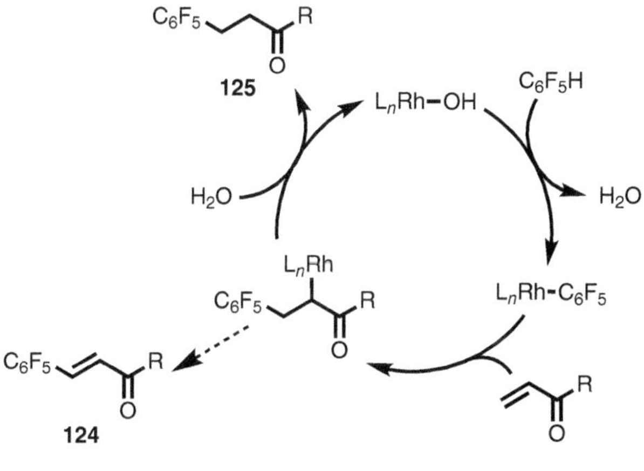

Figure 4.16 Mechanism for hydroarylation and C—C coupling using a $[Rh^I(OH)]$ complex.

unfunctionalized arenes [37–40], arenes with directing substituents [32, 41–62], and highly fluorinated arenes [63–66]. With these systems, high yields (or TONs) have been observed for the desired products with reasonable catalyst loadings (~5 mol%). Unfunctionalized arenes such as benzene have been used in a one-step process to generate the corresponding styrene product with over 800 TONs using a bisimine-ligated Rh complex [37]. For the hydroarylation of ethylene with benzene catalyzed by Rh complexes, in a process similar to that operating in many hydroarylation mechanisms [37, 40]. However, rather than generating a hydroarylation product by reductive elimination or σ-bond metathesis, a β-hydrogen elimination occurs to generate a hydride species with a bound styrene. With the use of an external oxidant, styrene is eliminated to regenerate a Rh^I species that can reenter the catalytic cycle.

The hydroarylation of arenes substituted with a coordinating heteroatom (imines, esters, amides, ketones, pyridines, etc.) has received substantial attention. It is important to note that with these arenes, only *ortho*-hydroarylation occurs due to the initial binding step during catalysis. This anchoring of the substrate places the *ortho*-aryl C—H bond in close proximity to the Rh^I center and allows for subsequent C—H activation to generate a five-membered chelate. Moreover, both *meta*- and *para*- C—H activation do not occur due to the geometry imparted by the directing group, which orients these C—H bonds away from the metal center. Additionally, C—H activation through these bonds would generate strained six- or seven-membered chelates. The use of these directing groups, therefore, selectively generates only *ortho*-hydroarylated products. This strategy has been successfully employed for both inter- [32, 41–44] and intramolecular [46, 47] hydroarylations.

Conveniently, a simple Rh^I species such as Wilkinson's catalyst serves as a catalyst for this *ortho*-hydroarylation reaction [32, 41–47]. Also, the commonly employed rhodium starting complex $[(coe)_2Rh^I(\mu\text{-}Cl)]_2$ is useful as a precatalyst, with addition of phosphorous-containing ligands [45]. The use of such ligands has demonstrated higher overall yields (vs PPh_3), and the flexibility of ligand choice allows for an element of catalyst design. One such benefit is that chiral catalysts are available for enantioselective hydroarylations. Chiral substituents on the aryl-directing group can also improve enantioselectivity [53]. Hydroarylations of α-olefins have also been investigated, and interestingly, only the anti-Markovnikov linear products were observed using Rh^I-based systems. The use of Rh^I in *ortho*-hydroarylation has therefore been used to control both regio- and enantioselectivity during the course of the reaction. Despite this control, dihydroarylation has often been observed with arenes containing multiple *ortho*-aryl C—H bonds.

4.4 Directed *ortho*-Hydroarylation Catalyzed by Ir^I Complexes

For comparative purposes, note that iridium has been featured in several reports by Periana and coworkers on catalytic hydroarylation with catalyst precursors of the type $[Ir^{III}(\mu\text{-acac-}O,O,C^3)(\text{acac-}O,O)(\text{acac-}C^3)]_2$ [67] or $[(acac)_2Ir^{III}(Ph)(L)]$ ($L = H_2O$, Py) [68, 69]. In these Ir^{III} examples (see Chapter 3 for further details), the catalytic hydroarylation of unactivated arenes (i.e., benzene) was achieved with various aliphatic olefins (i.e., ethylene, propylene, 1-hexene, etc.), with activities that exceed those observed for platinum-based catalysts (see Section 4.2). A mechanistic analysis of this catalysis by DFT calculations [70, 71] confirms that olefin insertion into the Ir–aryl bond is energetically significant and likely rate determining. The key C—H activation step was calculated to involve a concerted,

4-centered transition state whereby hydrogen is transferred from an arene ligand to the metal-bound carbon of the Ir(CH_2CH_2Ar) group (σ-bond metathesis or oxidative hydrogen migration). One might expect Ir^I catalysts to operate by a different mechanism involving C—H oxidative addition, and a few examples of Ir^I-catalyzed hydroarylation have been reported [72–79]. In contrast to that demonstrated by Ir^{III} catalysts, Ir^I complexes have not been reported to catalyze the hydroarylation of unactivated arenes. For the most part, Ir^I systems have been used for the directed *ortho*-hydroarylation of arenes bearing directing functionalities, similar to Rh^I complexes that have been extensively used for this purpose (see Section 4.3). Typically, strained cyclic olefins (i.e., norbornene) have been employed as hydroarylation substrates along with arenes bearing directing substituents such as amides or esters. As with many Rh^I systems, the active Ir^I species are usually generated *in situ* using commercial or readily synthesized Ir^I species.

In an initial report by Togni and coworkers [72], several Ir^I complexes with ancillary cyclopentadienyl (Cp) ligands were synthesized and isolated from the corresponding $[L_2Ir^I(\mu\text{-Cl})]_2$ dimer precursors (**126–129**). Treatment of the dimeric complex $[(coe)_2Ir^I(\mu\text{-Cl})]_2$ (**130**) with free diphosphine ligands (Figure 4.17; L_2, **131–134**) afforded complexes **126–129**. Using NaCp in THF, complexes **126–129** underwent disproportionation reactions to afford the corresponding $[L_2Ir^I(Cp)]$ complexes **135–138** along with an equivalent of NaCl.

Using 1 mol% of these complexes, the *ortho*-hydroarylation of norbornene with benzamide was observed (Eq. (4.11)). Upon heating at 100 °C for 72 h, complexes **126–129** and **135–138** gave either the hydroarylation product **139** or the hydroamination product **140** in moderate yields. High enantioselectivities for the *exo*-norbornane isomers of **139** and **140** were observed, with ee values greater than 70%. Catalysis with the Cp-ligated complexes **135–138** afforded **139** as the only product, although in low yields (about 20%). In contrast, the starting dimer species **126–129** generally produced the hydroamination product **140**. Since these Cp complexes are formally 18-electron species, it is likely that phosphine dissociation or Cp ring slippage occurs to generate the active catalytic species. It was suggested, therefore, that the ability of Cp ligands to change their binding motif from η^5 to η^3 (or even η^1) would provide better access to the Ir center allowing for amide coordination. The resulting Ir(amide) intermediate would then undergo C—H activation and subsequent olefin insertion to produce the hydroarylation product.

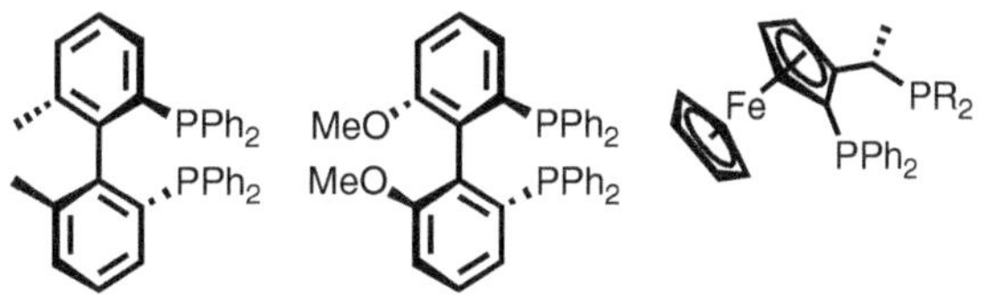

131 **132** **133**: R = Cy
134: R = Ph

Figure 4.17 Chiral phosphine ligands used to generate $[L_2CpIr^I]$ complexes.

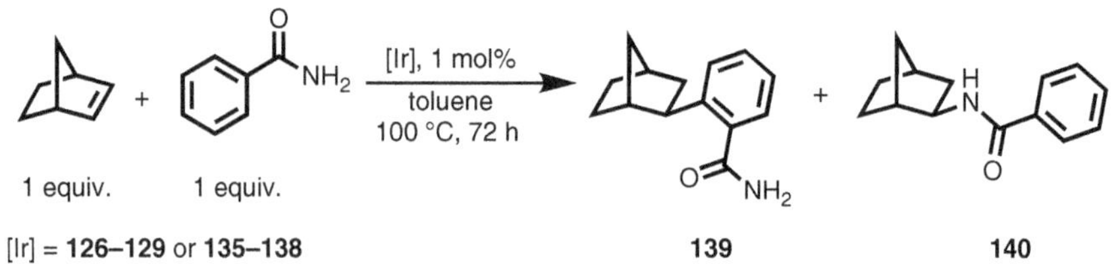

(4.11)

In contrast to benzamide, phenol as a substrate provided only hydroarylation products, even upon heating for several days [73]. However, dihydroarylation products were also observed (Eq. (4.12)). This reaction is associated with a low stereoselectivity for the *exo*- and *endo*-norbornane products, and a low ee of 4.5%.

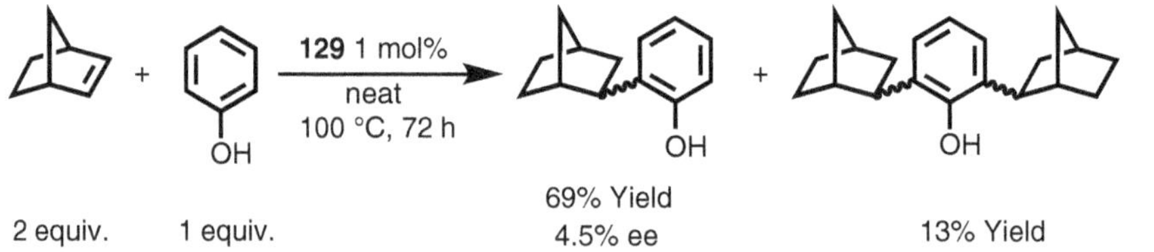

(4.12)

A mechanism was proposed for catalysis with the Ir^{I} dimeric precatalyst **129** [73] (Figure 4.18). An initial coordination of the phenol to the Ir dimer induces a disproportionation to generate the $[L_2Ir^{I}Cl(phenol)]$ intermediate, **141**. Due to the geometry imparted by this anchoring directing group, only *ortho*-aryl C—H bond activation can occur, to give an Ir species with the substrate bound in a bidentate fashion (**142**). Coordination and insertion of norbornene couples the substrates and generates an Ir^{III}–hydride species (**143**). Reductive elimination produces the hydroarylation product and regenerates the active Ir^{I} complex (**129**).

A report by Yamamoto and coworker [74] describes enantioselective *ortho*-hydroarylations of norbornene with ester-functionalized anisole derivatives (Table 4.13). For this system, cationic $[(cod)_2Ir^{I}]^+$ complexes with counterions such as SbF_6^- or $B(C_6F_5)_4^-$ were employed as precatalysts. The active species was generated by addition of a chiral phosphine, and initial ligand screening with $[(cod)_2Ir^{I}][B(C_6F_5)_4]$ (**144**) showed that the sulfur-linked bis(phosphoramidite) (*R*,*R*)-S-Me-BIPAM (**145**, Figure 4.19) provided the highest enantioselectivities and overall yields. With this system, a range of arene substrates were observed to give high enantioselectivities in greater than 90% yields for the *exo* product. Electron-withdrawing aryl substituents (R^2) exhibited reduced conversion to the product, and significantly increased yields were observed with amide-directing (vs carbonyl) groups. To probe the mechanism, the *N*-cyclohexylbenzamide-d_5 arene substrate was used to provide a KIE of 2.08. It was suggested that this small primary KIE indicates a C—H(D) cleavage step in the mechanism.

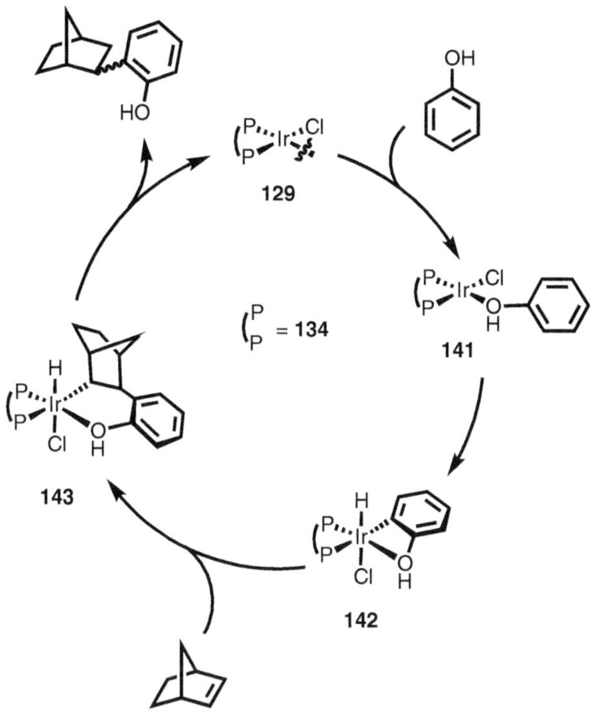

Figure 4.18 Proposed catalytic species for the *ortho*-hydroarylation of norbornene with phenol using dimeric (diphosphine)Ir[I] complexes.

Table 4.13 Substrate scope for *ortho*-hydroarylation of norbornene with **144** and added **145** [74].

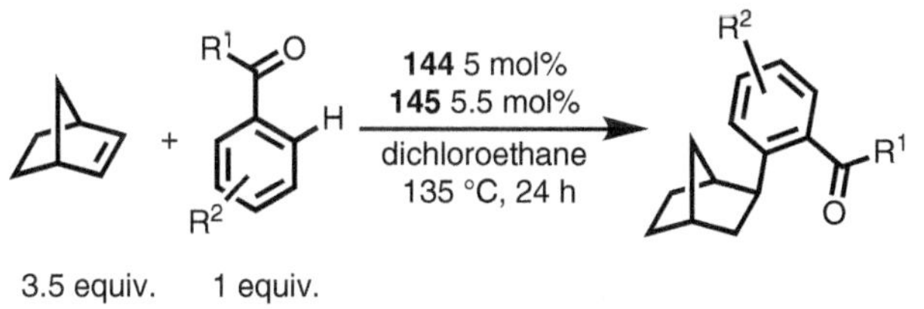

R^1	R^2	Isolated yield (%)	ee (%)
Me	H	40	92
	o-OMe	82	88
	o-F	42	93
	m-Me	50	88
Et	*o*-OMe	61	92
NMe_2	H	90	99
$N(^iPr)_2$	H	90	96
Piperidine	H	91	99
	p-Me	97	99

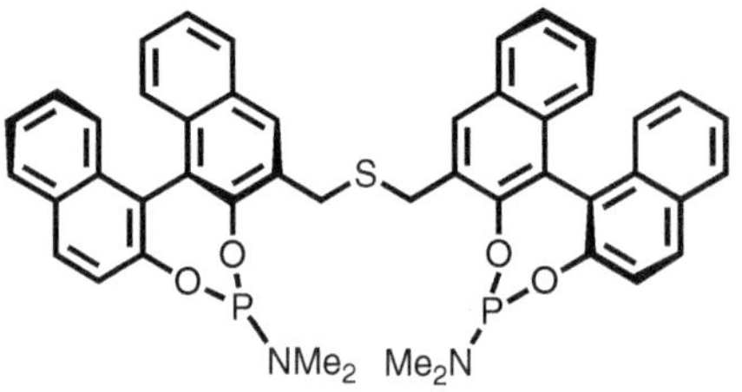

Figure 4.19 Structure of (*R*,*R*)-S-Me-BIPAM, **145**.

A similar system employing 2,2′-bis(diphenylphosphino)-1,1′-binaphthyl (BINAP) and **144** demonstrated the enantioselective *ortho*-hydroarylation of aryl ketones with norbornene and styrene derivatives [75]. With styrene-based substrates, branched and linear alkyl arene products were observed. With the bulky BINAP ligand, linear products were preferred for all styrene-containing products. Similarly, *exo*-norbornane based products were observed to be favored in the hydroarylation of norbornene with 2′-methylacetophenone, giving an ee of 70%. This system was also successfully applied toward the *ortho*-hydroarylation of asymmetric alkynes with substituted acetophenone derivatives.

Bower and coworkers [76, 77] examined a variety of chelating diphosphines to further investigate ancillary ligand effects on hydroarylation. These diphosphine ligands contain aliphatic linkers of varying lengths (i.e., CH_2, CH_2CH_2, etc.). It was hypothesized that the bite angle of the diphosphine could potentially influence the regioselectivity of *ortho*-hydroarylation (i.e., Markovnikov versus anti-Markovnikov) by changing the steric profile at the metal center. A catalytically active species could be generated *in situ* by addition of a diphosphine to complex **144**. These species catalyzed the hydroarylation of styrene with diethylbenzamide with 5 mol% loading in dioxane at 100 °C for 24 h. The overall yields and product ratios (branched:linear) were determined by ^1H NMR spectroscopy and the results are summarized in Table 4.14. While the racemic-BINAP ligand provided high yields, it gave a high selectively for the linear product **147**. Surprisingly, the dppm ligand with a small bite angle was very selective for the anti-Markovnikov product (product ratio of 8 : 92). By increasing the diphosphine bite angle, a switch in selectivity was observed. Thus, dppb (with a $(CH_2)_4$-linked diphosphine) provided complete selectivity for the branched product (branched:linear ratio of 100 : 0). However, ligands with longer alkyl tethers exhibited significantly reduced yield. With dppb, only a 28% overall yield was obtained. Improved yields of the branched products were obtained by using fluorinated versions of the diphosphines, as illustrated by d^Fppb and d^Fppe [$(C_6F_5)_2P(CH_2)_nP(C_6F_5)$; $n = 4$ and 2, respectively], which gave increased yields and retained high branched:linear product ratios compared to the "parent" ligands dppb and dppe.

To further investigate this system, d^Fppb and **144** were used to probe the substrate scope. High yields and Markovnikov product selectivities were observed for a series of arene substrates containing various carbonyl-directing groups (e.g., amides, ketones, esters) and other arene substitutions. Additionally, various substituted styrene substrates (i.e., *p*-Me, *p*-F, *m*-Cl, and *o*-F), as well as aliphatic olefins were tested. Styrene substituents had a minimal effect on the overall yield and selectivity, typically displaying yields above 70% (and in some cases quantitative conversion). Branch:linear selectivities of greater than 25 : 1 were

Table 4.14 Ligand scope with alkyl tethered diphospine ligands [76].

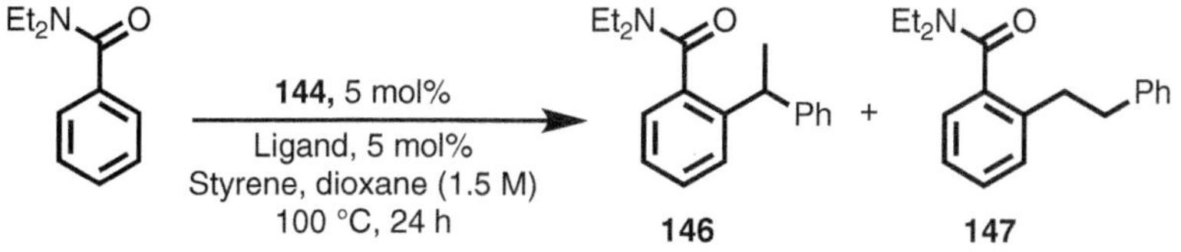

Ligand	Overall yield (%)	146 : 147
rac-BINAP	100	29 : 71
dppm	73	8 : 92
dppe	82	23 : 77
dppp	52	60 : 40
dppb	28	100 : 0
d^Fppe	100	36 : 64
d^Fppb	100	100 : 0

observed with all styrene substrates. In contrast, bulky aliphatic olefins gave reduced yields; however, regioselectivity for branched products was preserved. No reaction was observed with ethyl acrylate. These results suggest that this system is tolerant of many substrates, but likely cannot catalyze the hydroarylation of electron-deficient aliphatic olefins. Additionally, it demonstrates that the regioselectivity is likely dictated by the diphosphine bite angle.

To investigate the operative mechanism in this system, a labeling experiment was conducted using deuterated styrene analogue **148** and diethylbenzamide as substrates with **144**/d^Fppb as a precatalyst (Eq. (4.13)). Although this reaction ultimately gives mainly the branched hydroarylation product, the labeling experiment suggests that several reversible steps are involved in the mechanism. Both the methyl and methine positions incorporated deuterium atoms, which is consistent with a reversible olefin insertion step. Additionally, deuterium incorporation into the *ortho*-arene position implies a reversible C—H activation step. As a result of the H(D) exchange, it was suggested that reductive elimination was rate limiting.

(4.13)

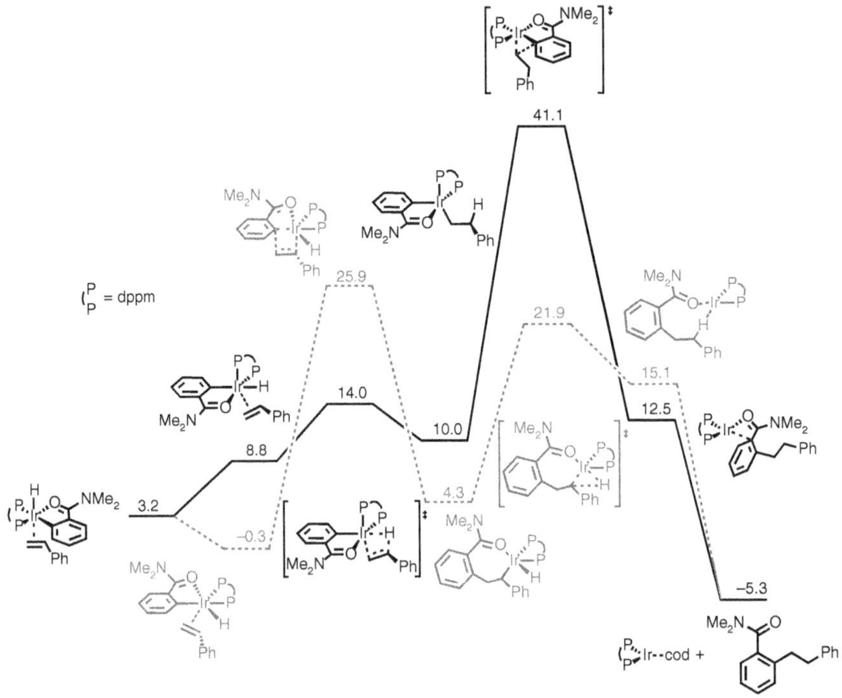

Figure 4.20 Calculated barriers for the two possible insertion and reductive elimination routes in *ortho*-hydroarylation using [(dppm)Ir^{I}(cod)]$^+$ as the added catalyst. Olefin insertion into the metal–hydride and subsequent C—C reductive elimination pathway (black, solid) was calculated to have a larger kinetic barrier than the pathway involving arene insertion followed by C—H reductive elimination (gray, dashed). Only the anti-Markovnikov pathways are depicted. All complexes exist as cationic species and the charge has been omitted for clarity. Energies are given in kcal/mol.

The original mechanism proposed by Bower and coworkers [76, 77] involved a mechanism whereby C—H oxidative addition generates an Ir–hydride species. Upon coordination of styrene, a migratory insertion into the Ir–hydride occurs to generate a new Ir–alkyl fragment and C—C reductive elimination then affords the desired alkyl arene. The operative mechanism has also been explored by Huang and Liu [78] using DFT calculations (Figure 4.20). The results of this study were surprising; the energetically preferred mechanism based on the DFT calculations for this hydroarylation involved several different steps from the mechanism proposed by Bower and coworkers [76, 77], namely a C—H reductive elimination. In the DFT calculated mechanism, a C—H activation of the arene substrate occurs to generate an Ir–hydride, followed by styrene coordination. Unlike the original mechanism, migration of the Ir–aryl bond into the olefin takes place. This insertion preserves the Ir–hydride and can occur either in a Markovnikov or in an anti-Markovnikov fashion. A C—H reductive elimination occurs to generate the resulting alkyl arene products.

While added ligands have been successfully employed to probe the selectivity and activity of Ir^{I}-based *ortho*-hydoarylation catalysts, a recent report by Nishimura [79] demonstrates that [(cod)Ir^{I}(μ-Cl)]$_2$ (**150**) with added NaB(C_6F_5)$_4$

Table 4.15 Substrate scope of ligand-free *ortho*-hydroarylation of vinyl ethers [79].

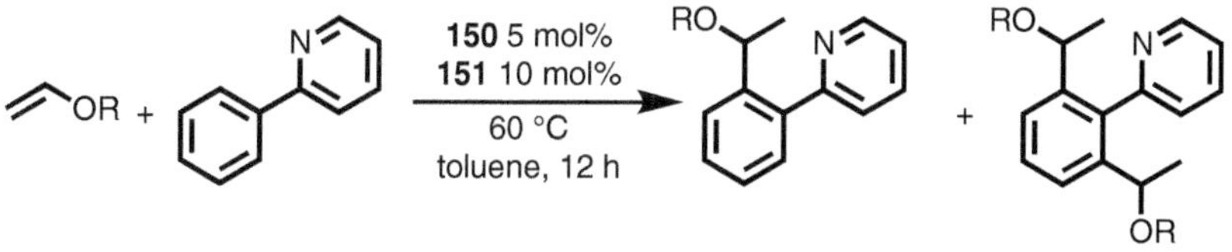

153
152

R	Isolated yield (%)	153 : 152
Et	90	100 : 0
^{n}Bu	99	100 : 0
^{i}Bu	96	95 : 5
Bn	91	95 : 5
Ph	91	100 : 0
p-MePh	98	100 : 0
p-MeOPh	93	95 : 5
p-FPh	90	94 : 6
p-ClPh	93	100 : 0
m-MePh	91	98 : 2
o-MePh	98	100 : 0
$SiMe_3$	88	100 : 0

(**151**) is an *ortho*-hydroarylation catalyst, without added ligand. The addition of **151** to **150** likely results in elimination of NaCl to produce a reactive, cationic iridium species. This catalyst couples vinyl ether substrates with pyridyl-substituted arenes, with the pyridine acting as a directing group. These reactions are selective for the branched product for a range of olefins and 2-phenylpyridine as the arene (Table 4.15). Linear products were not observed with any substrates. Instead, only branched hydroarylated products were isolated, generally in yields greater than 85%. While dihydroarylated species (**152**) were occasionally observed, selective conversion to only monohydroarylated products (**153**) occurred for most substrates. This ligand-free system was also applied to arene substrates with different types of directing groups (e.g., imine, oxazole, benzothiazole). A deuterium-labeling study using deuterated 2-phenylpyridine (C_6D_5-2-Py) as a substrate revealed scrambling of hydrogen into the *ortho*-aryl positions. This result, and deuterium incorporation into the methyl and methine positions of the product suggests reversible C—H activation and olefin insertion steps for hydroarylation catalyzed by Ir^I species. Nishimura and coworkers [80] have also demonstrated that similar $[(\text{diene})Ir^I(\mu\text{-OH})]_2$ complexes are competent catalysts for the asymmetric hydroarylation of vinyl ethers with benzamides.

The Ir^I-based hydroarylation systems in the literature have provided insights into several aspects of reaction selectivity. With chiral phosphine ligands, the *ortho*-hydroarylation of norbornene with various arenes possessing directing groups enantioselectively produces *exo*-norbornane [72–75]. The *ortho*-hydroarylation of styrenes and vinyl ethers with arenes bearing directing groups was also achieved [76–79]. These studies demonstrate the potential for ligand design to influence the regioselectivity of hydroarylation. Specifically, investigations with diphosphine ligands with varying bite angles illustrate a connection between catalyst structure and olefin insertion regioselectivity [76], allowing access to both Markovnikov and anti-Markovnikov products. Smaller bite angles lead to increased generation of linear alkyl arene products, while wider bite angles provide branched alkyl arenes. The mechanism involving these diphosphine complexes was probed further using DFT calculations [78]. An unexpected mechanism was calculated to be energetically favorable: migration of an Ir–aryl bond into an olefin occurs rather than migration of an Ir—H moiety.

While several examples of Ir^I-catalyzed hydroarylations exist, it is a relatively unexplored d^8 metal center for the hydroarylation of olefins. Moreover, the active catalytic species are typically generated *in situ*; it is possible, therefore, that these catalyst systems generate protic species that mediate hydroarylation (see Pt-catalyzed hydroarylation, Section 4.2). Further studies are required to identify likely mechanistic pathways available to these catalysts based on Ir^I precursors.

4.5 Hydroarylation with Ni^0 Complexes via Ni^{II} Intermediates

Hydroarylation has been reported for Ni^0 catalyst precursors and while these complexes do not have a d^8 electronic configuration, Ni^{II} species have been implied as intermediates in the catalytic cycle. The Hartwig group [81] recently investigated the hydroarylation of various olefins with electron-deficient bis(trifluoromethyl)benzene substrates using $[(cod)_2Ni^0]$ (**154**) and an added carbene ligand (1,3-bis(2,6-diisopropylphenyl)-1,3-dihydro-2*H*-imidazol-2-ylidene, IPr). While these reaction conditions also catalyze the hydroarylation of 1-hexene with benzene, low yields were observed determining that highly electron-deficient arenes are the preferred substrates in this system. To probe the substrate scope, a variety of terminal and internal olefins were tested (Table 4.16). The complex $[(IPr)_2Ni^0]$ (**155**) was used as a precatalyst in large catalyst loadings (20 mol%) with a slight excess of arene. Generally, high selectivities were observed for the linear anti-Markovnikov product **156**. While some olefin substrates (R = Me) afford a mixture of both **156** and the branched product **157**, most substrates exhibited greater than 99% selectivity for the linear products. Similarly, moderate yields were generally demonstrated for most olefin substrates.

Several labeling studies were performed with 5-deuterio-1,3-bis(trifluoromethyl)benzene (**158**) using complex **155** as a precatalyst (Figure 4.21) [81]. After each reaction, the position of the deuterium atom was determined by ^2H NMR spectroscopy. In the first experiment (**A**), 1-octene with added NaO^tBu generated a linear hydroarylation product, for which deuterium incorporation occurred primarily in the β-methylene position. Minimal

Table 4.16 Hydroarylation substrate scope using Ni^0 with electron-deficient arenes [81].

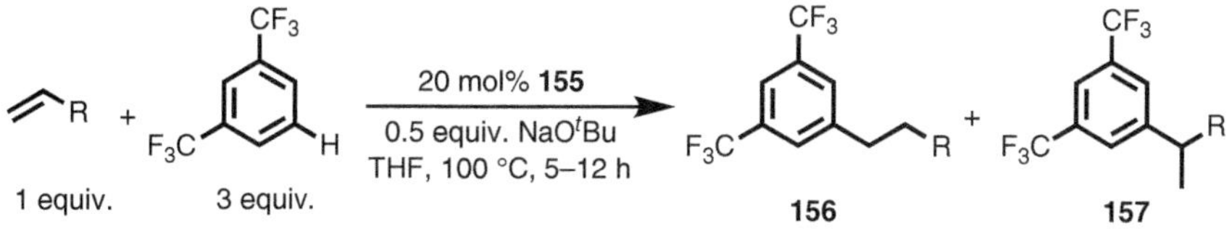

R	Isolated yield (%)	156 : 157
Me	57	89 : 11
^{n}Hex	65	97 : 3
^{t}Bu	87	>99 : 1
Cy	59	>99 : 1
CH_2CH_2Ph	73	97 : 3
CH_2SiMe_3	39	>99 : 1
$SiEt_3$	78	>99 : 1
CH_2NBn_2	68	>99 : 1
O^tBu [a)]	23	99 : 1
Cyclohexene	70	—

a) Reaction did not include NaO^tBu.

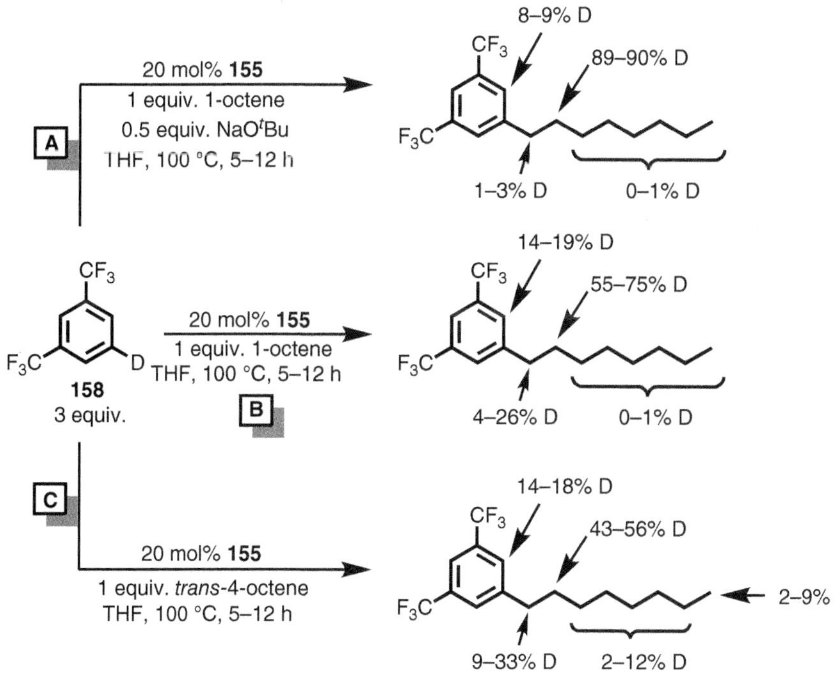

Figure 4.21 Deuterium-labeling study with and without NaO^tBu using various octene substrates.

isomerization of 1-octene was indicated by a lack of deuterium incorporation into the aliphatic chain. Also, observed *ortho*-aryl C—H scrambling suggests a reversible arene-activation step. In a similar reaction without base (**B**), deuterium scrambling into the α-methylene was observed to a greater extent. A similar increase in H(D) exchange into the *ortho*-aryl position was noted. Similarly, very little deuterium incorporation into the aliphatic chain was observed. With the internal olefin *trans*-4-octene (reaction **C**), an increase in deuterium incorporation into the aliphatic chain was observed, likely due to prior isomerization of the olefin. Differences in the initial rates of catalysis with proteo- and deutero-1,3-bis(trifluoromethyl)benzene gave a statistically insignificant KIE of 1.1 ± 0.1.

The mechanism of this hydroarylation was probed using DFT calculations, which gave activation energies for the proposed transition states and intermediates [81]. This analysis points to $[(IPr)Ni^0(olefin)_2]$ as the resting state. Arene coordinates to this species in an η^2-fashion, displacing an olefin. Proton transfer from the bound arene to the ancillary olefin ligand occurs to produce two new Ni—C bonds, affording a complex of the type $[(IPr)Ni^{II}(R)(Ar)]$. The proposed turnover-limiting step for the formation of both branched and linear products is reductive elimination from the corresponding $[(IPr)Ni^{II}(R)(Ar)]$ ($R = CH(CH_3)_2$ or $CH_2CH_2CH_3$, respectively) species, with a barrier of 13.1 kcal/mol. The selectivity of the reaction was attributed to the rate-limiting reductive elimination step; the energy difference between the transition states of the linear and branched products was calculated to be 1.1 kcal/mol. This difference corresponds to a product ratio of 82 : 18, which was in good agreement with the experimental data.

In a similar report by Hiyama and coworkers [82], the hydroarylation of alkynes and olefins with fluorinated arenes (e.g., C_6F_5H, 1,2,3,4-C_6F_4H, etc.) was achieved using $[(cod)_2Ni^0]$ with added $PCyp_3$ ligand (Cyp = cyclopentyl). The hydroarylation of 2-vinylnaphthalene with C_6F_5H yielded the corresponding branched diarylethane product whereby C—H activation occurs preferentially over C—F activation. The resulting $[(C_6F_5)Ni^{II}H]$ species may then coordinate an olefin, which then undergoes migratory insertion (into Ni—H) in a Markovnikov fashion. Reductive elimination of the aryl and alkyl ligands generates the alkyl arene product.

While Ni^{II} complexes have not been used directly as precatalysts for olefin hydroarylation, a computational study by Cundari and Gonzalez [83] concludes that species of the type $[(\kappa^2\text{-Tp})Ni^{II}(Ph)(NCMe)]$ (Tp = hydrido-tris(pyrazolyl)borate) could act as competent hydroarylation catalysts. In this study, the analogous Pt^{II} species was also investigated. The energies of various intermediates and transition states in the catalytic cycle for hydroarylation of ethylene with benzene were calculated by DFT. The mechanism is reminiscent of that for Pt^{II}-catalyzed hydroarylation (see Section 4.2), whereby olefin coordination and insertion occur. The resulting M—Ar species may then abstract hydrogen from benzene. It was determined that while the mechanisms for Ni^{II} and Pt^{II} are similar, the C—H-activation steps are different. The Pt^{II} species likely undergoes a two-step oxidative addition/reductive elimination pathway, whereas the Ni^{II} species undergoes a one-step σ-bond metathesis pathway, with a barrier of 31.6 kcal/mol.

4.6 Formal Hydroarylation Reactions with Pd^{II} Catalysts via Heck-Like Mechanisms

Catalysts based on Pd^{II} are known to couple olefins with arenes that possess a suitable leaving group (halide, triflate, etc.) or transmetallating substituent (SnR_3, $B(OH)_2$, etc.). In these cases, such as the Heck reaction [84], an olefin product is generated by the formal elimination of HX between the coupling partners. It has been demonstrated that with the addition of a suitable hydrogen source hydroarylation products are generated instead, with formal addition of a hydrogen and an aryl group across the olefin. It is important to note that unlike other hydroarylation catalysts that proceed via arene C—H bond activation, Pd^{II}-catalyzed hydroarylation reactions occur either through C—X activation or arene transmetallation. Several reviews have been published on such hydroarylation reactions [85, 86]. Two categories of conditions have been employed: formate-assisted hydroarylation [87–96] and oxidatively coupled hydroarylation [97–102]. Details of these two conditions for Pd^{II}-catalyzed hydroarylation are discussed below in this section.

4.6.1 Formate-Assisted Pd^{II} Catalyzed Hydroarylation

An initial report by Cacchi and coworkers [87] demonstrated that [$(PPh_3)_2Pd^{II}(OAc)_2$] (**159**) catalyzes the hydroarylation of α,β,γ,δ-dienones (**160a–d**) with various aryl halides (Eq. (4.14) reports hydroarylation with 1-iodo-4-methoxybenzene). In contrast to other d^8 metal-based hydroarylation catalysts, an aryl C—H activation step does not occur with this Pd^{II} system. Instead, the aryl group is introduced as a ligand by oxidative addition of a C—X bond to generate a Pd(X)(Ar) fragment, and stoichiometric formic acid serves as the source of hydrogen to generate the resulting hydroarylation products **161a–d**. This system was employed with several dienones and aryl iodide substrates. With some substrates, however, the substituted vinylic Heck products **162a–d**

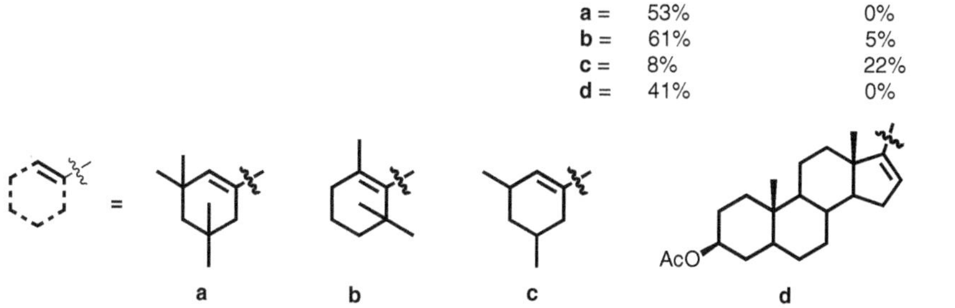

(4.14)

Table 4.17 Formal hydroarylation of norbornene derivatives with aryl iodides [89].

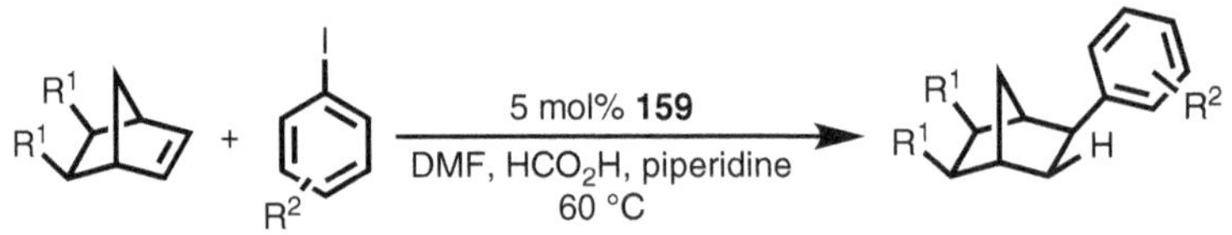

R^1	R^2	Reaction time	Isolated yield (%)
H	H	8	83
	p-NHCOMe	4	90
	p-NH_2	6	29
	p-OH	6	65
	m-Me	8	72
	o-CH_2OH	8	70
CO_2Me	H	3	85
	p-Cl	8	83
	o-OMe	8	73

were observed as a byproduct. This reaction type has also been applied to the formal hydroarylation of alkynes [88].

In a subsequent report by Cacchi and colleagues [89], the scope of this system was explored. In the presence of formic acid and base (either piperidine or tributylamine), modest loadings of **159** catalyzed the hydroarylation of norbornene with a wide variety of aryl iodides (Table 4.17) to selectively generate the aryl-substituted *exo*-norbornane product. Arenes containing potentially coordinating groups (*p*-NH_2 or *p*-OH) exhibited reduced yields when compared to non-coordinating substituents (*m*-Me or H). Additionally, electron-deficient norbornenes bearing *exo*-ester functionalities exhibited high activities under reaction conditions. With all substrates tested, only hydroarylation products were isolated and vinylic-substituted products (similar to **162a–d**) were not observed, likely due to the strained geometry required for β-hydride elimination.

To probe the mechanism, Tam and Mayo [90] used substituted norbornadiene substrates to probe stereo- and regioselectivity. Using $Pd^{II}(OAc)_2$ and added PPh_3 ligand, catalysis (Table 4.18) was achieved in the presence of formate (either added directly or generated *in situ*). With ester- or silyl-containing norbornene substrates, high yields were observed using formic acid and added piperidine base. Using these conditions, low yields were observed with 2,3-dibromonorbornadiene as a substrate. Adding potassium formate instead of generating formate *in situ* resulted in an increased yield. An additional increase in reactivity was observed upon addition of tetrabutylammonium chloride. Two potentially reactive olefins exist in norbornadiene, but surprisingly, only one product was identified. The functionalized olefin remained intact and hydroarylation occurred only across the unfunctionalized olefin.

A mechanism was proposed to account for the observed selectivity for preferential hydroarylations across unfunctionalized olefins (Figure 4.22). An

Table 4.18 Formal hydroarylation of norbornadiene derivatives with phenyliodide [90].

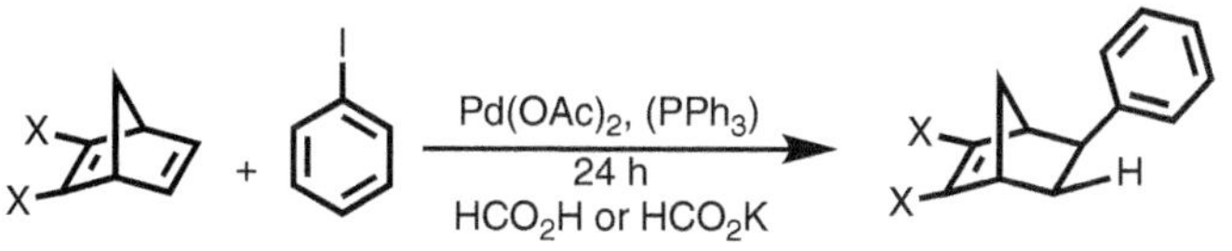

X	Conditions	Isolated yield (%)
CO_2Me	**A**	91
$SiMe_3$	**A**	80
Br	**A**	<10
	B	33
	C	58

Conditions: **A** = 20 mol% $Pd(OAc)_2$, 40 mol% PPh_3, HCO_2H, piperidine, THF, 60 °C; **B** = 20 mol% $Pd(OAc)_2$, 40 mol% PPh_3, HCO_2K, DMF, 25 °C; **C** = 20 mol% $Pd(OAc)_2$, 40 mol% PPh_3, HCO_2K, Bu_4NCl, DMF, 25 °C.

initial ligand substitution step generates an active Pd center that oxidatively adds phenyl iodide to generate a $Pd^{II}(Ph)(I)$ species. Norbornadiene (**163**) coordination can occur either through the unfunctionalized or functionalized olefin, generating complexes **164** and **165**, respectively. The steric profile of the metal center for these complexes dictates their overall stability; compound **164** directs the norbornadiene substituents away from the metal center while compound **165** orients them directly into the metal center. Thus, compound **164** is the favored intermediate during catalysis. From this intermediate, olefin insertion can occur in a *syn* fashion, positioning the arene and Pd in *exo*-positions on the norbornene product. Coordination of formate is then followed by a β-hydride abstraction and then decomplexation affords a Pd–hydride fragment along with an equivalent of CO_2. Reductive elimination affords the final *exo*-norbornene product **166**. The other regioisomer, **167**, is not observed.

While simple PPh_3 ligands have been used in this type of hydroarylation, Brunel and coworkers [91] have investigated several dimeric Pd species as catalysts. These dimeric complexes contain bridging acetate ligands and a bidentate phosphine ligand (see complexes **168–170**, Table 4.19). Complex **168** is a catalyst for hydroarylation of norbornene with aryl halides to generate *exo*-phenylnorbornanes. Formate was produced *in situ* from triethylamine and formic acid. Using this structural motif, the reaction conditions were optimized and other aryl substituents were explored as leaving groups (Table 4.19). Initial optimizations identified dimethylsulfoxide (DMSO) as the preferred solvent; similarly, heating the reaction to 120 °C gave more product. Using this temperature and solvent, catalyst loadings were reduced from 0.5 mol% to 5×10^{-9} mol% with little to no reduction in yield as identified by gas chromatography. Similarly, complexes **169** and **170** bearing heterocyclic nitrogen ligands were demonstrated to be competent catalysts. The Br, OTf, and I leaving groups gave similar yields, but a Cl leaving group required higher catalyst loadings.

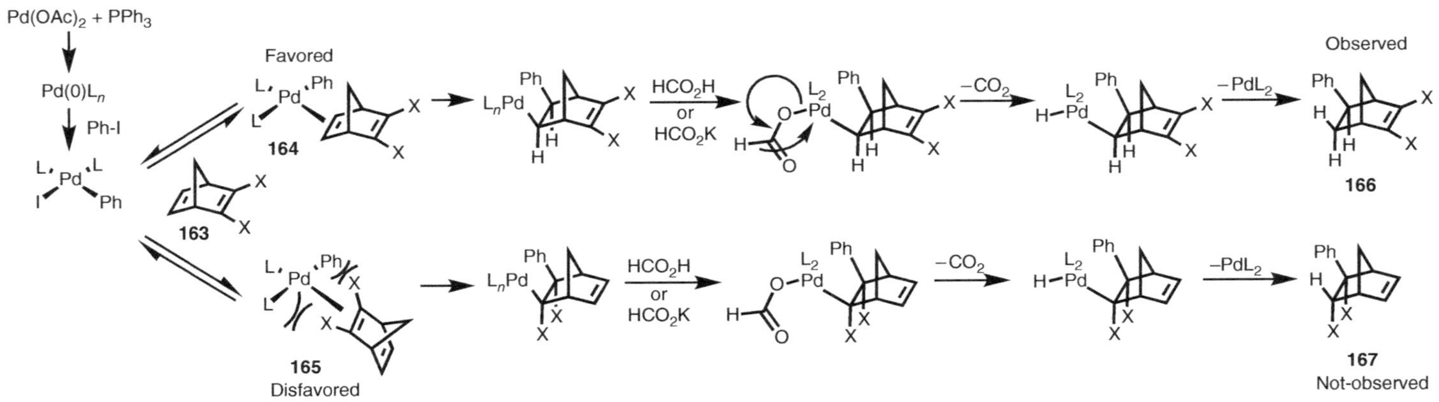

Figure 4.22 Proposed mechanism for formate-assisted Pd^{II} hydroarylation of substituted norbornenes.

Table 4.19 Substrate scope of norbornene hydroarylation with dimeric Pd^{II} complexes [91].

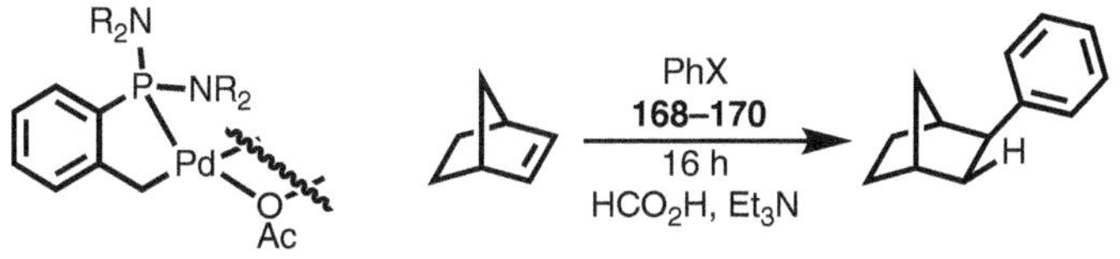

168: NR_2 = NMe_2
169: NR_2 = pyrrolidine$_6$
170: NR_2 = piperidine

X	Catalyst	Catalyst loading (mol%)	Solvent	Temperature (°C)	GC yield (%)
I	**168**	0.5	Toluene	25	0
		0.5	Toluene	110	80
		0.5	CH_3CN	80	12
		0.5	DMF	120	100
		0.5	DMSO	120	100
		5×10^{-2}	DMSO	120	100
		5×10^{-9}	DMSO	120	100
	169	5×10^{-6}	DMSO	120	100
	170	5×10^{-6}	DMSO	120	100
Cl	**168**	0.5	DMSO	120	100
		5×10^{-3}	DMSO	120	5
Br	**168**	5×10^{-4}	DMSO	120	100
OTf	**168**	5×10^{-4}	DMSO	120	100

Chiral diphosphines [92], oxazolines [93, 94], and (β-*N*-sulfonylaminoalkyl) phosphines [95] have been employed as ligands (see Figure 4.23) for enantioselective hydroarylation and the results are summarized in Table 4.20. All reactions employed DMSO as the solvent at elevated temperatures using formic acid and triethylamine. Notably, the diphosphine-based ligands **171** and **172** as well as oxazoline-containing ligands (**174–181**) provided lower yields than the (β-*N*-sulfonylaminoalkyl)phosphine ligands **182–184**. Similarly, compounds **182–184** gave higher enantioselectivities than the nitrogen- or phosphorous-based donors. A related ligand effect study on the hydroarylation of 7-oxybenzonorbornene with PhI or PhOTf was reported by Fiaud and coworker [96]. While PhI generally gave higher yields, low enantioselectivity was observed even with chiral ligands such as **171** and **173**. With PhOTf, however, significantly higher enantioselectivities for the *exo*-norbornane product were achieved. Ligand **171** exhibited the highest performance in terms of both yields and enantioselectivities. Overall, these reports demonstrate that chiral, bidentate ligands may be used to induce enantioselectivity.

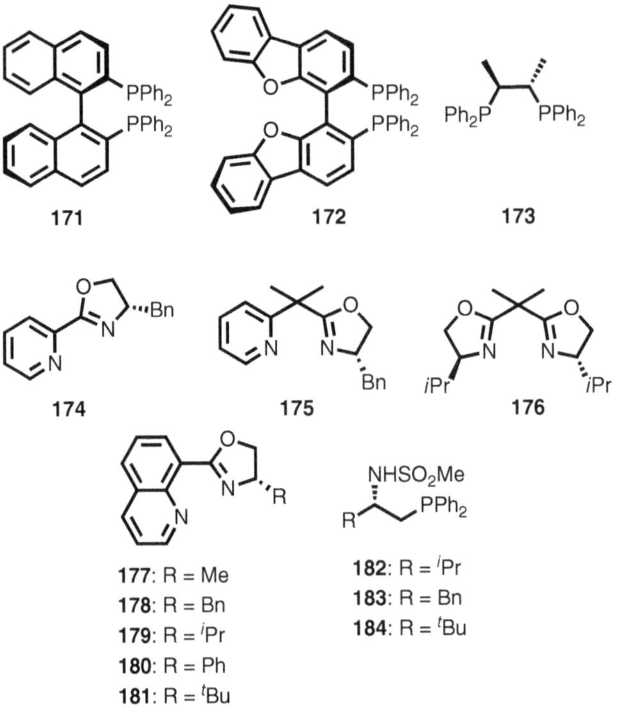

Figure 4.23 Chiral ligands for Pd hydroarylation.

4.6.2 Oxidatively Coupled Pd^{II}-Catalyzed Hydroarylation with Aryltin and Arylboronic Ester Substrates

While Pd^{II} species have been employed as hydroarylation catalysts with norbornene substrates, several reports describe hydroarylations using stannyl- or boryl-substituted arenes, as typically employed in cross-coupling reactions (such as the Stille or Suzuki). A recent report by the Sigman group [97] demonstrates that hydroarylation is achieved in alcohol-containing solvents. In the proposed mechanism (Figure 4.24), a [$L_nPd^{II}X_2$] species mediates oxidation of *iso*-propanol to generate a Pd^{II} hydride along with acetone and an equivalent of HX. Olefin insertion occurs in a Markovnikov fashion to produce a Pd–alkyl intermediate that undergoes transmetallation with arylboronic esters (or tributylphenyl tin). Reductive elimination affords the final hydroarylation product along with a Pd^0 species. Exposure to O_2 and "HX" regenerates the active Pd^{II} catalyst along with an equivalent of H_2O_2.

To further develop this system, Sigman and coworkers [97] employed Pd^{II} complexes with a variety of ligands in the hydroarylation of styrene derivatives with arylboronic esters. The results of the optimization experiments are summarized in Table 4.21. Both the Markovnikov hydroarylation product (**185**) and the classical Heck product (**186**) were observed. Several ligands were tested, containing either nitrogen- (**187**) or carbene-based (**188**, **189**) donors. Additionally, various conditions were employed for the *in situ* generation of the active transmetallation agent (i.e., $ArB(OR)_3^-$). Two different bases were used in this study: **187** or a mixture of **187** and K^{t}OBu. Finally, various arylboronic

Table 4.20 Hydroarylation of norbornene catalyzed by Pd^{II}-bearing chiral ligands.

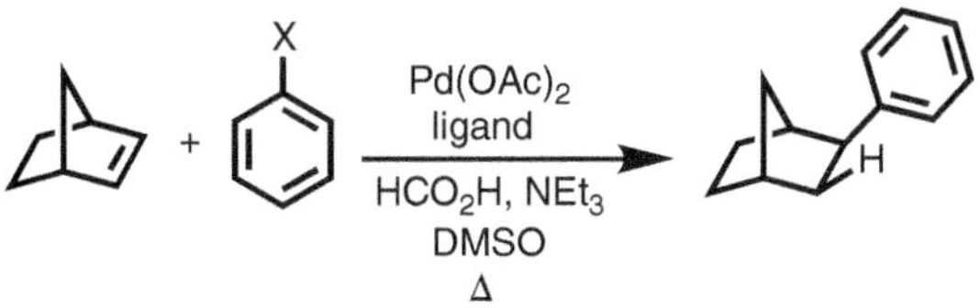

Conditions (References)	X	Ligand	Time (h)	Isolated yield (%)	ee (%)
A [92]	OTf	**171**	12	9	8
		172	36	18	26
		173	14	58	24
		182	2	37	81
	I	**171**	18	23	15
B [93]	I	**174**	46	44	29
		175	14	47	0
		176	24	51	0
B [94]	I	**177**	14	42	51
		178	58	60	73
		179	43	54	74
		180	14	57	28
		181	14	47	18
C [95]	OTf	**182**	20	74	71
		183	20	89	60
		184	20	75	62

Conditions: **A** = 65 °C, 2 mol% Pd, 4 mol% ligand, excess PhX; **B** = 25 °C, 5 mol% Pd, 10 mol% ligand, excess PhX; **C** = 65 °C, 1.2 mol% Pd, 2.4 mol% ligand, excess norbornene, Ar atmosphere.

esters were employed (**190–193**). With these conditions, it was determined that an equimolar mixture of **187** and K^{t}OBu achieved the highest activity and selectivity for **185**. Moreover, the ethylene glycol-derived (**191**) and the pinacol-derived (**192**) phenylboronic esters exhibited the highest activity, generally with quantitative conversion.

The Sigman group has also demonstrated that other arenes can be delivered to olefins using the corresponding arylboronic ester [97] or tributylaryl tin [98] reagents. Added MnO_2 was added to reactions using the aryltin reagents, to decompose any H_2O_2 generated during the reaction generated during the reaction. A summary of the substrate scope is given below in Table 4.22.

A deuterium-labeling study with Me_2CHOD solvent provided information regarding the mechanism of this reaction (Scheme 4.8) [98]. Under these conditions, no deuterium was incorporated into the products, suggesting that acidic protons are not involved in the mechanism. A similar experiment using Me_2CDOH as a solvent, however, demonstrated that a majority (92%) of the product contained deuterium, either at the methyl or methine position, implying a role for a Pd^{II}–hydride intermediate. The existence of multiple isotopomers

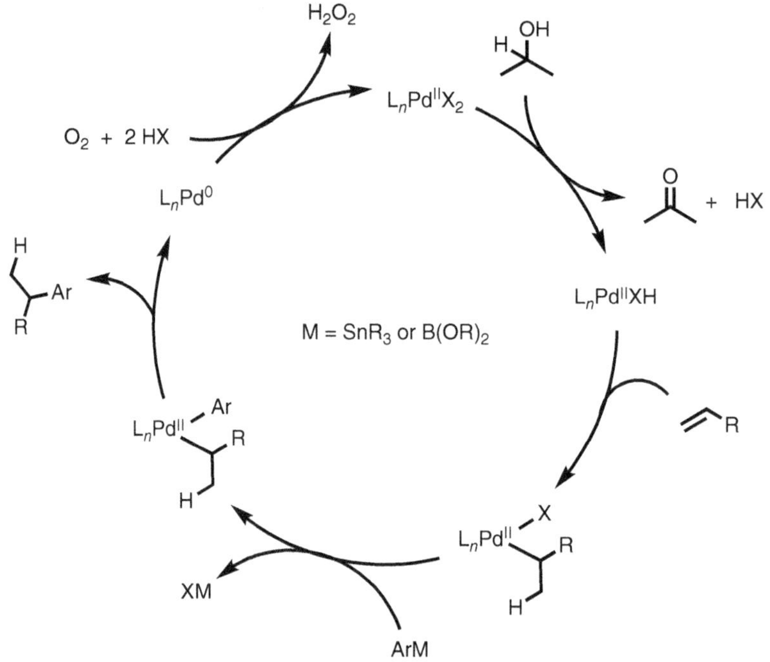

Figure 4.24 Mechanism for aerobic alcohol oxidation coupled to hydroarylation.

suggests operation of a reversible β-hydride elimination step, which allows deuterium scrambling into the olefin.

In a subsequent report by Sigman and coworkers [99], chiral ligands were employed for enatioselective hydroarylations using either arylboronic ester or aryltin reagents. With bulky bis(oxazoline)-based ligands, modest enantioselectivities were achieved for a series of substrates.

Recent reports by the Sigman group [100, 101] describe oxidatively coupled hydroarylations with alkenyl alcohols (Eqs (4.15) and (4.16)). While the oxidation of added ^{i}PrOH has previously been employed as a hydrogen source, the oxidation of an olefin-containing alcohol has also been explored toward oxidatively coupled hydroarylations. With alkenyl alcohols, an initial oxidation of the alcoholic group generates a carbonyl and a Pd^{II}–hydride. Subsequent coordination and insertion of the olefin can occur. Arylation (Eq. (4.15)) or transmetallation (Eq. (4.16)) followed by reductive elimination generates the desired hydroarylation product. The reaction was made enantioselective with the use of the chiral pyridyl-oxazoline ligand **194**. The hydroarylation of nitroalkenes with $PdCl_2$ using aryltin or tetraarylborate as arene sources has also been demonstrated by Uemura and coworker [102].

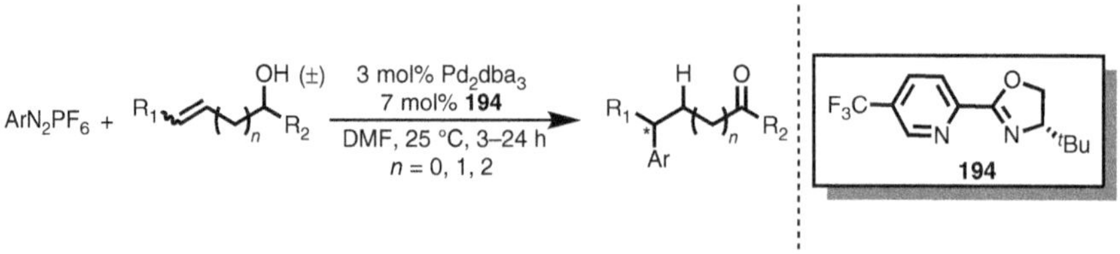

(4.15)

Table 4.21 Optimization of Pd hydroarylation using arylboronic esters [97].

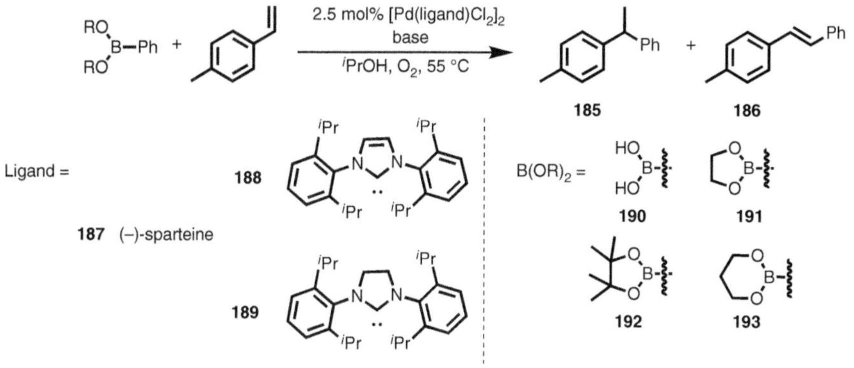

Ligand	PhB(OR)$_2$	Base (mol%)	Time (h)	Conversion (%)	185 : 186
187	**190**	**187** (40)	24	35	1.4 : 1
188	**190**	**187** (20)	24	63	3.2 : 1
	190	**187** (20) + K^{t}OBu (5)	24	18	6.4 : 1
	191	**187** (7.5) + K^{t}OBu (7.5)	4	>99	27 : 1
	192	**187** (5) + K^{t}OBu (5)	24	99	2.7 : 1
	192	**187** (7.5) + K^{t}OBu (7.5)	24	>99	17 : 1
189	**191**	**187** (7.5) + K^{t}OBu (7.5)	8	>99	>30 : 1
	193	**187** (7.5) + K^{t}OBu (7.5)	24	46	>30 : 1

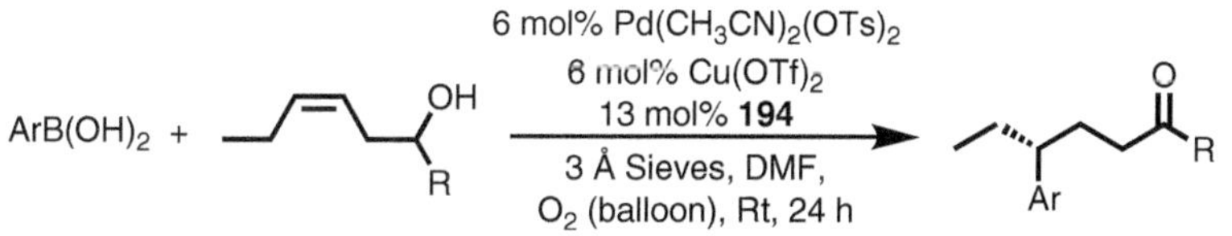

(4.16)

4.6.3 Summary of Pd^{II}-Catalyzed Formal Hydroarylation Reactions

It has been demonstrated that Pd^{II} complexes are catalytically active for hydroarylation [85–102]. In contrast to other commonly employed metals for hydroarylation, however, aryl C—H activation has not yet been utilized with Pd^{II}. Instead, a leaving group (I, Br, OTf) or transmetallating substituent such as SnR_3 or $B(OR)_2$ is required; additionally, an external hydrogen source is necessary. Using arenes-bearing leaving groups, formate has been employed as a hydrogen source (either *in situ* or as an added salt), to liberate CO_2 as a byproduct [85–96]. For arenes substituted with cross-coupling functionalities, the oxidation of an alcohol to a carbonyl group generates a Pd–hydride intermediate that can reductively eliminate with an alkyl fragment to generate the desired hydroarylation product [97–101].

While Pd^{II} complexes have shown promise as hydroarylation catalysts, further studies are required to explore the possibility of designing species that can

Table 4.22 Substrate scope for reductively coupled hydroarylation.

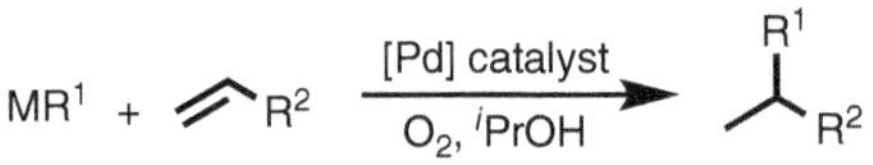

Conditions (Reference)	M	R^1	R^2	Product	Isolated yield (%)
A [97]	**191**	Ph	Ph	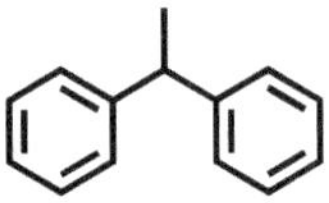	91
		p-^{i}PrPh	Ph	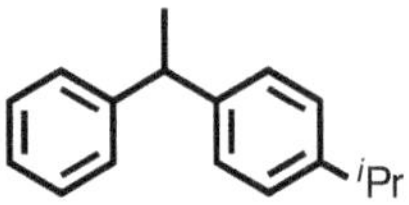	65
		p-(CO_2Me)Ph	Ph	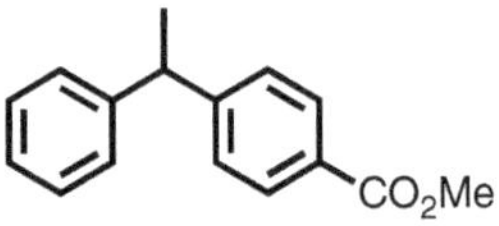	63
		Ph	*p*-MePh	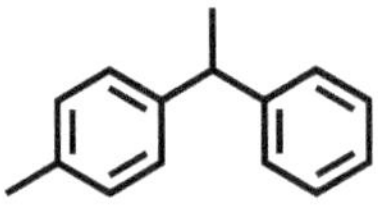	81
		p-FPh	*p*-MePh	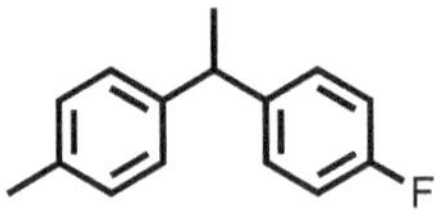	78
		p-OMePh	*p*-MePh	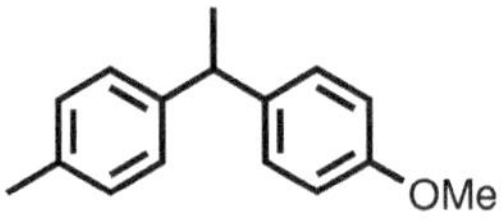	58
		Ph	*p*-NHBocPh	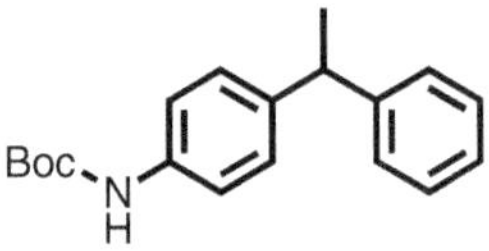	90

Table 4.22 (Continued)

Conditions (Reference)	M	R^1	R^2	Product	Isolated yield (%)
B [98]	$SnBu_3$	Ph	*p*-MePh	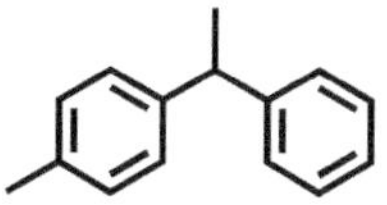	76
		p-OMePh	*p*-MePh	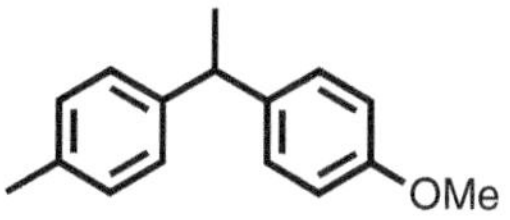	70
		p-CF_3Ph	*p*-MePh	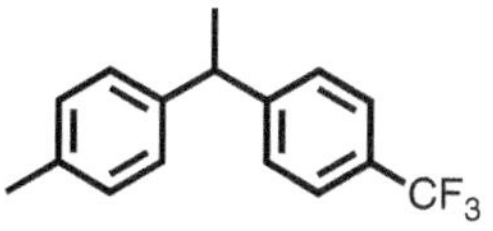	67
		p-OMePh	*p*-ClPh	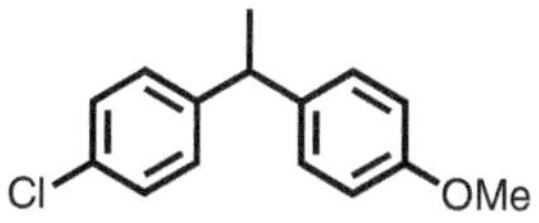	65

Conditions: **A** = 1 equiv. alkene, 3 equiv. of MR^1, 0.75 mol% [Pd(**189**)Cl_2]$_2$, 6 mol% **187**, 6 mol% K^{t}OBu, 55 °C, 24 h, ^{i}PrOH solvent, O_2 atmosphere; **B** = 1 equiv. alkene, 1.5 equiv. of MR^1, 2.5 mol% Pd(**187**)Cl_2, 40 mol% **187**, 75 mol% MnO_2, 60 °C, 18 h, ^{i}PrOH solvent, 25 psi O_2.

Scheme 4.8 Deuterium-labeling study in oxidatively coupled hydroarylation.

access hydroarylation products via a C—H activation pathway. Such complexes would allow for hydroarylation to occur without the need to prefunctionalize the arene. Moreover, a C—H activation pathway would eliminate the need for stoichiometric amounts of a hydrogen source (such as formate or ^{i}PrOH). Finally, such C—H activations are more difficult to achieve than C—X activations; as such, accessing this more difficult type of transformation with Pd and applying it toward productive hydroarylation is of interest.

4.7 Conclusions

The alkylation of arenes is a massive industrial process that generates millions of tons of alkyl arenes annually. Moreover, these products are typically used as polymer precursors (e.g., for synthesis of polystyrene). While industrial processes typically employ Lewis acids (such as zeolites or $AlCl_3$ with a Brønsted acid) to couple ethylene and benzene via a Friedel–Crafts mechanism, this chemistry can lead to over-alkylation. Additionally, for higher olefins, such catalysis affords the less-desired Markovnikov products and often leads to competitive olefin isomerization. As a result, there is current interest in the development of homogeneous transition metal catalysts for selective, direct hydroarylation of α-olefins with arenes.

Given the history of research in this area, it seems that d^8 precatalysts are promising candidates for future development. Many desirable features exist in systems employing d^8 complexes as catalysts; for example, hydroarylations of simple unfunctionalized arenes such as benzene or toluene have been demonstrated with Pt^{II} complexes [8–27]. Regio- and enantio-selectivity has been achieved in olefin hydroarylations with Rh^{I} [32, 37–66] and Ir^{I} [67–80] based catalysts, and have been successfully applied toward complex total syntheses [47, 53, 57, 58]. A wide range of substrates have been identified with Rh^{I} catalysts, elucidating the compatibility of the catalyst with various substrate functionalities. Additionally, several strategies for the hydroarylation of olefins employing Pd^{II} complexes with added external hydrogen sources have been explored [84–102].

While hydroarylations with d^8 metal centers offer potential benefits when compared to zeolites, several important limitations remain. Overall, d^8 catalysts appear to generally suffer from a lack of activity and catalyst stability issues. Several steps that may lead to catalyst decomposition have been identified, including β-hydride elimination and non-productive reductive elimination. For late transition metals, reduction to elemental states can often be an issue. To address catalytic stability, further research into combating possible decomposition routes is necessary. Currently, only expensive second- and third-row d^8 metals have been well studied. While a few examples of Ni^0 hydroarylations exist [81, 82], Ni^{II}-catalyzed hydroarylations have yet to be investigated. Computational studies employing DFT have suggested that Ni^{II} complexes could be competent toward the hydroarylation of ethylene and benzene [83]. As such, further development of first row systems is required to determine whether such systems are catalytically active. It is possible that the application of the successes from second- and third-row metal systems toward first row analogs could yield new, cheap catalysts for hydroarylation.

References

1 Welch, V.A., Fallon, K.J., and Gelbke, H.-P. (2005) Ethylbenzene, in *Ullmann's Encyclopedia of Industrial Chemistry* vol. 13, John Wiley and Sons, Inc., pp. 451–464.

2 James, D.H. and Castor, W.M. (2011) Styrene, in *Ullmann's Encyclopedia of Industrial Chemistry* vol. 34, John Wiley and Sons, Inc., pp. 529–544.
3 Maul, J., Frushour, B.G., Kontoff, J.R., Eichenauer, H., Ott, K.-H., and Schade, C. (2007) Polystyrene and styrene copolymers, in *Ullmann's Encyclopedia of Industrial Chemistry* vol. 29, John Wiley and Sons, Inc., pp. 475–522.
4 Perego, C. and Ingallina, P. (2002) Recent advances in the industrial alkylation of aromatics: new catalysts and new processes. *Catal. Today*, **73**, 3–22.
5 Hartwig, J.F. (ed.) (2010) *Organotransition Metal Chemistry: From Bonding to Catalysis*, University Science Books, USA.
6 Lersch, M. and Tilset, M. (2005) Mechanistic aspects of C—H activation by Pt complexes. *Chem. Rev.*, **105** (6), 2471–2526.
7 Chianese, A.R., Lee, S.J., and Gagné, M.R. (2007) Electrophilic activation of alkenes by platinum(II): so much more than a slow version of palladium(II). *Angew. Chem. Int. Ed.*, **46** (22), 4042–4059.
8 Karshtedt, D., Bell, A.T., and Tilley, T.D. (2004) Pt—Ag catalyst system for hydroarylations with unactivated arenes and olefins. *Organometallics*, **23** (18), 4169–4171.
9 Karshtedt, D., McBee, J.L., Bell, A.T., and Tilley, T.D. (2006) Stoichiometric and catalytic arene activations by Platinum complexes containing bidentate monoanionic nitrogen-based ligands. *Organometallics*, **25** (7), 1801–1811.
10 Suslick, B.A., Liberman-Martin, A.L., Wambach, T.C., and Tilley, T.D. (2017) Olefin hydroarylation catalyzedby (pyridyl-indolate)Pt(II) complexes: catalytic efficiencies and mechanistic aspects. *ACS Catal.*, **2017** (7), 4313–4322.
11 Luedtke, A.T. and Goldberg, K.I. (2007) Reductive elimination of ethane from five-coordinated platinum(IV) alkyl complexes. *Inorg. Chem.*, **46** (21), 8496–8498.
12 Luedtke, A.T. and Goldberg, K.I. (2008) Intermolecular hydroarylation of unactivated olefins catalyzed by homogeneous platinum complexes. *Angew. Chem. Int. Ed.*, **47** (40), 7694–7696.
13 Clement, M.L., Grice, K.A., Luedtke, A.T., Kaminsky, W., and Goldberg, K.I. (2014) Platinum(II) olefin hydroarylation catalysts: tuning selectivity for the anti-Markovnikov product. *Chem. Eur. J.*, **20** (52), 17287–17291.
14 Johansson, L., Ryan, O.B., Rømming, C., and Tilset, M. (2001) Unexpected selectivities in C—H activations of toluene and *p*-xylene at cationic platinum(II) diimine complexes. New mechanistic insight into product-determining factors. *J. Am. Chem. Soc.*, **123** (27), 6579–6590.
15 Johansson, L., Tilset, M., Labinger, J.A., and Bercaw, J.E. (2000) Mechanistic investigation of benzene C—H activation at a cationic platinum(II) center: direct observation of a platinum (II) benzene adduct. *J. Am. Chem. Soc.*, **122** (44), 10846–10855.
16 Bowring, M.A., Bergman, R.G., and Tilley, T.D. (2013) Isolation of a dicationic platinum complex with two accessible coordination sites. *Organometallics*, **32** (19), 5266–5268.

17 McKeown, B.A., Foley, N.A., Lee, J.P., and Gunnoe, T.B. (2008) Hydroarylation of unactivated olefins catalyzed by platinum(II) complexes. *Organometallics*, **27** (16), 4031–4033.
18 McKeown, B.A., Gonzalez, H.E., Friedfeld, M.R., Gunnoe, T.B., Cundari, T.R., and Sabat, M. (2011) Mechanistic studies of ethylene hydrophenylation catalyzed by bipyridyl Pt(II) complexes. *J. Am. Chem. Soc.*, **133** (47), 19131–19152.
19 McKeown, B.A., Prince, B.M., Ramiro, Z., Gunnoe, T.B., and Cundari, T.R. (2014) Pt^{II}-catalyzed hydrophenylation of α-olefins: variation of linear/branched products as a function of ligand donor ability. *ACS Catal.*, **4** (5), 1607–1615.
20 McKeown, B.A., Gonzalez, H.E., Friedfeld, M.R., Brosnahan, A.M., Gunnoe, T.B., Cundari, T.R., and Sabat, M. (2013) Platinum(II)-catalyzed ethylene hydrophenylation: switching selectivity between alkyl- and vinylbenzene production. *Organometallics*, **32** (9), 2857–2865.
21 McKeown, B.A., Gonzalez, H.E., Gunnoe, T.B., Cundari, T.R., and Sabat, M. (2013) Pt^{II}-catalyzed ethylene hydrophenylation: influence of dipyridyl chelate ring size on catalyst activity and longevity. *ACS Catal.*, **3** (6), 1165–1171.
22 McKeown, B.A., Gonzalez, H.E., Michaelos, T., Gunnoe, T.B., Cundari, T.R., Crabtree, R.H., and Sabat, M. (2013) Control of olefin hydroarylation catalysis via a sterically and electronically flexible platinum(II) catalyst scaffold. *Organometallics*, **32** (14), 3903–3913.
23 Bowring, M.A., Bergman, R.G., and Tilley, T.D. (2011) Disambiguation of metal and Brønstead acid catalyzed pathways for hydroarylation with platinum(II) catalysts. *Organometallics*, **30** (6), 1295–1298.
24 Szuromi, E. and Sharp, P.R. (2006) Alkene coupling through triflic acid elimination on platinum(II). *Organometallics*, **25** (3), 558–559.
25 McBee, J.L., Bell, A.T., and Tilley, T.D. (2008) Mechanistic studies of the hydroamination of norbornene with electrophilic platinum complexes: the role of proton transfer. *J. Am. Chem. Soc.*, **130** (49), 16562–16571.
26 Bowring, M.A., Bergman, R.G., and Tilley, T.D. (2013) Pt-catalyzed C—C activation induced by C—H activation. *J. Am. Chem. Soc.*, **135** (35), 13121–13128.
27 Brosnahan, A.M., Talbot, A., McKeown, B.A., Kalman, S.E., Gunnoe, T.B., Ess, D.H., and Sabat, M. (2015) Phosphine and *N*-heterocyclic carbene ligands on Pt(II) shift selectivity from ethylene hydrophenylation toward benzene vinylation. *J. Organomet. Chem.*, **793**, 248–255.
28 Cucciolito, M.E., D'Amora, A., Tuzi, A., and Vitagliano, A. (2007) Catalytic hydroarylation of olefins promoted by dicationic platinum(II) and palladium(II) complexes. The interplay of C—C bond formation and M—C bond cleavage. *Organometallics*, **26** (21), 5216–5223.
29 Partridge, M.G., Field, L.D., and Messerle, B.A. (1996) Activation of C—H bonds in thiophenes by $Rh(PMe_3)_2(CO)Cl$: a low-temperature NMR study. *Organometallics*, **15** (2), 872–877.

30 Kawamoto, T. (2000) Activation of C—H and C—Cl bonds in the reactions of 2-substituted benzothiazolines with $[RhCl(PPh_3)_3]$. *Inorg. Chim. Acta*, **300–302**, 512–516.

31 Ghosh, C.K. and Graham, W.A. (1989) A rhodium complex that combines benzene activation with ethylene insertion. Subsequent carbonylation and ketone formation. *J. Am. Chem. Soc.*, **111** (1), 375–376.

32 Jun, C.-H., Hong, J.-B., Kim, Y.-H., and Chung, K.-Y. (2000) The catalytic activation of aromatic imines by Wilkinson's complex: the domino reaction of hydroacylation and *ortho*-alkylation. *Angew. Chem. Int. Ed.*, **39** (19), 3440–3442.

33 Ritleng, V., Sirlin, C., and Pfeffer, M. (2002) Ru-, Rh-, and Pd-catalyzed C—C bond formation involving C—H activation and addition on unsaturated substrates: reactions and mechanistic aspects. *Chem. Rev.*, **102** (5), 1731–1769.

34 Colby, D.A., Bergman, R.G., and Ellman, J.A. (2010) Rhodium-catalyzed C—C bond formation via heteroatom-directed C—H bond activation. *Chem. Rev.*, **110** (2), 624–655.

35 Colby, D.A., Tsai, A.S., Bergman, R.G., and Ellman, J.A. (2012) Rhodium catalyzed chelation-assisted C—H bond functionalization reactions. *Acc. Chem. Res.*, **45** (6), 814–825.

36 Klei, S.R., Tan, K.L., Golden, J.T., Yung, C.M., Thalji, R.K., Ahrendt, K.A., Ellman, J.A., Tilley, T.D., and Bergman, R.G. (2004) Carbon–hydrogen bond activation by iridium and rhodium complexes: catalytic hydrogen/deuterium exchange and carbon–carbon bond-forming reactions, in *ACS Symposium Series 885, Activation and Functionalization of C—H Bonds* (eds K.I. Goldberg and A.S. Goldman), American Chemical Society, pp. 46–55.

37 Vaughan, B.A., Webster-Gardiner, M.S., Cundari, T.R., and Gunnoe, T.B. (2015) A rhodium catalyst for single-step styrene production from benzene and ethylene. *Science*, **348** (6233), 421–424.

38 Webster-Gardiner, M.S., Fu, R., Fortman, G.C., Nielsen, R.J., Gunnoe, T.B., and Goddard, W.A. III, (2015) Arene C—H activation using Rh(I) catalysts supported by bidentate nitrogen chelates. *Catal. Sci. Technol.*, **5**, 96–100.

39 Webster-Gardiner, M.S., Piszel, P.E., Fu, R., McKeown, B.A., Nielsen, R.J., Goddard, W.A. III, and Gunnoe, T.B. (2017) Electrophilic Rh^I catalysts for arene H/D exchange in acidic media: evidence for an electrophilic aromatic substitution mechanism. *J. Mol. Catal. A: Chem.*, **426**, 381–388.

40 Vaughan, A.B., Khani, S.K., Gary, J.B., Kammert, J.D., Webster-Gardiner, M.S., McKeown, B.A., Davis, R.J., Cundari, T.R., and Gunnoe, T.B. (2017) Mechanistic studies of single-step styrene production using a rhodium(I) catalyst. *J. Am. Chem. Soc.*, **139** (4), 1485–1498.

41 Jun, C.-H., Lee, H., and Hong, J.-B. (1997) Chelationassisted intermolecular hydroacylation: direct synthesis of ketone from aldehyde and 1-alkene. *J. Org. Chem.*, **62** (5), 1200–1201.

42 Jun, C.-H. and Hong, J.-B. (1999) Catalytic transformation of aldimine to ketimine by Wilkinson's complex through transamination. *Org. Lett.*, **1** (6), 887–889.

43 Lim, S.-G., Ahn, J.-A., and Jun, C.-H. (2004) Ortho alkylation of aromatic ketamine with functionalized alkene by Rh(I) catalyst. *Org. Lett.*, **6** (25), 4687–4690.
44 Vo-Thanh, G., Lahrache, H., Loupy, A., Kim, I.-J., Chang, D.-H., and Jun, C.-H. (2004) Rh(I)-catalyzed solvent-free *ortho*-alkylation of aromatic imines under microwave irradiation. *Tetrahedron*, **60**, 5539–5543.
45 Yoon, J.H., Park, Y.J., Lee, J.H., Yoo, J., and Jun, C.-H. (2005) Recyclable self-assembly-supported catalytic system for orthoalkylation. *Org. Lett.*, **7** (14), 2889–2892.
46 Thalji, R.K., Ahrendt, K.A., Bergman, R.G., and Ellman, J.A. (2001) Annulation of aromatic imines via directed C—H activation with Wilkinson's catalyst. *J. Am. Chem. Soc.*, **123** (39), 9692–9693.
47 Thalji, R.K., Ahrendt, K.A., Bergman, R.G., and Ellman, J.A. (2005) Annulation of aromatic imines via directed C—H bond activation. *J. Org. Chem.*, **70** (17), 6775–6781.
48 Lim, Y.-G., Han, J.-S., Yang, S.-S., and Chun, J.H. (2001) Rhodium-catalyzed *ortho*-alkylation of aromatic aldimines and ketimines via C—H bond activation. *Tetrahedron Lett.*, **42**, 4853–4856.
49 Lim, Y.-G., Han, J.-S., Koo, B.T., and Kang, J.-B. (2004) Regioselective alkylation of aromatic aldimines and ketimines via C—H bond activation by a rhodium catalyst. *J. Mol. Catal. A: Chem.*, **209**, 41–49.
50 Lim, Y.-G. and Koo, B.T. (2005) Rhodium-catalyzed alkylation of terephthaldialdimine with terminal alkenes via C—H bond activation. *Tetrahedron Lett.*, **46**, 7997–8001.
51 Thalji, R.K., Ellman, J.A., and Bergman, R.G. (2004) Highly efficient and enantioselective cyclization of aromatic imines via directed C—H bond activation. *J. Am. Chem. Soc.*, **126** (23), 7192–7193.
52 Harada, H., Thalji, R.K., Bergman, R.G., and Ellman, J.A. (2008) Enantioselective intramolecular hydroarylation of alkenes via directed C—H bond activation. *J. Org. Chem.*, **73** (17), 6772–6779.
53 Watzke, A., Wilson, R.M., O'Malley, S.J., Bergman, R.G., and Ellman, J.A. (2007) Asymmetric intramolecular alkylation of chiral aromatic imines via catalytic C—H bond activation. *Synlett*, **15**, 2383–2389.
54 Kakiuchi, F., Le Gendre, P., Yamada, A., Ohtaki, H., and Murai, S. (2000) Atropselective alkylation of biaryl compounds by means of transition metal-catalyzed C—H/olefin coupling. *Tetrahedron: Asymmetry*, **11** (13), 2647–2651.
55 Shibata, K. and Chatani, N. (2014) Rhodium-catalyzed alkylation of C—H bond in aromatic amides with α,β-unsaturated esters. *Org. Lett.*, **16** (19), 5148–5151.
56 Shibata, K., Yamaguchi, T., and Chatani, N. (2015) Rhodium-catalyzed alkylation of C—H bonds in aromatic amides with styrenes via bidentate–chelation assistance. *Org. Lett.*, **17** (14), 3584–3587.
57 O'Malley, S.J., Tan, K.L., Watzke, A., Bergman, R.G., and Ellman, J.A. (2005) Total synthesis of (+)-lithospermic acid by asymmetric intramolecular alkylation via catalytic C—H bond activation. *J. Am. Chem. Soc.*, **127** (39), 13496–13497.

58 Ahrendt, K.A., Bergman, R.G., and Ellman, J.A. (2003) Synthesis of a tricyclic mescaline analogue by catalytic C—H bond activation. *Org. Lett.*, **5** (8), 1301–1303.
59 Wentzel, M.T., Reddy, V.J., Hyster, T.K., and Douglas, C.J. (2009) Chemoselectivity in catalytic C—C and C—H bond activation: controlling intermolecular carboacylation and hydroarylation of alkenes. *Angew. Chem. Int. Ed.*, **48** (33), 6121–6123.
60 Brunet, J.-J., Neilbecker, D., and Philippot, K. (1992) Unexpected catalytic and stereoselective hydroarylation of norbornene during the attempted rhodium-catalysed hydroamination of norbornene with aniline or diphenylamine. *J. Chem. Soc. Chem. Commun.*, (17), 1215–1216.
61 Brunet, J.-J., Commenges, G., Neilbecker, D., and Philippot, K. (1994) Rhodium-catalysed hydroamination-hydroarylation of norbornene with aniline, toluidines, or diphenylamine. *J. Organomet. Chem.*, **469** (2), 221–228.
62 Lenges, C.P. and Brookhart, M. (1999) Addition of olefins to aromatic ketones catalyzed by Rh(I) olefin complexes. *J. Am. Chem. Soc.*, **121** (28), 6616–6623.
63 Sun, Z.-M., Zhang, J., Manan, R.S., and Zhao, P. (2010) Rh(I)-catalyzed olefin hydroarylation with electron-deficient perfluoroarenes. *J. Am. Chem. Soc.*, **132** (20), 6935–6937.
64 Bercaw, J.E., Hazari, N., and Labinger, J.A. (2009) Activation of a C—H bond in indene by $[(COD)Rh(\mu_2\text{-}OH)]_2$. *Organometallics*, **28** (18), 5489–5492.
65 Kloek, S.M., Heinekey, D.M., and Goldberg, K.I. (2007) C—H activation by rhodium(I) hydroxide and phenoxide complexes. *Angew. Chem. Int. Ed.*, **46** (25), 4736–4738.
66 Tsui, G.C., Menard, F., and Lautens, M. (2010) Regioselective rhodium(I)-catalyzed hydroarylation of protected allylic amines with arylboronic acids. *Org. Lett.*, **12** (11), 2456–2459.
67 Matsumoto, T., Taube, D.J., Periana, R.A., Taube, H., and Yoshida, H. (2000) Anti-Markovnikov olefin arylation catalyzed by an iridium complex. *J. Am. Chem. Soc.*, **122** (30), 7414–7415.
68 Periana, R.A., Liu, X.Y., and Bhalla, G. (2002) Novel bis-acac-*O,O*-Ir(III) catalyst for anti-Markovnikov, hydroarylation of olefins operates by arene CH activation. *Chem. Commun.*, (24), 3000–3001.
69 Bhalla, G., Oxgaard, J., Goddard, W.A. III, and Periana, R.A. (2005) Anti-Markovnikov hydroarylation of unreactive olefins catalyzed by a bis-tropolonato iridium(III) organometallic complex. *Organometallics*, **24** (13), 3229–3232.
70 Oxgaard, J., Muller, R. P., Goddard, W. A. III, Periana, R. A. (2004) Mechanism of homogeneous Ir(III) catalyzed regioselective arylation of olefins, *J. Am. Chem. Soc.*, **126**(1), 352–363.
71 Bhalla, G., Bischof, S.M., Ganesh, S.K., Liu, X.Y., Jones, C.J., Borzenko, A., Tenn, W.J. III, Ess, D.H., Hashiguchi, B.G., Lokare, K.S., Leung, C.H., Oxgaard, J., Goddard, W.A. III, and Periana, R.A. (2011) Mechanism of efficient anti-Markovnikov olefin hydroarylation catalyzed by homogeneous Ir(III) complexes. *Green Chem.*, **13**, 69–81.

72 Aufdenblatten, R., Diezi, S., and Togni, A. (2000) Iridium(I)-catalyzed asymmetric intermolecular hydroarylation of norbornene with benzamide. *Montash. Chem.*, **131**, 1345–1350.
73 Dorta, R. and Togni, A. (2003) Addition of the *ortho*-C—H bonds of phenol across an olefin catalyzed by a chiral iridium(I) diphosphine complex. *Chem. Commun.*, (6), 760–761.
74 Shirai, T. and Yamamoto, Y. (2015) Cationic iridium/S-Me-BIPAM-catalyzed direct asymmetric intermolecular hydroarylation of bicycloalkenes. *Angew. Chem. Int. Ed.*, **54** (34), 9894–9897.
75 Tsuchikama, K., Kasagawa, M., Hashimoto, Y.-K., Endo, K., and Shibata, T. (2008) Cationic iridium-BINAP complex-catalyzed addition of aryl ketones to alkynes and alkenes via directed C—H bond cleavage. *J. Organomet. Chem.*, **693** (26), 3393–3942.
76 Crisenza, G.E.M., McCreanor, N.G., and Bower, J.F. (2014) Branch-selective, iridium-catalyzed hydroarylation of monosubstituted alkenes via a cooperative destabilization strategy. *J. Am. Chem. Soc.*, **136** (29), 10258–10261.
77 Crisenza, G.E.M., Sokolva, O.O., and Bower, J.F. (2015) Branch-selective alkene hydroarylation by cooperative destabilization: Iridium-catalyzed *ortho*-alkylation of acetanilides. *Angew. Chem. Int. Ed.*, **54** (49), 14866–14870.
78 Huang, G. and Liu, P. (2016) Mechanism and origins of ligand-controlled linear versus branched selectivity of iridium-catalyzed hydroarylation of alkenes. *ACS Catal.*, **6**, 809–820.
79 Ebe, Y. and Nishimura, T. (2015) Iridium-catalyzed branch-selective hydroarylation of vinyl ethers via C—H activation. *J. Am. Chem. Soc.*, **137** (18), 5899–5902.
80 Hatano, M., Ebe, Y., Nishimura, T., and Yorimitsu, H. (2016) Asymmetric alkylation of *N*-sulfonylbenzamides with vinyl ethers via C—H bond activation catalyzed by hydroxoiridium/chiral diene complexes. *J. Am. Chem. Soc.*, **138** (12), 4010–4013.
81 Bair, J.S., Schramm, Y., Sargeev, A.G., Clot, E., Eisenstein, O., and Hartwig, J.F. (2014) Linear-selective hydroarylation of unactivated terminal and internal olefins with trifluoromethyl-substituted arenes. *J. Am. Chem. Soc.*, **136** (38), 13098–13101.
82 Nakao, Y., Kashihara, N., Kanyiva, K.S., and Hiyama, T. (2008) Nickel-catalyzed alkenylation and alkylation of fluoroarenes via activation of C—H bond over C—F bond. *J. Am. Chem. Soc.*, **130** (48), 16170–16171.
83 Cundari, T.R. and Gonzalez, H.E. (2012) A computational comparison of Ni^{II} and Pt^{II} hydrido-tris(pyrazolyl)borate supported hydroarylation catalysis. *J. Mol. Cat. A.: Chem.*, **353–354**, 1–6.
84 Cabri, W. and Candiani, I. (1995) Recent developments and new perspectives in the Heck reaction. *Acc. Chem. Res.*, **28** (1), 2–7.
85 Cacchi, S. (1990) The palladium-catalyzed hydroarylation and hydrovinylation of carbon–carbon multiple bonds: new perspectives in organic synthesis. *Pure Appl. Chem.*, **62** (4), 713–722.

86 Namyslo, J.C., Storsberg, J., Klinge, J., Gärtner, C., Yao, M.-L., Ocal, N., and Kaufmann, D.E. (2010) The hydroarylation reaction—scope and limitations. *Molecules*, **15**, 3402–3410.

87 Arcadi, A., Marinelli, F., and Cacchi, S. (1986) The reaction of aryl iodides with hindered α,β,γ,δ-dienones in the presence of the $[Pd(OAc)_2(PPh_3)_2]$-trialkylammonium formate reagent. *J. Organomet. Chem.*, **312** (2), c27–c32.

88 Acardi, A., Bernocchi, E., Burini, A., Cacchi, S., Marinelli, F., and Pietroni, B. (1988) The palladium-tributylammonium formate reagent in the stereoselective hydrogenation, and stereo- and regioselective hydroarylation of alkyl 4-hydroxy-2-alkynoates: a route to substituted butenolides. *Tetrahedron*, **44** (2), 481–490.

89 Arcadi, A., Marinelli, F., Bernocchi, E., Cacchi, S., and Ortar, G. (1989) Palladium-catalyzed preparation of *exo*-aryl derivatives of the norbornane skeleton. *J. Organomet. Chem.*, **368** (2), 249–256.

90 Mayo, P. and Tam, W. (2002) Palladium-catalyzed hydrophenylation of bicyclic alkenes. *Tetrahedron*, **58** (47), 9527–9540.

91 Brunel, J.M., Heumann, A., and Buono, G. (2000) Highly efficient phosphapalladacyclic catalysts for the hydroarylation of norbornene. *Angew. Chem. Int. Ed.*, **39** (11), 1946–1949.

92 Namyslo, J.C. and Kaufmann, D.E. (1997) Palladium-catalyzed enantioselective hydrophenylation and hydroheterarylation of bicyclo[2.2.1]hept-2-ene: influence of the chiral ligand, the leaving group, and the solvent. *Chem. Ber. Recl.*, **130**, 1327–1331.

93 Wu, X.-Y., Xu, H.-D., Tang, F.-Y., and Zhou, Q.-L. (2001) Asymmetric hydroarylation of norbornene derivatives catalyzed by palladium complexes of chiral quinolinyl-oxazolines. *Tetrahedron: Asymmetry*, **12** (18), 2565–2569.

94 Wu, X.-Y., Xu, H.-D., Zhou, Q.-L., and Chan, A.S.C. (2000) Enantioselective Heck-type hydroarylation of norbornene with phenyl iodide catalyzed by palladium/quinolinyl-oxazolines. *Tetrahedron: Asymmetry*, **11** (6), 1255–1257.

95 Sakuraba, S., Awano, K., and Achiwa, K. (1994) Asymmetric Heck-type hydroarylation of norbornene with phenyl triflate catalyzed by palladium complexes of chiral (β-*N*-sulfonylaminoalkyl)phosphines. *Synlett*, (4), 291–292.

96 Moinet, C. and Fiaud, J.-C. (1995) Palladium-catalyzed asymmetric hydrophenylation of 1,4-dihydro-1,4-epoxynapthalene. *Tetrahedron Lett.*, **36** (12), 2051–2052.

97 Iwai, Y., Gligorich, K.M., and Sigman, M.S. (2008) Aerobic alcohol oxidation coupled to palladium-catalyzed alkene hydroarylation with boronic esters. *Angew. Chem. Int. Ed.*, **47** (17), 3219–3222.

98 Gligorich, K.M., Cummings, S.A., and Sigman, M.S. (2007) Palladium-catalyzed reductive coupling of styrenes and organostannanes under aerobic conditions. *J. Am. Chem. Soc.*, **129** (46), 14193–14195.

99 Podhajsky, S.M., Iwai, Y., Cook-Sneathen, A., and Sigman, M.S. (2011) Asymmetric palladium-catalyzed hydroarylation of styrenes and dienes. *Tetrahedron*, **67** (24), 4435–4441.

100 Werner, E.W., Mei, T.-S., Burckle, A.J., and Sigman, M.S. (2012) Enantioselective Heck arylations of acyclic alkenyl alcohols using a redox-relay strategy. *Science*, **338** (6113), 1455–1458.
101 Mei, T.-S., Werner, E.W., Burckle, A.J., and Sigman, M.S. (2013) Enantioselective redox-relay oxidative heck arylations of acyclic alkenyl alcohols using boronic acids. *J. Am. Chem. Soc.*, **135** (18), 6830–6833.
102 Ohe, T. and Uemura, S. (2003) Palladium(II)-catalyzed Michael-type hydroarylation of nitroalkenes using aryltins and sodium tetraarylborates. *Bull. Chem. Soc. Jpn.*, **76** (7), 1423–1431.

5

Hydroarylation of C—C Multiple Bonds Using Nickel Catalysts

Yoshiaki Nakao

Kyoto University, Graduate School of Engineering, Department of Materials Chemistry, Katsura, Nishikyo-ku, Kyoto, 615-8510, Japan

5.1 Introduction

Nickel catalysts have rarely been employed for hydroarylation of C—C multiple bonds until recently [1] in spite of a very early report on a stoichiometric example of C—H metalation of azobenzene by nickelocene [2]. As related nickel-catalyzed C—H functionalization reactions, hydroacylation (Scheme 5.1) [3] and hydroalkynylation (Scheme 5.2) [4] of alkynes and alkylation of imidazolium salts (Scheme 5.3) [5], and a BF_3-adduct of thiazole (Scheme 5.4) [6] through insertion of alkynes and alkenes into acidic C—H bonds have been reported. Nickel-catalyzed hydroarylation of unsaturated compounds using arenes and heteroarenes has made significant progress in the last decade and is discussed in this chapter.

5.2 Hydroarylation of Alkynes

Hydroarylation of alkynes by nickel catalysis was first reported with fluorobenzenes [7]. The reaction of pentafluorobenzene with alkynes proceeds smoothly in the presence of catalytic amounts of $Ni(cod)_2$ and tricyclopentylphosphine [$PCyp_3$ (Cyp = cyclopentyl)] in toluene at 80 °C to give a variety of alkenylated pentafluorobenzenes stereo- and regioselectively (Scheme 5.5). (IMes)Ni(1,5-hexadiene) [IMes = 1,3-bis(2,4,6-trimethylphenyl)imidazol-2-ylidene] has also been introduced as a viable catalyst for the hydroarylation [8]. It is worth noting that the C—H bond of the fluoroarenes is exclusively functionalized over C—F bonds, whereas experimental [9] and theoretical [10] studies have revealed that electron-rich nickel(0) complexes favor oxidative addition of C—F bonds rather than C—H bonds. Exclusive *cis*-addition is observed, and the reaction across unsymmetrical alkynes gives alkenyl(pentafluoro)benzenes in which the aryl and a sterically less hindered substituent of alkynes are bound to the same carbon. This regioselectivity is observed generally with other nickel-catalyzed hydroarylation reactions

Catalytic Hydroarylation of Carbon-Carbon Multiple Bonds, First Edition.
Edited by Lutz Ackermann, T. Brent Gunnoe and Laurel Goj Habgood.

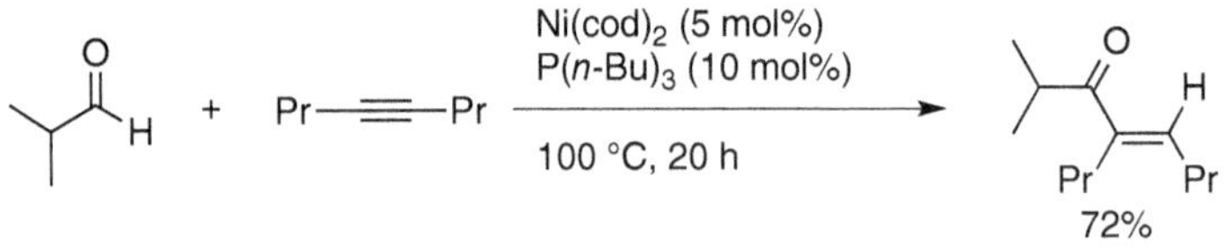

Scheme 5.1 Nickel-catalyzed hydroacylation of alkynes.

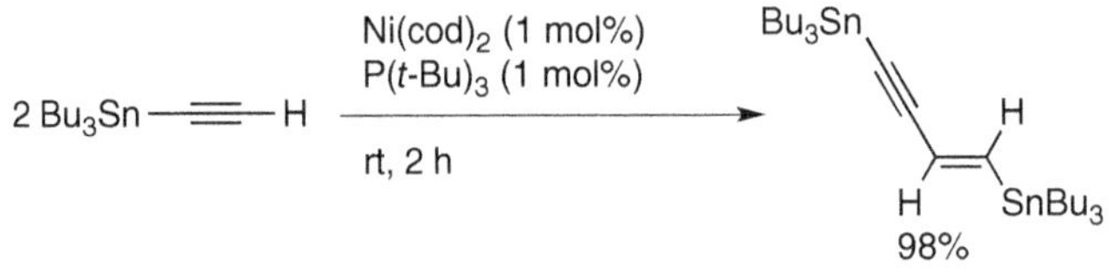

Scheme 5.2 Nickel-catalyzed hydroalkynylation of alkynes.

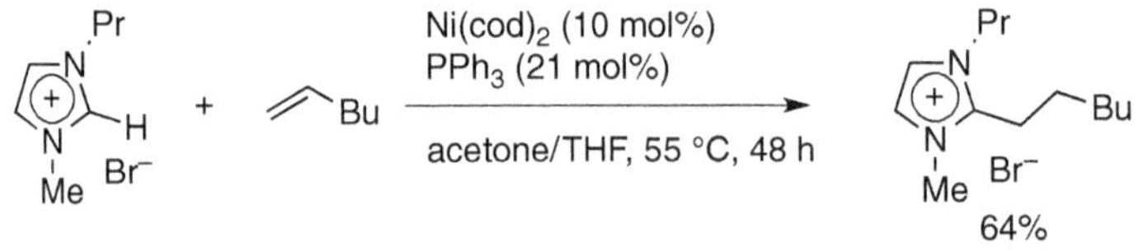

Scheme 5.3 Nickel-catalyzed alkylation of imidazolium salts.

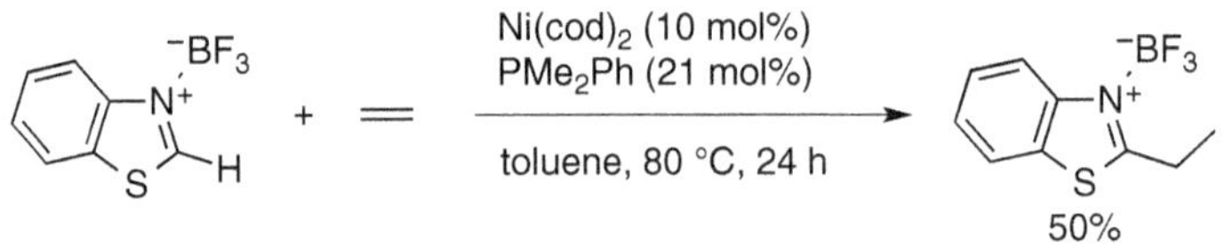

Scheme 5.4 Nickel-catalyzed alkylation of heteroarene–BF_3 complexes.

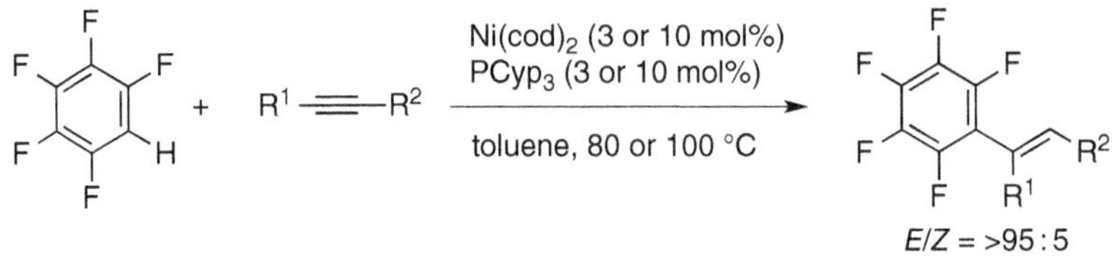

R^1, R^2 = Pr, Pr: 99% (3 h)
CH_2SiMe_3, CH_2SiMe_3: 75% (6 h)
Ph, Ph: 68% (3.5 h)
Me, *t*-Bu: 89% (3 h)
Me, $SiMe_3$: 47% (13 h)
Ph, $SiMe_3$: 63% (15 h)

Scheme 5.5 Nickel-catalyzed hydroarylation of alkynes with pentafluorobenzene.

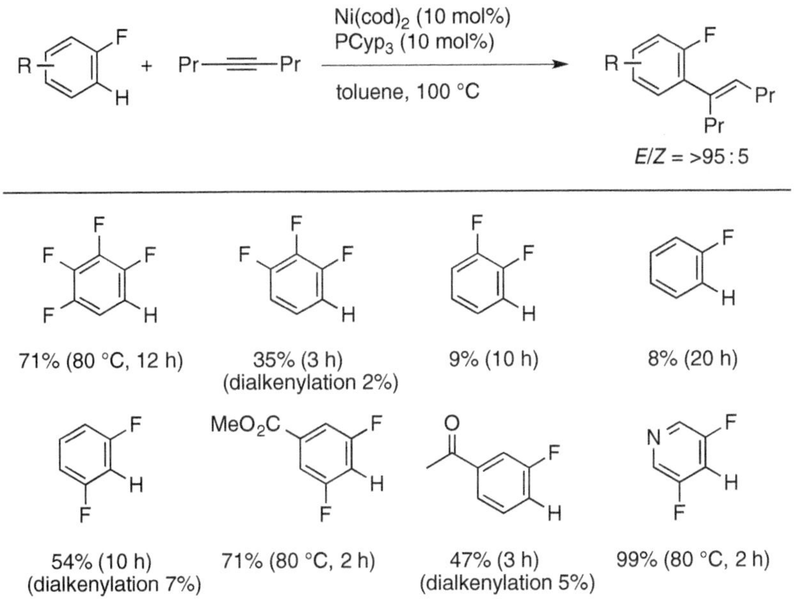

Scheme 5.6 Nickel-catalyzed hydroarylation of 4-octyne with fluoroarenes.

discussed in this chapter. Other fluorobenzene derivatives also participate in the hydroarylation across 4-octyne (Scheme 5.6). Excellent site-selectivity is also observed on the arene moieties bearing potential directing groups such as carbonyl, and C—H bonds *ortho* to a fluorine are exclusively alkenylated.

A catalytic cycle of the nickel-catalyzed hydrofluoroarylation of alkynes was originally proposed to be initiated by the formation of an η^2-arene–nickel(0) complex [11]. Oxidative addition of the C—H bond follows, and migratory insertion of an alkyne into the Ni—H bond and the subsequent C—C bond-forming reductive elimination give the products (Scheme 5.7). The oxidative addition of a pentafluorophenyl C—H bond was observed experimentally and was found to be reversible to compete with kinetically less favored but thermodynamically favored oxidative addition of C—F bonds to give a nickel–fluoride complex. The proposed catalytic cycle has been revised based on theoretical calculations [12]. The computed catalytic cycle involves the initial formation of an η^2-alkyne–nickel(0) complex with σ-coordination of the reacting Ar—H bond (Scheme 5.8). Cleavage of the Ar—H and formation of two C—Ni bonds as well as a new C—H bond proceed in a concerted manner (ligand-to-ligand hydrogen transfer) to give an alkenyl(aryl)nickel(II) complex, which gives the alkenylated fluoroarene after geometrical isomerization as a turnover-limiting step followed by C—C bond forming reductive elimination. A measured kinetic isotope effect (KIE) of 1.0 shows that the ligand-to-ligand hydrogen transfer process is indeed a facile process. The proposed cycle works well with a phosphine ligand equivalent to nickel, whereas the observed oxidative addition of the Ar—H bond proceeds to NiL_2. Indeed, the optimized catalytic reaction proceeds more efficiently with NiL than with NiL_2.

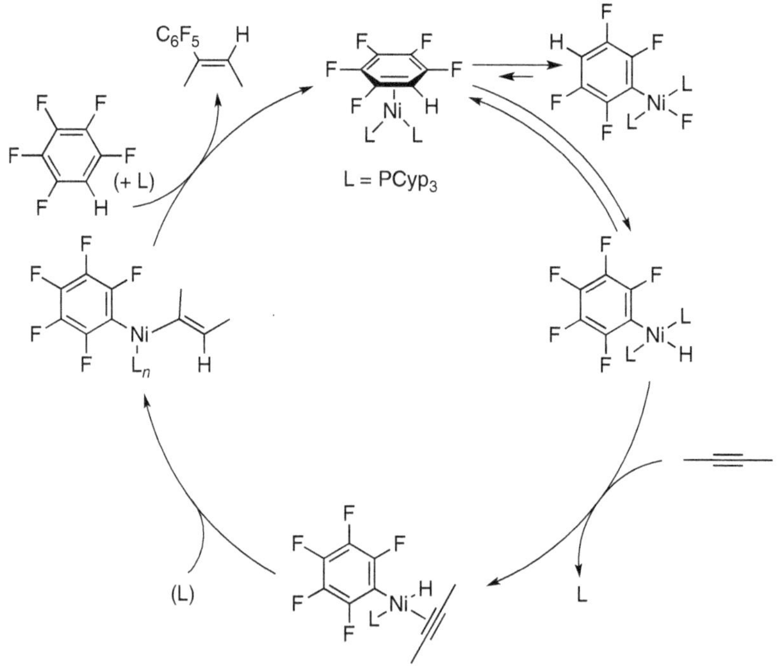

Scheme 5.7 Plausible mechanism of nickel-catalyzed hydroarylation of alkynes proposed based on stoichiometric experiments.

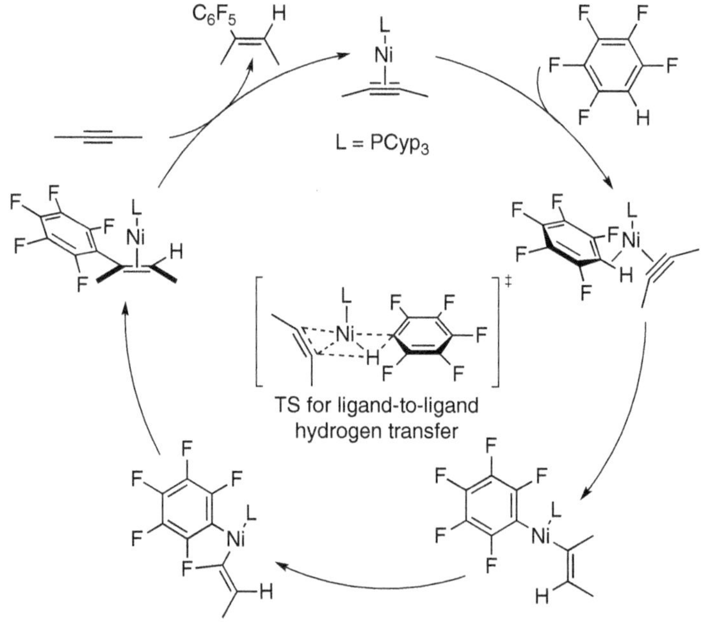

Scheme 5.8 A plausible mechanism of nickel-catalyzed hydroarylation of alkynes proposed based on theoretical calculations.

5.3 Hydroheteroarylation of Alkynes

5.3.1 Hydroheteroarylation of Alkynes with five-Membered Heteroarenes

The reaction conditions for the hydrofluoroarylation of alkynes also affect hydroheteroarylation of alkynes [13]. Indole derivatives bearing a relatively electron-withdrawing substituent at the C-3 position react with alkynes at the C-2 position selectively to give various 2-alkenylindoles with excellent stereo- and regioselectivity (Scheme 5.9). Additional five-membered heteroarenes, including benzimidazole, caffeine, purine, benzoxazole, benzofuran, benzothiophene, and thiazole, also participate in the alkenylation reaction with 4-octyne exclusively at the C-2 position to give the corresponding alkenylheteroarenes in a stereoselective manner (Scheme 5.10). A nickel(0) species generated *in situ* from air-stable $Ni(acac)_2$ [14] or (IMes)Ni(1,5-hexadiene) [8] also catalyzes the hydroheteroarylation of alkynes. It has also been shown that 1,3,4-oxadiazoles can be alkenylated by a Ni/PCy_3 catalyst to give the corresponding alkenylated heterocycles (Scheme 5.11) [15].

The alkenylation of *N*-methylimidazole under these reaction conditions proceeds sluggishly to afford a mixture of C-2 mono-alkenylation and C-2/C-5 dialkenylation products in low yields. The use of $AlMe_3$ as a co-catalyst, on the other hand, improved the transformation. In addition, the use of $P(t\text{-Bu})_3$ instead of $PCyp_3$ as a ligand further improves the selectivity to give a range of 2-alkenylimidazoles (Scheme 5.12) [16]. Phenyl-substituted alkynes give formal *trans*-adducts via isomerization of the initially formed *cis*-adducts under the reaction conditions.

C-2-substituted imidazoles undergo the alkenylation reaction at the C-5 position selectively by cooperative $Ni/PCyp_3/AlMe_3$ catalysis (Scheme 5.13). A variety of substituents in both the imidazole and alkyne substrates are tolerated

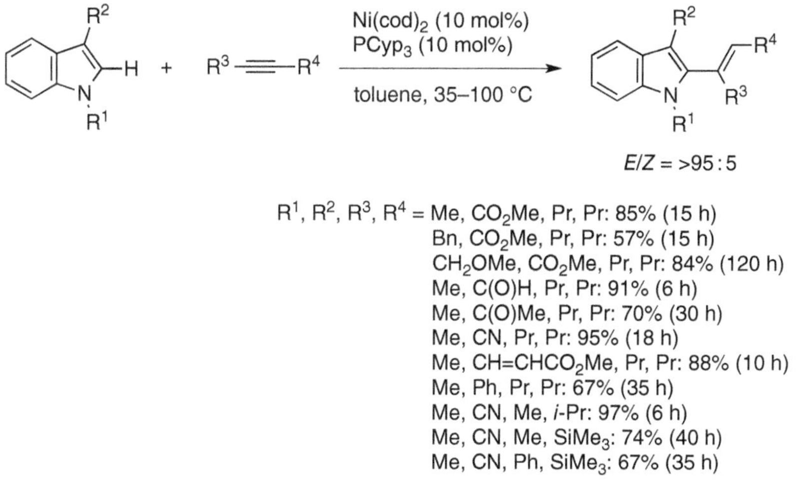

Scheme 5.9 Nickel-catalyzed hydroheteroarylation of alkynes with indoles.

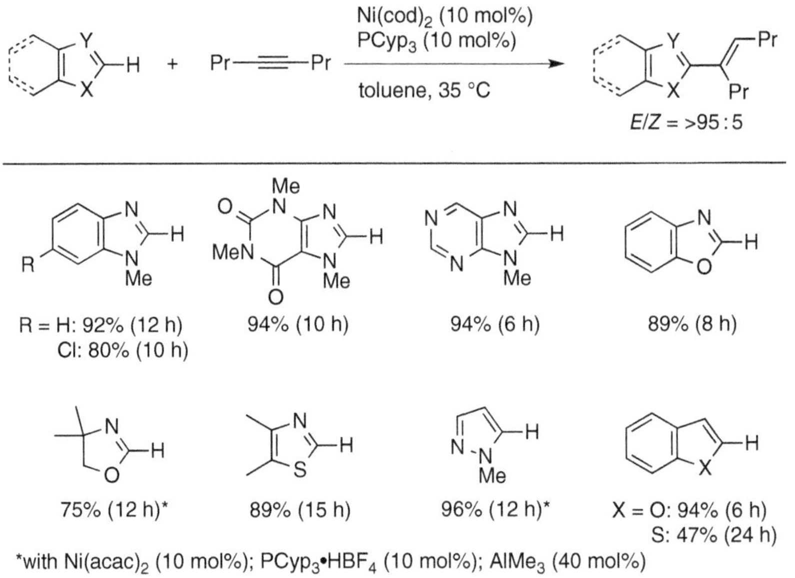

Scheme 5.10 Nickel-catalyzed hydroheteroarylation of 4-ocytne with five-membered heteroarenes.

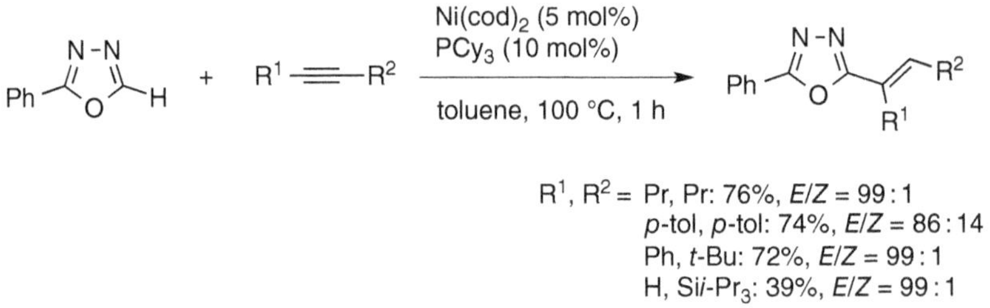

Scheme 5.11 Nickel-catalyzed hydroheteroarylation of alkynes with 2-phenyl-1,3,4-oxadiazole.

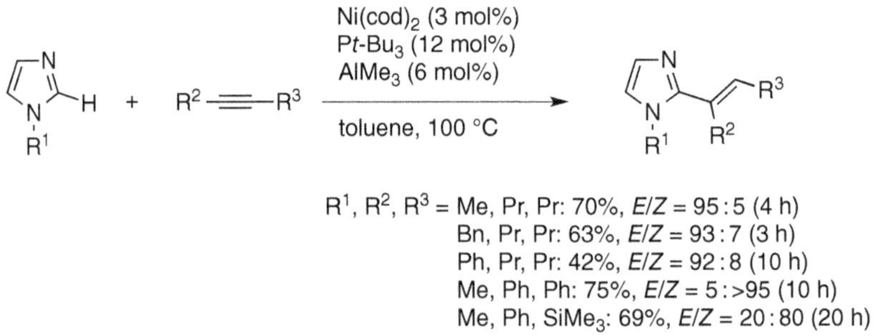

Scheme 5.12 Nickel/aluminum-catalyzed hydroheteroarylation of alkynes with imidazoles.

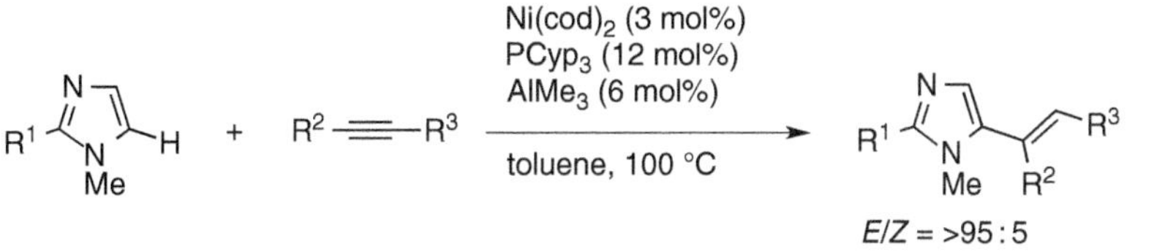

R^1, R^2, R^3 = Me, Pr, Pr: 55% (1 h)
Ph, Pr, Pr: 81% (2 h)
t-BuMe$_2$Si, Pr, Pr: 42% (1 h)
Me, Me, *i*-Pr: 75% (4 h)
Me, Me, *t*-Bu: 71% (4 h)

Scheme 5.13 Nickel/aluminum-catalyzed hydroheteroarylation of alkynes with 2-substituted imidazoles.

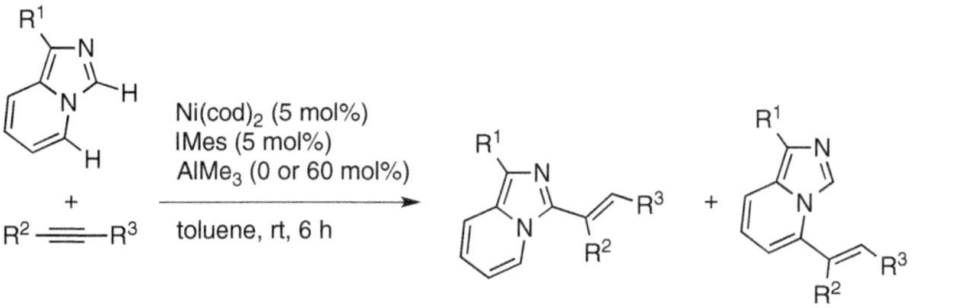

R^1, R^2, R^3 =	w/o AlMe$_3$	w/ AlMe$_3$
H, Pr, Pr:	96%, 4 : 1, *E*/*Z* = 98 : 2	98%, 1 : 10, *E*/*Z* = 99 : 1
Ph, Pr, Pr:	99%, 5 : 1, *E*/*Z* = 99 : 1	99%, 1 : 7, *E*/*Z* = 95 : 5
CH=CHPh, Pr, Pr:	97%, 5 : 1, *E*/*Z* = 99 : 1	83%, 1 : 11, *E*/*Z* = 80 : 20
H, Me, *t*-Bu:	—	93%, 1 : 11, *E*/*Z* = 99 : 1

Scheme 5.14 Nickel-catalyzed hydroheteroarylation of alkynes with imidazo[1,5-*a*]pyridines.

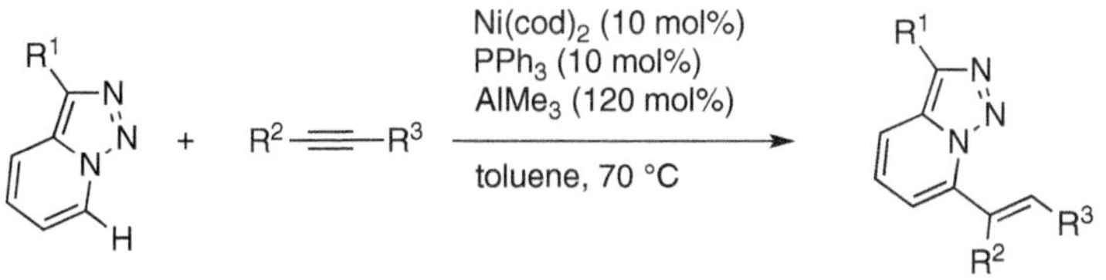

R^1, R^2, R^3 = H, Ph, Ph: 97%
Me, Ph, Ph: 93%
Ph, Ph, Ph: 94%
H, Pr, Pr: 76%
H, Me, *i*-Pr: 82% (regioselectivity >95 : 5)
H, Me, Ph: 95% (regioselectivity 85 : 15)

Scheme 5.15 Nickel-catalyzed hydroheteroarylation of alkynes with triazolo[1,5-*a*]pyridines.

to afford 2,5-di-substituted imidazoles stereo- and regioselectively in modest to good yields. Coordination of imidazoles to the Lewis acidic aluminum center of $AlMe_3$ at their sp^2-nitrogen likely makes their C(2)—H and C(5)—H bonds reactive enough to undergo C—H functionalization by electron-rich nickel(0) species.

Imidazo[1,5-*a*]pyridines can be alkenylated at the C-2 position by a nickel catalyst bearing a *N*-heterocyclic carbene (NHC) as a ligand (Scheme 5.14) [17]. The presence of $AlMe_3$ as a Lewis acid co-catalyst leads the alkenylation to proceed at the C-5 position, thus allowing divergent C—H functionalization of the bicyclic heteroaromatic core. Triazolopyridines are also alkenylated at the C-7 position by the cooperative nickel/aluminum system (Scheme 5.15) [18].

5.3.2 Hydroheteroarylation of Alkynes with Azine-*N*-oxides

The Ni/$PCyp_3$ system also catalyzes the C-2 alkenylation of azine-*N*-oxides through insertion of alkynes into the C(2)—H bond of the heterocycles [19]. Various 2-alkenylazine-*N*-oxides are obtained under mild reaction conditions stereo- and regioselectively (Scheme 5.16). The transformation provides an alternative to the direct C—H alkenylation of azines because the *N*-oxide moiety can readily be introduced/removed by several known protocols.

5.3.3 Hydroheteroarylation of Alkynes with Azines

Azine-*N*-oxides can be readily functionalized by the nickel catalysis, whereas the parent azines do not react with alkynes at all under these reaction conditions even at elevated temperatures. As in the case of imidazoles, however, the presence of a Lewis acid co-catalyst dramatically promotes the transformation. Of Lewis acid co-catalysts examined for the reaction of pyridine with 4-octyne in the presence of a Ni/P(i-Pr)$_3$ catalyst, diorganozinc derivatives such as $ZnMe_2$ and $ZnPh_2$ accelerate the alkenylation reaction to give the corresponding 2-alkenylpyridine [20]. A range of substituted pyridines and other azine derivatives are alkenylated stereo- and regioselectively to give 2-alkenylazine products by the cooperative Ni/P(i-Pr)$_3$/ZnR_2 catalysis (Scheme 5.17). The use of $AlMe_3$ as a Lewis acid co-catalyst gives 2-dienylpyridine through double incorporation of alkynes at

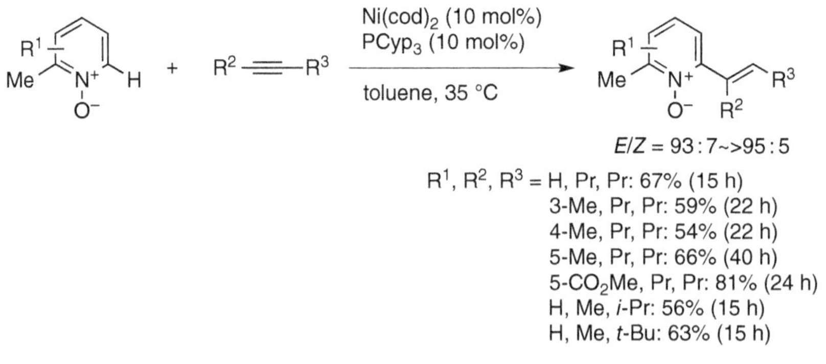

Scheme 5.16 Nickel-catalyzed hydroheteroarylation of alkynes with azine-*N*-oxides.

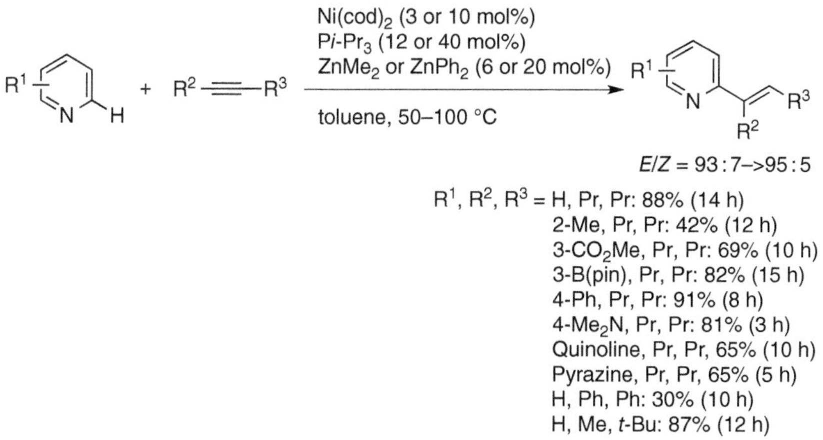

Scheme 5.17 Nickel/zinc-catalyzed hydroheteroarylation of alkynes with azines.

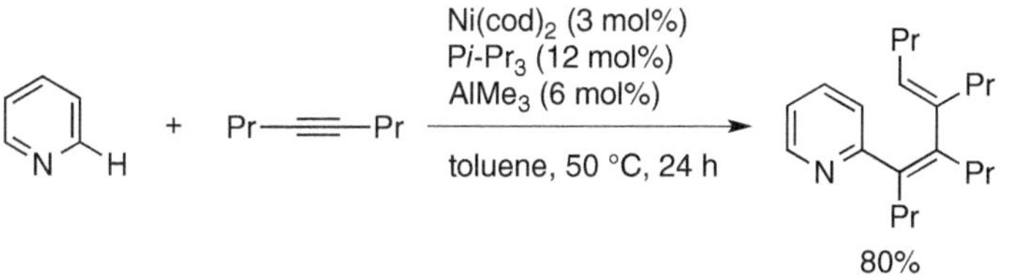

Scheme 5.18 Nickel/aluminum-catalyzed hydroheteroarylation of 4-octyne with pyridine.

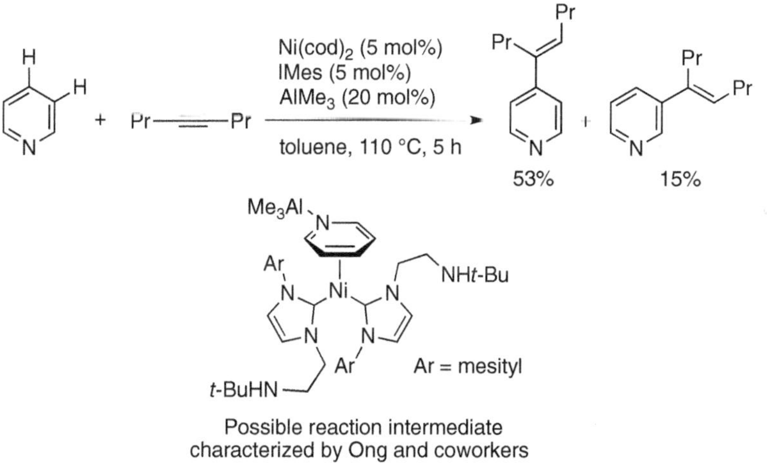

Scheme 5.19 Nickel/aluminum-catalyzed hydroheteroarylation of 4-octyne with pyridine.

the C-2 position of pyridine (Scheme 5.18). On the other hand, the use of IMes as a ligand and $AlMe_3$ as a Lewis acid catalyst switches the reaction site from C-2 to C-3 and C-4 positions, favoring particularly the C-4 functionalization (Scheme 5.19) [21, 22]. It can be proposed that pyridine coordinating to the Lewis acid co-catalysts at its sp^2-nitrogen has enhanced reactivity at the C(2)—H and C(4)—H bonds to undergo the ligand-to-ligand hydrogen transfer process to afford alkenylpyridines. The electronic and/or steric properties of ligands might be responsible for controlling the site-selectivity. A Ni/NHC complex bearing pyridine coordinating to $AlMe_3$ at the sp^2-nitrogen as a ligand in an η^2-fashion has been characterized, and exclusive coordination of the heteroarene ligand to the nickel center at the C3—C4 double bond is likely relevant to the observed site-selectivity of the alkenylation process [21].

5.4 Hydroarylation of Alkenes

Nickel-catalyzed hydroarylation of alkenes show different regioselectivity depending on an alkene substrate. Pentafluorobenzene adds across 2-vinylnaphthalene to selectively give a 1,1-diarylethane product (Scheme 5.20) [7]. On the other hand, hydroarylation of aliphatic 1-alkenes with trifluoromethyl-substituted benzenes has been reported to give linear alkylated arenes exclusively (Scheme 5.21) [23]. A catalytic cycle of the latter reaction is proposed based on theoretical calculations. The mechanism resembles the one proposed for the alkyne-hydrofluoroarylation involving the ligand-to-ligand hydrogen transfer process (Scheme 5.8) but with a bis(alkene)nickel complex as a resting state and C—C bond-forming reductive elimination as a rate-limiting step (Scheme 5.22). The observed KIE value (1.1) for the linear alkylation matches well with this mechanistic scenario.

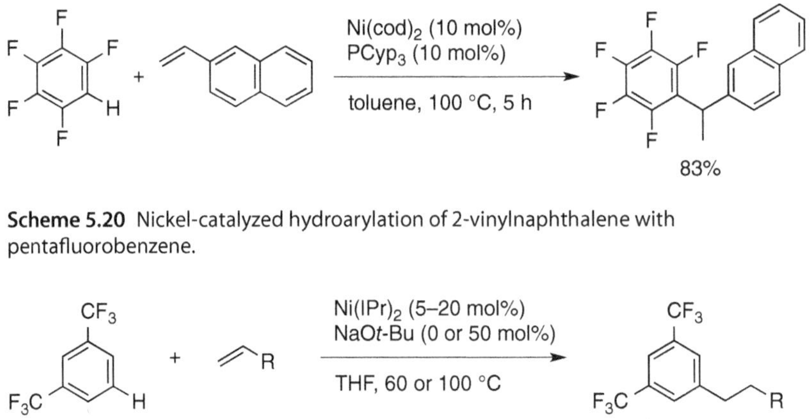

Scheme 5.20 Nickel-catalyzed hydroarylation of 2-vinylnaphthalene with pentafluorobenzene.

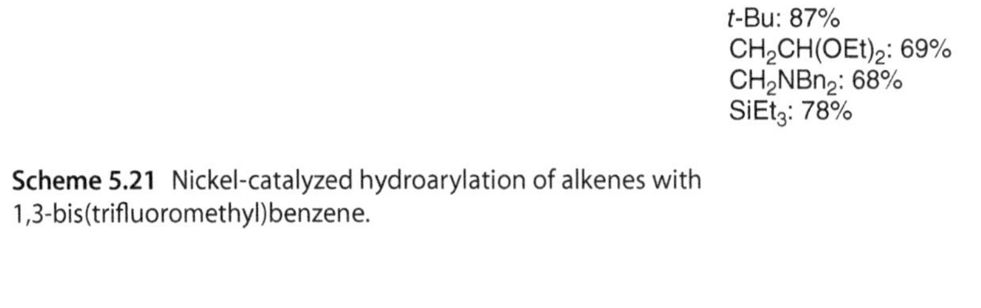

Scheme 5.21 Nickel-catalyzed hydroarylation of alkenes with 1,3-bis(trifluoromethyl)benzene.

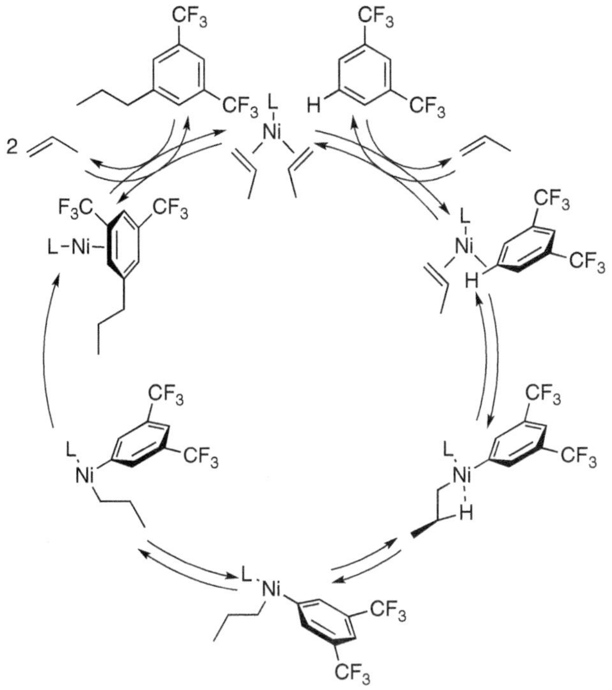

Scheme 5.22 Plausible mechanism of nickel-catalyzed hydroarylation of alkenes proposed based on theoretical calculations.

5.5 Hydroheteroarylation of Alkenes

5.5.1 Hydroheteroarylation of Alkenes with five-Membered Heteroarenes

Nickel catalysts bearing NHC ligands have been found to affect the alkylation of various heteroarenes through hydroheteroarylation. A range of functional groups in five-membered heteroarenes, including indoles (Scheme 5.23), benzofurans (Scheme 5.24), pyrroles (Scheme 5.25), and furans (Scheme 5.26), are tolerated to react with various aliphatic 1-alkenes with exclusive C-2 and linear selectivity to give a variety of 2-alkylheteroarenes [24]. An *E*/*Z* mixture of 2-octene reacts with the indole substrate with concomitant olefin isomerization to give 2-octylindole (Scheme 5.27). The olefin isomerization process can be hampered in the presence of a base additive, which is effective for some heteroarene and/or alkene substrates to obtain 2-alkylheterocycles in good yields.

While vinylarenes react with benzimidazole with a reversed regioselectivity to give 1,1-diarylethane products (Scheme 5.28), the presence of $AlMe_3$ as a Lewis acid co-catalyst turns the regioselectivity to give 1,2-diarylethanes (Scheme 5.29) [25]. A similar regio-divergency can be observed with the reaction of allylarenes with olefin isomerization accompanied to allow the access to both 1,1- (Scheme 5.30) and 1,3-diarylpropanes (Scheme 5.31) [26].

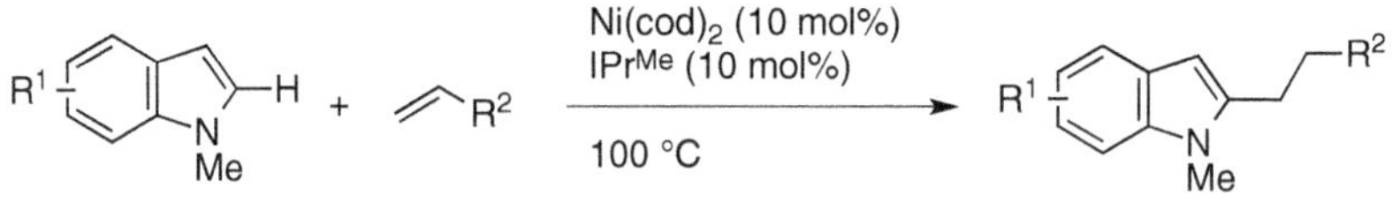

l/*b* = >95 : 5

R^1, R^2 = 4-CO_2Me, $C_{11}H_{23}$: 82% (1 h)
5-CO_2Me, $C_{11}H_{23}$: 82% (24 h w/ 2 mol% cat.)
5-C(O)Me, C_8H_{17}: 62% (14 h)
5-OMe, $C_{11}H_{23}$: 89% (24 h)
5-NMeAc, C_8H_{17}: 95% (18 h)
5-F, $C_{11}H_{23}$: 94% (24 h)
5-B(pin), C_8H_{17}: 45% (16 h)
5-Si(OEt)$_3$, C_8H_{17}: 72% (16 h)
6-CO_2Me, $C_{11}H_{23}$: 86% (1 h)
H, $CH_2CH(OEt)_2$: 61% (24 h, w/ 50 mol% NaO*t*-Bu)
H, $CMe_2CH_2CO_2Me$: 54% (21 h)
H, CH_2SiMe_2Ph: 74% (24 h)
H, cyclohexen-4-yl: 90% (1 h)
H, O*t*-Bu: 61% (18 h)
H, $SiEt_3$: 89% (24 h)

Scheme 5.23 Nickel-catalyzed hydroheteroarylation of alkenes with indoles.

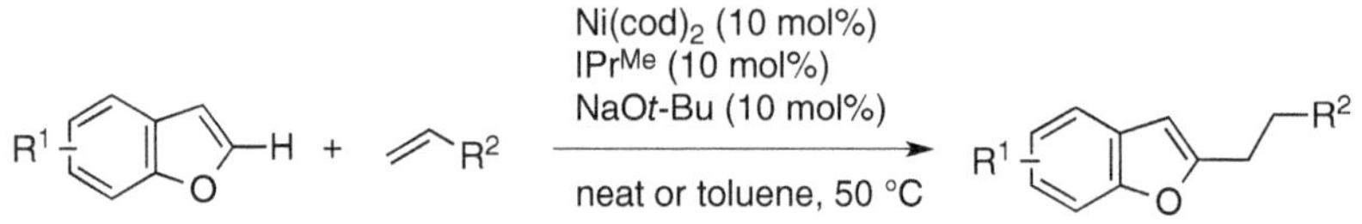

l/*b* = >95 : 5

R^1, R^2 = H, C_8H_{17}: 98% (16 h)
5-CO_2Me, C_8H_{17}: 98% (21 h w/o NaO*t*-Bu)
5-C(O)Me, C_8H_{17}: 52% (16 h w/o NaO*t*-Bu)
5-OMe, C_8H_{17}: 94% (21 h)
5-F, C_8H_{17}: 99% (21 h)
5-B(pin), C_8H_{17}: 77% (21 h)
5-SiMe(OSiMe$_3$)$_2$, C_8H_{17}: 77% (16 h w/ 1 mol% cat.)
6-OMe, C_8H_{17}: 98% (22 h at 65 °C)
H, *t*-Bu: 86% (16 h w/ 1 mol% cat.)

Scheme 5.24 Nickel-catalyzed hydroheteroarylation of alkenes with benzofurans.

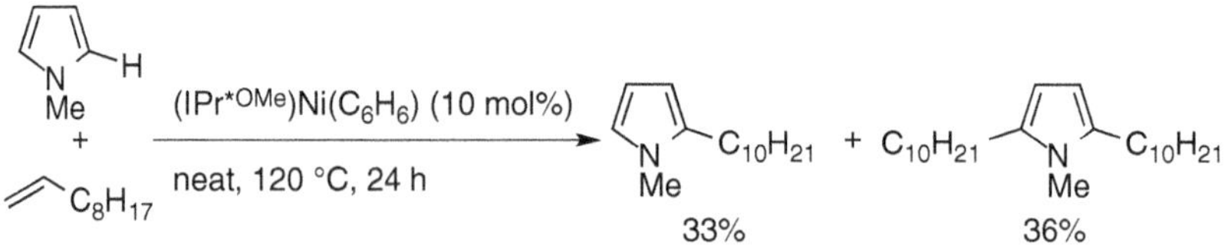

Scheme 5.25 Nickel-catalyzed hydroheteroarylation of 1-decene with 1-methylpyrrole.

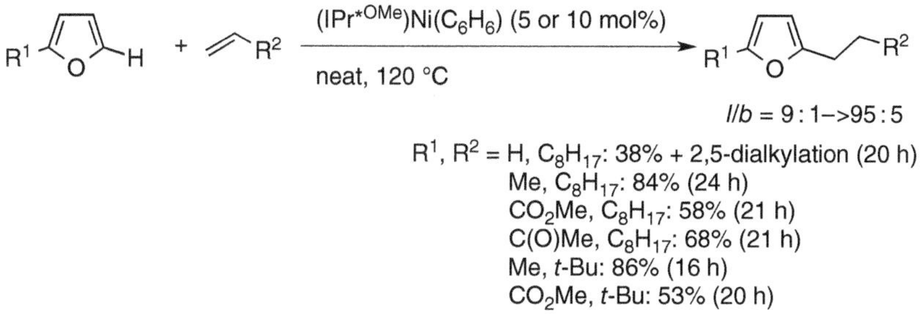

Scheme 5.26 Nickel-catalyzed hydroheteroarylation of alkenes with furans.

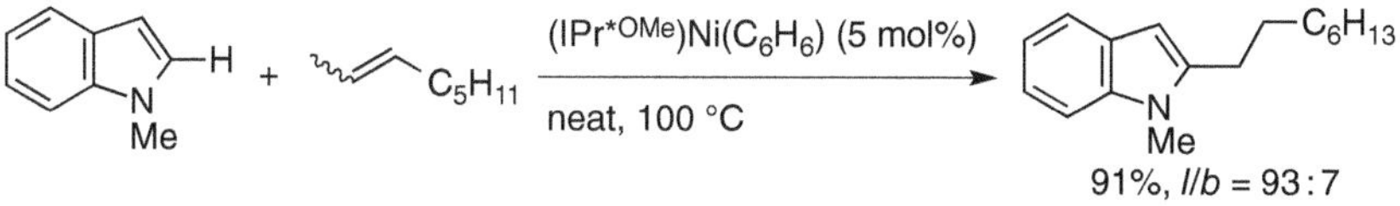

Scheme 5.27 Nickel-catalyzed hydroheteroarylation of 2-octene with 1-methylindole.

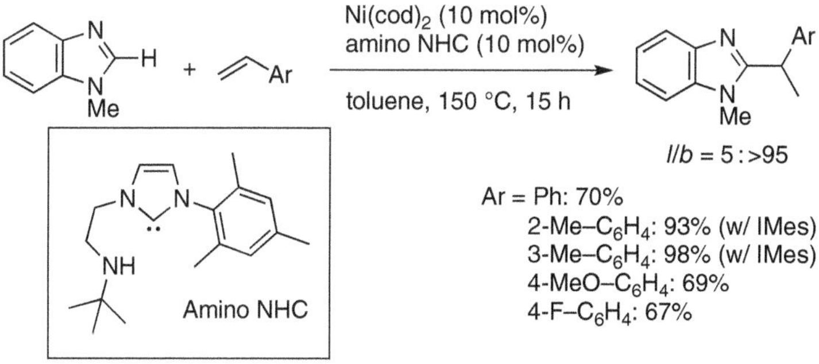

Scheme 5.28 Nickel-catalyzed hydroheteroarylation of vinylarenes with 1-methylbenzimidazole.

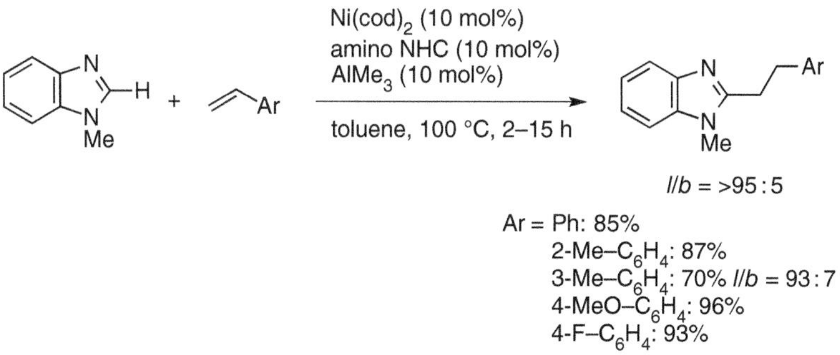

Scheme 5.29 Nickel/aluminum-catalyzed hydroheteroarylation of vinylarenes with 1-methylbenzimidazole.

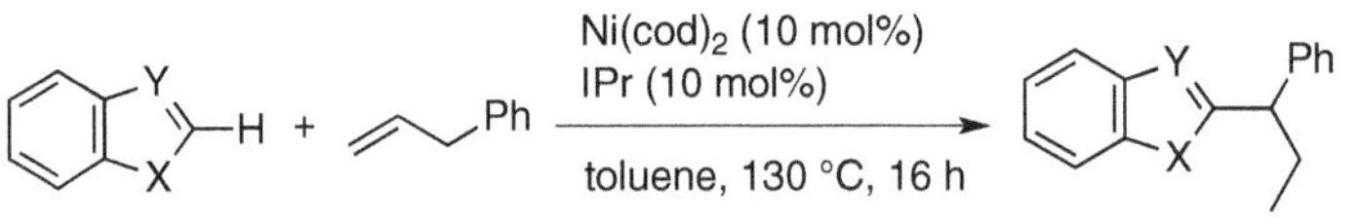

l/b = 15 : 85–5 : >95

X, Y = NMe, CH: 58%
NMe, N: 97%
NBn, N: 81%
N(4-MeO–C_6H_4), N: 85%
O, N: 73%

Scheme 5.30 Nickel-catalyzed hydroheteroarylation of allylbenzene.

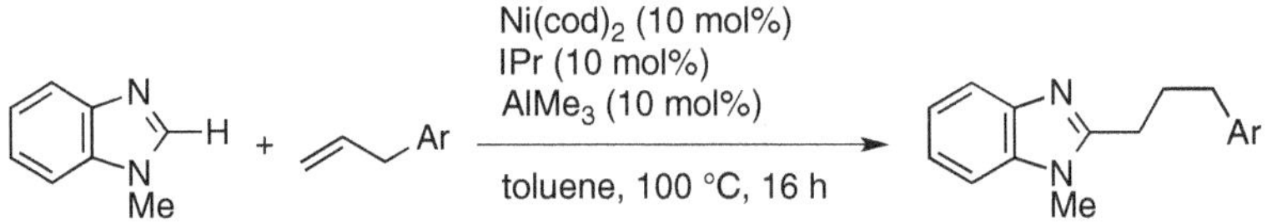

l/b = 92 : 8–>95 : 5

Ar = Ph: 92%
2-Me–C_6H_4: 90%
3-Me–C_6H_4: 88%
4-MeO–C_6H_4: 81%
4-F–C_6H_4: 77%

Scheme 5.31 Nickel/aluminum-catalyzed hydroheteroarylation of allylarenes with 1-methylbenzimidazole.

5.5.2 Hydroheteroarylation of Alkenes with Azines

Azines are alkylated selectively at the C-4 position by cooperative Ni/IPr and MAD catalysis [22]. The use of the very bulky Lewis acid co-catalyst is crucial to achieve the high C-4 selectivity. Irrespective of substitution patterns of azine substrates, excellent C-4 and linear selectivities are attained (Scheme 5.32),

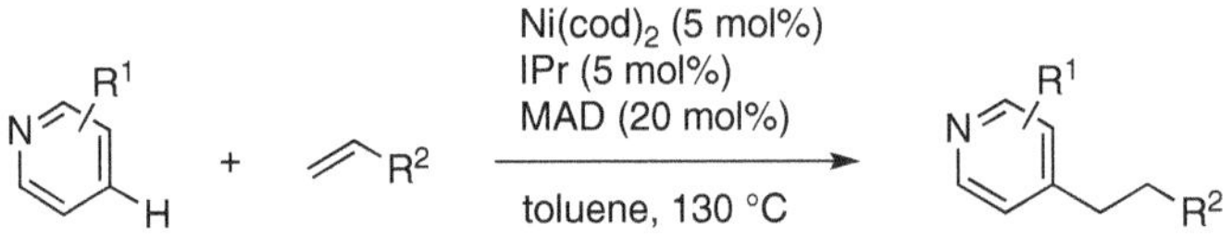

l/b = 90 : 10–>95 : 5

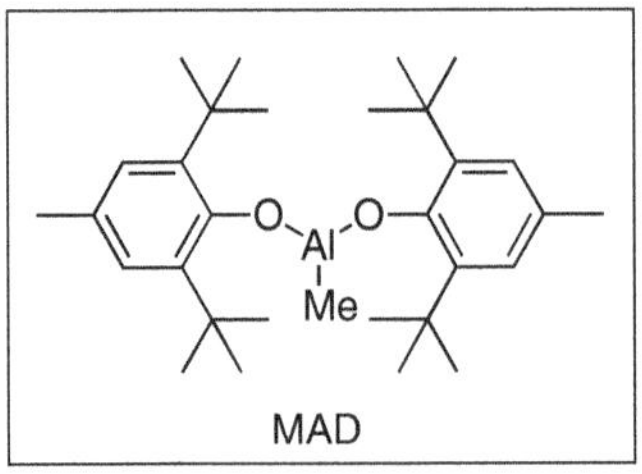

R^1, R^2 = H, $C_{11}H_{23}$: 87% (5 h)
2-Me, $C_{11}H_{23}$: 91% (10 h)
3-CO_2Me, $C_{11}H_{23}$: 52% (9 h)
Quinoline, $C_{11}H_{23}$: 89% (9 h)
H, CH_2Ph: 85% (9 h)
H, $(CH_2)_2CH{=}CH_2$: 48% (23 h)
H, $(CH_2)_3$OBoc: 61% (23 h)
H, cyclohexen-4-yl: 91% (10 h)
H, $SiMe_3$: 83% (19 h)

Scheme 5.32 Nickel/aluminum-catalyzed hydroheteroarylation of alkenes with azines.

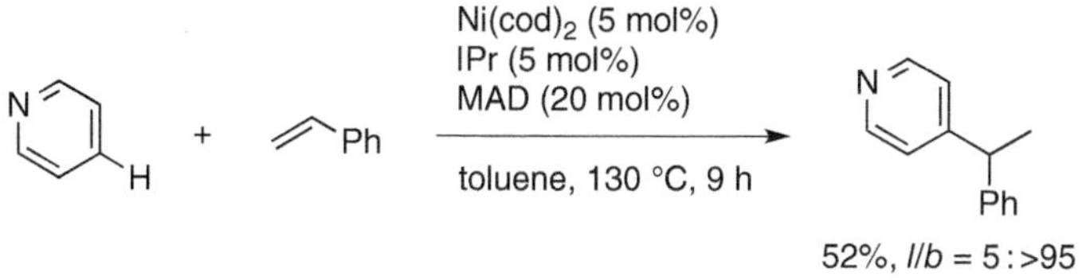

Scheme 5.33 Nickel/aluminum-catalyzed hydroheteroarylation of styrene with pyridine.

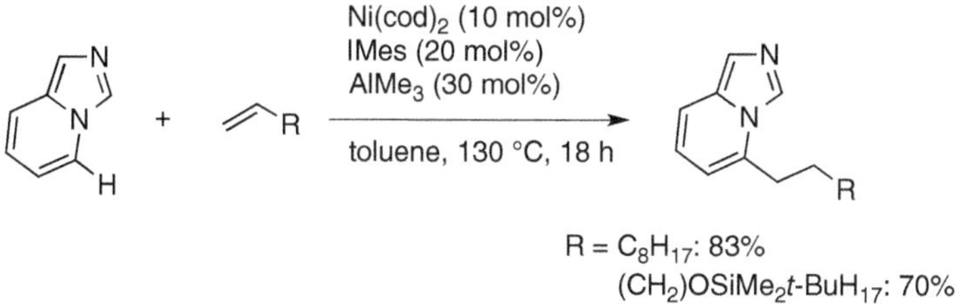

Scheme 5.34 Nickel/aluminum-catalyzed hydroheteroarylation of alkenes with imidazo[1,5-*a*]pyridine.

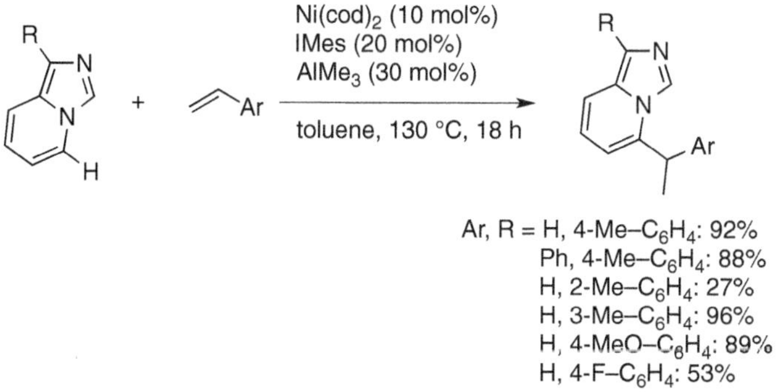

Scheme 5.35 Nickel/aluminum-catalyzed hydroheteroarylation of vinylarenes with imidazo[1,5-*a*]pyridines.

whereas styrene reacts with pyridine to give a C-4 branched alkylation product (Scheme 5.33).

C-5 selective alkylation of imidazo[1,5-*a*]pyridines can be catalyzed cooperatively by Ni/IMes/$AlMe_3$ [17]. Aliphatic 1-alkenes give linear alkylation products (Scheme 5.34), while vinylarenes give 1,1-diarylethanes (Scheme 5.35).

5.6 Summary and Outlook

Nickel-catalyzed hydroarylation and hydroheteroarylation of alkynes and alkenes have been discussed in this chapter. Some features of the nickel catalysis are worth mentioning. The regioselectivities of these nickel-catalyzed transformations are generally very high. Aryl and heteroaryl groups are introduced to less-hindered carbons of alkynes and alkenes except for the reaction across vinylarenes, with which the opposite regioselectivity is observed in most cases. Qualitatively, relatively acidic Ar—H bonds are selectively functionalized. Therefore, the second feature is that the nickel system works effectively with relatively electron-poor arene substrates. When the reactivity of arenes and heteroarenes is less sufficient, the use of Lewis acid co-catalysts often accelerates the C—H functionalization to expand the scope of the nickel-catalyzed protocols. Notably, Lewis acid co-catalysts electronically activate the arene substrates through Lewis acid–Lewis base interaction and contribute to control the reaction site on arene and heteroarene substrates and in some cases accelerate it. These characteristic features associated with the nickel systems would have potential to further establish hydroarylation and hydroheteroarylation reactions that cannot be achieved by other metal catalysis.

References

1 For recent reviews, see: (a) Nakao, Y. (2011) Hydroarylation of alkynes catalyzed by nickel. *Chem. Rec.*, **11** (5), 242–251; (b) Castro, L.C.M. and Chatani, N. (2015) Nickel catalysts/*N,N*-bidentate directing groups: an excellent partnership in directed C—H activation reactions. *Chem. Lett.*, **44** (4), 410–421; (c) Henrion, M., Ritleng, V., and Chetcuti, M.J. (2015) Nickel *N*-heterocyclic carbene-catalyzed C—C bond formation: reactions and mechanistic aspects. *ACS Catal.*, **5** (2), 1283–1302.
2 Kleiman, J.P. and Dubeck, M. (1963) The preparation of cyclopentadienyl [*o*-(phenylazo)phenyl]nickel. *J. Am. Chem. Soc.*, **85** (10), 1544–1545.
3 Tsuda, T., Kiyoi, T., and Saegusa, T. (1990) Nickel(0)-catalyzed hydroacylation of alkynes with aldehydes to α,β-enones. *J. Org. Chem.*, **55** (8), 2554–2558.
4 (a) Ishikawa, M., Oshita, J., Ito, Y., and Minato, A. (1988) Carbon–hydrogen bond activation by a nickel complex for the catalytic formation of dienyne systems. *J. Chem. Soc., Chem. Commun.*, 804–805; (b) Ogoshi, S., Ueta, M., Oka, M., and Kurosawa, H. (2004) Dimerization of terminal alkynes catalyzed by a nickel complex having a bulky phosphine ligand. *Chem. Commun.*, 2732–2733.
5 (a) Clement, N.D. and Cavell, K.J. (2004) Transition-metal-catalyzed reactions involving imidazolium salt/*N*-heterocyclic carbene couples as substrates. *Angew. Chem. Int. Ed.*, **43** (29), 3845–3847; (b) Normand, A.T., Yen, S.K., Huynh, H.V., Hor, T.S.A., and Cavell, K.J. (2008) Catalytic annulation of heterocycles via a novel redox process involving the imidazolium salt *N*-heterocyclic carbene couple. *Organometallics*, **27** (13), 3153–3160.
6 Normand, A.T., Hawkes, K.J., Clement, N.D., Cavell, K.J., and Yates, B.F. (2007) Atom-efficient catalytic coupling of imidazolium salts with ehylene involving Ni–NHC complexes as intermediates: a combined experimental and DFT study. *Organometallics*, **26** (22), 5352–5363.
7 Nakao, Y., Kashihara, N., Kanyiva, K.S., and Hiyama, T. (2008) Nickel-catalyzed alkenylation and alkylation of fluoroarenes via activation of C—H bond over C–F bond. *J. Am. Chem. Soc.*, **130** (48), 16170–16171.

8 Nett, A.J., Zhao, W., Zimmerman, P.M., and Montgomery, J. (2015) Highly active nickel catalysts for C—H functionalization identified through analysis of off-cycle intermediates. *J. Am. Chem. Soc.*, **137** (24), 7636–7639.

9 (a) Cronin, L., Higgitt, C.L., Karch, R., and Perutz, R.N. (1997) Rapid intermolecular carbon–fluorine bond activation of pentafluoropyridine at nickel(0): comparative reactivity of fluorinated arene and fluorinated pyridine derivatives. *Organometallics*, **16** (22), 4920–4928; (b) Braun, T., Foxon, S.P., Perutz, R.N., and Walton, P.H. (1999) Nickel-assisted carbon–fluorine bond activation of 2,4,6-trifluoropyrimidine: synthesis of new pyrimidine and pyrimidinone derivatives. *Angew. Chem. Int. Ed.*, **38** (22), 3326–3329; (c) Schaub, T., Fischer, P., Steffen, A., Braun, T., Radius, U., and Mix, A. (2008) C–F activation of fluorinated arenes using NHC-stabilized nickel(0) complexes: selectivity and mechanistic investigations. *J. Am. Chem. Soc.*, **130** (29), 9304–9317.

10 Reinhold, M., McGrady, J.E., and Perutz, R.N. (2004) A comparison of C–F and C—H bond activation by zerovalent Ni and Pt: a density functional study. *J. Am. Chem. Soc.*, **126** (16), 5268–5276.

11 Kanyiva, K.S., Kashihara, N., Nakao, Y., Hiyama, T., Ohashi, M., and Ogoshi, S. (2010) Hydrofluoroarylation of alkynes with fluoroarenes. *Dalton Trans.*, **39** (43), 10483–10494.

12 Guihaumé, J., Halbert, S., Eisenstein, O., and Perutz, R.N. (2012) Hydrofluoroarylation of alkynes with Ni catalysts. C—H activation via ligand-to-ligand hydrogen transfer, an alternative to oxidative addition. *Organometallics*, **31** (4), 1300–1314.

13 Nakao, Y., Kanyiva, K.S., Oda, S., and Hiyama, T. (2006) Hydroheteroarylation of alkynes under mild nickel catalysis. *J. Am. Chem. Soc.*, **128** (25), 8146–8147.

14 Kanyiva, K.S., Nakao, Y., and Hiyama, T. (2007) Practical approach for hydroheteroarylation of alkynes using bench-stable catalyst. *Heterocycles*, **72** (1), 677–680.

15 Mukai, T., Hirano, K., Satoh, T., and Miura, M. (2009) Nickel-catalyzed C—H alkenylation and alkylation of 1,3,4-oxadiazoles with alkynes and styrenes. *J. Org. Chem.*, **74** (16), 6410–6413.

16 Kanyiva, K.S., Löbermann, F., Nakao, Y., and Hiyama, T. (2009) Regioselective alkenylation of imidazoles by nickel/Lewis acid catalysis. *Tetrahedron Lett.*, **50** (26), 3463–3466.

17 Yu, M.-S., Lee, W.-C., Chen, C.-H., Tsai, F.-Y., and Ong, T.-G. (2014) Controlled regiodivergent C—H bond activation of imidazo[1,5-*a*]pyridine via synergistic cooperation between aluminum and nickel. *Org. Lett.*, **16** (18), 4826–4829.

18 Liu, S., Sawicki, J., and Driver, T.G. (2012) Ni-catalyzed alkenylation of triazolopyridines: synthesis of 2,6-disubstituted pyridines. *Org. Lett.*, **14** (14), 3744–3747.

19 Kanyiva, K.S., Nakao, Y., and Hiyama, T. (2007) Nickel-catalyzed addition of pyridine-*N*-oxides across alkynes. *Angew. Chem. Int. Ed.*, **46** (46), 8872–8874.

20 Nakao, Y., Kanyiva, K.S., and Hiyama, T. (2008) A strategy for C—H activation of pyridines: direct C-2 selective alkenylation of pyridines by nickel/Lewis acid catalysis. *J. Am. Chem. Soc.*, **130** (8), 2448–2449.

21 Tsai, C.-C., Shih, W.-C., Fang, C.-H., Ong, T.-G., and Yap, G.P.A. (2010) Bimetallic nickel aluminun mediated para-selective alkenylation of pyridine: direct observation of η^2,η^1-pyridine Ni(0)–Al(III) intermediates prior to C—H bond activation. *J. Am. Chem. Soc.*, **132** (34), 11887–11889.

22 Nakao, Y., Yamada, Y., Kashihara, N., and Hiyama, T. (2010) Selective C-4 alkylation of pyridine by nickel/Lewis acid catalysis. *J. Am. Chem. Soc.*, **132** (39), 13666–13668.

23 Bair, J.S., Schramm, Y., Sergeev, A.G., Clot, E., Eisenstein, O., and Hartwig, J.F. (2014) *J. Am. Chem. Soc.*, **136** (38), 13098–13101.

24 Schramm, Y., Takeuchi, M., Semba, K., Nakao, Y., and Hartwig, J.F. (2015) *J. Am. Chem. Soc.*, **137** (38), 12215–13101.

25 Shih, W.-C., Chen, W.-C., Lai, Y.-C., Yu, M.-S., Ho, J.-J., Yap, G.P.A., and Ong, T.-G. (2012) The regioselective switch for aamino-NHC mediated C—H activation of benzimidazole via Ni–Al synergistic catalysis. *Org. Lett.*, **14** (8), 2046–2049.

26 Lee, W.-C., Wang, C.-H., Lin, Y.-H., Shih, W.-C., and Ong, T.-G. (2013) Tandem isomerization and C—H activation: regioselective hydroheteroarylation of allylarenes. *Org. Lett.*, **15** (20), 5358–5361.

6

Hydroarylation of Alkynes and Alkenes using Group 7–9 First-Row Transition Metal Catalysts

Naohiko Yoshikai

Nanyang Technological University, School of Physical and Mathematical Sciences, Division of Chemistry and Biological Chemistry, Singapore, 637371, Singapore

6.1 Introduction

The introduction of alkyl and alkenyl groups into aromatic rings is a fundamentally important task in synthetic chemistry. The addition of arenes and heteroarenes to unsaturated hydrocarbons such as alkenes and alkynes represents an atom- and step-economical approach to this task. Such hydroarylation reactions are typically promoted either through activation of the aromatic C—H bond or through activation of the unsaturated reaction partner. While the latter mode of hydroarylation has been known for the classical Friedel–Crafts alkylation, the former mode of hydroarylation has been reported rather sporadically until the early 1990s [1]. Notable among such seminal studies is the work of Lewis and Smith on ortho-alkylation of phenol catalyzed by a ruthenium–triphenyl phosphite complex, which likely involves phenoxide exchange on phosphorus and cyclometalation of the phosphite ligand as key steps [2]. In 1993, Murai and coworkers reported a ruthenium(0)-catalyzed addition reaction of an aryl ketone to an alkene, where the carbonyl oxygen functions as a directing group to assist and ensure the regioselective reaction at the ortho-position (Scheme 6.1) [3]. The discovery of this reaction was followed by the development of a series of directing group-assisted hydroarylation reactions using low-valent transition metal catalyst, which are generally considered to proceed through three elementary steps, that is, oxidative addition of the *ortho* C—H bond, migratory insertion of alkene or alkyne into the M—H bond, and C—C bond-forming reductive elimination (Scheme 6.2).

Directing group-assisted hydroarylation reactions have been developed dominantly using noble transition metal catalysts, those based on ruthenium [4, 5], rhodium [6], and iridium [7] in particular. However, recent years have witnessed a significant progress in the development of alternative and complementary catalysts based on earth-abundant first-row transition metals. This chapter summarizes hydroarylation reactions catalyzed by the group 7–9 first-row transition metal catalysts, with particular focus on reactions involving directing group-assisted C—H activation. In Section 6.2, low-valent cobalt-catalyzed

Catalytic Hydroarylation of Carbon-Carbon Multiple Bonds, First Edition.
Edited by Lutz Ackermann, T. Brent Gunnoe and Laurel Goj Habgood.

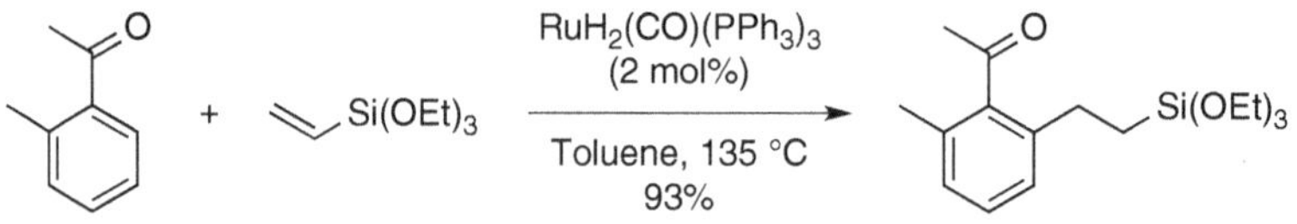

Scheme 6.1 Ruthenium-catalyzed, ketone-directed hydroarylation of alkene.

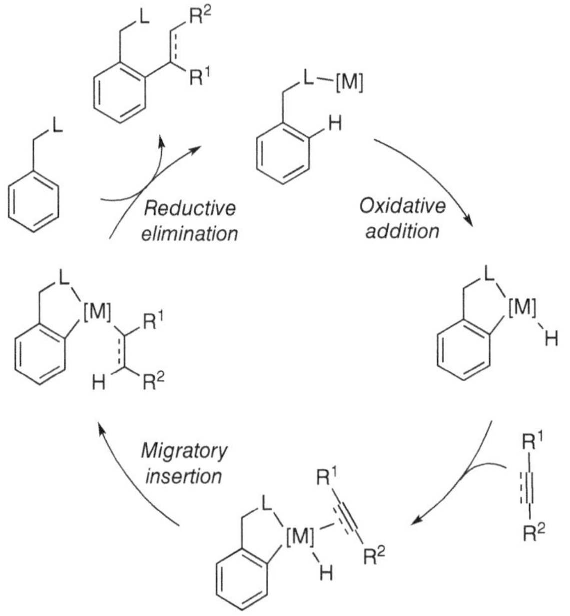

Scheme 6.2 A general catalytic cycle for directing group-assisted hydroarylation of alkenes and alkynes.

hydroarylation reactions of alkynes and alkenes, which feature mild conditions, broad substrate scopes, and notable regioselectivities, are reviewed [8–10]. In addition, hydroarylation reactions employing high-valent $Cp^{*}Co^{III}$ catalysts are also described. Iron- and manganese-catalyzed hydroarylation reactions, which are somewhat limited in scope at this time, are briefly discussed in Sections 6.3 and 6.4, respectively.

6.2 Hydroarylation of Alkynes and Alkenes using Cobalt Catalysts

6.2.1 Hydroarylation of Alkynes using Low-Valent Cobalt Catalysts

The ability of low-valent cobalt complexes to undergo chelation-assisted C—H activation of arenes was first reported by Murahashi in 1955, who discovered a $Co_2(CO)_8$-catalyzed ortho-carbonylation reaction of benzaldimine under CO pressure to afford an isoindolinone derivative [11]. Much later, in 1993, Klein and

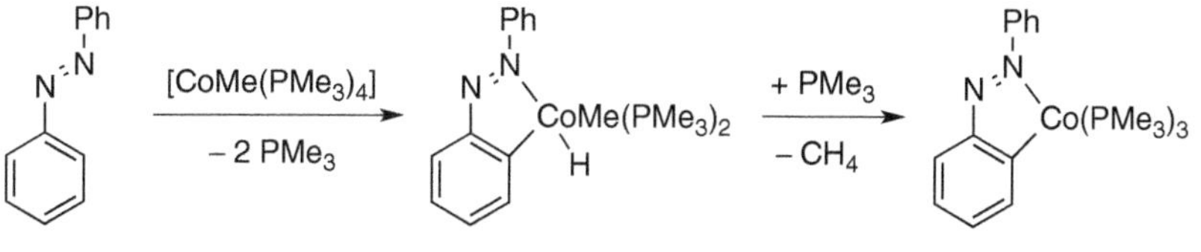

Scheme 6.3 Cyclometalation of azobenzene with a Co^{I} complex.

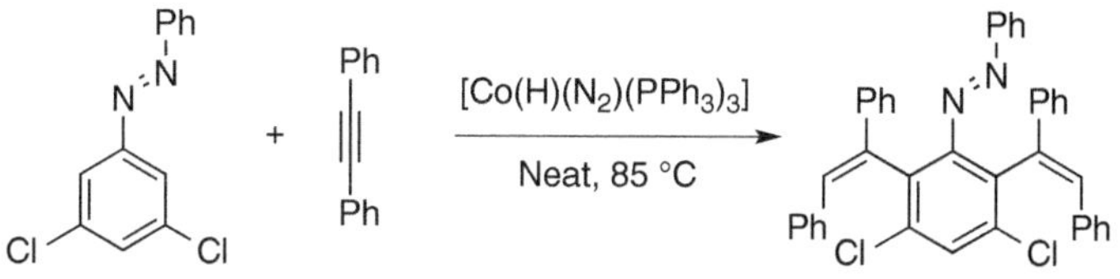

Scheme 6.4 Cobalt-catalyzed addition of azobenzene to diphenylacetylene.

coworkers reported the first example of a well-defined cyclometalation reaction using a Co^{I} complex $[CoMe(PMe_3)_4]$ and azobenzene, which is considered to proceed through nitrogen-assisted oxidative addition of the *ortho* C—H bond followed by reductive elimination of methane (Scheme 6.3) [12]. Since then, the same complex has been demonstrated to undergo cyclometalation of arenes bearing various other directing groups [13], representative examples being aryl imines and ketones to form five-membered metalacycles [14, 15] and aryl phosphines to form four-, five-, or six-membered metalacycles [16, 17]. Around the same time, Kisch and coworkers disclosed the first example of cobalt-catalyzed-directed hydroarylation of an alkyne using $Co(H)(N_2)(PPh_3)_3$ or $CoH_3(PPh_3)_3$ as the catalyst and an azobenzene derivative as the substrate (Scheme 6.4) [18]. The reaction affords an ortho-dialkenylation product in an anti-fashion, while the origin of the stereoselectivity remains unclear.

Inspired by the above early stoichiometric and catalytic studies, in 2010, Yoshikai and coworkers reported that an addition reaction of 2-arylpyridine to an internal alkyne, which was previously achieved by Lim *et al.* using a rhodium catalyst [19], can be promoted by a cobalt-based catalytic system comprising $CoBr_2$, $PMePh_2$, and MeMgCl as a precatalyst, a ligand, and a reductant, respectively (Scheme 6.5) [20]. This hydroarylation reaction features syn-stereoselectivity and complete transfer of the *ortho*-hydrogen atom to the vinylic position of the product, as confirmed by deuterium-labeling experiments. Thus, in light of mechanistic arguments on analogous hydroarylation reactions

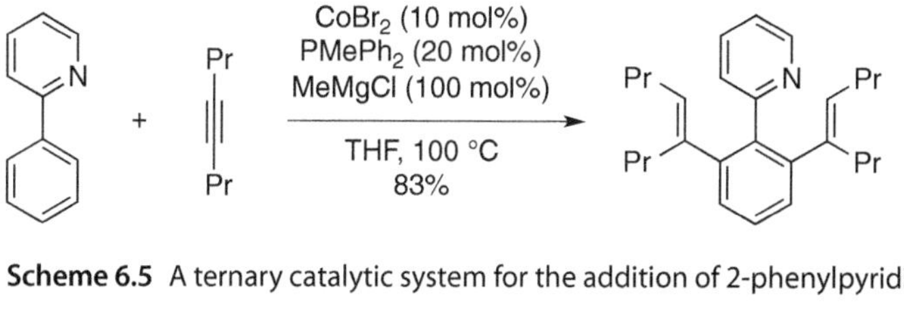

Scheme 6.5 A ternary catalytic system for the addition of 2-phenylpyridine to 4-octyne.

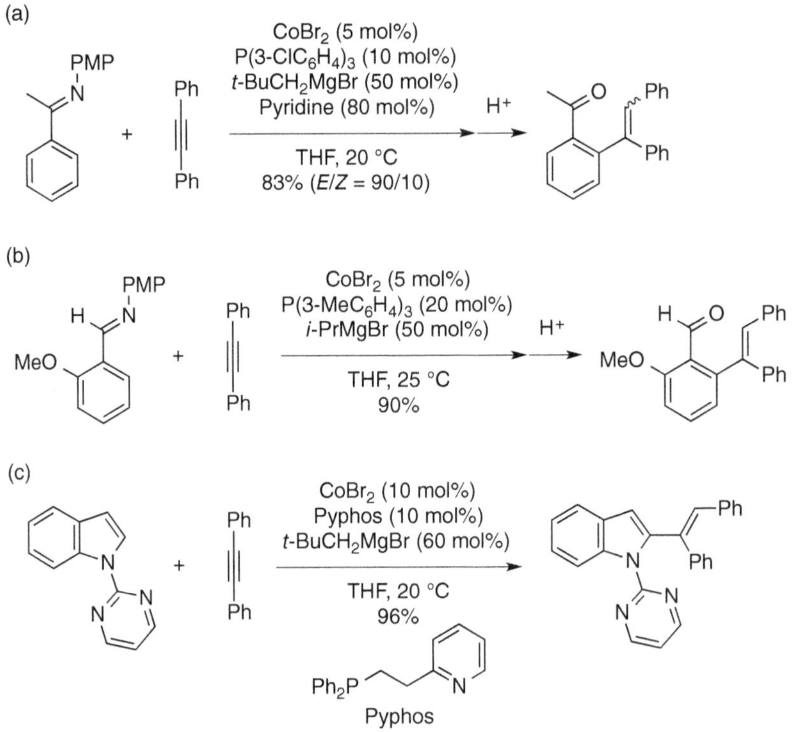

Scheme 6.6 Cobalt-catalyzed alkyne hydroarylation reactions.

catalyzed by rhodium(I) [6], a catalytic cycle involving pyridine-assisted C—H oxidative addition, migratory insertion of the alkyne into Co—H, and reductive elimination was conceived.

The discovery of the ternary catalytic system has opened the opportunity to emulate the reactivity of a Rh^{I} catalyst toward C—H activation using a cobalt catalyst, and has thus triggered further development of cobalt catalysts for alkyne hydro(hetero)arylation reactions involving directed C—H activation. Thus, the scope of aromatic substrates has been extended to aryl ketimines [21], aryl aldimines [22], and indoles bearing the N-pyrimidyl group [23] through careful optimization of the catalytic system for each substrate class (Scheme 6.6). These hydroarylation reactions are achieved efficiently at ambient temperature much lower than that typically required for the same type of reactions using Rh^{I} catalysts [24].

The trends in the regioselectivity with respect to the arenes and the alkynes in these hydroarylation reactions are worth mentioning. The reaction of aryl ketimines bearing methyl and trifluoromethyl groups on the meta-position reacted at the less hindered ortho position (Scheme 6.7). By contrast, alkoxy, cyano, and halogen (F, Cl, and Br) substituents on the meta-position caused C—H activation at their proximity, thus affording sterically more congested hydroarylation products. While such secondary directing effects of the alkoxy and fluorine substituents have been displayed in other transition metal-catalyzed

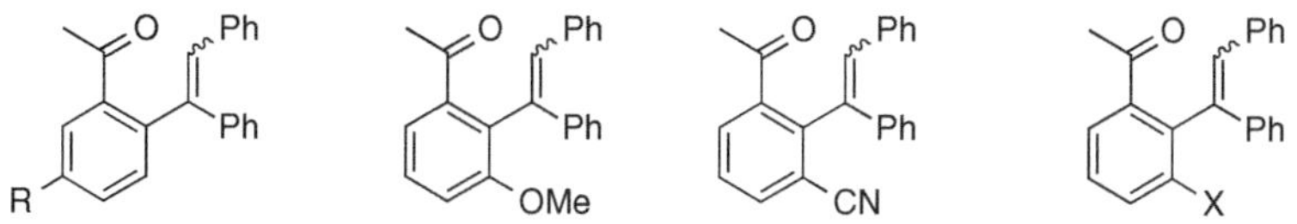

R = Me: 73% (*E*/*Z* = 90/10)
R = CF_3: 80% (*E*/*Z* = 88/12)

70% (*E*/*Z* = 92/8)

87% (*E*/*Z* = 58/42)

X = F: 64% (*E*/*Z* = 84/16)
X = Cl: 70% (*E*/*Z* > 99/1)
X = Br: 20% (*E*/*Z* > 99/1)

Scheme 6.7 Products of the cobalt-catalyzed addition of aryl ketimines to diphenylacetylene (Scheme 6.6a).

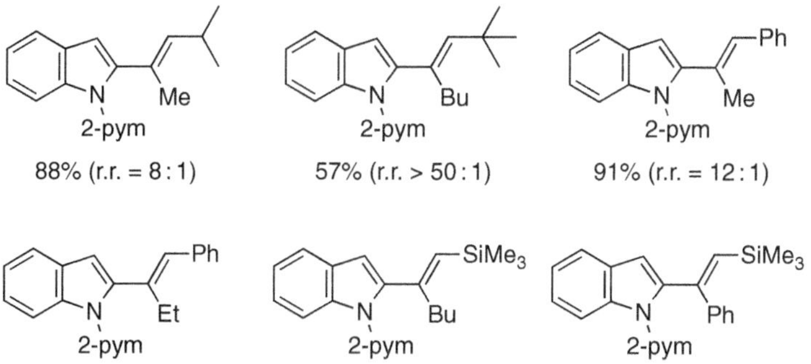

88% (r.r. = 8 : 1)   57% (r.r. > 50 : 1)   91% (r.r. = 12 : 1)

92% (r.r. = 1.6 : 1)   91% (r.r. > 50 : 1)   79% (r.r. > 50 : 1)

Scheme 6.8 Products of the cobalt-catalyzed addition of *N*-pyrimidylindole to unsymmetrically substituted alkynes (2-pym = 2-pyrimidyl; r.r. = regioisomer ratio; Scheme 6.6c).

hydroarylation reactions [25, 26], the same trend of regioselectivity with the cyano, chloro, and bromo substituents is rare. Regardless of the (hetero)arene substrates, the reaction of unsymmetrically substituted alkynes resulted in preferential C—C bond formation at the sterically less hindered side of the acetylenic carbons, as illustrated by the products of the reaction of *N*-pyrimidylindole (Scheme 6.8).

With the above observations and other mechanistic studies, a possible catalytic cycle for the cobalt-catalyzed addition of aryl ketimine to alkyne was proposed as shown in Scheme 6.9 [21]. Reduction of the cobalt precatalyst with neopentylmagnesium bromide leads to the formation of an active low-valent cobalt species, which is speculated to have the oxidation state of zero, with concomitant formation of a homocoupling product of the Grignard reagent. The active cobalt species undergoes reversible coordination of the alkyne, which is followed by chelation-assisted oxidative addition of the *ortho* C—H bond of the aryl imine. The resulting aryl(hydrido)cobalt species undergoes migratory insertion of the alkyne into the Co—H bond to produce an aryl(alkenyl)cobalt species. The general regioselectivity trend observed for unsymmetrical alkynes (see Scheme 6.8) may be attributed to the choice of the cobalt center to minimize the steric repulsion with the alkyne substituents in the migratory insertion. Finally, C—C reductive elimination results in the hydroarylation product while regenerating the active cobalt species. The reaction was found to exhibit a large

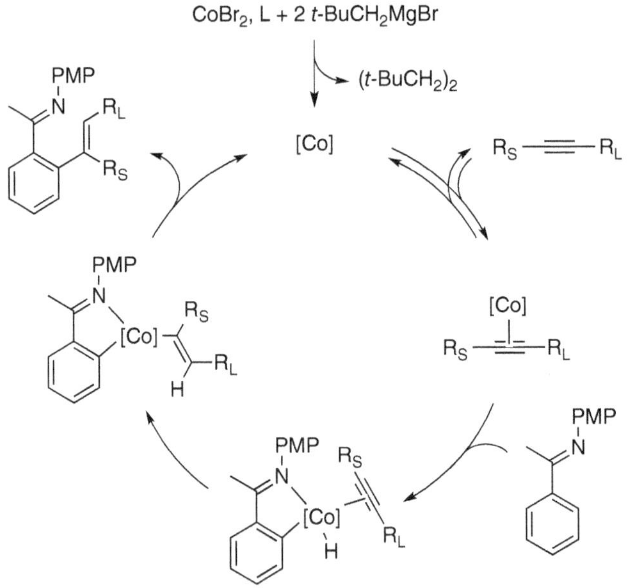

Scheme 6.9 Possible catalytic cycle for the addition of aryl ketimine to alkyne (R_L and R_S denote larger and smaller substituents, respectively).

H/D kinetic isotope effect ($k_H/k_D \approx 5$) with respect to the *ortho* C—H bond as well as a first-order dependence on the concentration of the imine, suggesting that the C—H activation is the rate-limiting step. The alkyne pre-coordination step was corroborated by a rate-saturation behavior observed with an increased alkyne concentration.

Petit and coworkers have recently demonstrated that a well-defined Co^0 complex $[Co(PMe_3)_4]$ serves as a single-component catalyst for the ketimine-directed hydroarylation of internal alkynes (Scheme 6.10) [27]. The reaction is performed with microwave irradiation at 170 °C, thus affording the anti-hydroarylation product in good yields. An analogous Co^I complex $[CoH(PMe_3)_4]$ also serves as a catalyst for the same transformation albeit with somewhat lower efficiency. Mechanistic experiments have shown that the anti-adduct is generated through isomerization of the initially formed syn-hydroarylation product. The well-defined catalyst also proved effective for β-alkenylation of α,β-unsaturated imines, which affords 1,2-dihydropyridine derivatives through 6π-electrocyclization of initially formed α,β,γ,δ-unsaturated imines [28, 29].

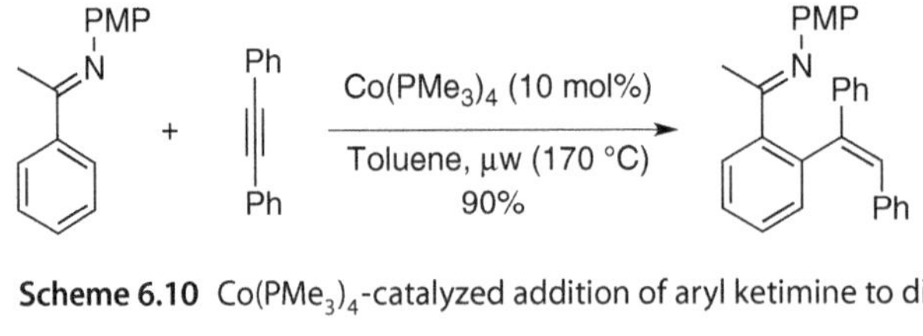

Scheme 6.10 $Co(PMe_3)_4$-catalyzed addition of aryl ketimine to diphenylacetylene.

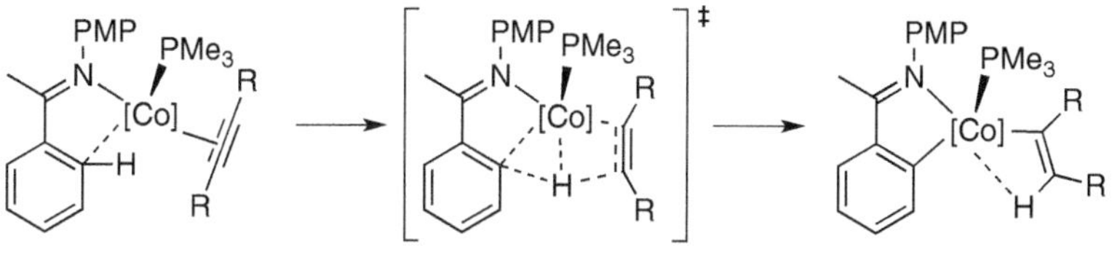

Scheme 6.11 Concerted C—H activation/hydrogen transfer mechanism.

The use of the well-defined catalyst in the above reaction prompted theoretical investigation of the reaction mechanism by DFT calculations. Interestingly, the DFT study suggested that the C—H bond cleavage and the insertion of the alkyne take place in a concerted manner (Scheme 6.11). This concerted mechanism is reminiscent of the ligand-to-ligand hydrogen transfer (LLHT) mechanism, which was coined for the concerted C—H activation/alkyne (or alkene) insertion process proposed for the nickel-catalyzed hydroarylation reactions of alkynes and alkenes [30, 31]. Regardless of the initially proposed stepwise oxidative addition/migratory insertion mechanism (Scheme 6.7), Yoshikai's *in situ*-generated cobalt catalysts may also operate according to a similar concerted mechanism, not only in the alkyne hydroarylation reactions but also in the alkene hydroarylation reactions described in the following subsection. It should be noted, however, that this type of concerted mechanisms awaits further experimental support.

Besides demonstrating the above chelation-assisted hydroarylation reactions of alkynes, Yoshikai also demonstrated that addition of azole derivatives to alkynes [32] could also be promoted by cobalt complexes. A catalyst generated from $CoBr_2$, DPEphos, and trimethylsilylmethylmagnesium chloride smoothly promotes the addition of (benz)oxazoles to internal alkynes, thus affording C2-alkenylated (benz)oxazole derivatives with high syn-stereoselectivity at ambient temperature (Scheme 6.12a) [33]. Modification of the catalytic system using Xantphos instead of DPEphos further enabled the use of (benzo)thiazoles as the substrates, while an elevated temperature of 60 °C was required (Scheme 6.12b) [34]. As was the case with the chelation-assisted hydroarylation, deuterium-labeling experiments confirmed complete transfer of the C2-hydrogen atom to the vinylic position of the product without any crossover. This observation is consistent with that noticed in a three-step catalytic cycle involving oxidative addition, migratory insertion, and reductive elimination, while the initial two steps may take place in a concerted manner (cf. Scheme 6.9).

6.2.2 Hydroarylation of Alkenes using Low-Valent Cobalt Catalysts

Building on the development of the chelation-assisted hydroarylation reactions of alkynes, Yoshikai and coworkers extended the scope of the cobalt catalysis to the hydroarylation of alkenes. Their study was initially focused on the addition of 2-phenylpyridine to styrene. A cobalt–tricyclohexylphosphine catalyst was found to promote the addition reaction in a highly branch-selective fashion to afford the corresponding 1,1-diarylethane derivative (Scheme 6.13a) [35]. By contrast, the use of a bulky *N*-heterocyclic carbene (NHC) ligand, IMes,

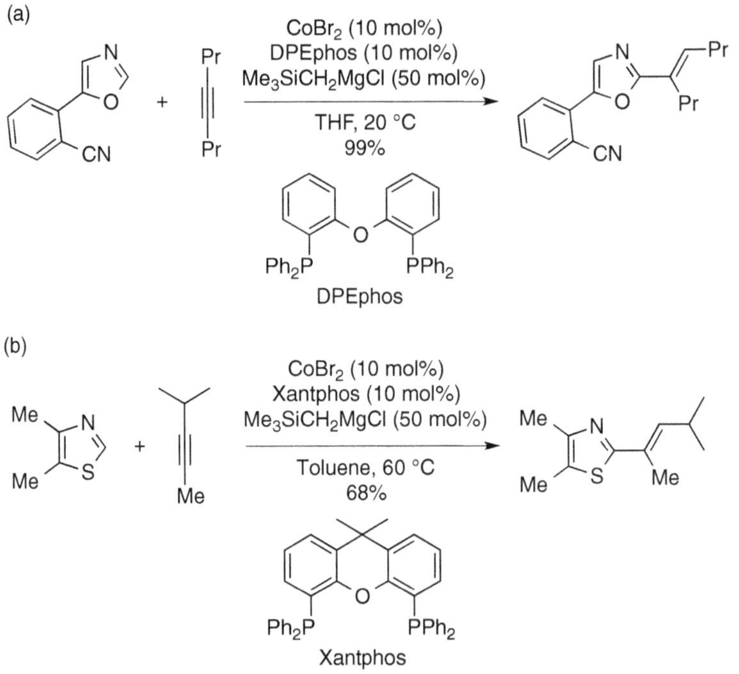

Scheme 6.12 Cobalt-catalyzed addition of oxazole and thiazole derivatives to internal alkynes.

instead of PCy_3, resulted in near complete reversal of the regioselectivity, thus leading to the formation of the 1,2-diarylethane isomer as the major product (Scheme 6.13b). The former, branched-selective addition is particularly notable because ruthenium- and rhodium-catalyzed styrene hydroarylations are known to mostly exhibit linear selectivity [1, 6, 36]. The regiodivergent catalytic systems showed reasonable levels of generality with respect to both the 2-arylpyridine and the styrene substrates, while in some cases the electronic nature of the substrates outweighed the ligand control.

The reaction of pentadeuterated 2-phenylpyridine and styrene resulted in the scrambling of the *ortho*-deuterium atoms and the α- and β-hydrogen atoms of styrene under both the Co–PCy_3 and Co–IMes catalytic systems. These observations suggest that (i) both the C—H activation and the styrene insertion steps are reversible; (ii) the styrene insertion can lead to either a branched or a linear alkylcobalt intermediate, and that these two pathways are competing with each other; (iii) the regioselectivity is determined in the reductive elimination step (Scheme 6.14). The branched selectivity of the Co–PCy_3 catalytic system may be ascribed to favorable η^3-benzyl type coordination in the branched alkylcobalt species. On the other hand, the linear selectivity of the Co–IMes catalytic system may be explained by the preference of the sterically more-shielding IMes ligand for the less-crowded linear alkylcobalt intermediate. These conjectures have been supported by computational model studies by Fu and coworkers [37]. It would be worthwhile to note that other examples of regiodivergent

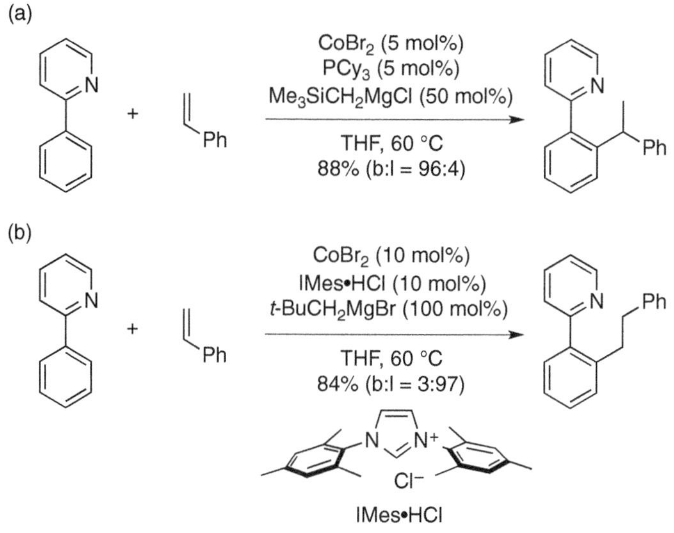

Scheme 6.13 Cobalt-catalyzed, ligand-controlled regiodivergent addition of 2-phenylpyridine to styrene.

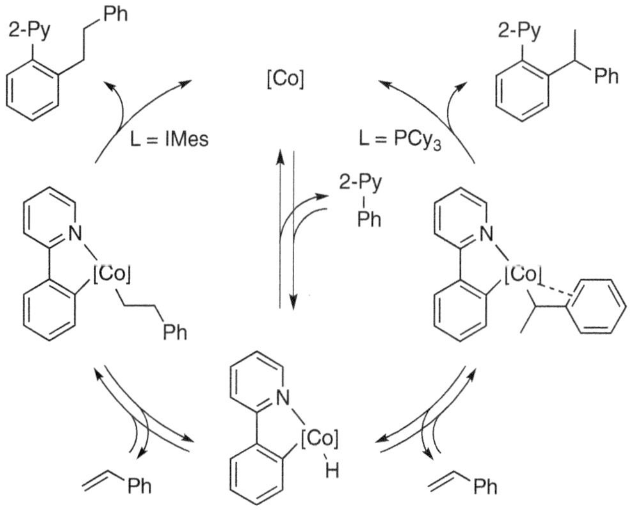

Scheme 6.14 Proposed catalytic cycles for the regiodivergent addition of 2-phenylpyridine to styrene.

hydroarylation of styrenes have recently been reported using different transition metal catalysts and arene substrates [36, 38–41].

The intrinsic branched selectivity of cobalt catalysts enabled only a facile extension of the scope of the arene substrates to aryl aldimines and ketimines (Scheme 6.15) [42, 43]. Catalytic systems consisting of $CoBr_2$, a triarylphosphine, and a Grignard reagent (t-$BuCH_2MgBr$ or CyMgBr) promote the addition of these aromatic substrates to various styrene derivatives with good to excellent

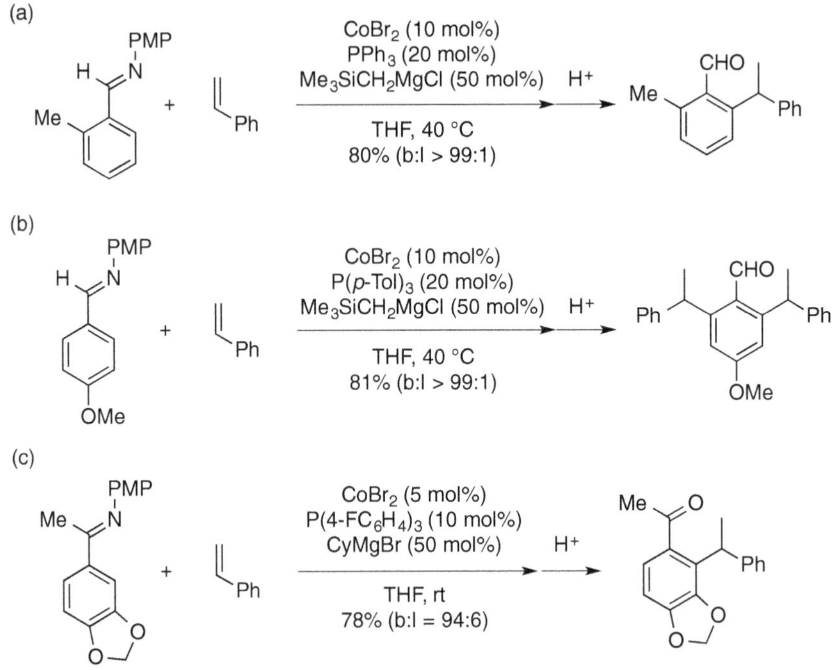

Scheme 6.15 Branch-selective addition of aryl aldimine and ketimine to styrene under cobalt–triarylphosphine catalysis.

branched-selectivity (>95:5 in many cases). The reaction of ortho-unsubstituted aldimines resulted in the formation of dialkylation products even with limited loading of styrene, which suggests that the coordination of the aldimine to the cobalt center is relatively strong and that the alkylation of the first *ortho* C—H bond is followed by the activation of the second *ortho* C—H bond rather than the dissociation of the cobalt catalyst.

Attempts to make the imine-directed, branched-selective styrene hydroarylation enantioselective have met with only partial success, because of the difficulty to identify monodentate chiral phosphorus ligands that can give rise to both reactive and enantioselective cobalt catalysts. The reaction of 1-Boc-3-iminoindoles with styrene derivatives proceeded in the presence of a cobalt catalyst bearing a chiral phosphoramidite ligand, thus affording the corresponding 1,1-diarylethane products with enantioselectivities up to 86% ee (Scheme 6.16) [44]. The catalytic system, however, does not promote the reaction of aryl imines other than the indole substrates.

Regardless of the intrinsic branched selectivity of the cobalt catalyst, elaboration of the ligand structure allowed the extension of the scope of linear-selective styrene hydroarylation using aryl ketimines. A catalytic system featuring the combination of a triarylphosphine ligand containing 2,4-dimethoxyphenyl groups and a Lewis basic additive such as 2-methoxypyridine or DBU promotes the addition of a series of aryl ketimines to styrene derivatives with high level of linear selectivity (Scheme 6.17; l:b > 90:10 in most cases) [45]. The

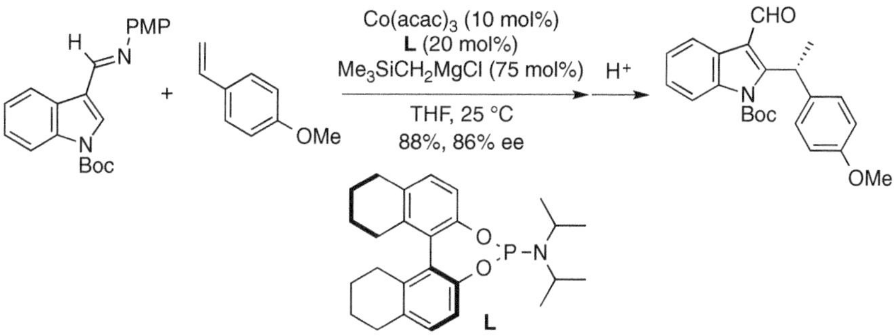

Scheme 6.16 Enantioselective addition of 1-Boc-3-iminoindole to styrene derivative with a cobalt–phosphoramidite catalyst.

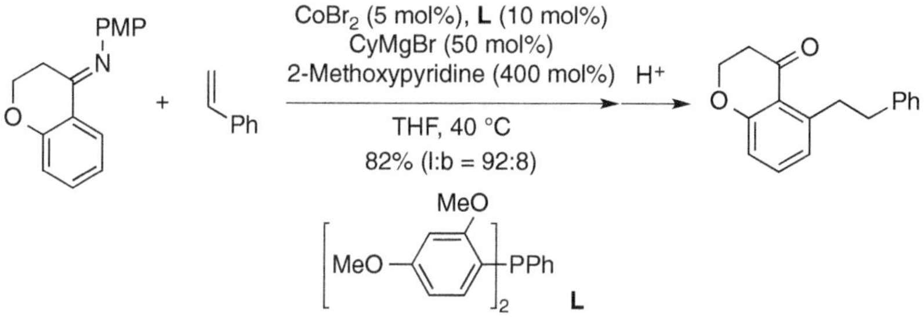

Scheme 6.17 Linear-selective addition of aryl ketimine to styrene under cobalt–triarylphosphine catalysis.

triarylphosphine ligand plays an essential role in controlling the regioselectivity, while the Lewis base enhances the catalytic efficiency.

Deuterium-labeling experiments again indicated that the regioselectivity is predetermined in the C—C bond-forming reductive elimination step. Unlike in the case of the Co–NHC catalyzed reaction (Scheme 6.13b), it may not be possible to explain the origin of the linear selectivity merely by the steric nature of the ligand, because bulkier triarylphosphines, such as tris(2,6-diethylphenyl)phosphine, promote the reaction with high branched selectivity (b:l > 90:10). Instead, a possible rationale is the role of the 2-methoxy group as a hemilabile ligand to the cobalt center (Scheme 6.18). Thus, the triarylphosphine would serve as a temporary PO bidentate ligand to stabilize the linear alkylcobalt intermediate and its reductive elimination transition state, while such stabilization is not feasible in the branched, η^3-benzyl intermediate and transition state due to the difference in the coordination number.

Cobalt catalysis also enables chelation-assisted hydroarylation of simple alkenes apart from that of styrene derivatives. Nakamura and coworkers developed an addition reaction of benzamide derivatives to alkenes with the aid of a cobalt catalyst generated *in situ* from $Co(acac)_2$ and CyMgCl in the presence of DMPU as an additive (Scheme 6.19) [46]. The catalytic system promotes the ortho-alkylation of secondary benzamides with various alkenes under ambient temperature conditions in excellent yields. Interestingly, reactions of both

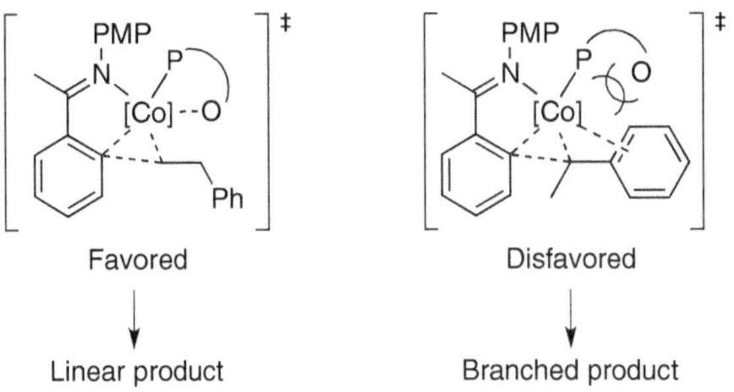

Scheme 6.18 Transition state models for the linear-selective addition of aryl ketimine to styrene.

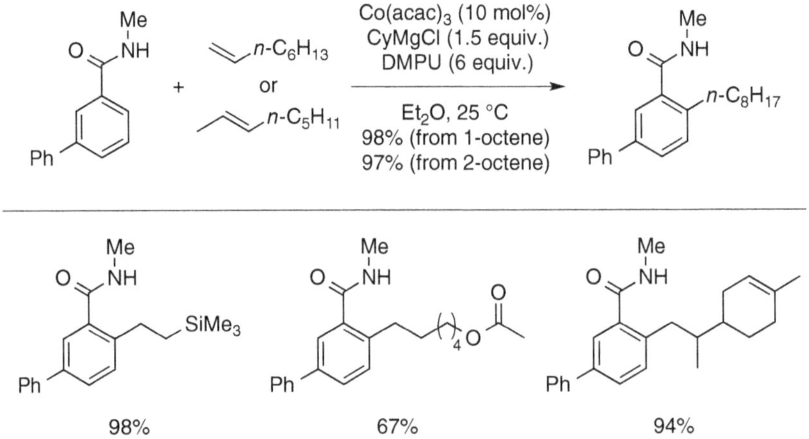

Scheme 6.19 Cobalt-catalyzed, amide-directed hydroarylation of alkenes.

1-octene and 2-octene resulted in the introduction of *n*-octyl group into the ortho position. Thus, the latter reaction is considered to proceed through olefin isomerization followed by hydroarylation. The reaction of limonene resulted in hydroarylation of the 1,1-disubstituted alkene moiety with the trisubstituted alkene moiety kept intact.

Besides the styrene hydroarylation reactions, Yoshikai and coworkers developed catalytic systems suitable for the addition of aryl ketimines to vinylsilanes and alkyl olefins [47, 48]. Thus, catalytic systems comprising $CoBr_2$, 1,10-phenanthroline or neocuproine, and *t*-$BuCH_2MgBr$ promote imine-directed hydroarylation reactions to vinylsilanes or alkyl olefins (Scheme 6.20). As was the case with the related styrene hydroarylation, deuterium-labeling experiments indicated the reversibility of the C—H activation and the olefin insertion steps, while the olefin insertion step predominantly leads to a linear aryl(alkyl)cobalt intermediate.

The imine-directed hydroarylation of an alkyl olefin has also been achieved in intramolecular settings. An indole derivative bearing a homoallyl group and an aldimine-directing group on the N and the C3 atoms, respectively, undergoes intramolecular cyclization in the presence of cobalt–NHC catalysts

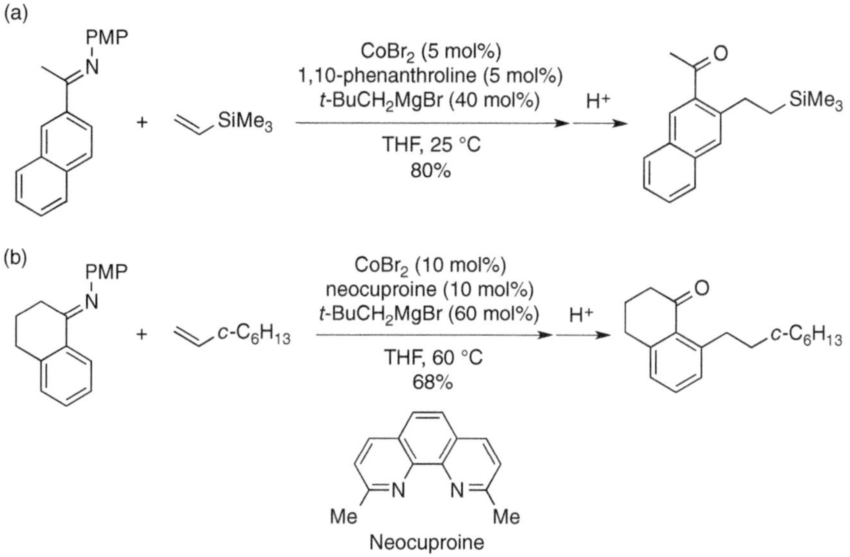

Scheme 6.20 Imine-directed hydroarylation of vinylsilane and alkyl olefin using cobalt–phenanthroline-type catalysts.

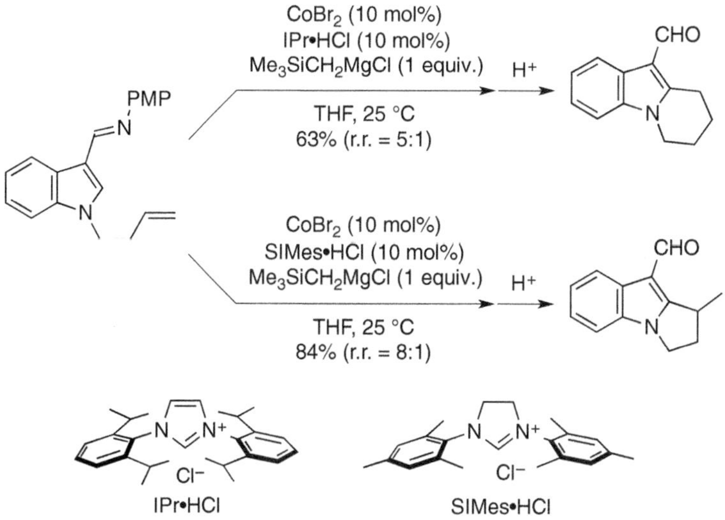

Scheme 6.21 Imine-directed intramolecular olefin hydroarylation on an indole platform.

(Scheme 6.21) [49]. Notably, the mode of cyclization may be controlled by the choice of the NHC ligand. Thus, the IPr ligand promotes 6-*endo*-type cyclization to afford a tetrahydropyridoindole derivative with decent regioselectivity, while the SIMes ligand prefers 5-*exo*-type cyclization and thus furnishes a dihydropyrroloindole derivative as the major product. Experiments using a C2-deuterated substrate suggested that the regioselectivity is predetermined in the olefin insertion step.

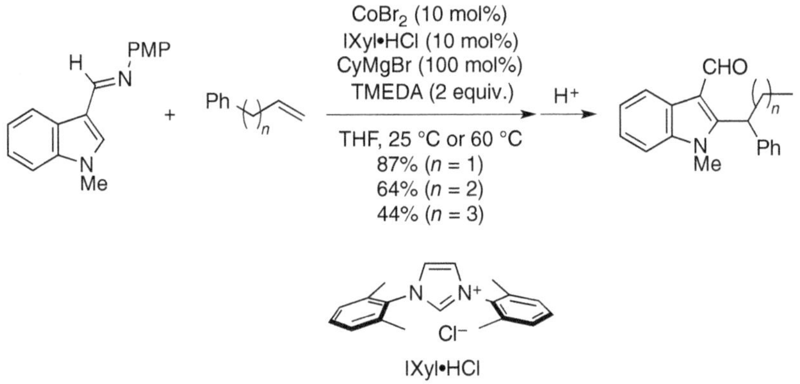

Scheme 6.22 Tandem alkene isomerization–hydroarylation reaction.

Low-valent cobalt species are known to be capable of promoting isomerization of alkenes [50–53]. Such reactivity is manifested in the above discussed amide-directed hydroarylation of an internal alkene (Scheme 6.19), where the formation of the linear alkylation product may be rationalized by kinetic preference for the reductive elimination of an aryl(*n*-alkyl)cobalt species rather than of an aryl(*sec*-alkyl)cobalt species. As a different example of tandem alkene isomerization–hydroarylation, Yoshikai reported on cobalt–NHC-catalyzed addition reactions of a 3-iminoindole derivative to allyl, homoallyl, and bishomoallylbenzene, which uniformly led to the formation of 1,1-diarylalkane-type products (Scheme 6.22) [54]. As was the case with the branched-selective styrene hydroarylation, the regioselectivity may be ascribed to the preference for η^3-benzylcobalt intermediates. A related tandem isomerization–hydroarylation reaction was achieved using a nickel catalyst and azole derivatives [55].

Besides the above alkene hydroarylation reactions involving C—H activation, Matsunaga and Kanai reported a different type of cobalt-catalyzed hydroarylation using pyridines and quinolines as the substrates [56, 57]. Thus, a catalytic system comprising $CoBr_2$, $LiBEt_3H$, and Et_3B promoted the addition of pyridines and quinolines to styrenes and alkyl olefins to afford the corresponding C4-alkylated products (Scheme 6.23). The reaction of styrenes led to branched alkylation products, while alkyl olefins gave rise to linear alkylation products. The reaction is proposed to involve a cobalt hydride as a catalytically active species. Hydrometalation of the olefin and C4-selective nucleophilic attack of the resulting alkylcobalt species to pyridine (or quinoline) generate a dihydropyridine intermediate, which undergoes rearomatization to afford the alkylation product while regenerating the cobalt hydride.

6.2.3 Hydroarylation of Alkynes and Alkenes using Cp^*Co^{III} Catalysts

In 2013, Matsunaga and Kanai disclosed the first example of the use of a cationic Cp^*Co^{III} complex in directing group-assisted C—H bond functionalization [58], which ignited subsequent developments of various Cp^*Co^{III}-catalyzed

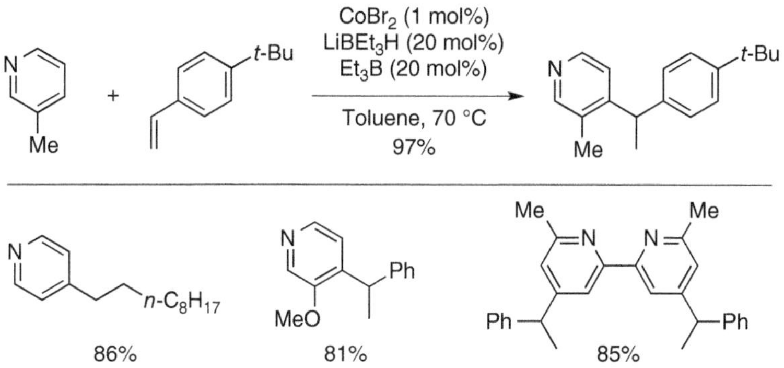

Scheme 6.23 Cobalt-catalyzed C4-alkylation of pyridine with alkenes.

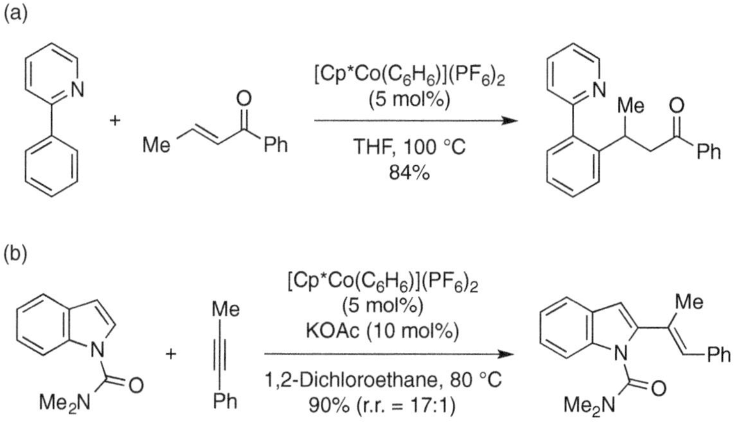

Scheme 6.24 Cp*Co^{III}-catalyzed, directing group-assisted hydroarylation of alkene and alkyne.

C—C and C–heteroatom bond-forming C—H functionalization reactions from them as well as from other research groups [10, 59]. Examples of hydroarylation reactions using the Cp*Co^{III} catalyst are summarized in Scheme 6.24. In the presence of a catalytic amount of a dicationic complex $[Cp^*Co(C_6H_6)](PF_6)_2$, 2-arylpyridine undergoes the addition of the *ortho* C—H bond to an α,β-unsaturated ketone to give a β-arylated ketone product (Scheme 6.24a) [58]. The same catalyst promotes alkenylation of the C2-position of *N*-dimethylcarbamoylindole with an internal or a terminal alkyne with high regio- and stereoselectivities (Scheme 6.24b) [60]. Interestingly, replacement of the dimethylamino group with a morpholino group in the latter reaction results in deaminative annulation rather than hydroarylation, thus affording a pyrroloindolone derivative. These hydroarylation reactions are considered to proceed through (carboxylate-assisted) deprotonative metalation, insertion of the unsaturated substrate into the Co—C bond, and protonation of the resulting organocobalt species.

6.3 Hydroarylation of Alkynes and Alkenes using Iron Catalysts

6.3.1 Hydroarylation of Alkynes and Alkenes using Low-Valent Iron Catalysts

Klein *et al.* demonstrated that Fe^0 and Fe^{II} complexes $[Fe(PMe_3)_4]$ and $[FeMe_2(PMe_3)_4]$, like the analogous Co^I complex $[CoMe(PMe_3)_4]$, are capable of undergoing nitrogen-assisted cyclometalation (Scheme 6.25) [14]. The former complex undergoes oxidative addition of the *ortho* C—H bond of an aryl imine, while the latter complex metalates the *ortho* position of another aryl imine with concomitant elimination of methane. In particular, the former C—H activation reaction points to the possibility of catalytic hydroarylation through the Lewis/Murai-type mechanism (Scheme 6.2), which involves directed C—H oxidative addition as one of the key elementary steps.

Stimulated by the above cyclometalation reactions as well as their own studies on the cobalt-catalyzed hydroarylation, Yoshikai and coworkers developed iron-based catalytic systems for imine-directed hydroarylation reactions (Scheme 6.26) [61]. An iron–NHC catalyst generated from $Fe(acac)_3$, imidazolium salt SIXyl·HCl, and CyMgCl in the presence of TMEDA promotes the addition of 3-iminoindole derivative to parent styrene and β-substituted styrenes to afford 1,1-diarylalkane products with exclusive regioselectivity. A similar iron–NHC catalytic system is effective for the addition of the same indole substrate to aryl-substituted internal alkynes, where C—C bond formation occurs selectively at the less hindered acetylenic carbon atom. Aryl imines other than the indole substrate, however, exhibit rather poor reactivities under these reaction conditions.

6.3.2 Hydroarylation of Alkenes using Lewis Acidic Iron Catalysts

Beller and coworkers reported that a simple iron salt $FeCl_3$ serves as a convenient catalyst for the hydroarylation of styrene derivatives with electron-rich arenes (Scheme 6.27) [62]. The reaction takes place simply by mixing a solvent amount of arene with styrene in the presence of catalytic $FeCl_3$, and affords the corresponding 1,1-diarylethane product with exclusive branched selectivity. However, as is the case with typical Friedel–Crafts reactions, the positional selectivity with respect to the arene substrate is often imperfect and modest.

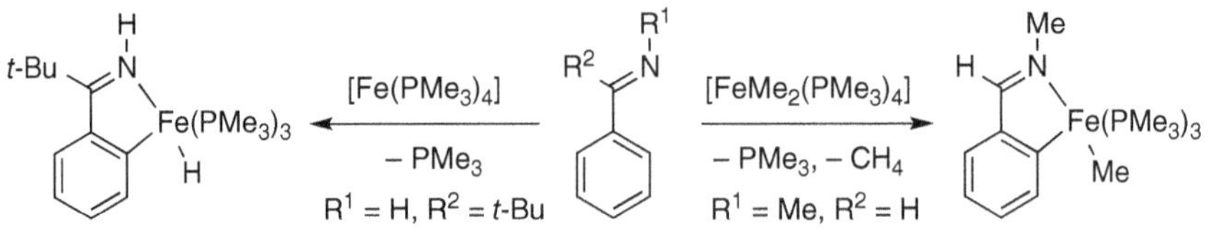

Scheme 6.25 Cyclometalation of aryl imines with Fe^0 and Fe^{II} complexes.

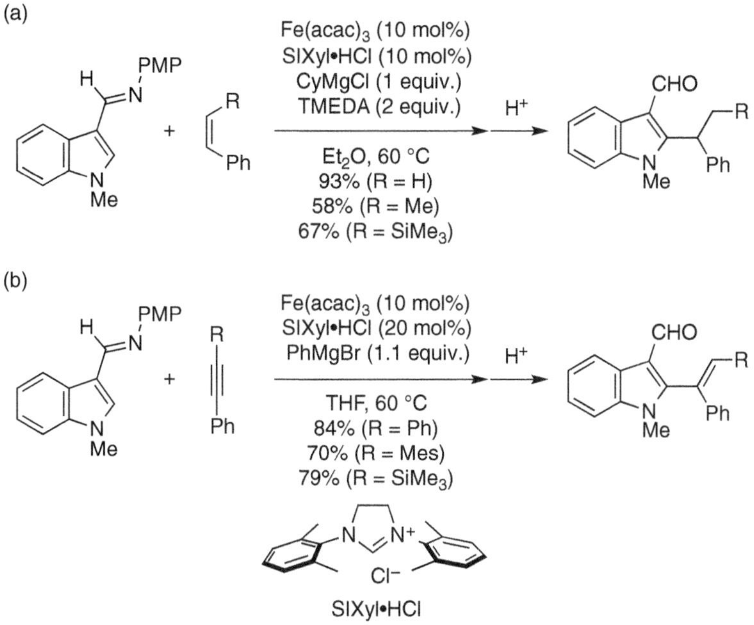

Scheme 6.26 Addition of 3-iminoindole to aryl olefins and alkynes under iron–NHC catalysis.

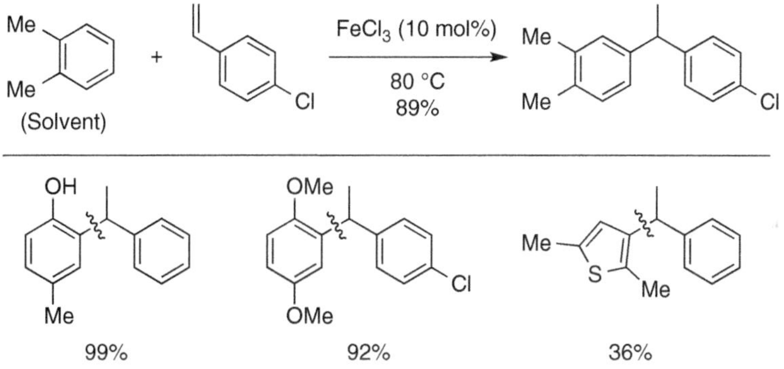

Scheme 6.27 $FeCl_3$-catalyzed hydroarylation of styrenes with electron-rich arenes.

6.4 Hydroarylation of Alkynes using Low-Valent Manganese Catalyst

The ability of a low-valent manganese complex such as $MnR(CO)_5$ (R = Me, Bn) to undergo cyclometalation of arenes bearing directing groups has been known since the 1970s [13]. The resulting manganacycle complex has also been demonstrated to undergo insertion of an alkyne or an alkene, since the late 1980s (Scheme 6.28) [63, 64]. However, the cyclometalation and the migratory insertion reactions have not been integrated into a catalytic hydroarylation

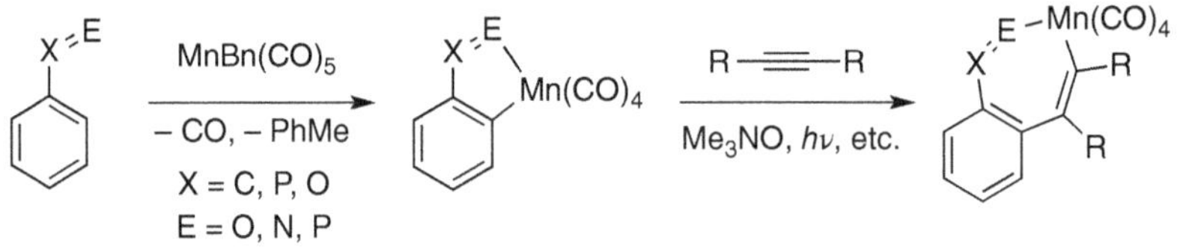

Scheme 6.28 Stoichiometric cyclometalation of $MnBn(CO)_5$ and subsequent alkyne insertion.

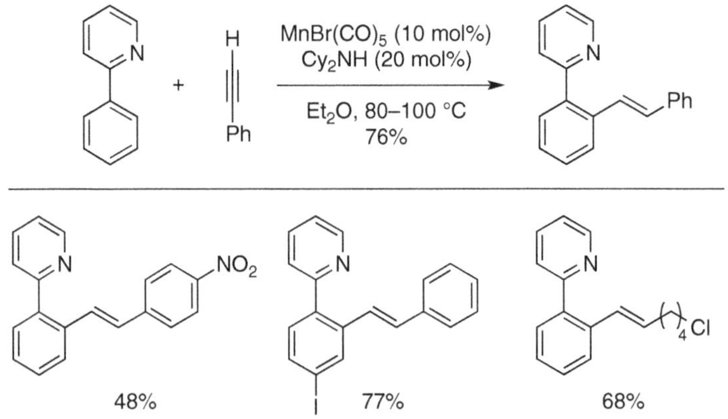

Scheme 6.29 Manganese-catalyzed directed hydroarylation of terminal alkynes.

process for a long time. A part of the reason for the difficulty of such catalysis might be the requirement for an amine N-oxide or irradiation to remove the CO ligand and thus promote the migratory insertion.

In 2013, Zhou, Chen, and Wang reported on the first example of a catalytic hydroarylation reaction of an alkyne via manganese-mediated cyclometalation (Scheme 6.29) [65, 66]. Thus, a catalytic system consisting of $MnBr(CO)_5$ and a secondary amine such as Cy_2NH promotes the addition of 2-phenylpyridine and related substrates to terminal alkynes to afford disubstituted alkene products with trans configuration. The catalytic system tolerates a variety of functional groups, while it does not show any activity toward internal alkynes. On the basis of experimental and computational mechanistic studies, a catalytic cycle involving deprotonative cyclometalation of the arene, insertion of the alkyne into the Mn–aryl bond, and protonation of the Mn–alkenyl bond with the terminal alkyne is proposed. The secondary amine is considered to serve as a base in the initial cyclometalation, while the catalytic turnover may be achieved through cyclometalation with a manganese acetylide species.

Wang and coworkers extended the scope of the Mn/amine catalytic system to hydroarylation of α,β-unsaturated carbonyl compounds (Scheme 6.30) [67]. A variety of 2-arylpyridines undergo addition to acrylate esters to afford the corresponding monoalkylation products in moderate to good yields. Heteroaromatic substrates such as N-pyrimidylindole can also be alkylated by methyl acrylate albeit in a modest yield. Furthermore, α,β-unsaturated ketones such as ethyl vinyl ketone and cyclohexenone can also be used as the alkylating agents.

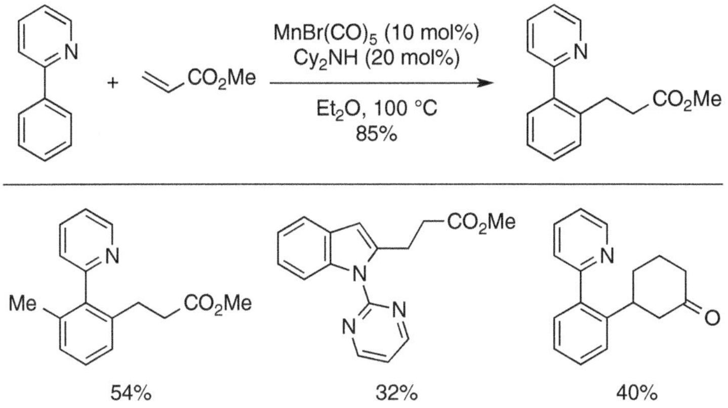

Scheme 6.30 Manganese-catalyzed directed hydroarylation of α,β-unsaturated carbonyl compounds.

The reaction is proposed to proceed through Cy_2NH-assisted cyclometalation, conjugate addition of the aryl–Mn species to acrylate, and C-protonation of the Mn enolate with $[Cy_2NH_2]^+$, as supported by DFT calculations.

6.5 Conclusions

Various catalytic systems based on the group 7–9 first-row transition metals have been developed for hydroarylation of alkynes and alkenes. In particular, cobalt catalysis has demonstrated a broad scope for hydroarylation through directing group-assisted C—H bond activation. Low-valent cobalt catalysts not only serve as cost-effective alternatives to well-established rhodium(I) catalysts in some reactions, but also feature mild C—H activation conditions and unique regioselectivities that have not been attained with rhodium(I). The branched-selective hydroarylation of styrenes is a notable example in this respect. Cp^*Co^{III} catalysis is attractive for the broad functional group compatibility, and further expansion of its scope is anticipated. The development of hydroarylation reactions using iron- and manganese catalysts is still in an exploratory stage, and the potential of these catalysts remains to be seen. On the whole, further exploration is required to establish group 7–9 first-row transition metal catalysts for practical and sustainable hydroarylation reactions.

6.6 Abbreviations

acac	acetylacetonato
Boc	*tert*-butoxycarbonyl
Cp*	pentamethylcyclopentadienyl
Cy	cyclohexyl
DBU	1,8-diazabicyclo[5.4.0]undec-7-ene

DFT	density functional theory
DMPU	1,3-dimethyl-3,4,5,6-tetrahydro-2(1*H*)-pyrimidinone
DPEphos	bis[(2-diphenylphosphino)phenyl] ether
IMes	1,3-bis(2,4,6-trimethylphenyl)imidazol-2-ylidene
IPr	1,3-bis(2,6-diisopropylphenyl)imidazol-2-ylidene
IXyl	1,3-bis(2,6-dimethylphenyl)imidazol-2-ylidene
NHC	*N*-heterocyclic carbene
PMP	*p*-methoxyphenyl
Py	pyridyl
pyphos	2-[2-(diphenylphosphino)ethyl]pyridine
SIMes	1,3-bis(2,4,6-trimethylphenyl)-4,5-dihydroimidazol-2-ylidene
SIXyl	1,3-bis(2,6-dimethylphenyl)-4,5-dihydroimidazol-2-ylidene
TMEDA	*N,N,N′,N′*-tetramethylethylenediamine
Xantphos	4,5-bis(diphenylphosphino)-9,9-dimethylxanthene

References

1 Kakiuchi, F. and Kochi, T. (2008) Transition-metal-catalyzed carbon-carbon bond formation via carbon-hydrogen bond cleavage. *Synthesis*, **2008**, 3013–3039.
2 Lewis, L.N. and Smith, J.F. (1986) Catalytic C-C bond formation via ortho-metalated complexes. *J. Am. Chem. Soc.*, **108**, 2728–2735.
3 Murai, S. *et al.* (1993) Efficient catalytic addition of aromatic carbon-hydrogen bonds to olefins. *Nature*, **366**, 529–531.
4 Kakiuchi, F. and Murai, S. (2002) Catalytic C—H/olefin coupling. *Acc. Chem. Res.*, **35**, 826–834.
5 Foley, N.A., Lee, J.P., Ke, Z., Gunnoe, T.B., and Cundari, T.R. (2009) Ru(II) catalysts supported by hydridotris(pyrazolyl)borate for the hydroarylation of olefins: reaction scope, mechanistic studies, and guides for the development of improved catalysts. *Acc. Chem. Res.*, **42**, 585–597.
6 Colby, D.A., Bergman, R.G., and Ellman, J.A. (2010) Rhodium-catalyzed C-C bond formation via heteroatom-directed C—H bond activation. *Chem. Rev.*, **110**, 624–655.
7 Pan, S. and Shibata, T. (2013) Recent advances in iridium-catalyzed alkylation of C—H and N—H bonds. *ACS Catal.*, **3**, 704–712.
8 Gao, K. and Yoshikai, N. (2014) Low-valent cobalt catalysis: new opportunities for C—H functionalization. *Acc. Chem. Res.*, **47**, 1208–1219.
9 Yoshikai, N. (2014) Development of cobalt-catalyzed C—H bond functionalization reactions. *Bull. Chem. Soc. Jpn.*, **87**, 843–857.
10 Moselage, M., Li, J., and Ackermann, L. (2016) Cobalt-catalyzed C—H activation. *ACS Catal.*, **6**, 498–525.
11 Murahashi, S. (1955) Synthesis of phthalimidines from schiff bases and carbon monoxide. *J. Am. Chem. Soc.*, **77**, 6403–6404.
12 Klein, H.-F., Helwig, M., Koch, U., Flörke, U., and Haupt, H.-J. (1993) Coordination and reactions of diazenes in trimethylphosphinecobalt(I) complexes - syntheses and structures of complexes containing

Mu-2-(N,N')-benzo[C]cinnoline and Eta-2-azobenzene ligands. *Z. Naturforsch., B*, **48**, 778–784.
13 Albrecht, M. (2010) Cyclometalation using d-block transition metals: fundamental aspects and recent trends. *Chem. Rev.*, **110**, 576–623.
14 Klein, H.-F., Camadanli, S., Beck, R., Leukel, D., and Flörke, U. (2005) Cyclometalation of substrates containing imine and pyridyl anchoring groups by iron and cobalt complexes. *Angew. Chem., Int. Ed.*, **44**, 975–977.
15 Camadanli, S., Beck, R., Flörke, U., and Klein, H.-F. (2008) First regioselective cyclometalation reactions of cobalt in arylketones: C—H versus C—F activation. *Dalton Trans.*, **2008**, 5701–5704.
16 Klein, H.-F., Beck, R., Flörke, U., and Haupt, H.-J. (2003) Five- and six-membered cobaltocycles with C,P-chelating ligands through regiospecific cyclometalation of mono(2-substituted) triphenylphosphane. *Eur. J. Inorg. Chem.*, **2003**, 1380–1387.
17 Klein, H.-F., Beck, R., Flörke, U., and Haupt, H.-J. (2003) Regiospecific cyclometalation of diphenyl(2-substituted phenyl)phosphane with methyltetrakis(trimethylphosphane)cobalt(I). *Eur. J. Inorg. Chem.*, **2003**, 853–862.
18 Halbritter, G., Knoch, F., Wolski, A., and Kisch, H. (1994) Transition-metal complexes of diazenes. 32. Functionalization of aromatic azo-compounds by the cobalt-catalyzed, regioselective double addition of tolane - 2,6-distilbenylazobenzenes and 2,3-dihydrocinnolines. *Angew. Chem., Int. Ed.*, **33**, 1603–1605.
19 Lim, Y.-G., Lee, K.-H., Koo, B.T., and Kang, J.-B. (2001) Rhodium(I)-catalyzed ortho-alkenylation of 2-phenylpyridines with alkynes. *Tetrahedron Lett.*, **42**, 7609–7612.
20 Gao, K., Lee, P.-S., Fujita, T., and Yoshikai, N. (2010) Cobalt-catalyzed hydroarylation of alkynes through chelation-assisted C—H bond activation. *J. Am. Chem. Soc.*, **132**, 12249–12251.
21 Lee, P.-S., Fujita, T., and Yoshikai, N. (2011) Cobalt-catalyzed, room-temperature addition of aromatic imines to alkynes via directed C—H bond activation. *J. Am. Chem. Soc.*, **133**, 17283–17295.
22 Yamakawa, T. and Yoshikai, N. (2013) Cobalt-catalyzed ortho-alkenylation of aromatic aldimines via chelation-assisted C—H bond activation. *Tetrahedron*, **69**, 4459–4465.
23 Ding, Z. and Yoshikai, N. (2012) Mild and efficient C2-alkenylation of indoles with alkynes catalyzed by a cobalt complex. *Angew. Chem., Int. Ed.*, **51**, 4698–4701.
24 Lim, S.-G., Lee, J.H., Moon, C.W., Hong, J.-B., and Jun, C.-H. (2003) Rh(l)-catalyzed direct ortho-alkenylation of aromatic ketimines with alkynes and its application to the synthesis of isoquinoline derivatives. *Org. Lett.*, **5**, 2759–2761.
25 Sonoda, M., Kakiuchi, F., Chatani, N., and Murai, S. (1997) Ruthenium-catalyzed addition of carbon-hydrogen bonds in aromatic ketones to olefins. The effect of various substituents at the aromatic ring. *Bull. Chem. Soc. Jpn.*, **70**, 3117–3128.

26 Tsuchikama, K., Kasagawa, M., Endo, K., and Shibata, T. (2010) Sequential catalytic reactions for the synthesis of benzofulvenes using an iridium complex with dual function. *Synlett*, **2010**, 97–100.
27 Fallon, B.J. *et al.* (2015) C—H activation/functionalization catalyzed by simple, well-defined low-valent cobalt complexes. *J. Am. Chem. Soc.*, **137**, 2448–2451.
28 Fallon, B.J. *et al.* (2015) Synthesis of 1,2-dihydropyridines catalyzed by well-defined low-valent cobalt complexes: C—H activation made simple. *ACS Catal.*, **5**, 7493–7497.
29 Yamakawa, T. and Yoshikai, N. (2013) Annulation of alpha,beta-unsaturated imines and alkynes via cobalt-catalyzed olefinic C—H activation. *Org. Lett.*, **15**, 196–199.
30 Guihaume, J., Halbert, S., Eisenstein, O., and Perutz, R.N. (2012) Hydrofluoroarylation of alkynes with Ni catalysts. C—H activation via ligand-to-ligand hydrogen transfer, an alternative to oxidative addition. *Organometallics*, **31**, 1300–1314.
31 Bair, J.S. *et al.* (2014) Linear-selective hydroarylation of unactivated terminal and internal olefins with trifluoromethyl-substituted arenes. *J. Am. Chem. Soc.*, **136**, 13098–13101.
32 Nakao, Y., Kanyiva, K.S., Oda, S., and Hiyama, T. (2006) Hydroheteroarylation of alkynes under mild nickel catalysis. *J. Am. Chem. Soc.*, **128**, 8146–8147.
33 Ding, Z. and Yoshikai, N. (2010) Cobalt-catalyzed addition of azoles to alkynes. *Org. Lett.*, **12**, 4180–4183.
34 Ding, Z. and Yoshikai, N. (2011) Cobalt-catalyzed alkenylation of thiazoles with alkynes via C—H bond functionalization. *Synthesis*, **2011**, 2561–2566.
35 Gao, K. and Yoshikai, N. (2011) Regioselectivity-switchable hydroarylation of styrenes. *J. Am. Chem. Soc.*, **133**, 400–402.
36 Crisenza, G.E.M. and Bower, J.F. (2016) Branch selective mural-type alkene hydroarylation reactions. *Chem. Lett.*, **45**, 2–9.
37 Yang, Z., Yu, H., and Fu, Y. (2013) Mechanistic study on ligand-controlled cobalt-catalyzed regioselectivity-switchable hydroarylation of styrenes. *Chem. - Eur. J.*, **19**, 12093–12103.
38 Shih, W.-C. *et al.* (2012) The regioselective switch for amino-NHC mediated C—H activation of benzimidazole via Ni-Al synergistic catalysis. *Org. Lett.*, **14**, 2046–2049.
39 Pan, S., Ryu, N., and Shibata, T. (2012) Ir(I)-catalyzed C—H bond alkylation of C2-position of indole with alkenes: selective synthesis of linear or branched 2-alkylindoles. *J. Am. Chem. Soc.*, **134**, 17474–17477.
40 Crisenza, G.E.M., McCreanor, N.G., and Bower, J.F. (2014) Branch-selective, iridium-catalyzed hydroarylation of monosubstituted alkenes via a cooperative destabilization strategy. *J. Am. Chem. Soc.*, **136**, 10258–10261.
41 Crisenza, G.E.M., Sokolova, O.O., and Bower, J.F. (2015) Branch-selective alkene hydroarylation by cooperative destabilization: iridium-catalyzed ortho-alkylation of acetanilides. *Angew. Chem., Int. Ed.*, **54**, 14866–14870.
42 Lee, P.-S. and Yoshikai, N. (2013) Aldimine-directed branched-selective hydroarylation of styrenes. *Angew. Chem., Int. Ed.*, **52**, 1240–1244.

43 Dong, J., Lee, P.-S., and Yoshikai, N. (2013) Cobalt-catalyzed branched-selective addition of aromatic ketimines to styrenes under room-temperature conditions. *Chem. Lett.*, **42**, 1140–1142.
44 Lee, P.-S. and Yoshikai, N. (2015) Cobalt-catalyzed enantioselective directed C—H alkylation of indole with styrenes. *Org. Lett.*, **17**, 22–25.
45 Xu, W. and Yoshikai, N. (2014) Highly linear selective cobalt-catalyzed addition of aryl imines to styrenes: reversing intrinsic regioselectivity by ligand elaboration. *Angew. Chem., Int. Ed.*, **53**, 14166–14170.
46 Ilies, L., Chen, Q., Zeng, X., and Nakamura, E. (2011) Cobalt-catalyzed chemoselective insertion of alkene into the ortho C—H bond of benzamide. *J. Am. Chem. Soc.*, **133**, 5221–5223.
47 Gao, K. and Yoshikai, N. (2011) Cobalt-phenanthroline catalysts for the ortho alkylation of aromatic imines under mild reaction conditions. *Angew. Chem., Int. Ed.*, **50**, 6888–6892.
48 Ding, Z. and Yoshikai, N. (2012) C2-alkylation of N-pyrimidylindole with vinylsilane via cobalt-catalyzed C—H bond activation. *Beilstein J. Org. Chem.*, **8**, 1536–1542.
49 Ding, Z. and Yoshikai, N. (2013) Cobalt-catalyzed intramolecular olefin hydroarylation leading to dihydropyrroloindoles and tetrahydropyridoindoles. *Angew. Chem., Int. Ed.*, **52**, 8574–8578.
50 Kobayashi, T., Yorimitsu, H., and Oshima, K. (2009) Cobalt-catalyzed isomerization of 1-alkenes to (*E*)-2-alkenes with dimethylphenylsilylmethyl-magnesium chloride and its application to the stereoselective synthesis of (*E*)-alkenylsilanes. *Chem. - Asian J.*, **4**, 1078–1083.
51 Chen, C., Dugan, T.R., Brennessel, W.W., Weix, D.J., and Holland, P.L. (2014) Z-selective alkene isomerization by high-spin cobalt(II) complexes. *J. Am. Chem. Soc.*, **136**, 945–955.
52 Punner, F., Schmidt, A., and Hilt, G. (2012) Up the hill: selective double-bond isomerization of terminal 1,3 dienes towards *Z*-1,3-dienes or 2*Z*,4*E*-dienes. *Angew. Chem., Int. Ed.*, **51**, 1270–1273.
53 Schmidt, A., Nodling, A.R., and Hilt, G. (2015) An alternative mechanism for the cobalt-catalyzed isomerization of terminal alkenes to (*Z*)-2-alkenes. *Angew. Chem., Int. Ed.*, **54**, 801–804.
54 Yamakawa, T. and Yoshikai, N. (2014) Alkene isomerization-hydroarylation tandem catalysis: indole C2Alkylation with aryl- substituted alkenes leading to 1,1-diarylalkanes. *Chem. - Asian J.*, **9**, 1242–1246.
55 Lee, W.-C., Wang, C.-H., Lin, Y.-H., Shih, W.-C., and Ong, T.-G. (2013) Tandem isomerization and C—H activation: regioselective hydroheteroarylation of allylarenes. *Org. Lett.*, **15**, 5358–5361.
56 Andou, T., Saga, Y., Komai, H., Matsunaga, S., and Kanai, M. (2013) Cobalt-catalyzed C4-selective direct alkylation of pyridines. *Angew. Chem., Int. Ed.*, **52**, 3213–3216.
57 Yamamoto, S., Saga, Y., Andou, T., Matsunaga, S., and Kanai, M. (2014) Cobalt-catalyzed C-4 selective alkylation of quinolines. *Adv. Synth. Catal.*, **356**, 401–405.
58 Yoshino, T., Ikemoto, H., Matsunaga, S., and Kanai, M. (2013) A cationic high-valent Cp*Co-III complex for the catalytic generation of nucleophilic

organometallic species: directed C—H bond activation. *Angew. Chem. Int. Ed.*, **52**, 2207–2211.

59 Yoshikai, N. (2015) Cp*Co-III-catalyzed C—H activation of (hetero)arenes: expanding the scope of base-metal-catalyzed C—H functionalizations. *ChemCatChem*, **7**, 732–734.

60 Ikemoto, H., Yoshino, T., Sakata, K., Matsunaga, S., and Kanai, M. (2014) Pyrroloindolone synthesis via a Cp*Co-III-catalyzed redox-neutral directed C—H alkenylation/annulation sequence. *J. Am. Chem. Soc.*, **136**, 5424–5431.

61 Wong, M.Y., Yamakawa, T., and Yoshikai, N. (2015) Iron-catalyzed directed C2-alkylation and alkenylation of indole with vinylarenes and alkynes. *Org. Lett.*, **17**, 442–445.

62 Kischel, J., Jovel, I., Mertins, K., Zapf, A., and Beller, M. (2006) A convenient FeCl3-catalyzed hydroarylation of styrenes. *Org. Lett.*, **8**, 19–22.

63 Liebeskind, L.S., Gasdaska, J.R., Mccallum, J.S., and Tremont, S.J. (1989) Ortho-functionalization of aromatic ketones Via manganation - a synthesis of indenols. *J. Org. Chem.*, **54**, 669–677.

64 Robinson, N.P., Main, L., and Nicholson, B.K. (1989) Reactions of ortho-manganated aryl-ketones, aldehydes and amides with alkynes - a New synthesis of inden-1-Ols and indenones. *J. Organomet. Chem.*, **364**, C37–C39.

65 Zhou, B., Chen, H., and Wang, C. (2013) Mn-catalyzed aromatic C—H alkenylation with terminal alkynes. *J. Am. Chem. Soc.*, **135**, 1264–1267.

66 Wang, C. (2013) Manganese-mediated C-C bond formation via C—H activation: from stoichiometry to catalysis. *Synlett*, **24**, 1606–1613.

67 Zhou, B., Ma, P., Chen, H., and Wang, C. (2014) Amine-accelerated manganese-catalyzed aromatic C—H conjugate addition to alpha,beta-unsaturated carbonyls. *Chem. Commun.*, **50**, 14558–14561.

7

Hydroarylation of Alkynes using Cu, Ag, and Au Catalysts

Mariia S. Kirillova, Fedor M. Miloserdov and Antonio M. Echavarren

Institute of Chemical Research of Catalonia (ICIQ), Barcelona Institute of Science and Technology, Av. Països Catalans 16, 43007, Tarragona, Spain

7.1 Introduction

The hydroarylation of alkynes (in other words alkenylation of (hetero)arenes with alkynes) is formally the insertion of an alkyne into the C—H bond of the arene leading to the rehybridization of the alkyne sp carbons to sp^2 (Scheme 7.1). This hydroarylation results in regio- and stereoselective formation of synthetically useful (*Z*)-alkenes (Scheme 7.1, $R^1 = H$) whereas the formally equivalent Mizoroki–Heck reaction generally affords (*E*)-alkenes. Several catalytic methods employing Brønsted [1] or Lewis acids [2–4] have been used to achieve the hydroarylation of alkynes. Various transition metals including Pt [5], Ni [6], Fe [7], Ru [8], W [9], and Re [10] have also shown interesting catalytic activity in this type of transformations.

An early report (perhaps the first) describing a copper-mediated intramolecular alkyne hydroarylation was published in 1962 [11]. However, only after more than 30 years, catalysis by 11th group metals has become more widespread in use in this context. This was mainly due to the discovery of the remarkable affinity of gold(I) to alkynes and allenes, and its great efficiency at promoting hydroarylation processes [12]. During the past decade, diverse catalytic systems (copper, silver, and gold complexes with phosphine or NCH–carbene ligands, Figure 7.1) were developed for the synthesis of substituted alkenes from terminal and internal alkynes, which proved to be effective on a range of activated or unactivated arenes and heteroarenes. For metal complexes, both ligand and counter-anion are important and poorly coordinating anions, such as OTf, NTf_2, PF_6, SbF_6, BAr^F, are commonly required [13]. Although the intramolecular alkyne hydroarylation has been studied thoroughly, the intermolecular version of this reaction is much more challenging and to date, only a few examples have been reported.

The main goal of this chapter is to review the advancements made in the field of metal-catalyzed intramolecular and intermolecular hydroarylation of alkynes promoted by gold, silver, and copper salts. Within the context of these two different reactions, the reactivity of alkynes (activated or non-activated), the selectivity, and mechanism of the transformation will be discussed. Particular

Catalytic Hydroarylation of Carbon-Carbon Multiple Bonds, First Edition.
Edited by Lutz Ackermann, T. Brent Gunnoe and Laurel Goj Habgood.

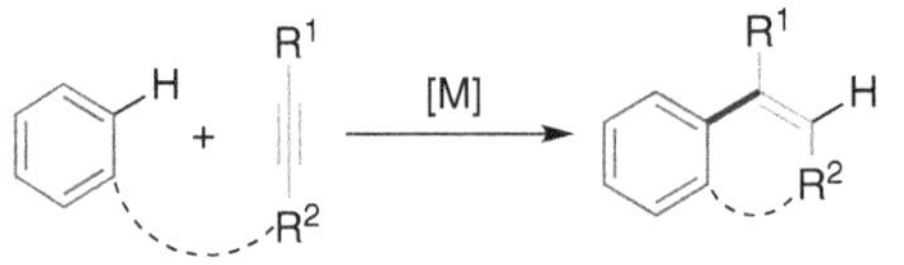

Scheme 7.1 Metal-catalyzed hydroarylation of alkynes.

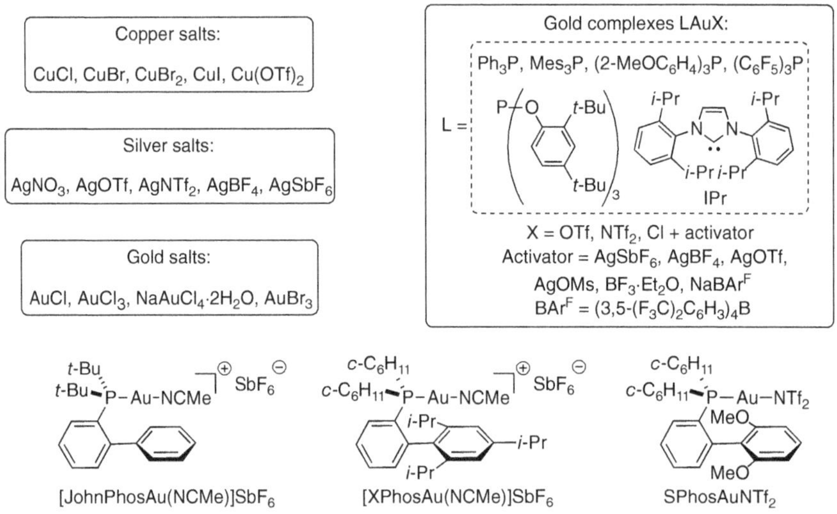

Figure 7.1 Salts and complexes of gold, silver, and copper used in alkyne hydroarylation.

attention will be given to illustrate the reactivity of electron-rich (hetero)arenes and unactivated arenes as arylating agents with alkynes. Examples of supported gold-catalyzed alkyne hydroarylation as well as the application of the hydroarylation strategy in the field of target-oriented synthesis will also be briefly discussed.

7.2 Intramolecular Hydroarylation of Alkynes

7.2.1 Alkyne Hydroarylation with Electron-Rich Arenes

7.2.1.1 Alkyne Hydroarylation with Aniline Derivatives

In both synthetic and medicinal chemistry, quinolines and dihydroquinolines (DHQs) are important building blocks whose structural features are present in a number of biologically active molecules [14]. The intramolecular hydroarylation of aniline derivatives is a rapid and efficient strategy to assemble these scaffolds. This transformation was reported to proceed in the presence of several metals and Lewis acids including platinum salts [15] and $BF_3 \cdot Et_2O$ [16], typically under harsh conditions. To our knowledge, the first example of copper-catalyzed hydroarylation was described in the context of the 2,2-dimethyl DHQ synthesis from *N*-(2,2-dimethylpropargyl) anilines, employing CuCl as the catalyst [11].

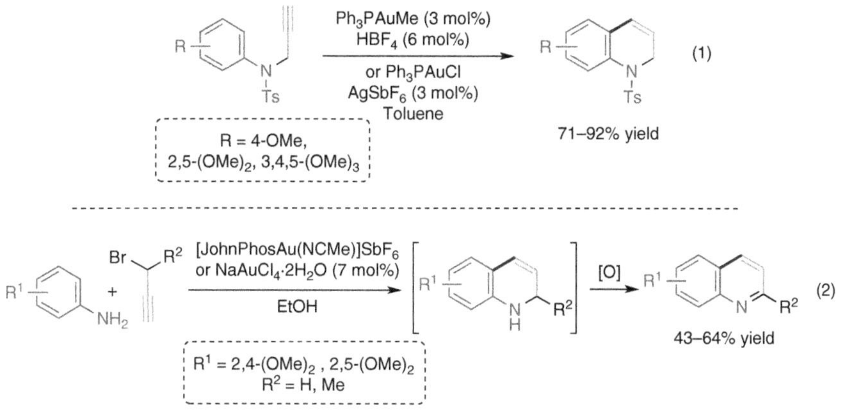

Scheme 7.2 Gold-catalyzed hydroarylation for the synthesis of dihydroquinolines (DHQs) and quinolones (JohnPhos = (2-biphenyl)di-*tert*-butylphosphine)

The evolution in the field of carbophilic π-Lewis acid catalysis resulted in the development of milder and broadly applicable processes promoted by gold as well as silver and copper salts [17].

The advantages of the gold-based π-acid catalysts are illustrated by one of the first reports that described gold(I)-catalyzed cyclization of *N*-propargylated *N*-tosylanilines. Under gold catalysis, the desired DHQs were obtained under milder conditions and often with greater yields than in the corresponding Pt(II)-mediated process (Scheme 7.2, Eq. 1) [18]. Contrary to the hydroarylation of protected *N*-propargylanilines, which yields DHQs, the gold-catalyzed cyclization of free propargylic (and homopropargylic) aniline derivatives takes place with concomitant oxidation of the DHQ intermediates to the corresponding quinolines. As an extension of this approach, a one-pot procedure for the synthesis of functionalized quinolines from anilines and propargylic bromides was developed through alkylation/gold-catalyzed hydroarylation cascade, by employing $NaAuCl_4{\cdot}2H_2O$ or cationic [JohnPhosAu(NCMe)]SbF_6 (Figure 7.1) as gold sources (Scheme 7.2, Eq. 2) [19].

Additionally, this strategy was applied for the preparation of various substituted 2-trifluoromethyl quinolines in good to excellent yields (Scheme 7.3, Eq. 1) [20]. The direct formation of quinoline derivatives was developed via an oxidative cyclization of *in situ*-generated *N*-aryl-2-propynylamines promoted by copper or gold salts in the presence of an imine as an oxidant (Scheme 7.3, Eq. 2) [21]. To gain access to similar structures, a one-pot A^3 coupling reaction (aldehyde, alkyne, amine coupling)/alkyne hydroarylation process was developed by employing gold(III) chloride (Scheme 7.3, Eq. 3) [22], copper(II) triflate [23] as well as some other silver [24] and copper [25, 26] salts.

A somewhat similar strategy was applied for the preparation of 3-spiroheterocyclic 2-oxindoles, employing a dual Cu(I)/Cu(II)-catalytic system. In this case, copper(I)-mediated KA^2 reaction (ketone, alkyne, amine coupling) is used for the *in situ* generation of propargylaniline derivatives. The subsequent Cu(II)-triggered hydroarylation completes the domino process,

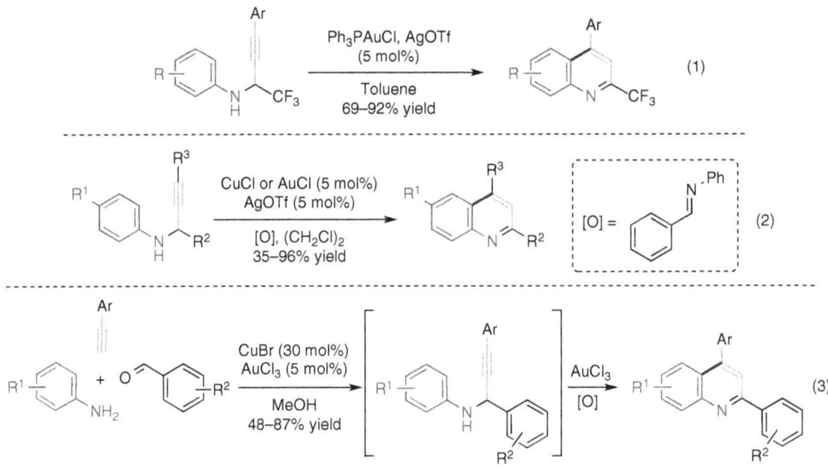

Scheme 7.3 Synthesis of substituted quinolines via gold- and copper-catalyzed intramolecular hydroarylation.

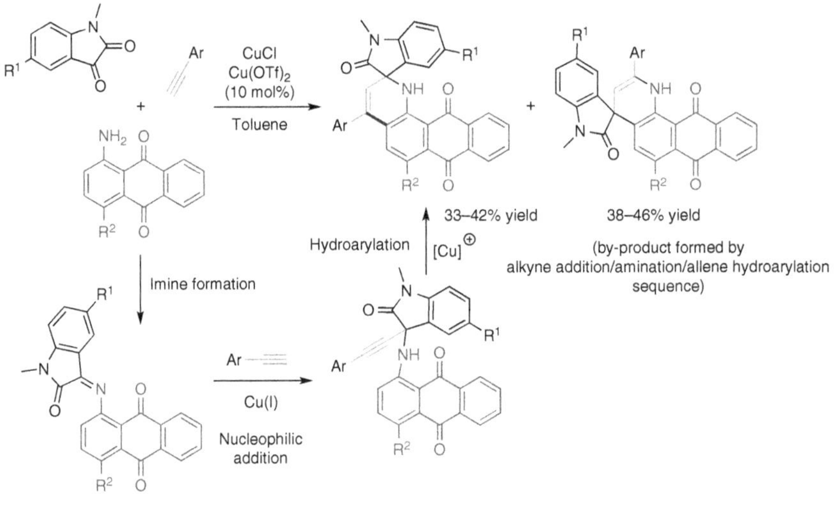

Scheme 7.4 Copper-catalyzed KA^2-hydroarylation for the preparation of 3-spiroheterocyclic 2-oxindoles.

yielding spiroheterocycles together with the regioisomer formed by the alkyne addition/amination/allene hydroarylation pathway (Scheme 7.4) [27].

A number of reports have been published, which describe the hydroarylation of alkynes promoted by silver and copper salts demonstrating their ability to act as carbophilic π-Lewis acid catalysts. In the presence of CuCl, *N*-(2,2-dimethylpropargyl) (hetero)anilines were successfully converted into the corresponding 2,2-dimethyl DHQs, generally requiring high temperatures (Scheme 7.5) [11, 28]. A dual Cu(I)/Cu(II)-catalyzed protocol for the synthesis of DHQs was developed through a one-pot dehydrative condensation of anilines with propargyl alcohols, followed by intramolecular hydroarylation [29].

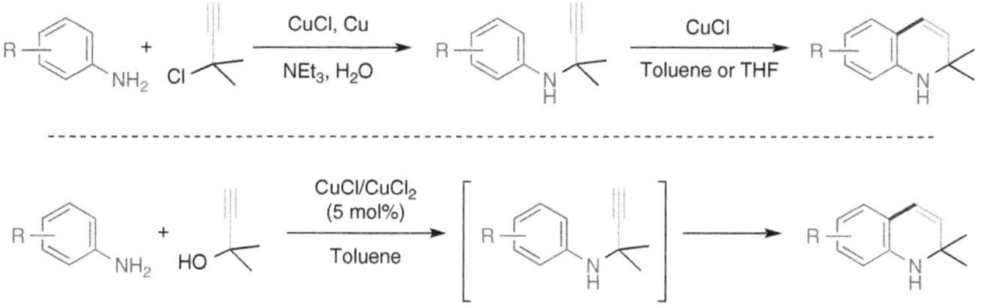

Scheme 7.5 Copper-catalyzed hydroarylation for the synthesis of dihydroquinolines.

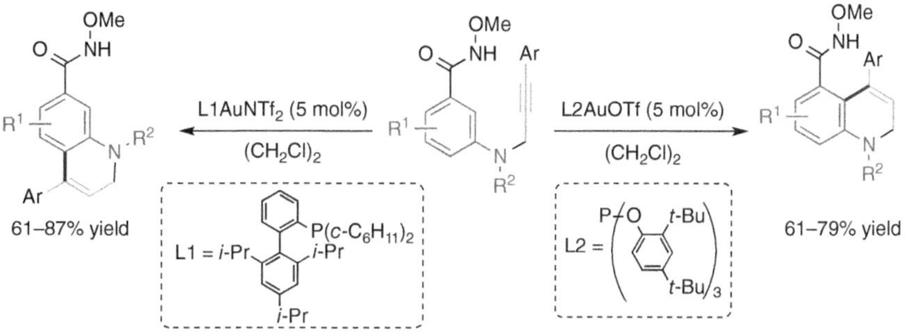

Scheme 7.6 Gold-catalyzed regiodivergent hydroarylation of propargylic anilines.

In the case of substrates bearing *N*-methoxy amide as a directing group, the regioselectivity of the gold-catalyzed hydroarylation could be controlled by both steric and electronic properties of the ligand. The high regioselectivity of the sterically hindered ortho-cyclization is achieved by the use of electrophilic Au(I) salts {such as phosphite gold(I) complex}, whereas more electron-rich gold(I) sources (e.g., those bearing phosphine ligands) led to the formation of the corresponding DHQs exclusively via para-cyclization (Scheme 7.6) [30].

Various *N*-propargylated aminonaphthoquinones undergo copper(II) triflate-catalyzed 6-*endo*-dig cyclization to give rise to 1,2-dihydrobenzo[*g*]quinoline-5, 10-diones in good yields (Scheme 7.7, Eq. 1) [31]. The efficiency of the reaction was not affected by the nature of the aryl substituent at the aniline nitrogen. In a somewhat similar strategy, *N*-propargyl aminoquinoxalines were converted into triazaphenanthrenes, using either copper or silver salts to trigger this transformation. Employing a stoichiometric amount of metal salts was essential to achieve high yields of the desired polyheterocycles (Scheme 7.7, Eq. 2). Interestingly, the 6-*endo*-dig cyclization results in the formation of only one regioisomer. According to computational studies, the 5-position of quinoxaline is more electron-rich than the 7-position, and thus more reactive toward an electrophilic alkyne moiety [32].

Silver-based π-acids also promote the hydroarylation of alkynes, which was illustrated by the regioselective formation of annulated pyridine and quinoline derivatives [33]. Another interesting example of silver-catalyzed cycloisomerization is described in the context of 3-methylimidazo[1,2-*a*]pyridines synthesis,

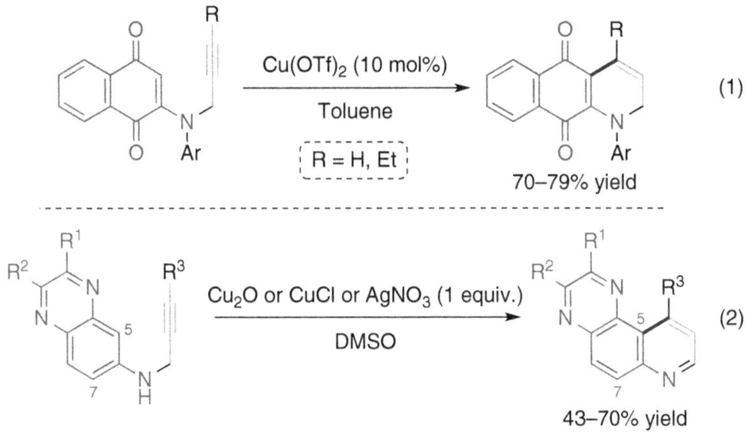

Scheme 7.7 Copper-catalyzed alkyne hydroarylation of *N*-propargylated aminonaphthoquinones and aminoquinoxalines.

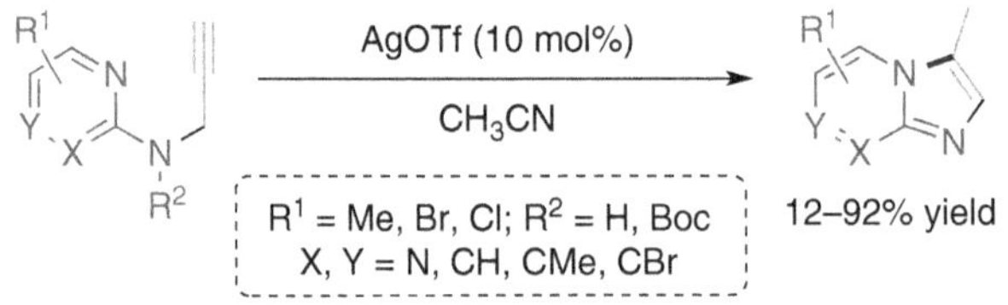

Scheme 7.8 Silver-catalyzed nitrogen heterocycle formation using propargylated 2-aminopyridines.

wherein the reaction proceeds via the *N*-atom of the heterocycle. A wide range of pyridine-derived propargylamines were converted into corresponding bicycles, although in moderate yields (Scheme 7.8). This reaction was applied in the synthesis of the sedative and anxiolytic drug saripidem [34].

The potential utility of the gold(I)-catalyzed alkyne hydroarylation for the construction of non-central chirality was illustrated by the synthesis of axially chiral 4-aryl-2-quinolinones and 4-arylcoumarins. A diverse array of *N*-aryl-arylethynylamides (propiolic anilides) were successfully converted into axially chiral heterobiaryls, by employing electrophilic gold(I) salts. Gold(I) complexes, bearing chiral phosphine ligands ((*R*)-DTBM-SEGPHOS or (*R*)-BINAP), were highly selective and efficient catalysts for the preparation of the 4-aryl-2-quinolinones in excellent yields and with up to 61% ee (Scheme 7.9) [35]. Further elaboration of this methodology allowed an enantioselective synthesis of aza[10]helicenes via an intramolecular quadruple hydroarylation of tetraynes in moderate yields and with up to 93% enantiomeric excess (ee) (Scheme 7.9) [36].

Similar to the aforementioned propargylic anilines, *N*-butynyl anilines also undergo gold-catalyzed hydroarylation, yielding *exo*-methylene-substituted tetrahydroquinolines via 6-*exo*-dig cyclization (Scheme 7.10) [37]. Under the developed reaction conditions, the isomerization of the exocyclic vinyl moiety to the more stable trisubstituted alkene led to the formation of DHQs. DHQs

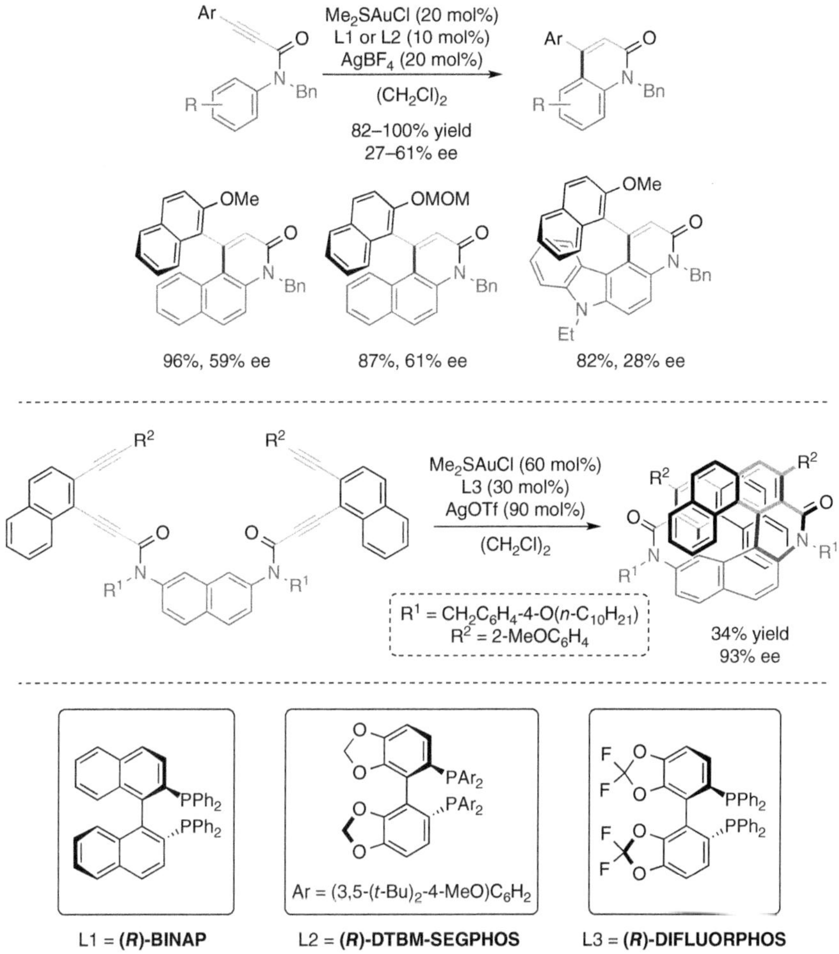

Scheme 7.9 Gold-catalyzed atroposelective intramolecular hydroarylation of internal alkynes with anilines.

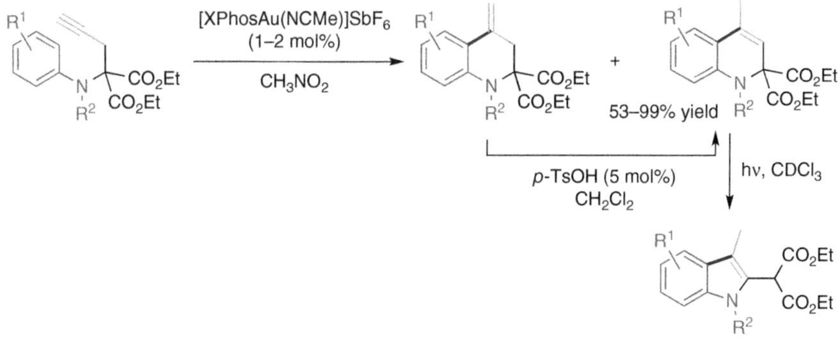

Scheme 7.10 Gold-catalyzed cycloisomerization of *N*-aminophenyl propargyl malonates.

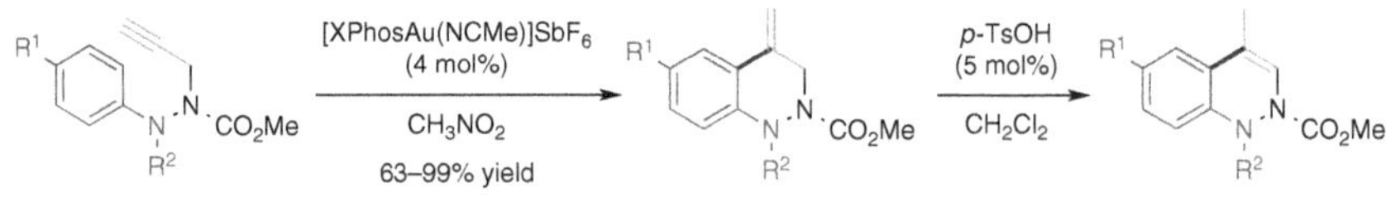

Scheme 7.11 Gold-catalyzed cycloisomerization of *N*-propargyl-*N*′-arylhydrazines.

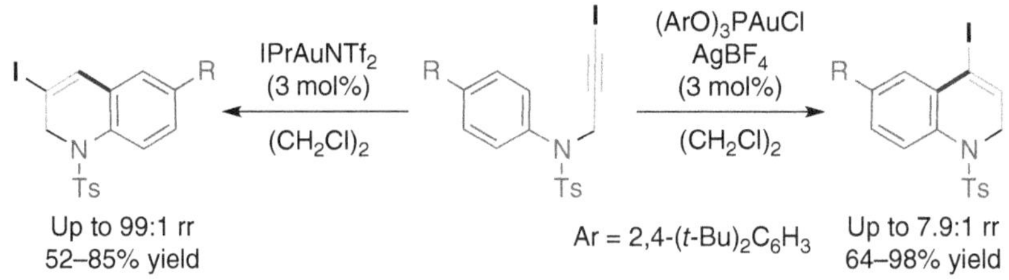

Scheme 7.12 Regioselectivity in the gold-catalyzed intramolecular hydroarylation of iodoalkynes with anilines.

could be exclusively obtained by treating the mixture of regioisomers with *para*-toluenesulfonic acid. Interestingly, further irradiation with visible light induces a photochemical rearrangement to form indoles.

Similarly, various dihydrocinnolines were prepared from *N*-arylhydrazine derivatives via gold-catalyzed hydroarylation, although the efficiency and selectivity of this transformation strongly depends on the nature of the protecting group (R^2) on the *N*-aryl nitrogen (Scheme 7.11) [38]. Halide-substituted alkynes also undergo hydroarylation with electron-rich arenes [39]. However, the nature of the catalyst has a strong effect on the outcome of the reaction. Non-rearranged 4-iododihydroquinolines were predominantly formed by employing phosphite gold(I) complexes, whereas more electron-rich NHC-Au(I) salts favored a 1,2-iodine migration, leading to rearranged 3-iododihydroquinoline (Scheme 7.12).

Several complex cascade transformations featuring alkyne hydroarylation of aniline derivatives have been also disclosed. An efficient preparation of dihydroquinolines from terminal alkynes and anilines or indolines was achieved by a one-pot hydroamination/alkynylation/hydroarylation cascade promoted by cationic NHC-Au(I) salts (Scheme 7.13) [40]. The proposed mechanism of this transformation involves hydroamination of the first molecule of the alkyne by an aniline, furnishing an enamine, whose tautomeric ketimine then reacts with a molecule of the alkyne. The *in situ*-generated propargylic aniline undergoes a gold(I)-catalyzed hydroarylation to form DHQs in excellent yields, considering the complexity of this transformation. When the same transformation was carried out with $AgBF_4$ as the catalyst, the dihydroquinoline underwent an additional hydroarylation to furnish vinylated DHQ derivatives [41].

In an analogous fashion, substrates featuring alkyne tethered to aniline [42] or anilide [43] were used as precursors for pyrrolo[1,2-*a*]quinolines and pyrrolo[1,2-*a*]quinolin-1(2*H*)-ones, respectively. This approach is based on a one-pot hydroamination/alkynylation/hydroarylation sequence. Unprotected *N*-(pent-4-ynyl)anilines, when subjected to Au(I) or Au(III) catalysts, undergo

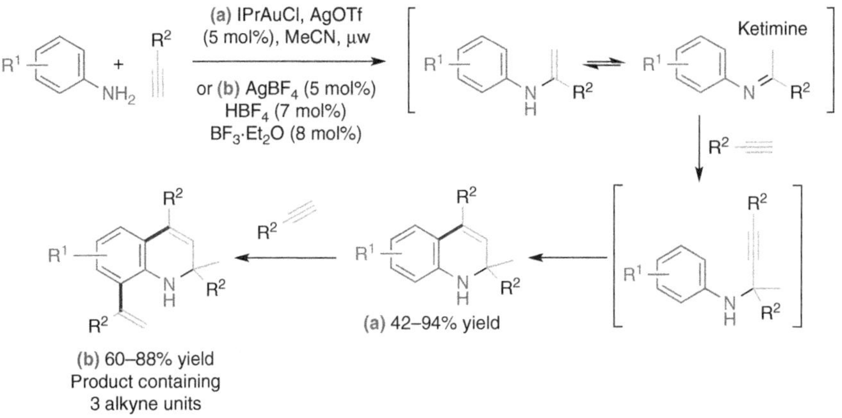

Scheme 7.13 Gold- and silver-catalyzed hydroamination/alkynylation/hydroarylation cascade.

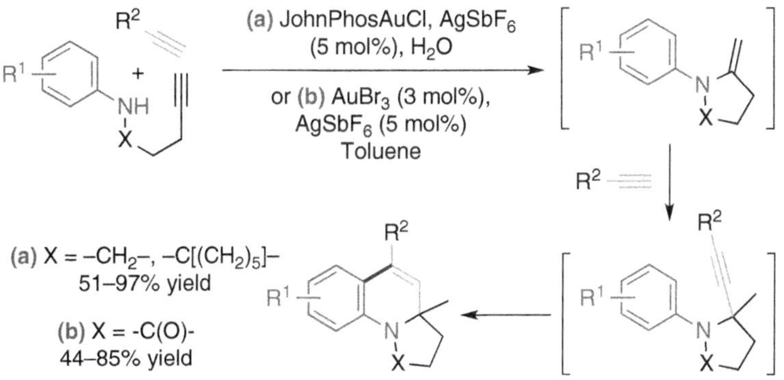

Scheme 7.14 Gold-catalyzed tandem hydroamination/alkynylation/hydroarylation in the synthesis of pyrroloquinolines.

intramolecular hydroamination in a 5-*exo*-dig manner, yielding dihydropyrroles bearing an *exo*-vinyl group. Subsequent addition of a terminal alkyne to an enamine intermediate generates a propargylated aniline, which, in the presence of an electrophilic gold catalyst, is converted to the corresponding pyrrolo[1,2-*a*]quinoline via an intramolecular hydroarylation (Scheme 7.14).

Additionally, the synthesis of tetrahydro-2*H*-pyrano[3,4-*c*]quinolines was developed via an alkyne hydroarylation/Prins cyclization cascade (Scheme 7.15) [44]. The gold(I)/indium (III) multi-catalyst system provided the desired tricycles in good to excellent yields, whereas moderate efficiency of the formation of the coupling product in the presence of indium(III) triflate solely was achieved only on employing elevated temperatures.

7.2.1.2 Alkyne Hydroarylation with Phenols and Phenol Ether Derivatives

In addition to the previously discussed propargylic anilines, other electron-rich arenes participate in hydroarylation reactions. The intramolecular hydroarylation of phenol ethers and esters is a direct approach to the synthesis of

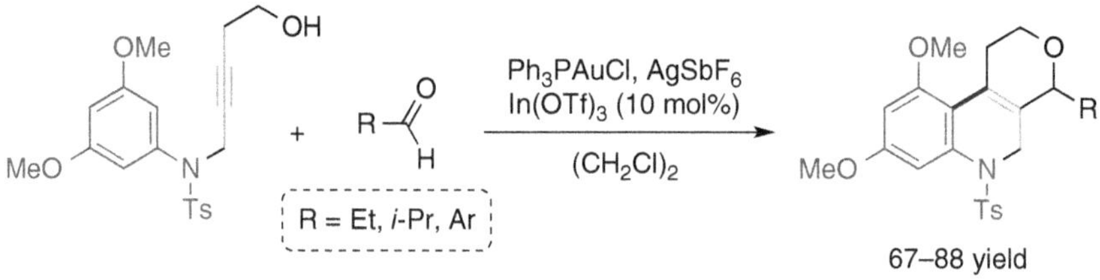

Scheme 7.15 Gold- and indium-catalyzed tandem hydroarylation/Prins reaction in the synthesis of pyranoquinolines.

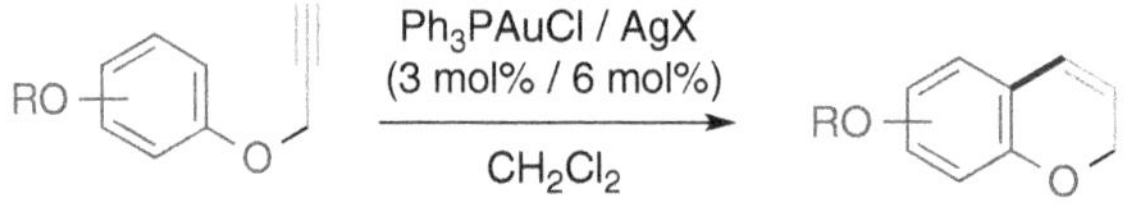

Scheme 7.16 Gold-catalyzed cycloisomerization of polyalkoxyaryl propargyl ethers.

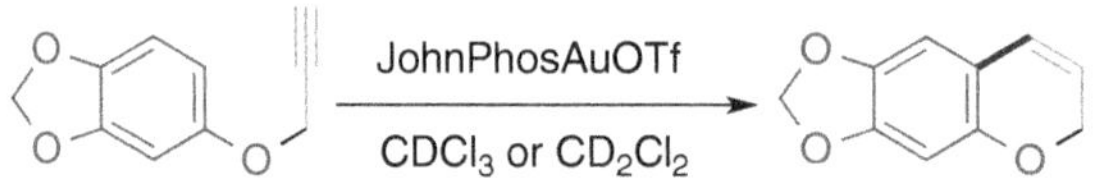

[Au] (%)	Time (min)	Yield (%)
2	5	98
0.6	30	98
0.2	60	0
0.2*	45	98

* with AgOTf (0.6 mol%)

Scheme 7.17 Effect of AgOTf-additive on efficiency of gold-catalyzed intramolecular hydroarylation of aryl propargyl ether.

2*H*-chromenes and coumarins, respectively. The cyclization of phenol ethers may be triggered by In(III) [45] or induced by harsh thermal conditions [46]. On the contrary, the gold-catalyzed hydroarylation of phenyl propargyl ethers usually occurs under mild reaction conditions (Scheme 7.16) [18b, 47].

The deactivation of cationic gold complexes, and the factors that cause it, have been studied in the context of the intramolecular hydroarylation of aryl propargyl ethers (Scheme 7.17) [48]. These studies revealed that contamination of the solvent ($CDCl_3$, CD_2Cl_2) with chlorine/chloride-containing compounds is responsible for the catalyst poisoning. This can be avoided through the purification of solvents and addition of AgOTf or triflic acid, which are often beneficial for the catalytic activity and catalyst loading. Taking advantage of the hydroarylation strategy, several scaffolds featuring a benzopyran skeleton were assembled by employing $Ph_3PAuNTf_2$ at elevated temperature (Scheme 7.18) [49]. Additionally, the gold-mediated alkenylation of phenyl propargyl ethers was also applied in the synthesis of a $5HT_4$ receptor agonist [50]. Further evolution of this method resulted in extension of the substrate scope to aryl propiolates and phenol derivatives bearing a nitro, cyano, carboxylic ester, or halogen substituent (Scheme 7.19) [18c, 51]. The formation of benzofuran as a by-product was observed with arenes bearing electron-withdrawing groups. A fragmentation of gold-vinyl intermediate to allene was proposed as a plausible explanation for the benzofuran formation. The gold(I)-catalyzed hydroarylation in combination with an Overman rearrangement was used for one-pot synthesis of allylic amine-derived

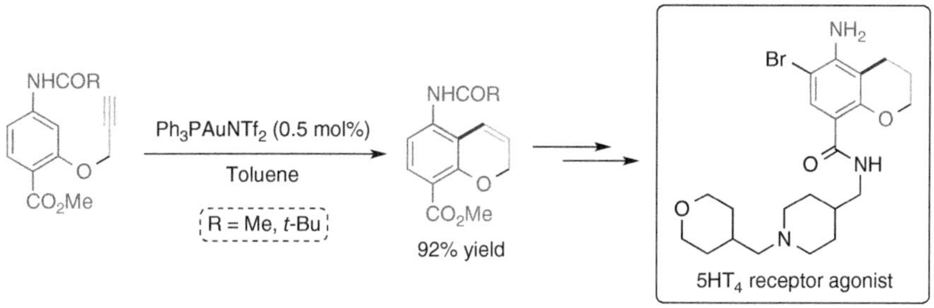

Scheme 7.18 Target-oriented gold-catalyzed intramolecular hydroarylation of phenyl propargyl ethers.

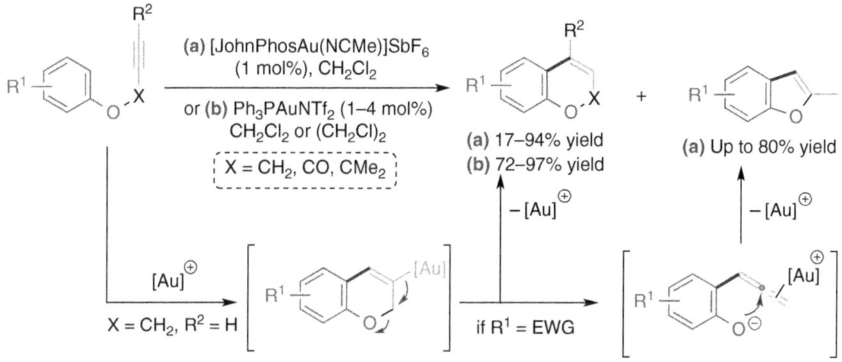

Scheme 7.19 Gold-catalyzed intramolecular hydroarylations of aryl propargyl ethers and aryl propiolates.

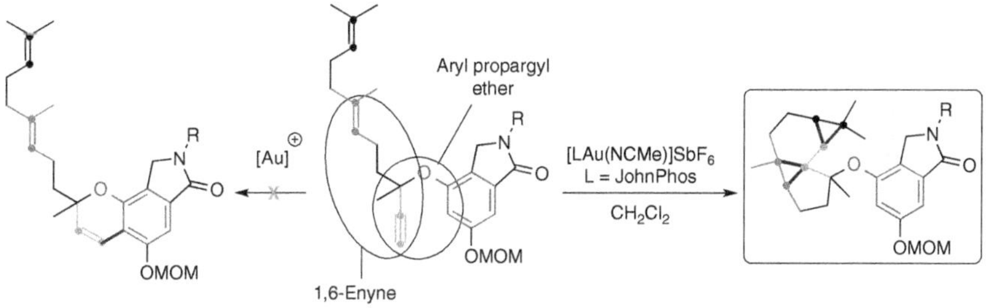

Scheme 7.20 Gold-catalyzed aryl propargyl ether hydroarylation vs. 1,6-enyne cycloisomerization.

2*H*-chromenes [52]. Attempts to apply the intramolecular alkenylation protocol on a substrate containing both an aryl propargyl ether and a 1,6-enyne fragment failed [53]. The cycloisomerization of enynes is largely favored over hydroarylation and no desired product could be obtained using gold catalysis (Scheme 7.20).

Propargylic phenol ethers [18b, 51, 52] as well as their internal counterparts are suitable precursors for the synthesis of 2*H*-chromenes via the intramolecular alkenylation. The examination of the product selectivity, scope, and limitations of alkyne hydroarylation reactions triggered by gold(I) or silver(I) salts was carried

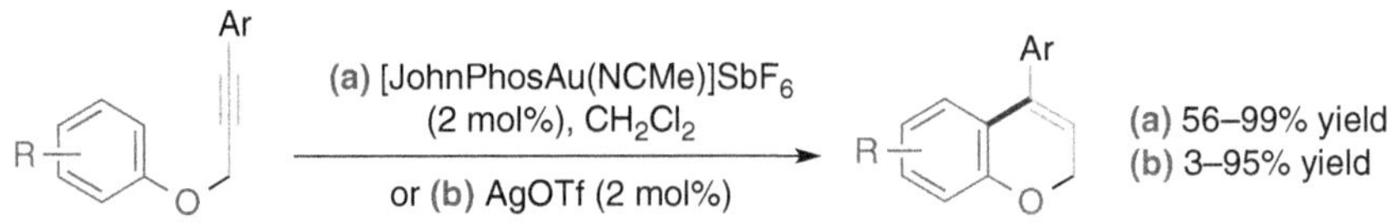

Scheme 7.21 Gold- and silver-catalyzed intramolecular hydroarylation of aryl alkyne tethered propargyl ethers.

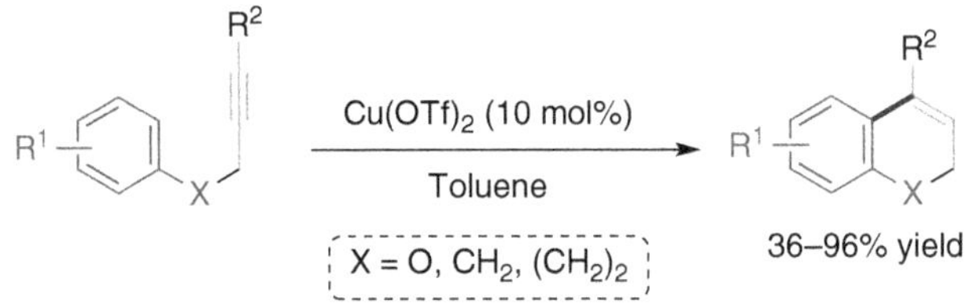

Scheme 7.22 Copper-catalyzed intramolecular hydroarylation of alkyne tethered to aryl ether.

out on both electron-rich and electron-deficient propargyl phenyl ethers. These studies revealed that cationic gold(I) catalysts show a remarkable efficiency and high regioselectivity for the para-alkenylation, even in case of meta-substituted aryl ethers; whereas silver-mediated process proceeds only for electron-rich substrates (Scheme 7.21) [54].

Copper-based π-acids also promote the intramolecular hydroarylation of alkynes. Internal alkynes, tethered to benzene ring via all-carbon or oxygen linkers, undergo endo-cyclization in the presence of copper(II) triflate, furnishing 2*H*-chromene derivatives in good to excellent yields (Scheme 7.22). Both electron-rich and electron-deficient arenes were good arylating agents. Moreover, various functional groups at the alkyne moiety, including different aryls, alkyls, and oxazolidinyl, were well tolerated [55]. Further extension of the intramolecular alkenylation strategy to diyne diaryl ethers resulted in the efficient preparation of 4,4′-bi(2*H*-chromene) derivatives (as well as bi-dihydroquinoline analogs and hybrid systems) via a gold-catalyzed double hydroarylation (Scheme 7.23) [56].

Similar to the halide-substituted propargyl anilines [39], iodoalkynes tethered to phenol ethers were converted into halogenated heterocycles via a gold-catalyzed migratory cycloisomerization (Scheme 7.24) [57]. The electron-rich cationic NHC-Au(I) complexes promote the formation of

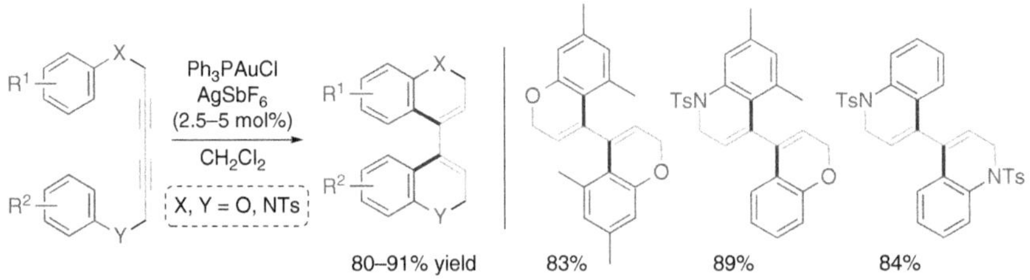

Scheme 7.23 Gold-catalyzed cycloisomerization of diynes toward the preparation of bi(2*H*-chromene), bi-dihydroquinoline, and their hybrid derivatives.

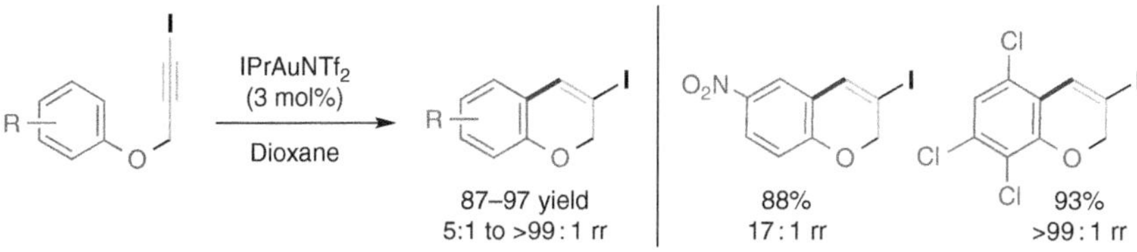

Scheme 7.24 Gold-catalyzed intramolecular hydroarylation of iodoalkyne tethered to aryl ether.

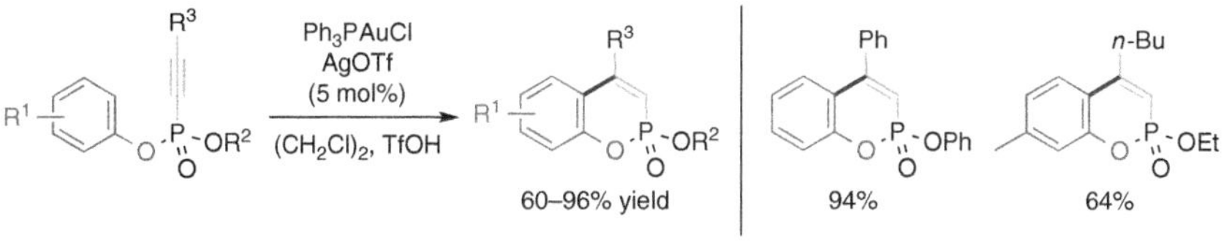

Scheme 7.25 Gold-catalyzed cycloisomerization of aryl alkynylphosphonates.

rearranged 3-iodo-(2*H*)-chromene derivatives with selectivities ranging from 5:1 to >99:1. This isomerization is favored even in the case of arenes bearing strong electron-withdrawing groups, which are known to make iodine migration less favorable. By taking advantage of the efficiency of the intramolecular hydroarylation reaction, a new synthetic route to a phosphacoumarin scaffold was developed through the cycloisomerization of aryl alkynylphosphonates. Careful optimization of the reaction conditions revealed the critical role of triflic acid as an additive, which served to prevent the hydration of the alkyne and favor the desired alkenylation (Scheme 7.25) [58]. The method tolerates a wide range of functional groups on alkyne and aryl moieties, enabling a direct construction of diversely substituted phosphacoumarin nucleus.

Aryl alkynoate esters were also competent substrates for alkyne hydroarylation in the presence of a π-Lewis acid catalyst [59]. This transformation is well illustrated by the synthesis of serotonin analogs in the context of structure–activity examination of fluorescent false neurotransmitter (FFN) neuroimaging probes (Scheme 7.26) [60]. Two protocols, employing $PtCl_4$ or $Ph_3PAuCl/AgSbF_6$, were developed in order to enhance the substrate scope. Comparing with $PtCl_4$, the cationic gold(I) complex was found to be superior in terms of reactivity, substrate scope, and reaction conditions used. A high selectivity for alkenylation at the 4-position of the indole nucleus was observed in the case of O-propargylated serotonin derivatives. However, the cyclization of 3-aminophenol derivatives proceeded with only moderate or no regioselectivity.

The advantages of intramolecular alkyne hydroarylation for the generation of functionalized polycyclic systems have been illustrated in the total synthesis of coumarin-containing natural products pimpinellin, fraxetin, and purpurasol [61] as well as the assembling of the tetracyclic core of the natural product berkelic acid (Scheme 7.27) [62].

A sensitive fluorescence probe for Au(III) ions was developed based on their ability to promote intramolecular alkenylation of amino-substituted phenyl propiolate derivatives (Scheme 7.28, Eq. 1) [63]. Furthermore, a novel class

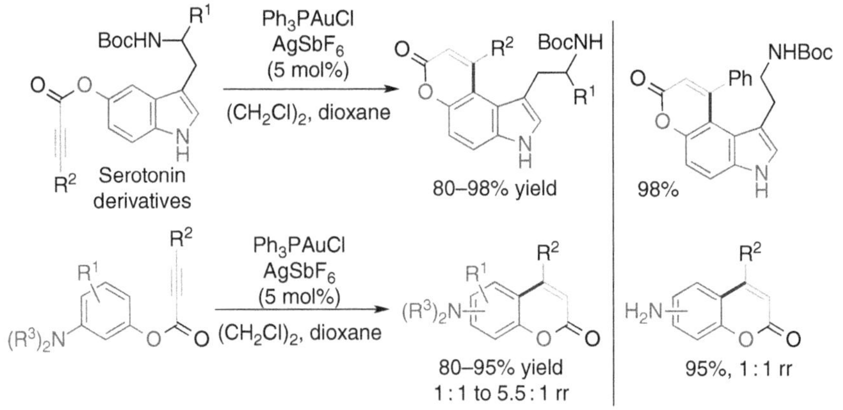

Scheme 7.26 Gold-catalyzed intramolecular cycloisomerization of aryl propiolates for the synthesis of coumarin-containing analogs of serotonin.

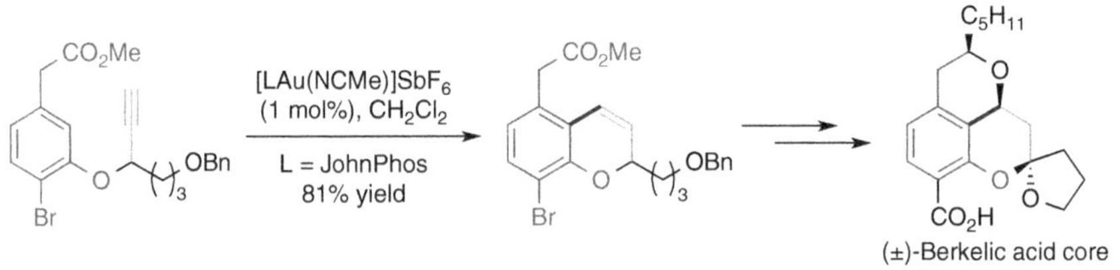

Scheme 7.27 Application of gold-catalyzed hydroarylation of alkynes tethered to aryl ethers in the syntheses of berkelic acid tetracyclic core.

of luminogens, based on pyrrolo-quinoxaline-associated coumarin scaffold, was prepared, using hydroarylation reaction to assemble the required rigid polyheretocycles (Scheme 7.28, Eq. 2) [64]. In addition to the aforementioned formation of the coumarine derivatives, the intramolecular hydroarylation of aryl alkynoate esters affords spirocyclic lactones (Scheme 7.29) [65]. This switch in regioselectivity toward ipso-cyclization was achieved by the addition of water to the standard reaction conditions, which could induce a hydrolytic demethylation. This reaction is limited to *para*-methoxy-substituted phenyl propiolate derivatives. A similar intramolecular ipso-alkenylation of phenols with terminal alkynes, featuring three-carbon tethers, gives ready access to spirocyclohexadienones via 5-*exo*-dig carbocyclization (Scheme 7.30). When para-substituted phenol derivatives are treated with a gold(I)-catalyst, in the presence of a stoichiometric amount of non-coordinating base and MsOH, the desired dienone products are obtained in good to excellent yields [66]. The same strategy was applied for the intramolecular dearomatization of two-carbon-tethered naphthol–ynes. A wide range of spiro[4,5]carbocycles was prepared in excellent yields using $Ph_3PAuOMs$ as catalyst (Scheme 7.31). A variety of internal alkynes bearing electron-rich and electron-deficient aryl substituents or halogen atoms are well tolerated [67]. Additionally, spirocyclization of anisole-substituted ynones was accomplished by employing copper(II) triflate as a catalyst [68].

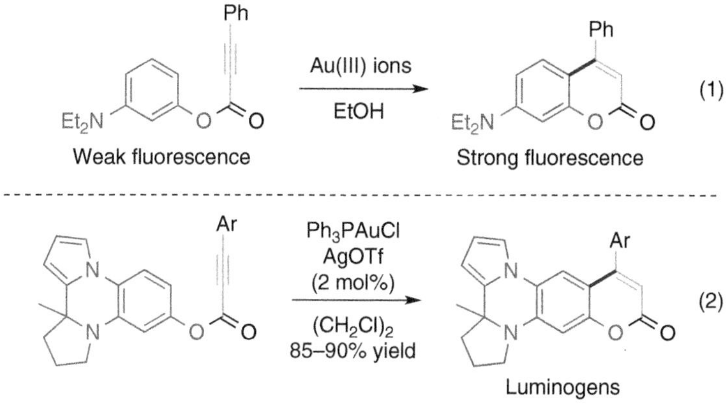

Scheme 7.28 Application of gold-catalyzed hydroarylation of alkynes tethered to aryl esters in sensing of Au(III) ions and synthesis of luminogens.

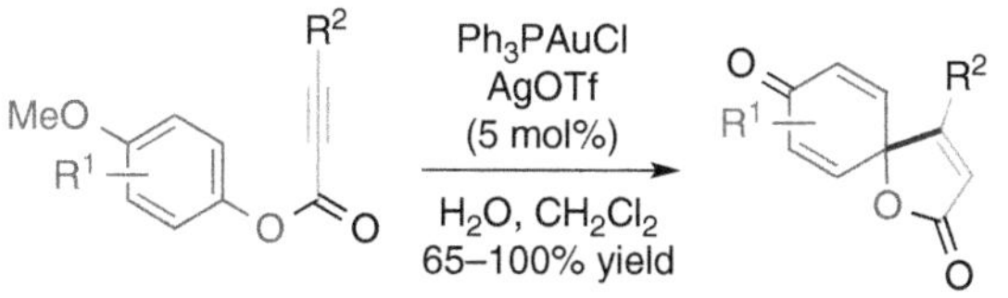

Scheme 7.29 Gold-catalyzed hydroarylation in the synthesis of spirocyclic lactones.

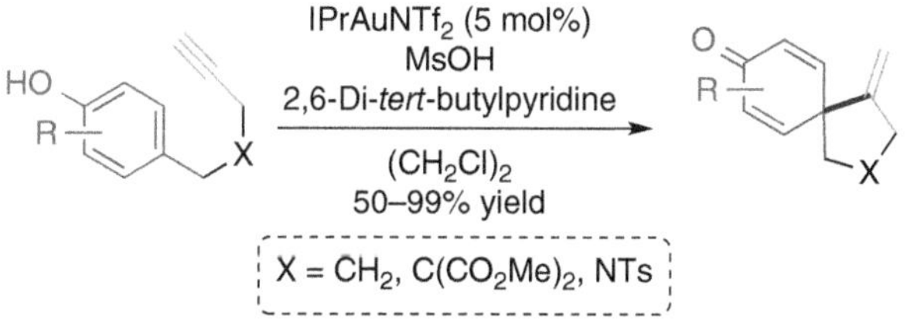

Scheme 7.30 Gold-catalyzed spirocyclization of alkyne-tethered phenols.

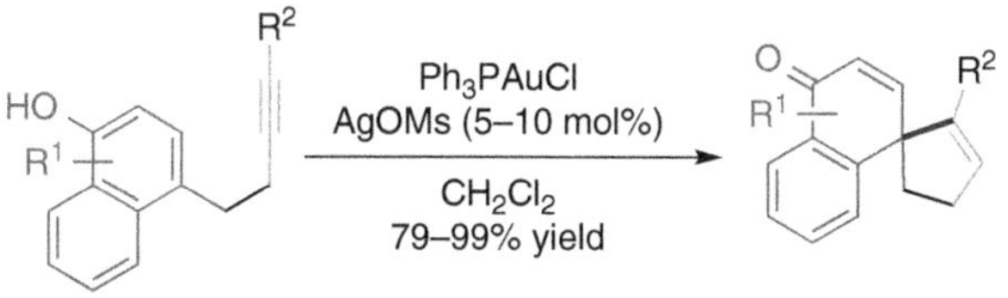

Scheme 7.31 Gold-catalyzed intramolecular dearomative spirocyclization of naphthols.

7.2.2 Alkyne Hydroarylation with Other Arenes

Aniline and phenol derivatives are not the only substrates used for hydroarylation reactions. Simple electron-neutral arenes have also been used as substrates in alkenylation reactions catalyzed by platinum [69] or ruthenium [8] and gallium(III) [3a,e, 70]. By employing biarylated substrates, this type of

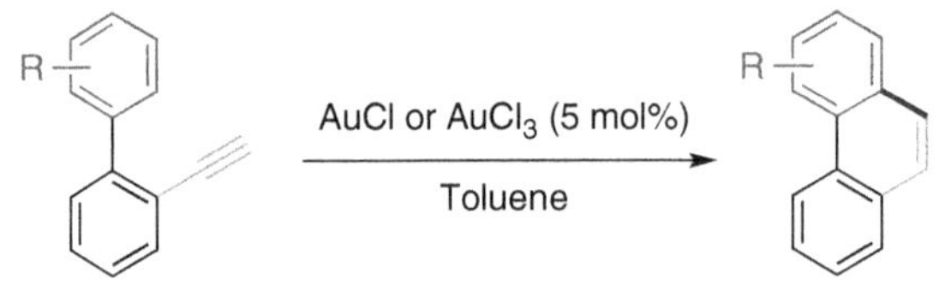

Scheme 7.32 Gold-catalyzed intramolecular cycloisomerization of *o*-alkynylbiaryls.

transformation was applied to the synthesis of polyarenes and polyheteroarenes. Interestingly, the cycloisomerization of *o*-alkynylbiaryls to form phenanthrenes proceeds with similar efficiency in the presence of Pt(II), Au(I), or Au(III) catalyst (Scheme 7.32) [5k, 71–73]. Computational studies addressing the mechanism and selectivity of the alkyne hydroarylation promoted by π-acids revealed that the preference for exo- or endo-cyclization pathways depends on the substitution pattern of the substrate and catalyst used, which is in good agreement with the experimental data [74].

The development of a new, strongly π-acidic phosphine gold catalyst gave an opportunity to expand the scope of this transformation with *o,o*-disubstituted *o*-alkynylbiaryls as well as with *o*-aryl-substituted *o*-alkynylbiaryls [75]. Remarkably, this new catalytic system switches the selectivity toward the formation of substituted phenanthrenes from *o*-aryl *o*-alkynylbiaryls; when under usual conditions the cyclization affords seven-membered ring-containing products (Scheme 7.33). Additionally, the total synthesis of a highly cytotoxic alkaloid (−)-cryptopleurine was achieved by employing a 6-*endo*-dig cyclization of an arylalkyne to form the phenanthrene ring system as a key step [76].

The use of internal aryl-substituted alkynes for alkenylation of non-activated arenes was first demonstrated in the synthesis of benzo[*a*]phenanthridines (Scheme 7.34, Eq. 1) [77a]. The desired 6-*endo*-dig carbocyclization of isoquinoline derivatives occurred efficiently in the presence of an electrophilic gold(I) catalyst upon heating, yielding fused helical tetracycles in good yields. Later, a similar method was developed for the preparation of aryl-substituted phenanthrenes employing copper(II) triflate as the catalyst (Scheme 7.34, Eq. 2)

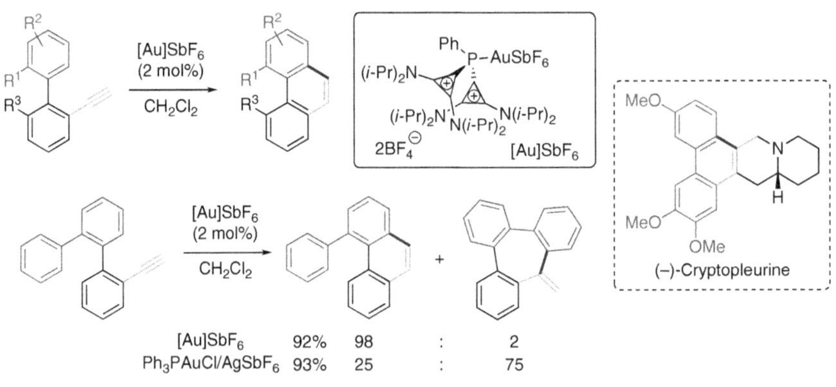

Scheme 7.33 The application of cyclopropenium-substituted phosphine ligands in gold-catalyzed hydroarylation of *o*-alkynylbiaryls.

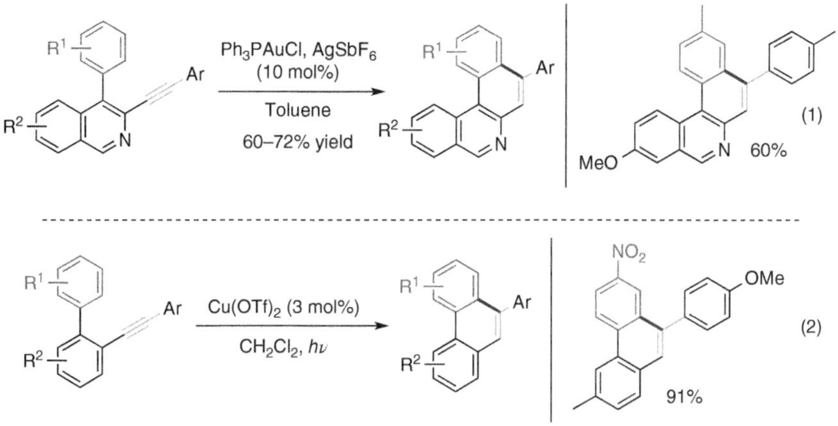

Scheme 7.34 Gold- and copper-catalyzed hydroarylation in the synthesis of aryl-substituted phenanthrenes and benzo[*a*]phenanthridines.

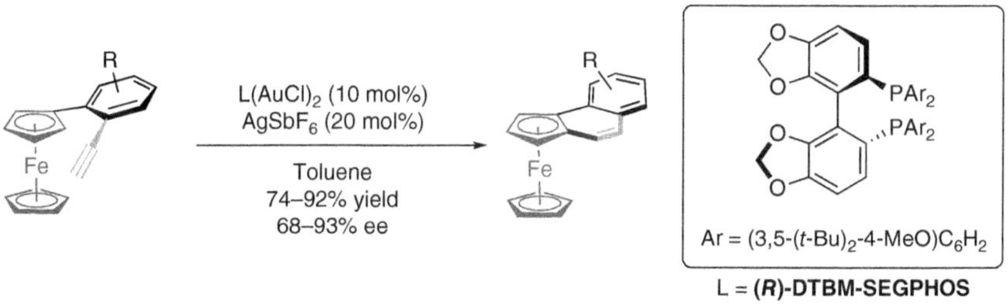

Scheme 7.35 Gold-catalyzed intramolecular hydroarylation of *ortho*-alkynylaryl ferrocenes.

[77b]. In contrast to the gold-catalyzed transformation, the copper-mediated process requires irradiation with UV light, and is proposed to occur via a 6-π-photocyclization. An extension of this concept to *ortho*-alkynylaryl ferrocenes allowed an efficient synthesis of planar-chiral tri- and tetracyclic *ortho*-condensed aromatic ferrocenes, in good to excellent yields and 68–93% ee, by employing digold complex, bearing the bidentate phosphine ligand (*R*)-DTBM-SEGPHOS (Scheme 7.35) [78].

A new entry to anthracene derivatives was elaborated taking advantages of π-acid-mediated alkenylation of *o*-alkynyldiarylmethanes (Scheme 7.36) [79].

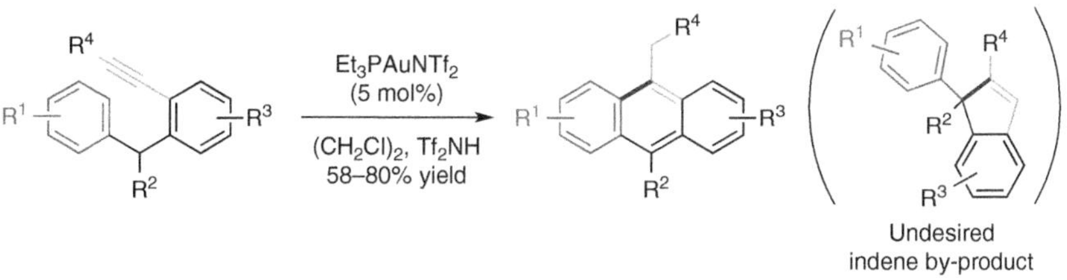

Scheme 7.36 The application of gold-catalyzed intramolecular hydroarylation for the synthesis of anthracene derivatives.

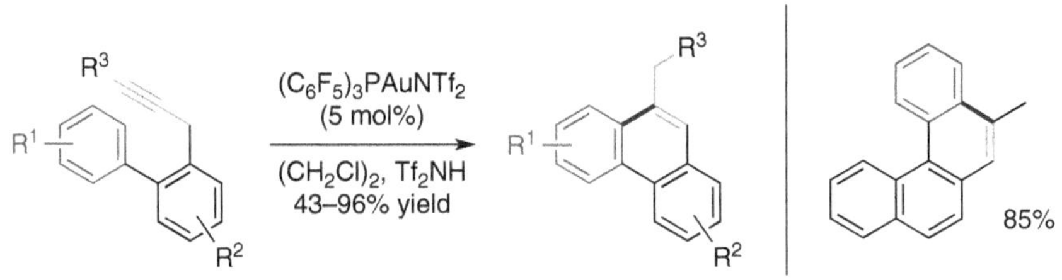

Scheme 7.37 Gold-catalyzed hydroarylation of *ortho*-propargylated biaryls in the synthesis of phenanthrenes.

An extensive optimization of reaction condition was performed in order to prevent competing hydration of the alkyne and formation of undesired indene by-products. A similar method was applied for the preparation of functionalized phenanthrenes from propargylbiphenyl precursors (Scheme 7.37) [80]. A wide range of functionalized biaryls as well as terminal and substituted alkynes are well tolerated and phenanthrene derivatives are obtained in mostly good to excellent yields. Regarding the mechanism, the reaction proceeds via a gold-catalyzed 6-*exo*-dig cycloisomerization followed by an *exo-endo* olefin migration.

Alkyne-tethered fluorenes also proved to be suitable substrates for the intramolecular hydroarylation leading to fluoranthenes and more complex decacyclenes (Scheme 7.38) [81]. In these transformations, the catalytic activity of Au(I) was comparable with $GaCl_3$. The formation of a polycyclic conjugated system containing a seven-membered ring was achieved via a 7-*endo*-dig-selective gold(I)-catalyzed cycloisomerization of *N*-(alkynylphenyl) anilines (Scheme 7.39) [82]. The hydroarylation of terminal and aryl-substituted alkynes resulted in the isolation of desired pentacyclic products in good to excellent yields.

The gold-catalyzed 7-*exo*-dig hydroarylation of terminal alkynes leads to dibenzocycloheptatrienes (Scheme 7.40) [83–85]. This method was also applied in the total synthesis of reticuol and (±)-dihydroisosubamol. It is worth mentioning that only substrates bearing very electron-rich aromatic groups smoothly underwent the desired transformation, whereas in other cases low yields or no conversions were observed.

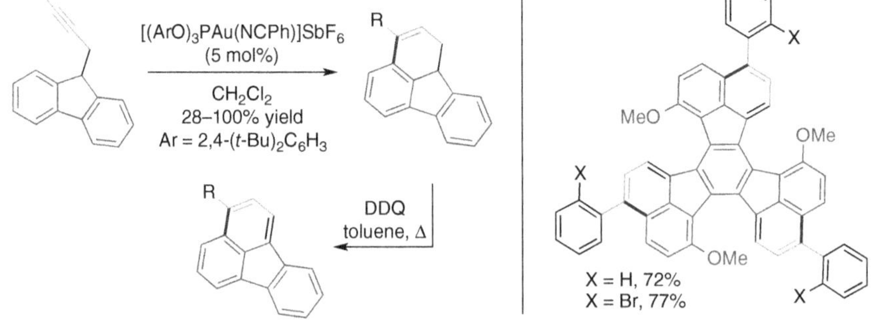

Scheme 7.38 Gold-catalyzed intramolecular hydroarylation in the synthesis of fluoranthenes.

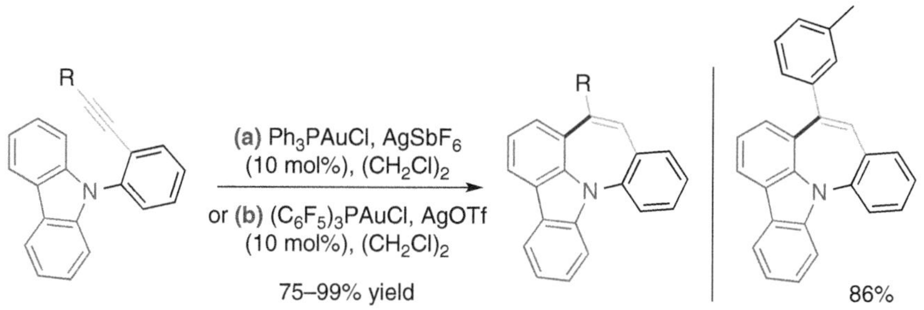

Scheme 7.39 Gold-catalyzed intramolecular hydroarylation of alkynes in the synthesis of dibenzazepine derivatives.

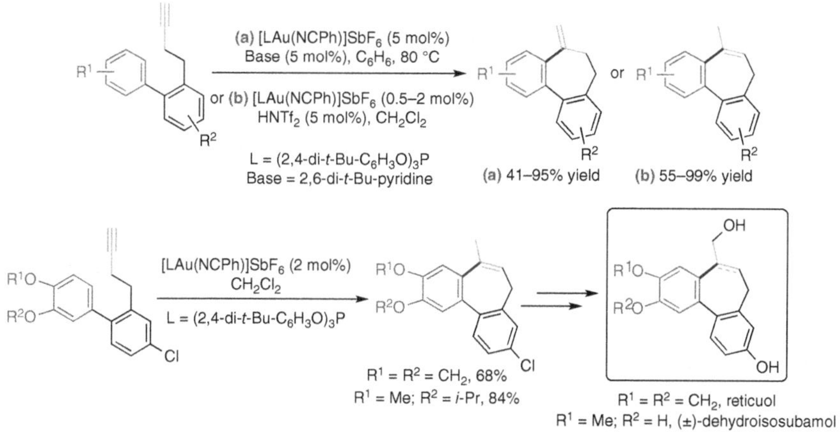

Scheme 7.40 Gold-catalyzed intramolecular hydroarylation of terminal alkynes in the synthesis of dibenzocycloheptatrienes.

Halide-substituted alkynes partake in alkenylation reactions, furnishing vinyl iodides using aniline [39] and aryl propargyl ether [57] substrates. In a similar fashion, biaryls bearing a haloalkyne at the ortho-position undergo a cycloisomerization with or without migration of a halogen atom in the presence of π-acid catalysts AuCl and $InCl_3$, respectively (Scheme 7.41, Eqs 1 and 2) [71]. The extension of this strategy to diiodoethynyl terphenyl derivatives resulted in the elaboration of an efficient method for the preparation of diiodinated picenes and dibenzo[*a*,*h*]anthracenes through a gold-catalyzed double hydroarylation (Scheme 7.41, Eqs 3 and 4) [86]. The initially proposed mechanism involved the formation of a gold vinylidene intermediate via a halide migration followed by the hydroarylation step (Scheme 7.42, pathway 1) [74]. However, additional computational studies, which were performed in the case of bromo-substituted alkynes, concluded that the most favored reaction pathway is an initial 6-*endo*-dig cyclization of an *o*-alkynylbiphenyl precursor followed by consecutive 1,2-*H*/1,2-halogen migrations (Scheme 7.42, pathway 2) [87].

An efficient method for the preparation of diaryl selenides was developed by hydroarylation of heteroatom-substituted alkynes (Scheme 7.43) [88]. The

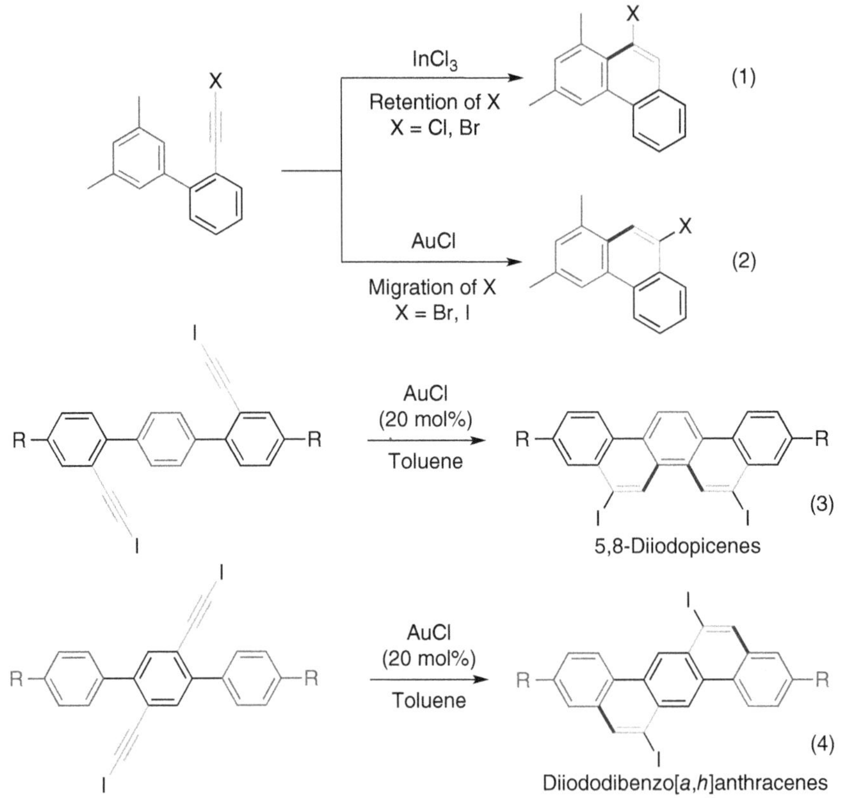

Scheme 7.41 Intramolecular hydroarylation of (*o*-arylphenyl)alkynyl halides catalyzed by π-acids. Application of a gold-catalyzed transformation to the synthesis of iodinated picene and dibenzo[*a*,*h*]anthracene derivatives.

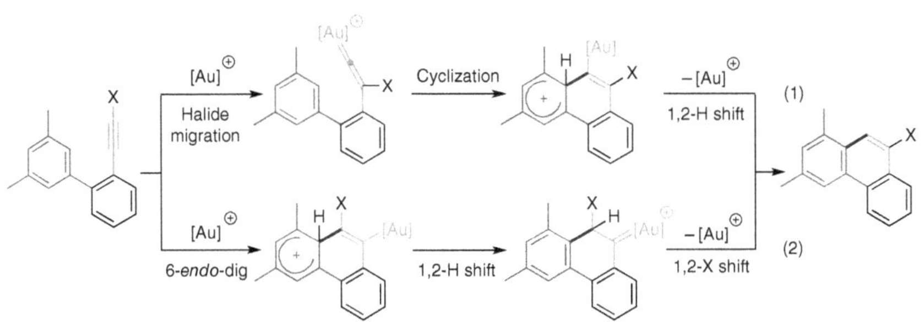

Scheme 7.42 Proposed mechanisms of halogen migration during gold-catalyzed hydroarylation of [*o*-(*m*-xylyl)phenyl]alkynyl halides.

rearranged phenanthryl selenides were obtained when (*o*-arylphenyl)alkynyl selenides were treated with gold(I) and gold(III) salts, causing the migration of the selenide from the terminal to the internal alkyne position. However, the regioselectivity of the transformation could be switched toward a non-rearranged hydroarylation product by using $In(OTf)_3$ as the catalyst.

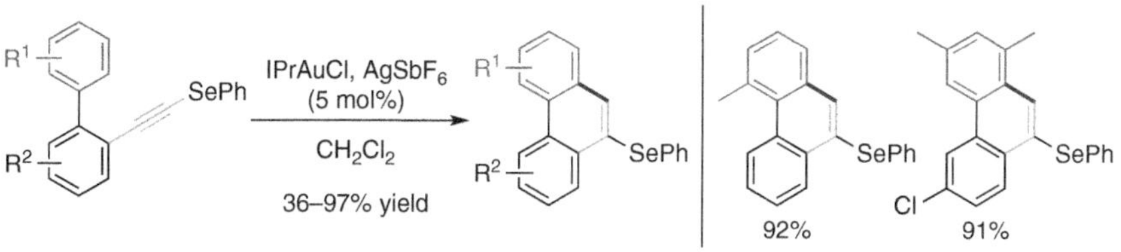

Scheme 7.43 Gold-catalyzed cycloisomerization of (*o*-arylphenyl)alkynyl selenides in the synthesis of seleno-substituted phenanthrenes.

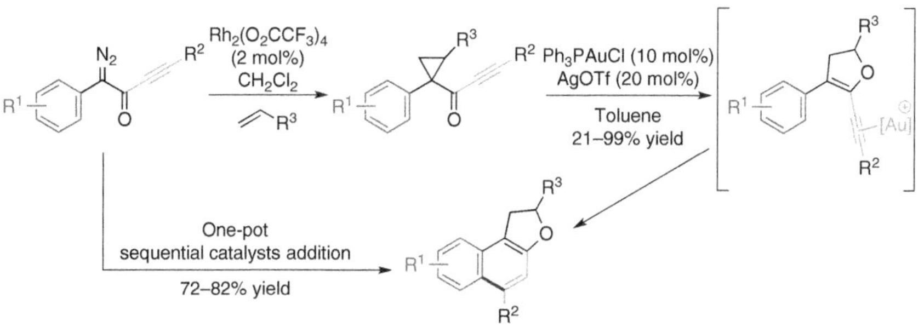

Scheme 7.44 Synthesis of benzannulated dihydrofurans via Rh-catalyzed cyclopropanation, Lewis acid-promoted cyclopropane opening, gold-catalyzed hydroarylation sequence.

Several complex cascade transformations featuring the hydroarylation of alkynes by simple arenes have been also reported. An efficient one-pot procedure for the synthesis of polycyclic dihydrofuran derivatives was developed by a combination of a rhodium-catalyzed cyclopropanation with a gold-mediated benzannulation (Scheme 7.44) [89]. The proposed mechanism of the annulation reaction involves a Lewis acid-promoted rearrangement of the cyclopropyl ketone to a dihydrofuran followed by a gold(I)-catalyzed cycloisomerization. Another example of a tandem transformation involving the hydroarylation of alkynes is described in the context of a single catalyst-mediated synthesis of benzocaradienes from styrenes and propargyl ester-containing diyne [90]. Upon heating with a gold(I) source, the diyne rearranges into a gold(I) carbene, which reacts with styrene, yielding an aryl-alkynyl-substituted cyclopropane that then undergoes a facile hydroarylation to form the benzonorcaradiene skeleton (Scheme 7.45).

7.2.3 Alkyne Hydroarylation with Indoles

As discussed earlier, both electron-rich and simple arenes are good nucleophiles for alkyne hydroarylation reactions. The scope of intramolecular alkenylation with alkynes is not limited by the simple arenes as substrates, and the hydroarylation of alkynes with electron-rich heteroarenes, such as indoles, is also well documented. Several metals and Lewis acidic metal salts, such as mercury [4d] and platinum [5j] salts are competent in catalyzing this process. Indoles are usually excellent nucleophiles for alkyne hydroarylation reactions. In the

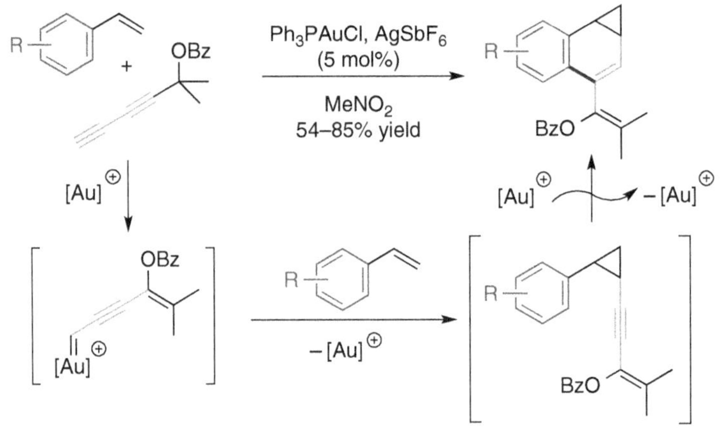

Scheme 7.45 Gold-catalyzed synthesis of benzonorcaradiene skeleton involving an alkyne hydroarylation.

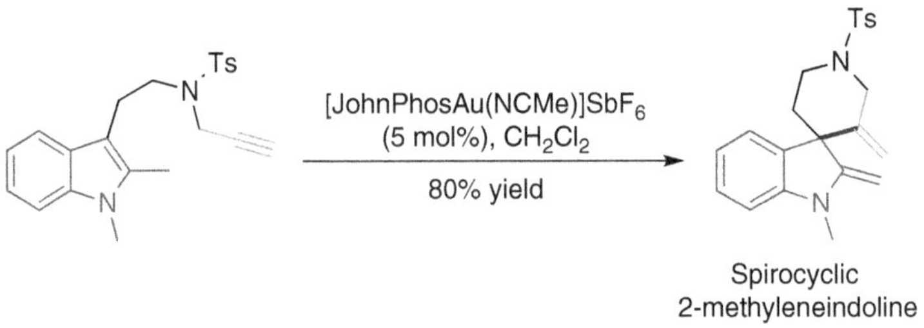

Scheme 7.46 Gold-catalyzed cyclization of propargylated 1,2-dimethyltryptamine into spirocyclic 2-methyleneindoline.

presence of gold(I) catalysts this transformation is assumed to proceed through an initial attack by the most nucleophilic 3-position of indole onto alkyne (Scheme 7.46) [91]. The isolation of a spirocyclic 2-methyleneindoline product in the case of the cycloisomerization of a 2-substituted tryptamine derivative provides evidence to support this proposal. Recent computational studies addressing the mechanism and selectivity of the propargyl-β-tetrahydrocarboline cyclization in the presence of gold revealed that reaction pathways may bifurcate after the formation of an alkyne-gold complex (Scheme 7.47) [92]. In other words, the hydroarylation of indole–ynes may start to occur simultaneously at the 2- and 3-position of indole, and at bifurcation point the cyclization pathway splits to α- or β-alkenylation routes. Moreover, the intermediate cations resulting from α- or β-alkenylation might interconvert to each other by corresponding alkenyl shifts. Thus, providing a general mechanism for the transformation containing a bifurcating potential energy surface as well as the establishing of a reasonable trend for the product distribution prediction is rather difficult. The

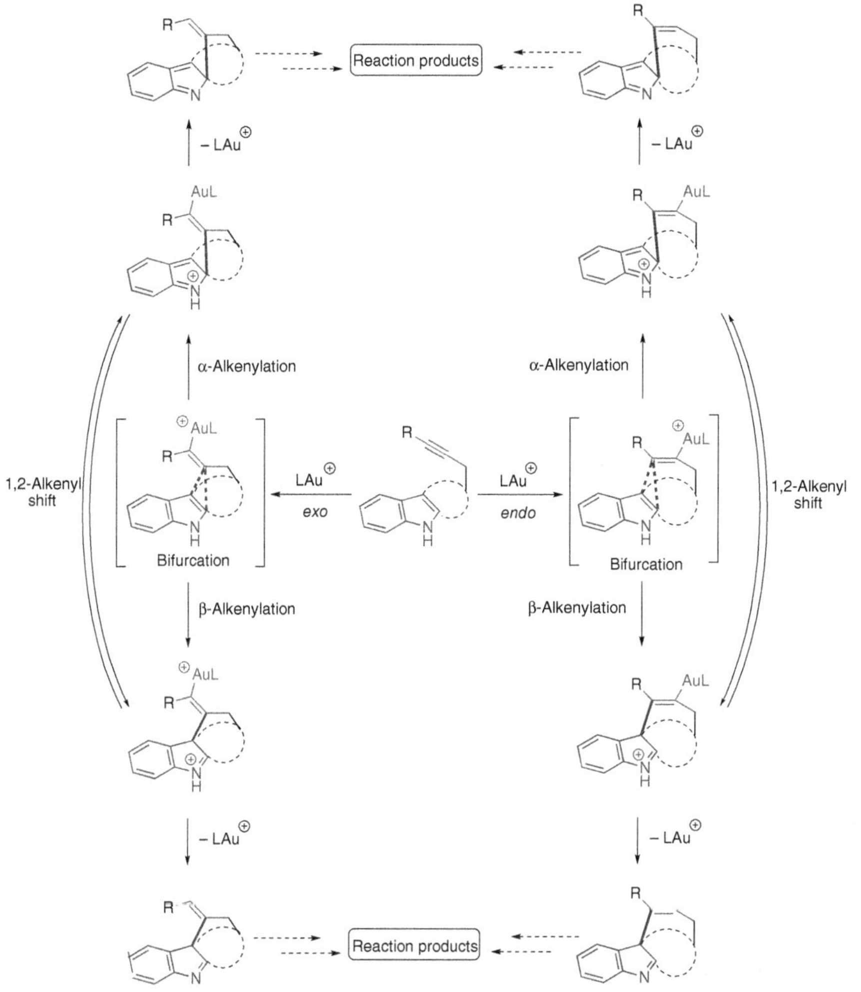

Scheme 7.47 General mechanism of gold-catalyzed intramolecular hydroarylation of alkynes with indole.

additional complexity of this transformation arises from the possibility of the indole-yne alkenylation to occur in exo- or endo-cyclization mode.

7.2.3.1 Alkenylation of Indoles at the 2-Position

The intramolecular hydroarylation of alkynes with indoles was first disclosed for *N*-propargyl tryptamine derivatives [91, 93]. The outcome of this transformation depends on the nature of both the alkyne and the catalyst used. In the presence of an electrophilic [JohnPhosAu(NCMe)]SbF_6 catalyst, the alkenylation of indole with terminal alkynes leads to seven-membered azepinoindoles presumably via a 6-*exo*-dig pathway and subsequent 1,2-alkene shift (Scheme 7.48).

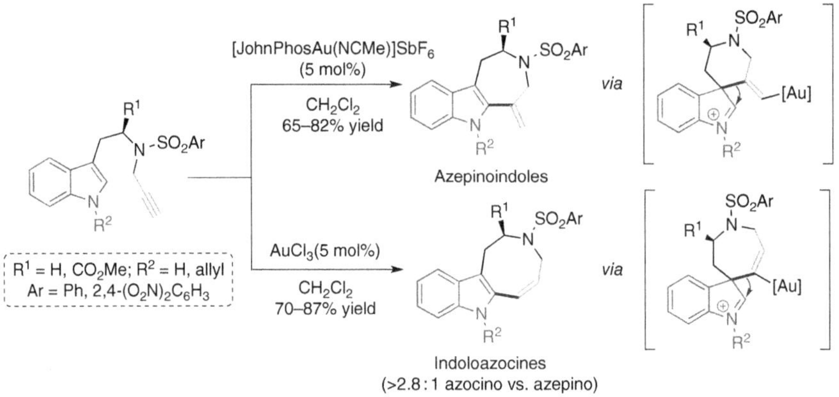

Scheme 7.48 Gold-catalyzed cyclization of tryptamines, bearing terminal alkyne into azepinoindoles and indoloazocines, through spirocyclic indolenine intermediate.

On the contrary, the formation of an eight-membered indoloazocines by 7-*endo*-dig pathway is favored in the presence of AuCl or $AuCl_3$, although a mixture of azepine/azocine compounds is generally observed. The alkenylation of the 2-position of indole proceeds satisfactorily for a series of *N*-protected-*N*-propargyl tryptamines and tryptophan derivatives either with free or protected indole nitrogen (Scheme 7.48). A similar gold-catalyzed cyclization was later developed for a variety of propargylated tryptophol derivatives [91]. The advantages of this strategy are illustrated by the regioselective formation of the azocine ring in a course of the synthesis of the lundurine family of alkaloids [94, 95].

In contrast to terminal alkynes, the hydroarylation of substrates featuring internal alkyne represents a more complex transformation leading to 2-allenylindole and/or indolino tetracyclic products (Scheme 7.49) [91, 93]. Contrary to the aforementioned *N*-propargyl tryptamine cyclization, the hydroarylation of diverse tryptamine-derived propargylamides featuring internal alkyne is a highly regioselective process and yields azocines via endo-cyclization (Scheme 7.50) [96]. In contrast, the alkenylation of indole-ynes-bearing terminal alkynes proceeds with poorer selectivity yielding exo-cyclization products (azepinoindoles and spirocycles). Indole derivatives, bearing terminal alkynes with all-carbon tethers undergo 6-*exo* or 6-*endo*-dig cyclization leading to the formation of dihydrocarbazoles (Scheme 7.51) [93].

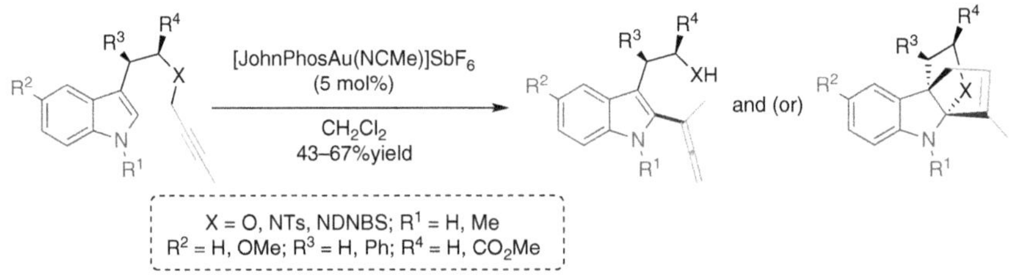

Scheme 7.49 Gold-catalyzed cyclization of internal propargylated tryptamines.

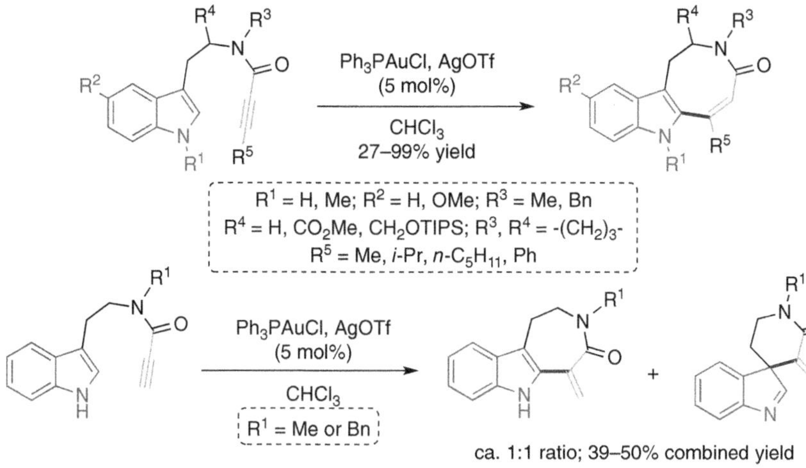

Scheme 7.50 Gold-catalyzed intramolecular hydroarylation of tryptamine-derived propargylamides.

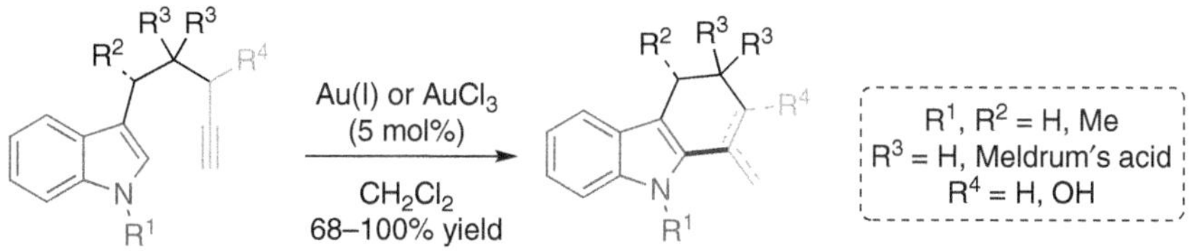

Scheme 7.51 Gold-catalyzed formation of dihydrocarbazoles.

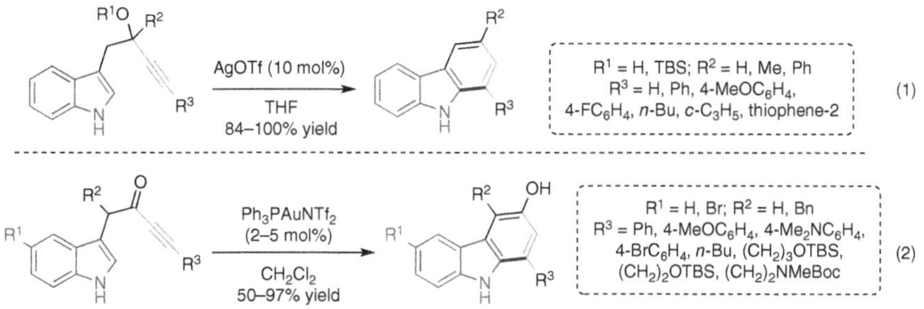

Scheme 7.52 Silver- and gold-catalyzed syntheses of carbazoles.

In contrast to the aforementioned gold-catalyzed intramolecular hydroarylation of alkynes with indoles, the instances of silver(I) salts being used to promote this transformation are scarce. Nevertheless, a silver(I)-promoted carbazole formation from indole-tethered propargyl alcohols has been reported (Scheme 7.52, Eq. 1) [97]. A proposed mechanism of this transformation is similar to that catalyzed by gold and consists of a dearomatizing spirocyclization followed by a subsequent 1,2-migration, protodemetalation, and aromatization through water elimination. Interestingly, either carbazoles or spirocyclic indolenines are exclusively formed depending on the silver salt and solvent used (see Scheme 7.59). In a similar fashion, indolyl ynones could be converted into the corresponding carbazole derivatives by using gold catalysis (Scheme 7.52, Eq. 2) [98].

7.2.3.2 Alkenylation of Indoles at the 3-position

Indoles bearing alkynyl chains at the 2-position are also suitable substrates for cycloisomerization reactions that lead to a large variety of substituted heterocycles. The intramolecular hydroarylation of indole-2-(*N*-propargylcarboxamides) gives access to diverse β-carboline derivatives (Scheme 7.53) [99]. The prevention of heterocyclization of the amide onto the alkyne is achieved by a simple protection of the amide nitrogen. An alkyne hydroarylation promoted by copper salts was used for the synthesis of β-carboline analogs (Scheme 7.54) [100], although harsher conditions were required when compared to gold-catalyzed processes. Depending on the substitution pattern, the cyclization proceeds in a 6-*exo*-dig or 7-*endo*-dig fashion.

A similar approach allows to efficiently generate substituted carbazoles, either from the corresponding 3-alkenyl-tethered indoles [101] or from 2-homopropargylic indoles (through hydroarylation and dehydration) [102]. The hydroarylation strategy gives ready access to useful heteroaromatic building blocks (present in natural products [103], biologically relevant compounds [104], and products of interest in material science [105]) with a significant potential to vary the substitution pattern and the structural features (Scheme 7.55). A slightly

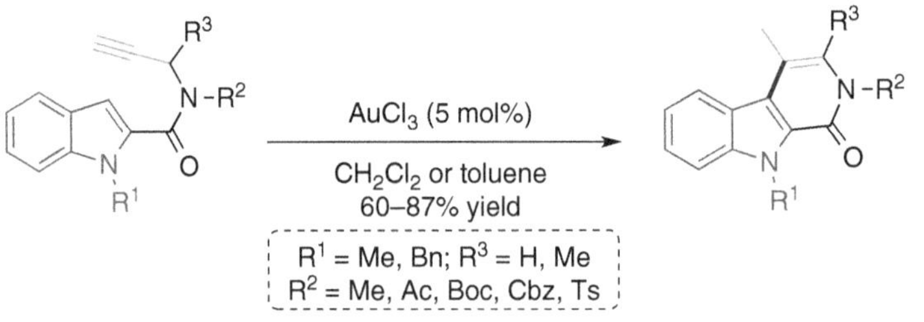

Scheme 7.53 Gold-catalyzed hydroarylation of indole-2-(*N*-propargylcarboxamides).

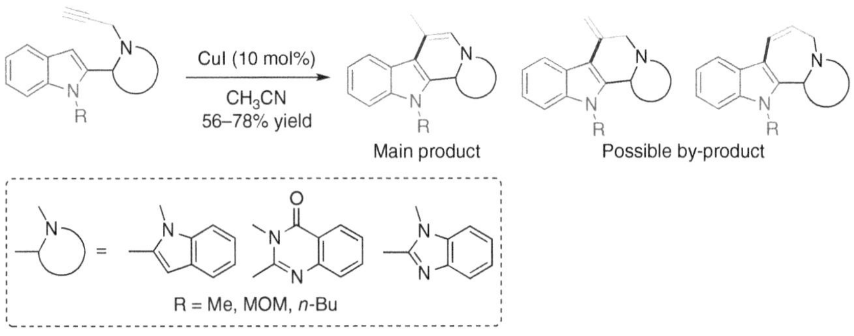

Scheme 7.54 Copper-catalyzed intramolecular hydroarylation in the synthesis of β-carboline-containing compounds.

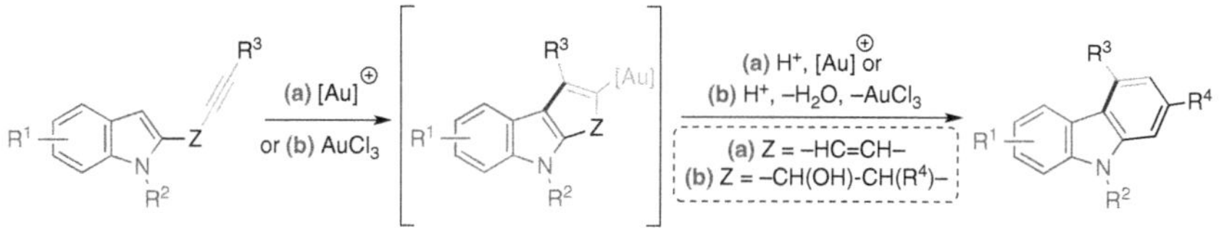

Scheme 7.55 Synthesis of carbazoles by gold-catalyzed intramolecular hydroarylation of internal alkynes with indoles.

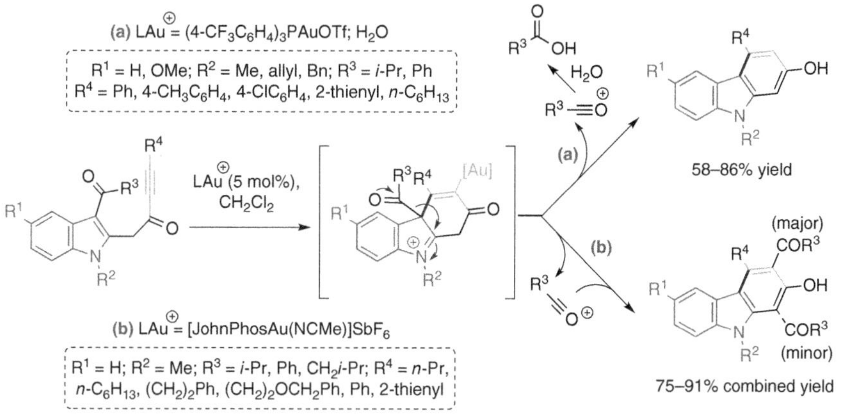

Scheme 7.56 Gold-catalyzed deacylative hydroarylation of internal alkynes with indoles.

different strategy, deacylative hydroarylation of 3-acylindoles, was applied for the preparation of a broad scope of carbazole derivatives (Scheme 7.56) [106]. This transformation involves a gold-catalyzed 6-*endo*-dig cyclization followed by a heterolytic fragmentation of a 3-alkenylated 3-acylindole and a final tautomerization of the conjugated ketone to yield a hydroxycarbazole. The eliminated acylium ion is converted into a carboxylic acid or trapped with the formation of a 1- and 3-acylcarbazole mixture depending on the alkyne substitution pattern and the catalytic system used. The hydroarylation of alkynes with indoles proceeding through 7-*exo*-dig [84, 107] and 8-*exo*-dig [108] pathways (in some cases followed by olefin isomerization) has been reported only for exceptional cases.

Various propargyl-β-tetrahydrocarbolines have been shown to undergo an intramolecular alkenylation with alkynes in the presence of a gold(I) catalyst providing nitrogen-containing heterocycles and complex scaffolds. A ring expansion of 2-propargyl-β-tetrahydrocarboline bearing internal alkyne proceeds in the presence of an electrophilic gold(I) catalyst yielding azocinoindole derivatives (Scheme 7.57, Eqs 1 and 2). A variety of protecting groups at the indole nitrogen atom as well as a wide range of substituents at the alkyne moiety [92, 109] were well tolerated. Alternatively, the dearomatizing spirocyclization occurs in case of terminal alkyne derivatives, furnishing bridged tetracyclic indolines (Scheme 7.57, Eq. 3). DFT calculations were performed to clarify the mechanism of the reaction and rationalize the relative preference for spirocyclization or ring-expansion pathways, depending on the substrate used (Scheme 7.58) [92]. In the case of terminal alkynes, the exo-cyclization at the 3-position of indole takes

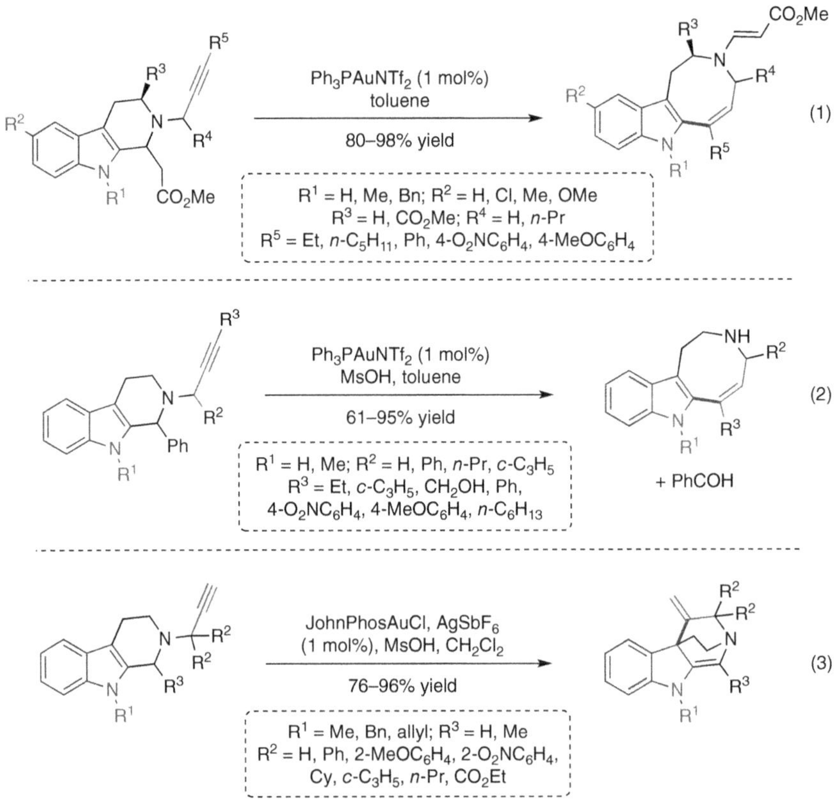

Scheme 7.57 Gold-catalyzed ring expansion and spirocyclization of propargyl-β-tetrahydrocarbolines.

place preferentially and results in the formation of β-alkenylated spirocycles. Contrarily, the hydroarylation of internal alkynes proceeds in an endo-fashion leading to the α-alkenylated intermediates, followed by a cationic Grob fragmentation that generates an iminium cation, which either isomerizes to the corresponding enamine or undergoes hydrolysis with the formation of amine.

7.2.3.3 Spirocyclizations

Several copper and silver salts are competent catalysts for the dearomatizing cyclization reaction of indole–ynones forming functionalized spirocyclic scaffolds including an asymmetric version (Scheme 7.59, Eq. 1) [68]. The stabilizing effect of the carbonyl group facilitates the formation of spirocycles versus 1,2-alkenyl migration products (Scheme 7.52, Eq. 1). This method was extended to indole-tethered propargyl alcohol precursors (Scheme 7.59, Eq. 2) [97, 98]. Interestingly, the selectivity of the reaction toward the formation of either carbazoles (Scheme 7.52, Eq. 1) or spirocyclic indolenines (Scheme 7.59, Eq. 2) could be controlled by the selection of the silver catalyst and the solvent.

Bridged tetracyclic indolenines could be prepared via a gold-catalyzed desilylative cyclization of indole–ynes (Scheme 7.60) [110]. The use of an NHC-gold

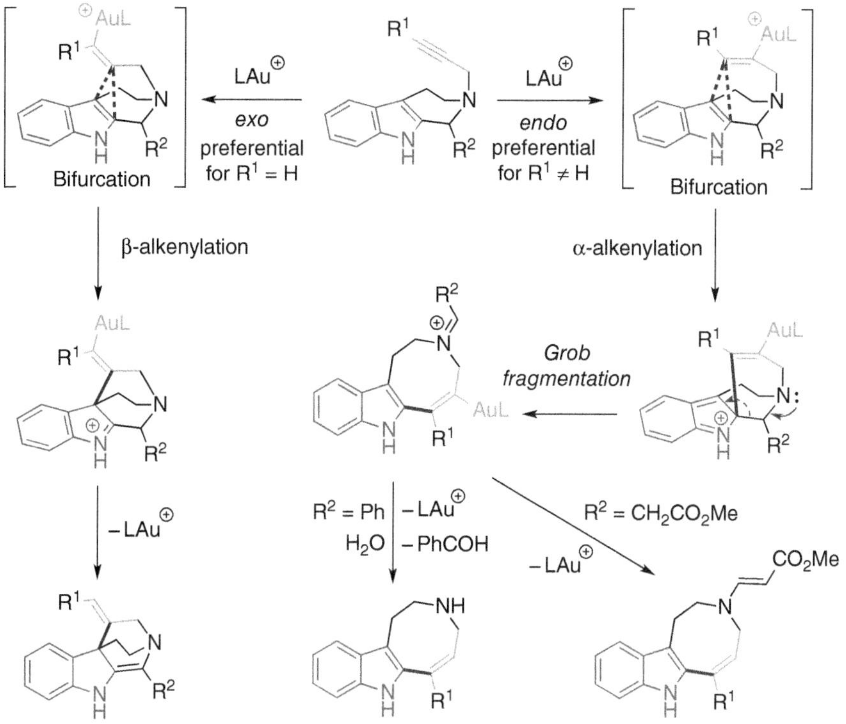

Scheme 7.58 Mechanism of gold-catalyzed intramolecular hydroarylation of propargyl-β-tetrahydrocarbolines.

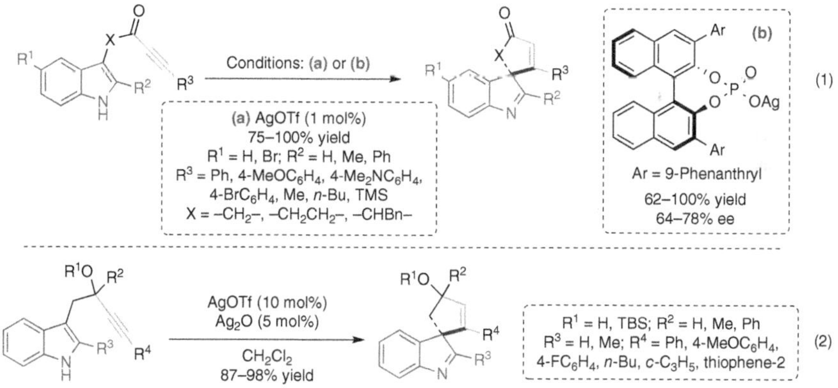

Scheme 7.59 Silver-catalyzed intramolecular spirocyclization reactions of indole–ynes.

catalyst and methanol, as a silyl scavenger and proton source, is essential for the formation of the desired polycyclic framework. A variety of protecting groups at the tryptamine nitrogen atom and indole substituents are well tolerated, whereas indole N-silylation is crucial to prevent C–N-cyclization. In addition, this methodology was also applied for the synthesis of the alkaloid (±)-strictamine [111].

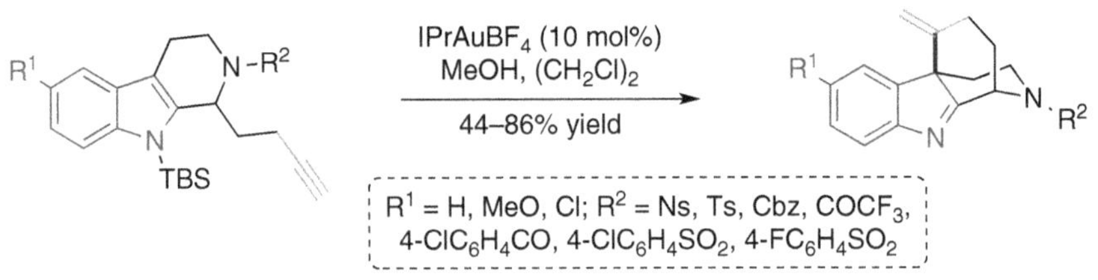

Scheme 7.60 Gold-catalyzed desilylative spirocyclization.

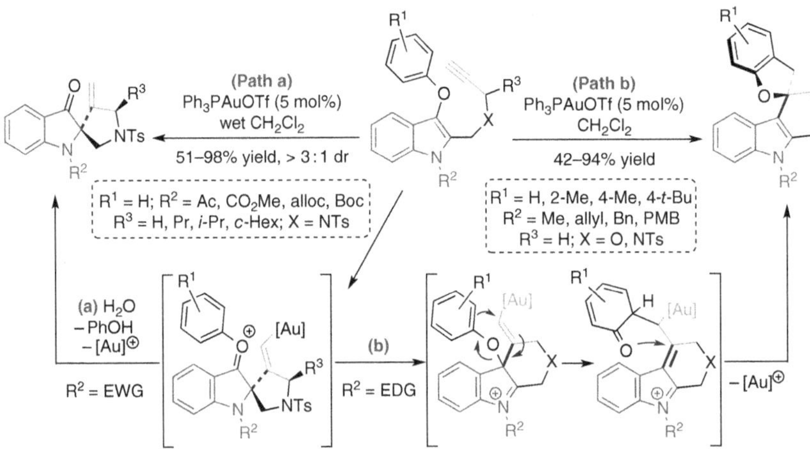

Scheme 7.61 Gold-catalyzed intramolecular alkyne hydroarylation with 3-aryloxy indoles.

Indoles bearing a phenoxy group at the 3-position and a 2-alkynyl tether undergo a regiodivergent annulation when exposed to a gold catalyst (Scheme 7.61) [112, 113]. The outcome of the reaction is strongly dependent on the indole *N*-protecting group, with an electron-withdrawing group (e.g., carbamate) at the indole nitrogen the cyclization leading to spiro-pseudoindoxyl structures (path a). Computational studies of the mechanism established that in the presence of water and triflic acid, hydrolysis of an oxonium intermediate is more energetically favored than rearrangements. In contrast, tetrahydro β-carboline derivatieves (THBC) are obtained when electron-donating protecting groups are used (path b). The mechanistic pathway, in this case, features a 5-*exo*-dig cyclization taking place simultaneously upon an alkyne activation, a subsequent 1,2-vinyl shift, and a C—O bond cleavage via an aromatic Claisen rearrangement. A nucleophilic attack by the transient carbonyl followed by a water-assisted rearomatization completes the benzo[*b*]dihydrofuran formation (Scheme 7.61).

7.2.3.4 More Complex Transformations Featuring a Hydroarylation of Alkynes

Alkyne Hydroarylation/Intramolecular Nucleophile Addition Sequence The 3-alkenylation of indoles followed by nucleophilic trapping of forming iminium

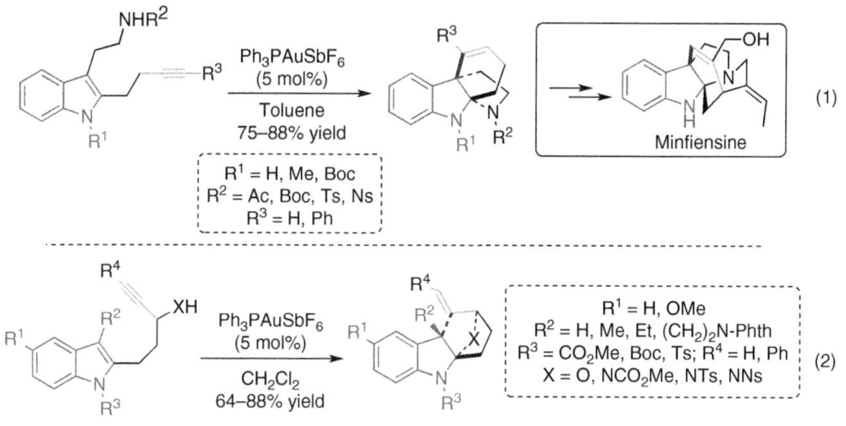

Scheme 7.62 Gold-catalyzed intramolecular hydroarylation/iminium trapping cascade.

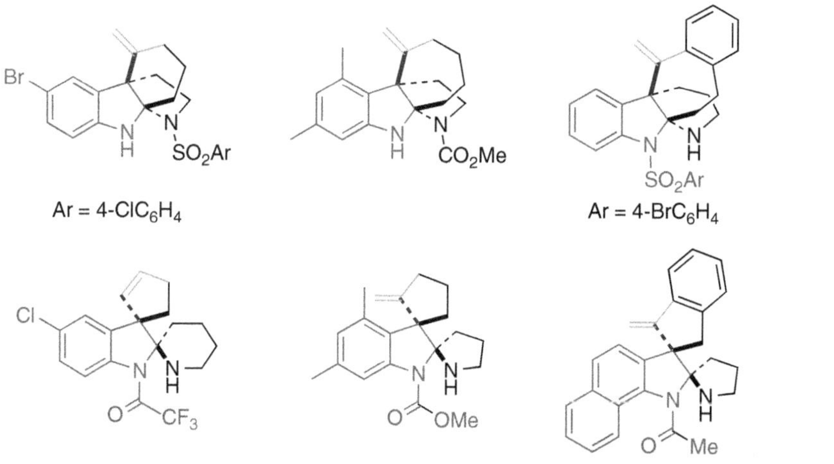

Figure 7.2 Examples of fused and spirocyclic compounds, which were assembled through gold-catalyzed intramolecular hydroarylation/iminium trapping cascade strategy.

ions leads to polycyclic indolines bearing two tetrasubstituted carbon centers. Various fused (Scheme 7.62, Eq. 1) and spirocyclic indolino tetracycles (Scheme 7.62, Eq. 2) are rapidly constructed from 2- and 3-substituted indoles and a nucleophile, either on the 3-substituent or on the alkyne tether (Figure 7.2) [114]. Additionally, this methodology was applied to the formal synthesis of the akuammiline alkaloid minfiensine [114b]. The proposed mechanism was supported by computational studies [115], which well elucidated the importance of the electron-withdrawing group on the indole nitrogen (favoring the formation of the hemiaminal/aminal) and the origin of the diastereoselectivity in the formation of spirocyclic indolines (Scheme 7.62, Eq. 2).

An efficient method to construct polyheterocycles, containing quaternary stereocenters presented in natural product frameworks, is based on a gold(I)-catalyzed hydroarylation/hemi-acetalization cascade of indole propargylic

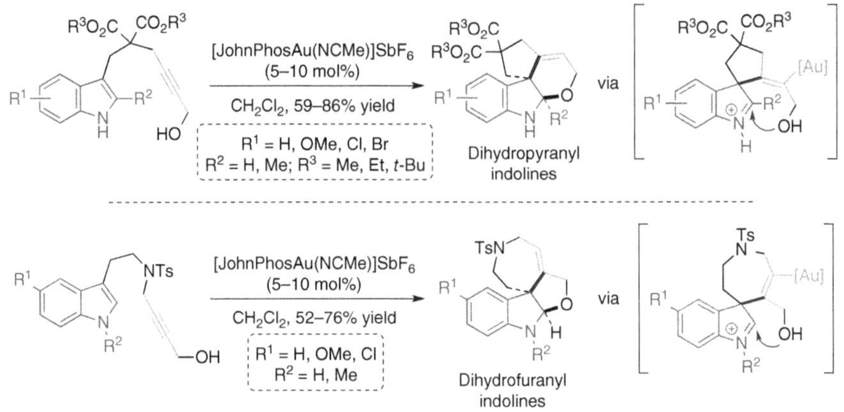

Scheme 7.63 Gold-catalyzed synthesis of pyran- and furan-fused indolines via hydroarylation/heteronucleophilic iminium trapping sequence.

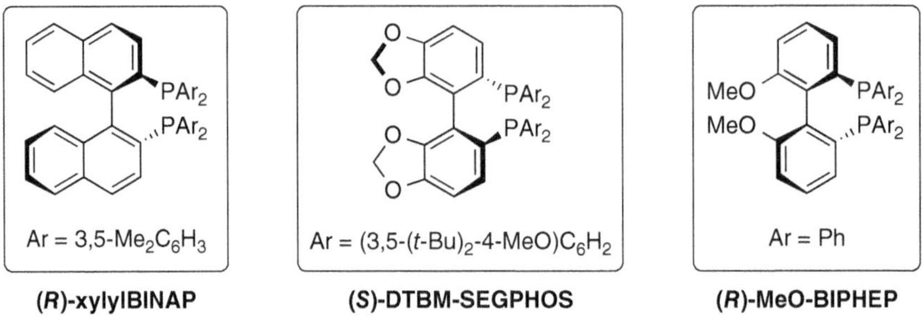

Figure 7.3 Ligands employed in enantioselective gold-catalyzed synthesis of dihydropyranyl indolines and dihydrofuranyl indolines.

alcohols [116]. This tandem transformation is assumed to proceed via an initial gold(I)-assisted alkenylation at the 3-position of indole followed by a nucleophilic attack of the hydroxy group onto the transient indolinium ion (Scheme 7.63). This method allows the rapid and efficient preparation of a wide range of fused tetracyclic dihydropyranyl and dihydrofuranyl indolines. Furthermore, the use of chiral phosphine digold catalysts [(*R*)-xylylBINAP($AuBF_4$)$_2$] and [(*S*)-DTBM-SEGPHOS(AuOTf)$_2$)] (Figure 7.3) resulted in the development of enantioselective processes [117]. Under these conditions, a gold(I)-catalyzed hydroindolination of propargylic alcohols proceeds in a moderate to good yield, generating the fused tetracyclic dihydropyranyl indolines with up to 87% ee and dihydrofuranyl indolines with up to 85% ee.

Extension of this process to indole-ynamides provided another method to construct structurally complex polyheterocycles featuring a quaternary carbon center (Scheme 7.64) [118]. A variety of protecting groups at tryptamine and indole nitrogen atoms as well as electron-donating and electron-withdrawing substituents at the indole moiety are well tolerated. The cyclization of the indole-ynamide ($R^1 = R^2 = R^5 = H$, $R^3 = Me$, $R^4 = Ts$) with a digold catalyst

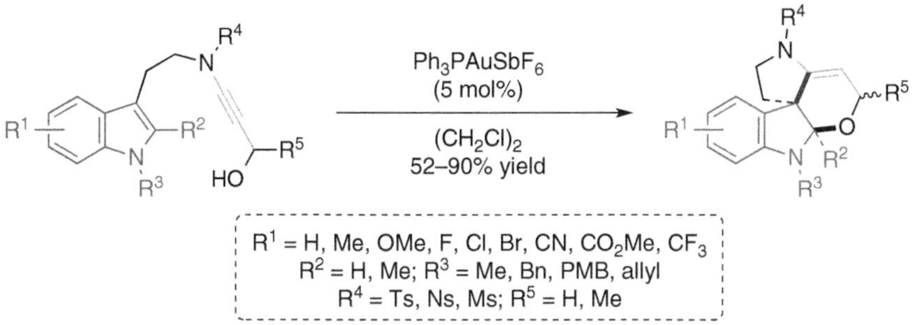

Scheme 7.64 Gold-catalyzed intramolecular hydroarylation/iminium sequence for indole-ynamides.

bearing a chiral phosphine ligand (*R*)-MeO-BIPHEP (Figure 7.3) furnished the annulation product with 60% ee.

Hydroamination/Alkyne Hydroarylation Cascade Transformations The ability of gold salts to promote complex multistep transformations and significantly increase molecular complexity was put to use for the synthesis of aryl-fused carbazole derivatives. This cascade reaction includes 5-*endo*-dig hydroamination/6-*endo*-dig hydroarylation of aniline-substituted diethynylarenes, ready available by the Sonogashira cross-coupling reaction (Scheme 7.65) [119]. This methodology was further extended to polyenyne aniline systems that feature a series of *o,o*-alkynylarenes, and are used for the generation of extended polyaromatic ribbon-like structures (Scheme 7.66) [120]. So far, this strategy has been applied to up to four consecutive 6-*endo*-dig cyclization events to form multiple benzene rings as a mixture of two stereoisomers (due to the non-planar nature of the extended polyarene). In a similar fashion, simpler polyarenes such as dihydrobenzoindole and dihydronaphthofuran could be prepared in good to excellent yields from corresponding 1,5-diyne precursors. A seven-membered ring formation was also achieved through hydroamination/hydroarylation of 1,6-diynes featuring a linear tether (Scheme 7.67) [121]. The further extension of the hydroamination/hydroarylation methodology to thiophene and thiophene/benzothiophene as alkynyl linkers illustrates the potential of this strategy in the context of material science (Scheme 7.68) [120, 122]. This strategy constitutes a simple entry to the preparation of various donor–acceptor–donor (D–A–D) molecules, which present an interesting UV–vis absorption and electrochemical properties.

Transformations Involving Friedel–Crafts-Type Reactions A synthesis of complex, fused polyheterocyclic skeletons was achieved via a cascade transformation consisting of an intermolecular Friedel–Crafts α-amidoalkylation of indoles and other heterocycles with subsequent intramolecular gold-catalyzed hydroarylation [123, 124]. Indole-containing scaffolds are constructed from *N*-propargylic *N,O*-acetals in good yields, although with moderate *exo*/*endo*-selectivity (Scheme 7.69). The single catalyst-mediated intermolecular reaction of indoles with (*Z*)-enynols leads to the formation of dihydrocyclohepta[*b*]indoles

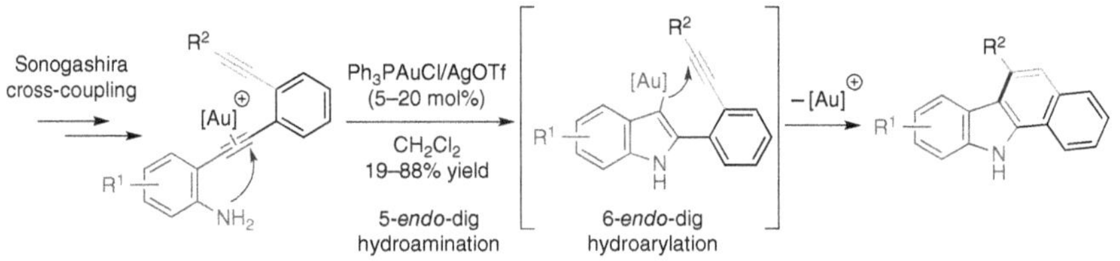

Scheme 7.65 Gold-catalyzed intramolecular hydroamination/hydroarylation cascade for the synthesis of aryl-fused carbazoles.

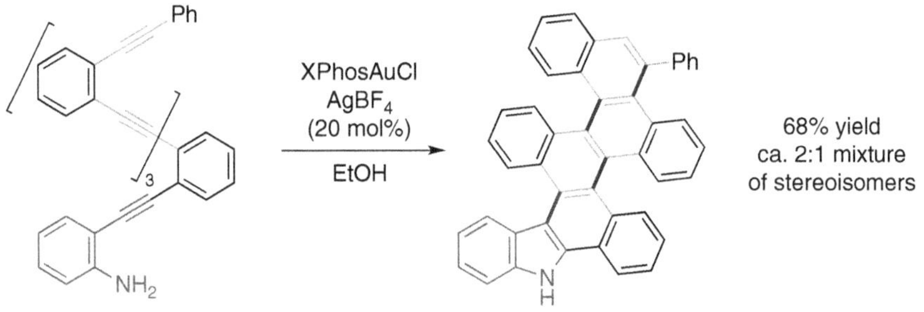

Scheme 7.66 Application of gold-catalyzed hydroamination/hydroarylation cascade for the preparation of polyaromatic compounds.

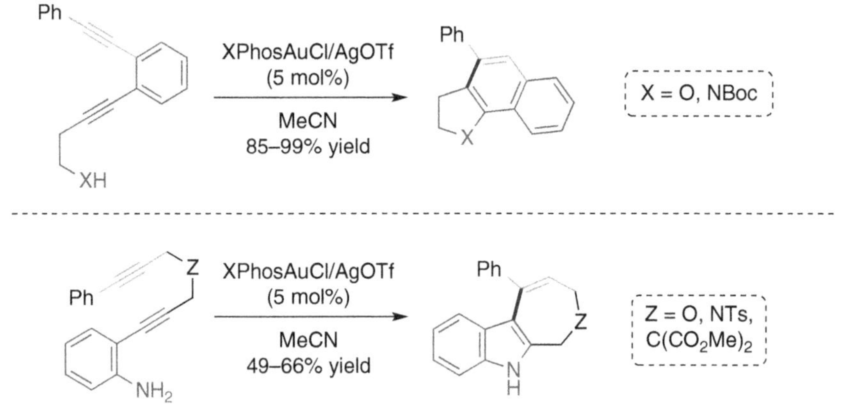

Scheme 7.67 Synthesis of dihydronaphthofuran, azepino-, oxepino-, and dihydrobenzoindoles through gold-catalyzed hydroamination/hydroarylation cascade.

[125]. This tandem transformation involves an initial Friedel–Crafts reaction to form an indole–yne intermediate, which is perfectly set to undergo gold(I)-catalyzed hydroarylation (6-*endo*-dig) yielding a new seven-membered ring after 1,2-alkenyl shift (Scheme 7.70). In a similar fashion, gold(III)-mediated intermolecular reaction of indoles with enones bearing two alkynyl substituents, furnishes polycyclic indole-fused scaffolds (Scheme 7.71) [126]. An intramolecular carbonyl-yne cycloisomerization of diyne-enone followed by a Friedel–Crafts reaction with the electron-rich indole partner generates a furan-tethered

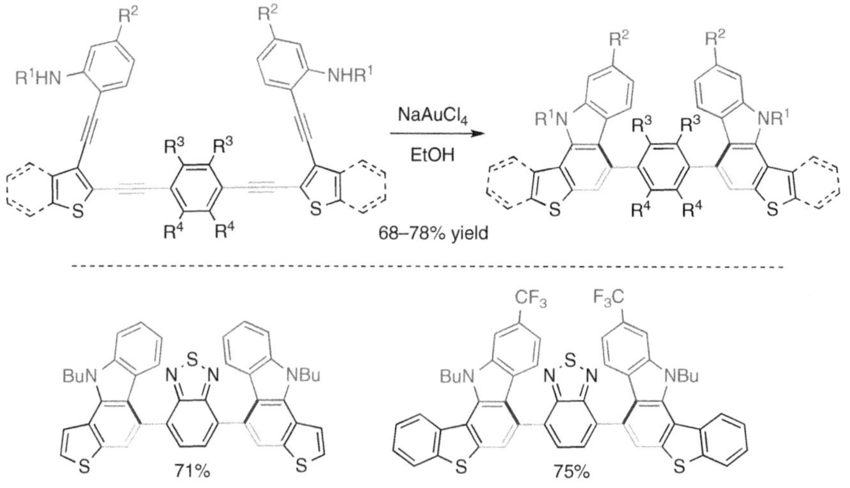

Scheme 7.68 Approach to donor-acceptor-donor molecules via gold-catalyzed hydroamination/hydroarylation cascade.

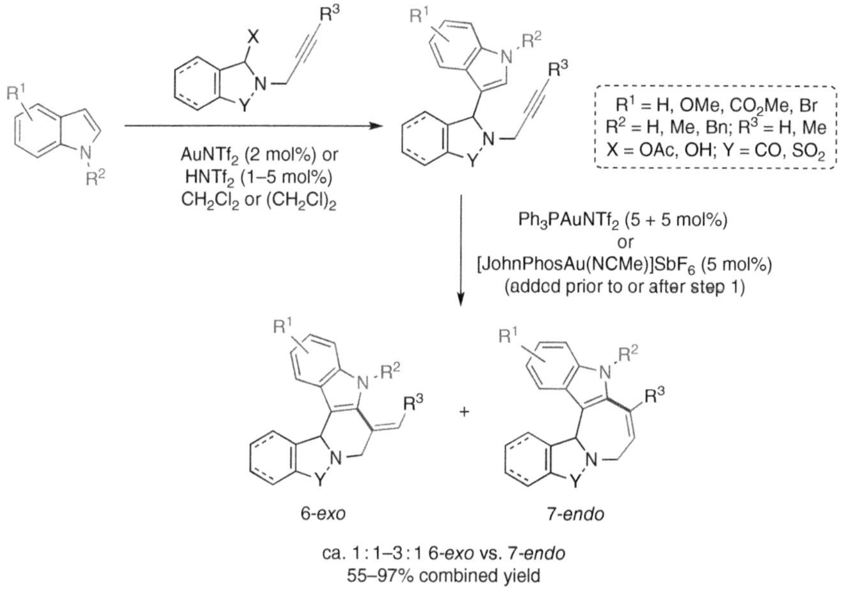

Scheme 7.69 Friedel–Crafts α-amidoalkylation/hydroarylation sequence.

alkynyl indole. The intermediate indole–yne then undergoes gold(III)-catalyzed hydroarylation to efficiently provide the furan-fused dihydrocyclohepta[*b*]indoles (Scheme 7.71). Additionaly, a [5–7]-fused tricyclic skeleton is assembled via a Michael addition/hydroarylation tandem transformation (Scheme 7.72) [127]. Various indole derivatives and vinyl propargyl ketones react in the presence of $NaAuCl_4$ hydrate to provide annulation products in a relatively simple manner.

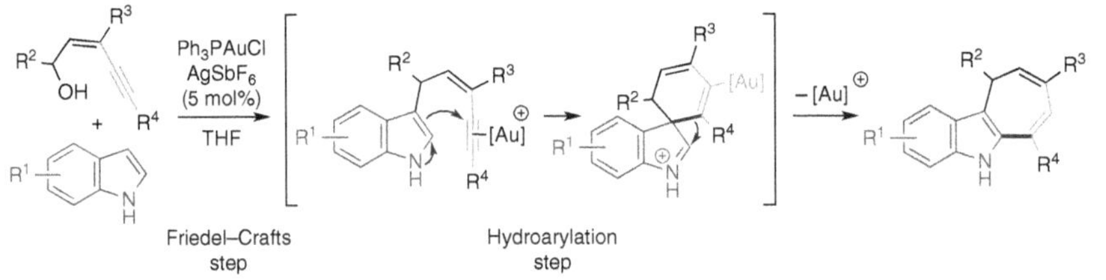

Scheme 7.70 Gold-catalyzed Friedel–Crafts/hydroarylation cascade reaction of indoles with (*Z*)-enynols.

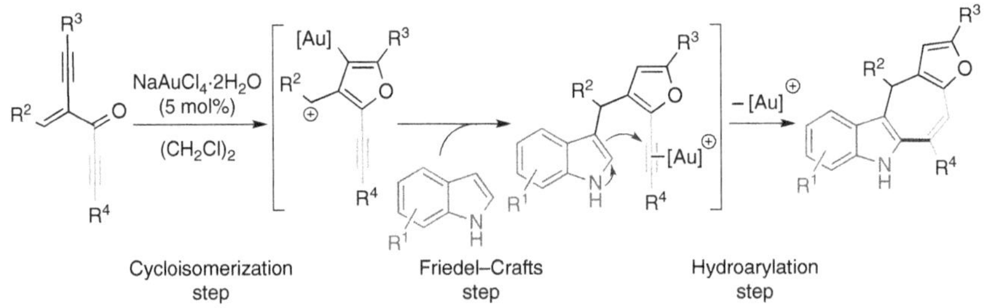

Scheme 7.71 Gold-catalyzed tandem cycloisomerization/Friedel–Crafts/hydroarylation reaction of diyne–enones with indoles.

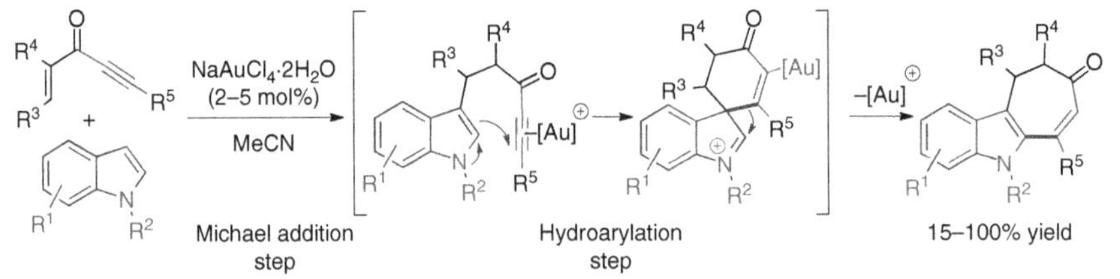

Scheme 7.72 Gold-catalyzed Michael addition/hydroarylation cascade reaction of indoles with enynones.

Another example of tandem Michael addition/alkyne hydroarylation was reported for *ortho*-alkynyl nitrostyrenes and indole derivatives (Scheme 7.73) [128]. This process takes place in water under microwave irradiation. The combination of an organocatalyzed asymmetric Michael addition with gold(I)-catalyzed hydroarylation of an alkyne moiety resulted in an enantioselective process for the synthesis of chiral indoloazepine derivatives with up to 98% ee (one-pot sequential catalysis, Scheme 7.73) [129].

A simple approach for the preparation of various indoloazepinobenzimidazoles was developed through a one-pot gold(I)-catalyzed domino oxidative condensation of N^1-alkynyl diamines with indole-3-carboxaldehydes followed by an intramolecular hydroarylation (Scheme 7.74) [130]. Generally, polycycles were obtained in moderate/good yields with an exception of the substrates bearing electron-withdrawing groups, for which no cyclized product was observed.

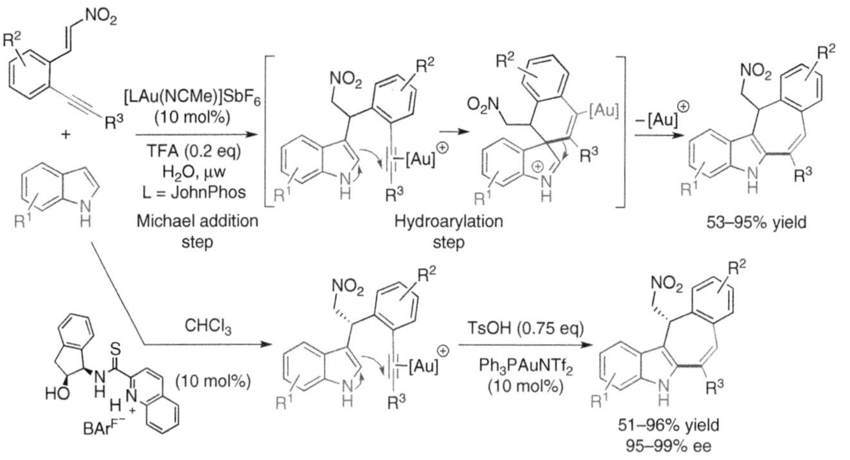

Scheme 7.73 Racemic and enantioselective versions of tandem gold-catalyzed Michael addition/hydroarylation reaction of indoles and *o*-alkynylnitrostyrenes.

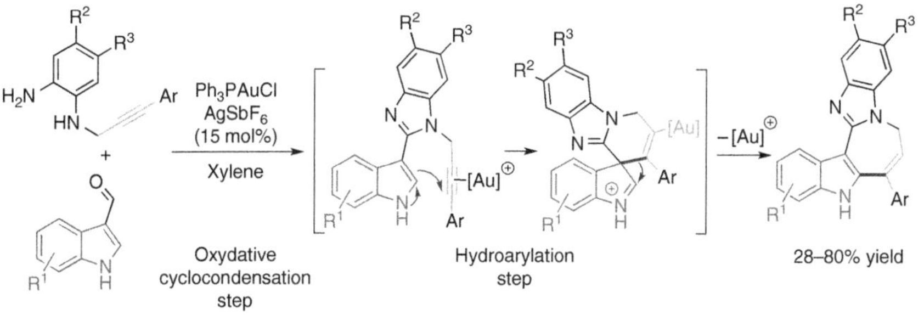

Scheme 7.74 Gold-catalyzed oxidative cyclocondestation/hydroarylation cascade.

A copper-mediated formation of ynamides from *gem*-dibromovinyls and sulphonamides followed by a one-pot intramolecular silver(I)-catalyzed hydroarylation is an efficient procedure for the synthesis of pyrrolo-/indolo[1,2-*a*] quinolines and naphtho[2,1-*b*]thiophenes (Scheme 7.75) [131].

Post-Ugi Alkyne Hydroarylation Indole-tethered propargylic amines, adducts of Ugi 4-component reaction (Ugi-4CR), undergo hydroarylation in the presence of gold(I) chloride as a catalyst [132]. However, the outcome of the reaction strongly depends on an alkyne substitution pattern: terminal alkynes are converted into corresponding dihydropyridoindoles, whereas propargyl amines with an internal triple bond give azepinoindoles (Scheme 7.76). This methodology was also extended to pyrroles [132]. Substrates obtained by condensation of indole-3-carboxaldehyde and propargyl amine in the presence of isocyanide and carboxylic acid are also suitable for a post-Ugi cyclization. The treatment of indole–ynes with a cationic triphenyl phosphine gold(I) complex results in the formation of tetracyclic spiroindolines [133]. The diastereoselective domino cyclization promoted by gold complexes proceeds via a 5-*exo*-dig

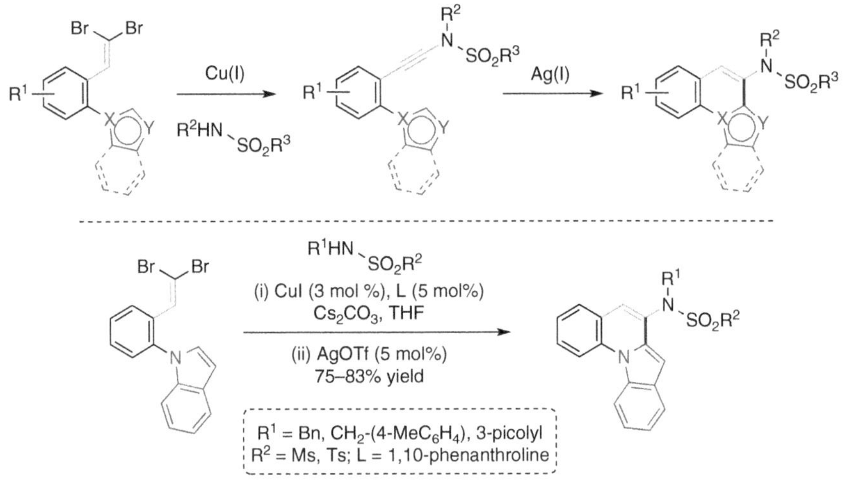

Scheme 7.75 Copper-catalyzed formation of ynamides and following silver-catalyzed hydroarylation.

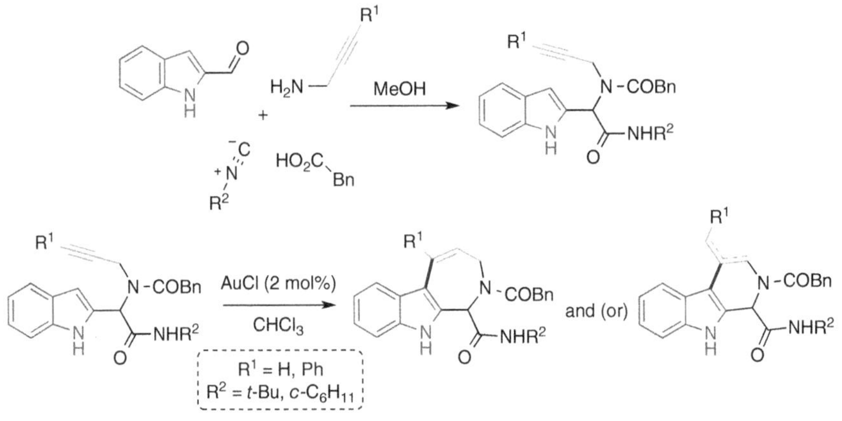

Scheme 7.76 Gold-catalyzed hydroarylation in C-2-tethered products of Ugi 4-component reaction.

hydroarylation of the activated alkyne by the indole 3-position (Scheme 7.77). The following diastereoselective attack of the nucleophilic amide (NH) onto the indolinium electrophile and subsequent protodeauration furnish spiroindolines (Scheme 7.77). The same reactivity was observed for indole propargylic amides, synthesized by Ugi 4-CR between substituted indole-3-carboxaldehydes, amine derivatives, isocyanides, and substituted propiolic acids (Scheme 7.77) [134]. In addition, the indoloazocine skeleton is easily accessible by *endo*-dig hydroarylation of the propargylamides (adducts of Ugi 4-CR), obtained by condensation of various aldehydes, tryptamine derivatives, an isocyanide, and 3-butynoic acid [135]. In a similar fashion, for substrates featuring a 3-(aminomethyl)indole instead of a tryptamine moiety, the cyclization resulted in the formation of

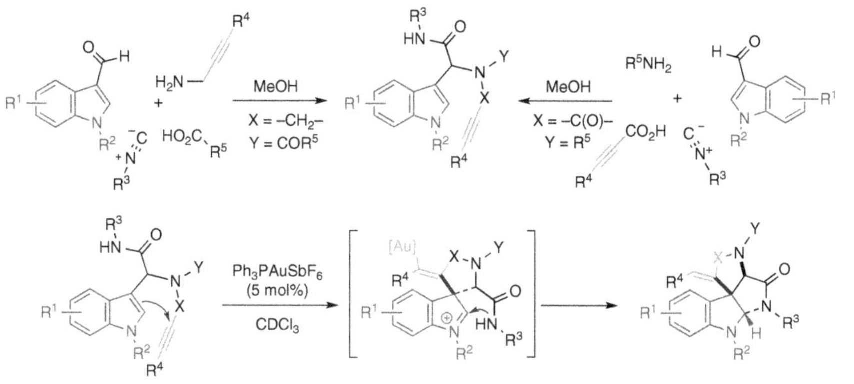

Scheme 7.77 Ugi reaction in the synthesis of the precursors for the gold-catalyzed hydroarylation/amide-iminium cyclization.

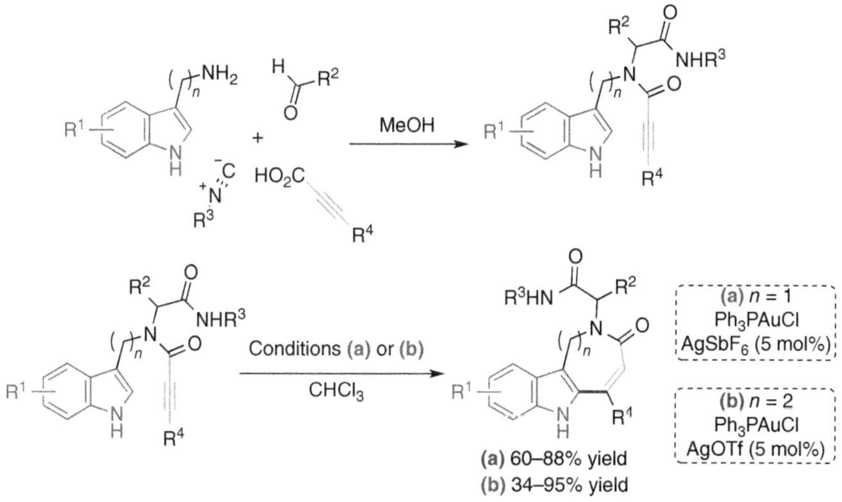

Scheme 7.78 Gold-catalyzed hydroarylation in C-3-tethered products of Ugi 4-component reaction.

indoloazepines (Scheme 7.78) [136]. It is noteworthy that in both cases the alkenylation with internal alkynes proceeds in an *endo*-dig manner, whereas terminal alkynes react via an *exo*-dig pathway, yielding indoloazepine or indolopyridine derivatives respectively.

In one case it was noted that with an alkyne tethered at the 4-position of indole the hydroarylation occurs at the 3-position of indole in the presence of $In(OTf)_3$ or electrophilic gold(I) sources. Remarkably, when the reaction was promoted by In(III) catalysts, the alkenylation proceeded via a 7-*exo*-dig pathway to yield azepino-[*c,d*]indolones, whereas treating the substrates with a gold(I) catalyst furnished azocino-[*c,d*]indolones, formed via 8-*endo*-dig cyclization (Scheme 7.79) [137].

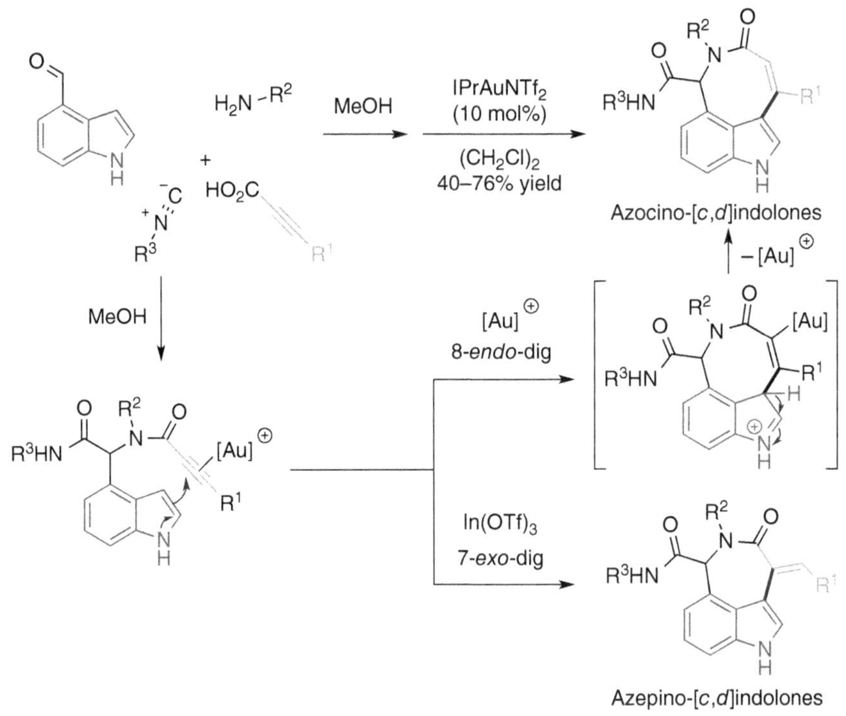

Scheme 7.79 A post-Ugi divergent intramolecular hydroarylation of internal alkynes tethered at the indole 4-position.

Alkyne Hydroarylation Reactions with Subsequent C—C Bond Cleavage It is commonly suggested that alkenylation at the 2-position of indole proceeds via an initial spirocyclization at the 3-position with a subsequent 1,2-alkenyl migration. However, during the course of the cyclization of alkyne-substituted indole-3-carboxamides, the 3,2-acylamino shift occurs preferentially and a rearranged product alkenylated at the 3-position is formed (Scheme 7.80) [138].

The silver-catalyzed hydroarylation of terminal alkynes with indoles was demonstrated in a tandem transformation, such as during a synthesis of substituted tetrahydrocarbazoles. In this case, silver(I) triflate was used as both σ-Lewis acid, to activate aldehyde and generate a 1,1-bis(indolyl)-4-alkyne intermediate, and as π-Lewis acid to promote 5-*exo*-dig hydroarylation through the indole 3-position (Scheme 7.81) [139]. Subsequent to the 1,2-alkyl shift,

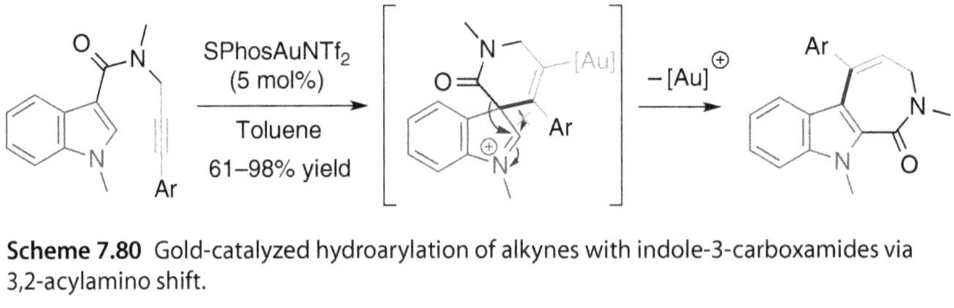

Scheme 7.80 Gold-catalyzed hydroarylation of alkynes with indole-3-carboxamides via 3,2-acylamino shift.

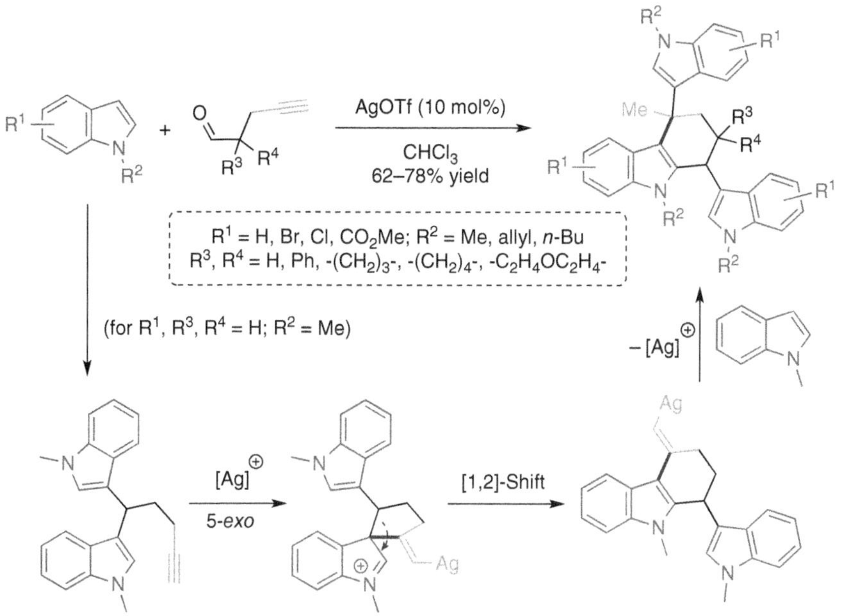

Scheme 7.81 Silver-catalyzed synthesis of tetrahydrocarbazoles.

aromatization/demetalation and an alkene hydroarylation complete the cascade process.

In the presence of gold catalysts 1,1-bis(indolyl)-5-alkynes undergo an intriguing rearrangement, featuring 6-*exo*-dig hydroarylation, followed by a pseudo 1,5-indole shift (Scheme 7.82) [140]. A nucleophilic attack by the *in situ*-generated vinyl–gold species on vinylogous iminium electrophile terminates the cascade and thereby furnishes the corresponding bis(indole) structures. Moreover, the use of a chiral phosphine gold catalyst (L = (*R*)-DM-SEGPHOS) resulted in an enantioselective process. The desired cyclic products were obtained with moderate to excellent *ee* (48–96% ee).

The cyclization of several indolyl alkynols, instead of the typical 1,2-alkenyl shift, features a 1,2-alkyl migration. Formed spirocyclic intermediates undergo heterolytic fragmentation/elimination sequence, yielding 3-allenyl indoles (Scheme 7.83) [141], or evolve into [*a*]carbazole, [*b*]carbazole derivatives, or carbazole ethers [142], depending on the substrate-substitution pattern and reaction conditions.

3-Propargylic indoles, in the presence of a gold(I) catalyst, undergo a complex rearrangement, which includes a formal 1,2-indole migration (Scheme 7.84) [143]. This transformation does not represent the typical hydroarylation of a triple bond, although it proceeds via a nucleophilic attack by the 3-position of indole onto the alkyne (Scheme 7.84, Eq. 1). This cycloisomerization involves an arylation of the alkyne and 1,2-migration of the indole nuclear, producing a gold(I) carbene intermediate. Different classes of indenes were selectively generated through subsequent Friedel–Crafts-type reaction or aura-Nazarov-type

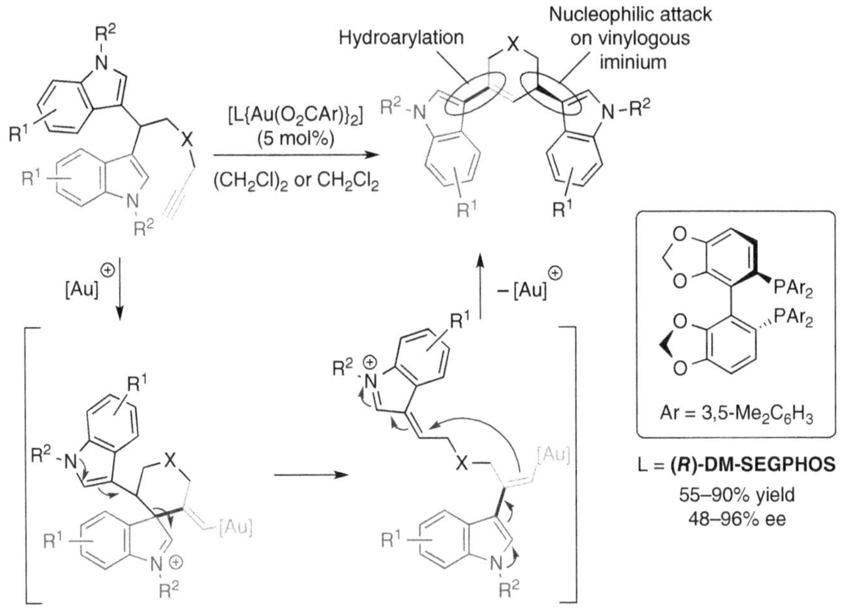

Scheme 7.82 Gold-catalyzed rearrangement of 1,1-bis(indolyl)-5-alkynes.

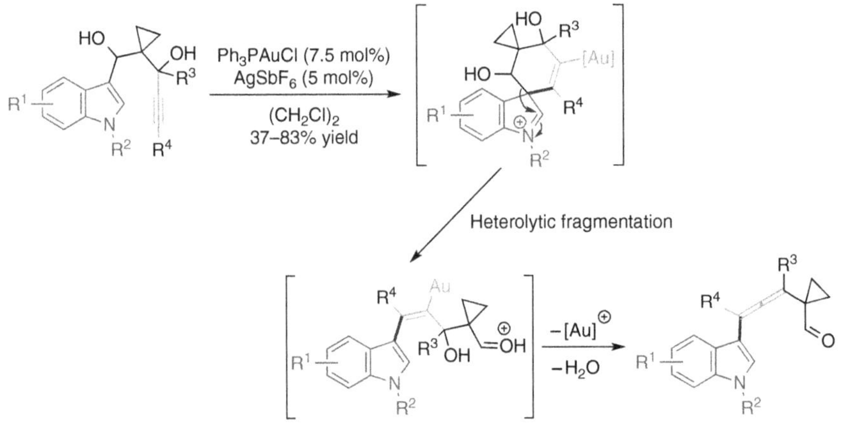

Scheme 7.83 Gold-catalyzed alkyne hydroarylation with subsequent heterolytic fragmentation in the synthesis of 3-allenyl indoles.

cycloaddition, depending on the substitution pattern of the alkyne fragment (Scheme 7.84, Eq. 1 vs Eq. 2). The proposed mechanism is supported by computational studies performed on a model system [144].

7.2.4 Alkyne Hydroarylation with Pyrroles

Alkyne hydroarylation with pyrroles is less common than with indoles, presumably because of selectivity issues considering the known high reactivity and lability of pyrroles. However, several transformations, discussed in the

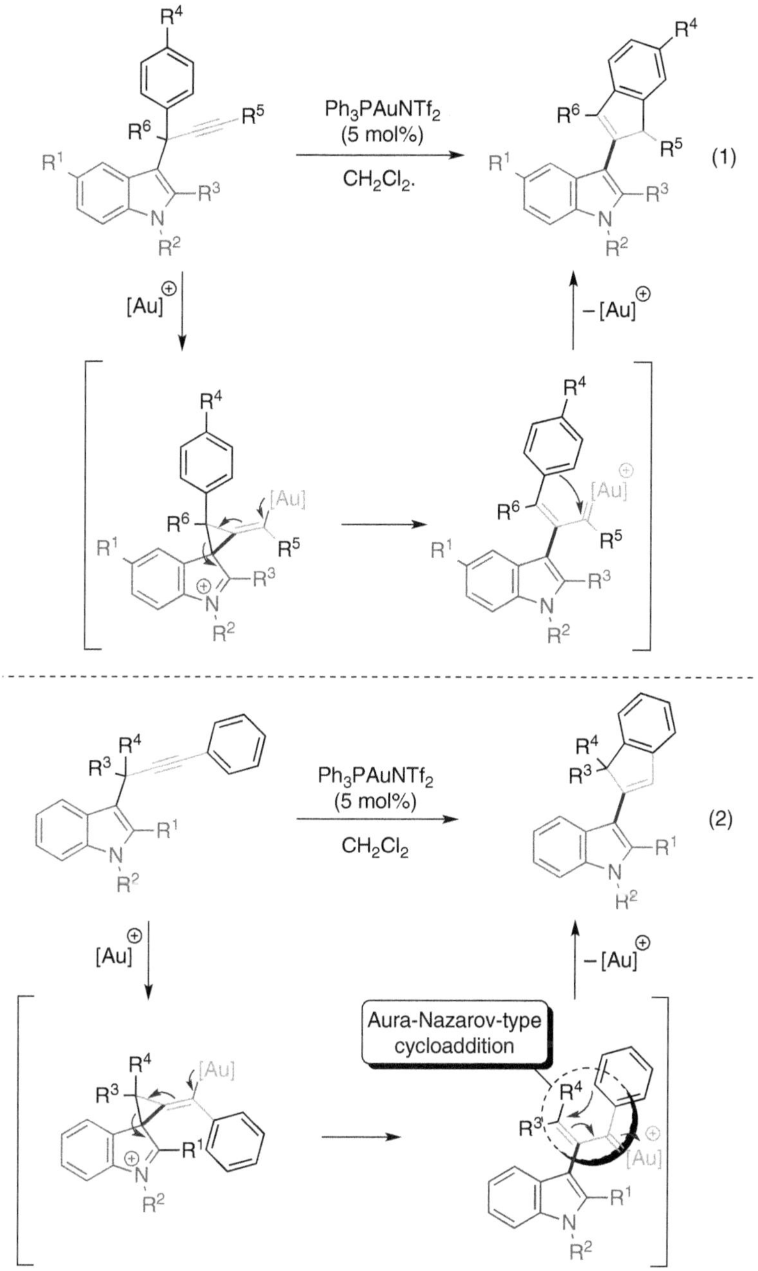

Scheme 7.84 Gold-catalyzed arylation of alkynes as a key step in the cycloisomerization of 3-propargylic indoles into indenes.

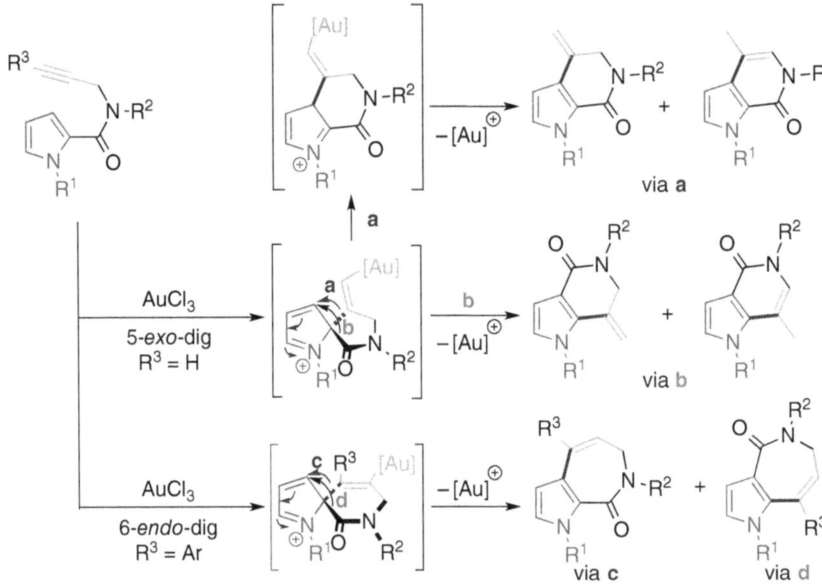

Scheme 7.85 Gold-catalyzed hydroarylation of 2-propargylamide-substituted pyrroles.

context of hydroarylations of indoles and furans, were extended to pyrroles [68, 84, 124, 131, 145, 146]. Indeed, difficulties in controlling the selectivity were observed during the course of the intramolecular hydroarylation of *N*-propargyl-pyrrole-2-carboxamides, which, generally, leads to a mixture of several regioisomers (Scheme 7.85) [147]. In the case of terminal alkynes, the spirocyclic intermediate that is generated after 5-*exo*-dig cyclization, undergoes either 1,2-alkenyl or 1,2-acyl shift, furnishing two regioisomeric [5,6]-fused bicycles. Interestingly, the polarity of a solvent has a strong influence on the ratio of products formed.

In contrast to the aforementioned hydroarylation of terminal alkynes, which yields [5,6]-fused bicycles, cycloisomerization of internal alkynes in the presence of gold catalysts furnishes solely seven-membered ring-containing products in moderate yields [147, 148]. The same tendency is observed when imidazolylpropargyl amides are used as substrate [149]. Mechanistically, this transformation is similar to the hydroarylation of terminal alkynes, although, the 6-*endo*-dig cyclization pathway occurs preferentially and no product of *exo*-dig cyclization is observed (Scheme 7.85). The proposed mechanism was supported by DFT calculations, which shows that an initial attack of pyrrole's α-carbon takes place at the most π-electron-deficient carbon center of the activated alkyne [150]. A similar strategy is applied for the preparation of a wide range of the substituted pyrroloazepine derivatives. Various β-yne-pyrroles, featuring all-carbon tethers, undergo hydroarylation with perfect *endo*-selectivity when treated with electrophilic [JohnPhosAu(NCMe)]SbF_6 catalyst (Scheme 7.86) [151]. In agreement with the previously discussed mechanistic studies, a terminal-alkyne-substrate reacts with 6-*exo*-selectivity (α-carbon of pyrrole attacks at the internal carbon

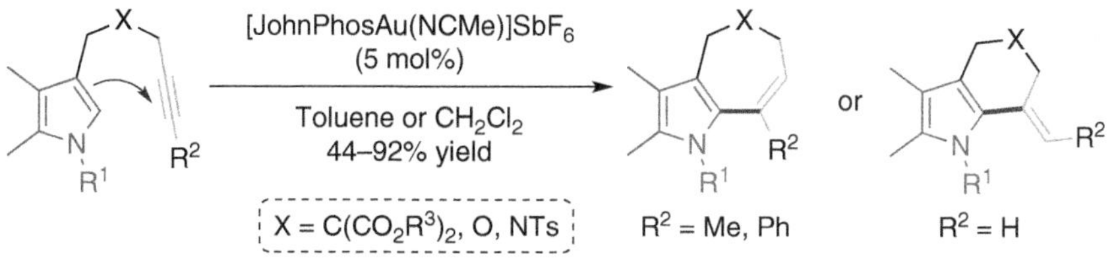

Scheme 7.86 Gold-catalyzed cycloisomerization of β-yne-pyrroles.

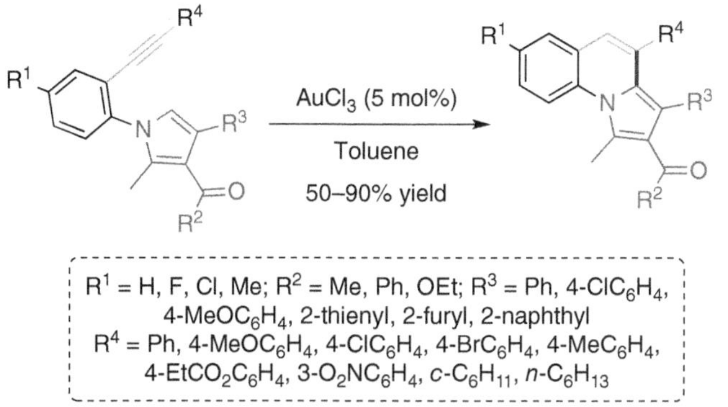

Scheme 7.87 The application of gold-catalyzed intramolecular hydroarylation to the synthesis of pyrrolo[1,2-*a*]quinolines.

carbon of a triple bond). In case of *N*-(2-alkynylphenyl)pyrroles, when the alkynyl moiety is tethered through the pyrrole nitrogen, cyclization is limited to *endo*/*exo* modes. The exclusive formation of pyrrolo[1,2-*a*]quinolines was achieved through 6-*endo*-dig cyclization by employing $AuCl_3$ (Scheme 7.87) [152]. The described method demonstrates high functional group tolerance.

A one-pot protocol was developed for a rapid and efficient asymmetric synthesis of substituted pyrroloazepines. In this tandem transformation pyrrole acts as a double nucleophile, first in organocatalyzed Friedel–Crafts–Michael-type reaction and then in gold(I)-catalyzed hydroarylation (6-*endo*-dig), to form a non-aromatic spirocyclic intermediate. A subsequent 1,2-alkenyl shift and rearomatization provides a seven-membered ring containing polycycle (Scheme 7.88) [153]. Similarly, the combination of an asymmetric copper-catalyzed alkylation of pyrroles with β,γ-unsaturated α-ketoesters, followed by a one-pot gold(I)-catalyzed hydroarylation of the alkyne moiety, allows a highly enantioselective synthesis of annulated pyrrole derivatives in moderate to good yields (Scheme 7.89) [154].

Post-Ugi Alkyne Hydroarylation A wide range of pyrrolocarboxamide derivatives, prepared via Ugi four-component reaction, have been used as substrates for intramolecular gold-catalyzed alkyne hydroarylation. When Ugi-4CR adducts, obtained from 2-formyl-*N*-methylpyrrole, propargyl amine, carboxylic acid, and isocyanide, were subjected to gold(I) chloride, 5-*exo*-dig cyclization/1,2-alkenyl

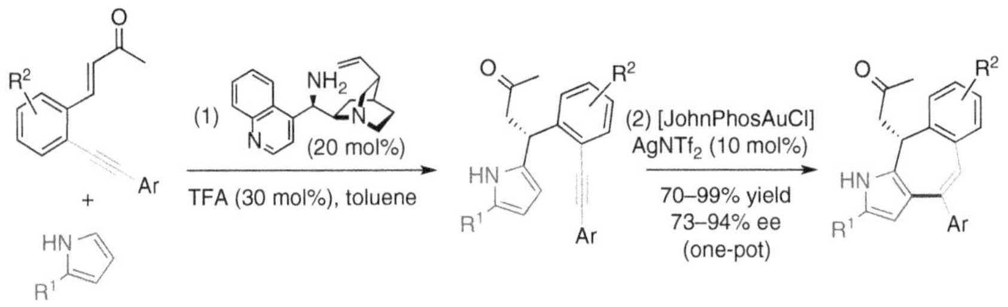

Scheme 7.88 Organocatalytic asymmetric Michael addition, gold-catalyzed intramolecular hydroarylation sequence.

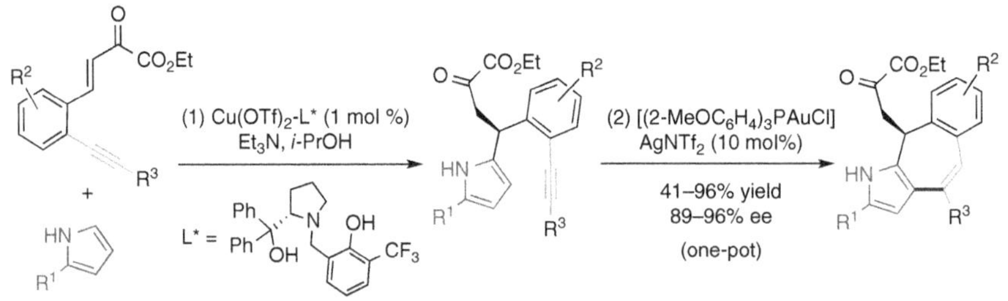

Scheme 7.89 Copper-catalyzed asymmetric Michael addition, gold-catalyzed intramolecular hydroarylation sequence.

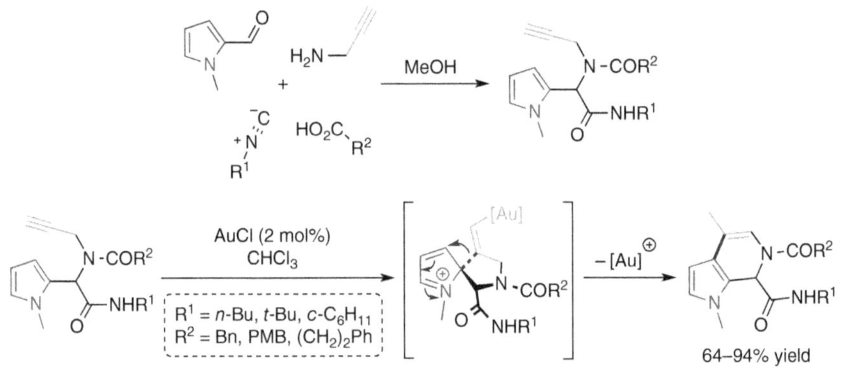

Scheme 7.90 Gold-catalyzed alkyne hydroarylation reaction of products of Ugi four-component condensation between 2-formyl *N*-methylpyrrole, propargyl amine, isonitrile and carboxylic acid.

migration took place, leading to the formation of various dihydro-azaindoles (Scheme 7.90) [132]. In a similar fashion, treatment of the product of Ugi-4CR between 2-formyl-pyrrole, primary amine, 2-alkynoic acid, and isocyanide with a gold(I) complex yields pyrrolopyridinones [155]. Remarkably, in the presence of an electrophilic Au(I) catalyst $Ph_3PAuOTf$, hydroarylation occurs exclusively with exo-selectivity, whereas the use of $PtCl_2$ favors the formation of an endo-cyclization product, in a mixture with the *exo*-isomer (Scheme 7.91) [156]. Similarly, this approach was applied to adducts of Ugi-4CR between

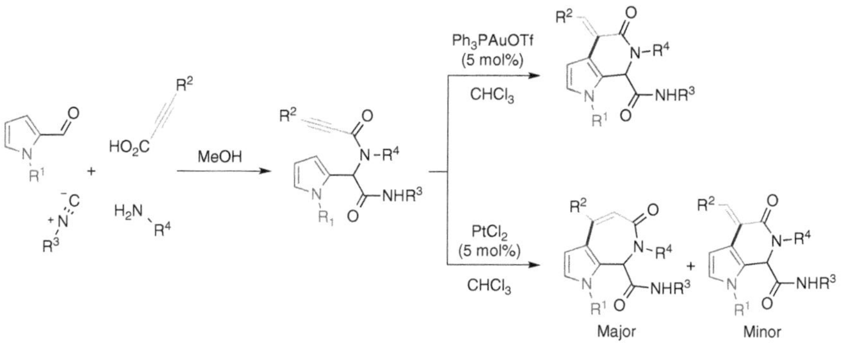

Scheme 7.91 Gold- and platinum-catalyzed hydroarylation reactions of products of Ugi four-component condensation between 2-formyl-pyrrole, 2-alkynoic acid, isonitrile, and primary amine.

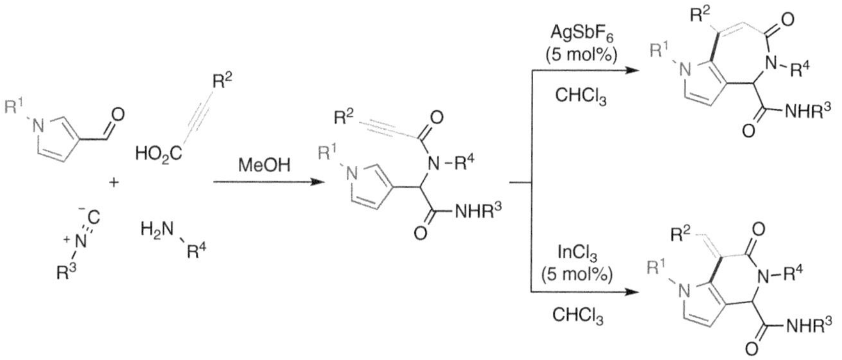

Scheme 7.92 Silver- and indium-catalyzed hydroarylation reactions of products of Ugi four-component condensation between 3-formyl-pyrrole, 2-alkynoic acid, isonitrile, and primary amine.

3-formyl-pyrrole, primary amine, 2-alkynoic acid and isocyanide. For these substrates, the exo/endo-selectivity of hydroarylation could be also controlled by the catalyst used: cyclization promoted by $AgSbF_6$ proceeded with perfect endo-selectivity, while employment of $InCl_3$ led to pyrrolopyridinones via the exo-cyclization pathway (Scheme 7.92) [157].

The synthetically challenging construction of nine-membered rings was achieved by gold(I)-catalyzed hydroarylation of pyrrolocarboxamide derivatives, prepared via Ugi four-component reaction between 2-(pyrrol-1-yl)-benzaldehyde, benzylic amine, 2-alkynoic acid, and an isocyanide (Scheme 7.93) [158].

7.2.5 Alkyne Hydroarylation with Furans and Benzofurans

As described above, indoles and pyrroles are excellent nucleophiles in alkyne hydroarylation reactions. Other electron-rich heteroarenes, such as furans and benzofurans, undergo facile intramolecular alkenylations with alkynes in the presence of π-Lewis acids. However, the reactivity profile of furans is

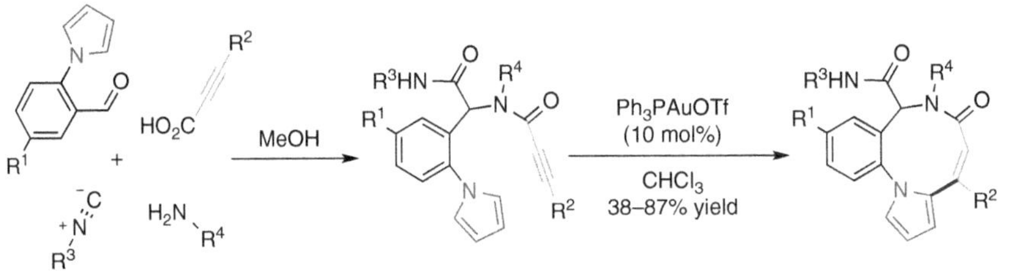

Scheme 7.93 Gold-catalyzed hydroarylation reaction of products of Ugi four-component condensation between 2-(pyrrol-1-yl)-benzaldehyde, 2-alkynoic acid, isonitrile, and benzylic amine.

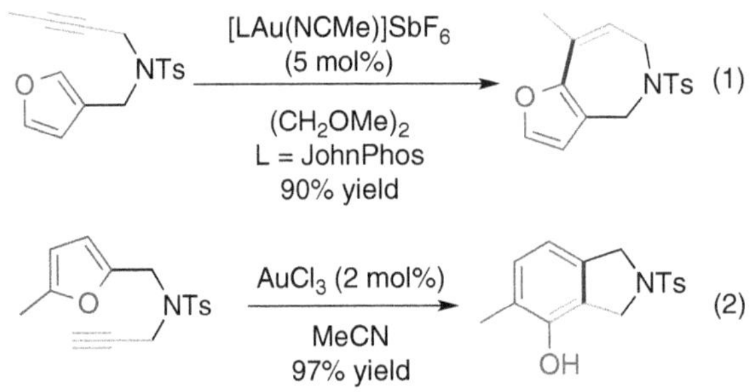

Scheme 7.94 Normal and abnormal reactivity of furans in intramolecular gold-catalyzed hydroarylation of tethered alkynes.

more complex due to their ability to suffer bond-cleavages and skeletal rearrangements. Formation of benzofuran derivatives via simple intramolecular alkenylation of furans with alkynes have been described employing ruthenium [159], tungsten [9], platinum [5d, e], mercury, and mercury/scandium salts [160] as catalysts. Gold salts have also been shown to promote hydroarylation with furans, although the outcome of the reaction depends on the substitution pattern of alkyne, and products of skeletal rearrangements could be obtained. In case of internal alkynes, a simple hydroarylation reaction generally occurs (Scheme 7.94, Eq. 1), whereas α-yne-furans, bearing terminal alkyne, often undergo rearrangement, and phenols are obtained, instead of the anticipated 3-alkenylated furans (Scheme 7.94, Eq. 2) [161].

7.2.5.1 Alkenylation at the 2-Position of Furan

The intramolecular α-alkenylation of furans with alkynes was developed for the synthesis of various fused heterocycles (Scheme 7.95) [162]. Remarkably, hydroarylation of substrates featuring internal alkyne with electron-donating substituent occurs with excellent endo-selectivity, while the exo-cyclization pathway takes place preferentially in the case of terminal alkynes. This method was extended to substrates, in which a furan moiety was replaced by other electron-rich arenes, such as benzofuran and thiophene. DFT analysis, performed to clarify the mechanism and regioselectivity issues, revealed that for substrates featuring internal alkyne with electron-donating groups, 7-*endo* transition state is favored. In contrast, the 6-*exo* cyclization pathway is preferred

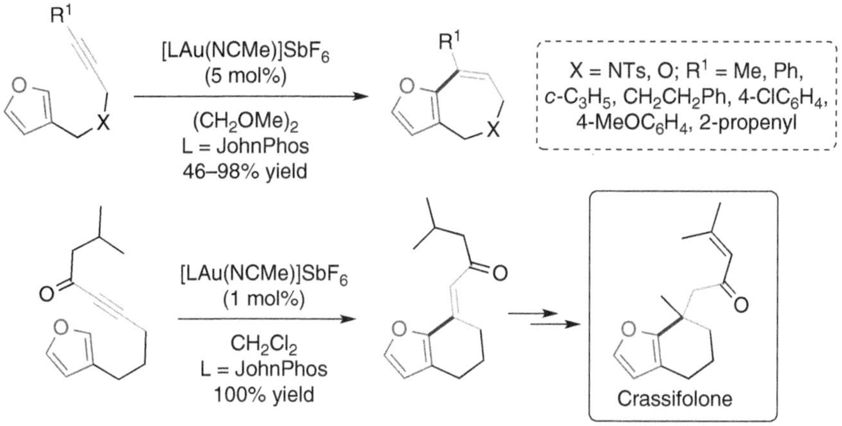

Scheme 7.95 Gold-catalyzed cycloisomerization of 3-furyl tethered alkynes in the synthesis of seven-membered ring containing polyheterocycles and crassifolone skeleton.

in case of the terminal alkynes or substrates bearing an electron-withdrawing group. A similar strategy for the construction of tetrahydrobenzofuran skeleton was applied in the syntheses of crassifolone and dihydrocrassifolone skeleton [163].

7.2.5.2 Alkenylation at the 3-Position of Furan

Although the gold-catalyzed hydroarylation of terminal alkynes usually results in the formation of phenols, several examples of simple alkenylation, yielding benzofurans, have been also disclosed. In the presence of electrophilic gold(I) salts, cyclization of propargylic alcohol-tethered furans leads to benzofuran derivatives in generally, a moderate yield. The short tether, of only 2 carbon atoms, favors a nucleophilic attack onto alkyne in an *endo*-fashion, preventing the formation of phenols (Scheme 7.96) [164, 165].

7.2.5.3 More Complex Transformations Featuring Hydroarylation of Alkynes

Hydroalkoxylation/Alkyne Hydroarylation Cascade A two-step synthesis of polyaromatic ribbons was developed via palladium-catalyzed Sonogashira cross-coupling followed by gold(I)-catalyzed alkyne hydroalkoxylation and alkyne hydroarylation. After an initial Pd-catalyzed cross-coupling, the *o*-alkynylphenol undergoes a one-pot hydroalkoxylation to furnish a benzofuran. The *endo*-selective alkenylation of the benzofuran moiety with the neighboring

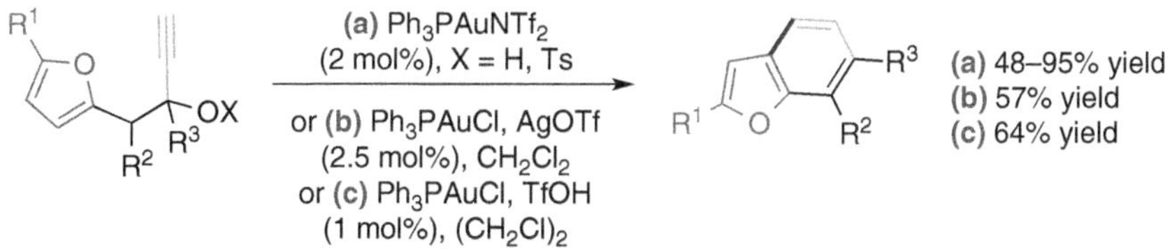

Scheme 7.96 Gold-catalyzed benzannulation of furans.

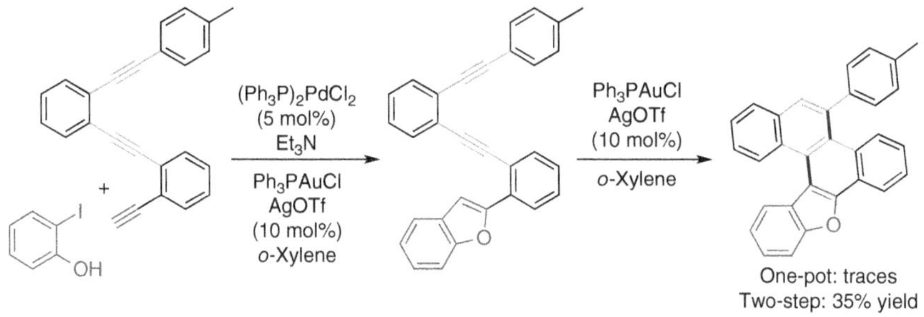

Scheme 7.97 Application of gold-catalyzed hydroarylation to the preparation of polyaromatic ribbons.

alkyne promoted by $Ph_3PAuCl/AgOTf$ leads to a benzofuran-fused naphthalene derivative that may give rise to polyarenes by further all-*endo* intramolecular hydroarylations (Scheme 7.97) [166]. Although, the overall cascade might be carried out in a one-pot fashion, in some cases the efficiency of polycyclization significantly improves by performing these reactions sequentially.

Fragmentation and Migration Furans featuring silyl-protected homopropargylic alcohols at the 2-position undergo cyclization in the presence of electrophilic gold(I) complexes to furnish benzofuran derivatives, which correspond to products of formal 2-alkenylation (Scheme 7.98) [167]. The proposed mechanism of this transformation involves the formation of spirocyclic intermediate formed via a nucleophilic attack onto an activated alkyne by the 2-position of furan in a 5-*exo*-dig fashion. Subsequent 1,2-alkyl shift, followed by protodemetalation and aromatization through silanol elimination, yields the 7-aryl benzofurans. This method was extended to homopropargylic benzofurans, which afford 4-substituted dibenzofuran derivatives [102d]. A similar strategy was later applied for the synthesis of the diverse di- and tri-substituted benzo[*b*]furans (Scheme 7.99) [168]. The gold(I)-catalyzed cyclization of furan–ynes in the presence of an external nucleophile proceeded via the formation of the

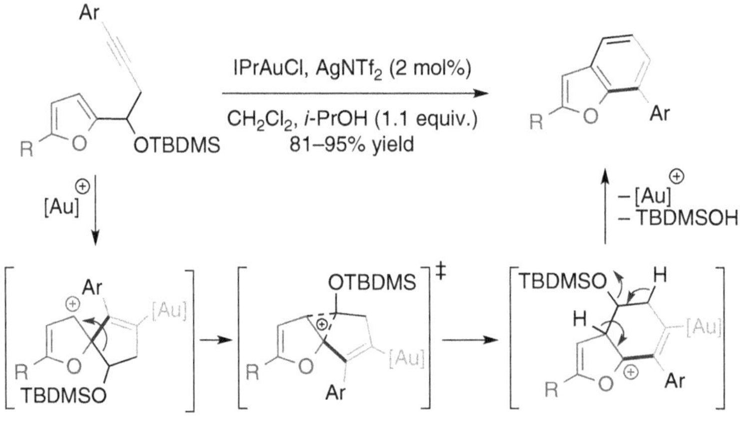

Scheme 7.98 Gold-catalyzed hydroarylation in the synthesis of benzofurans.

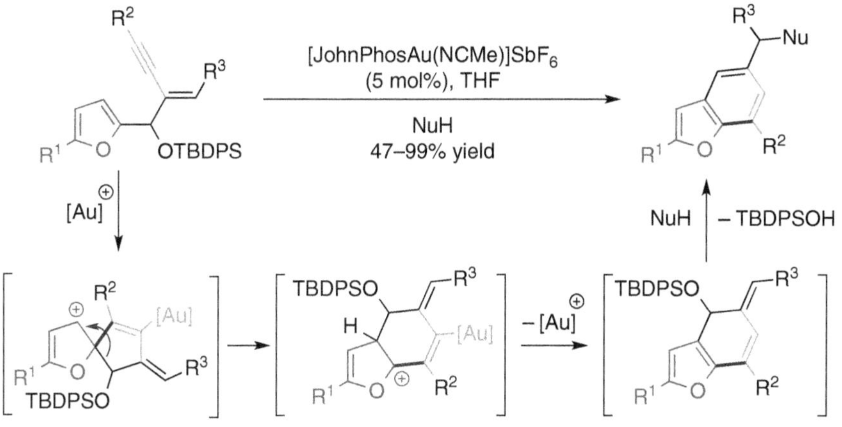

Scheme 7.99 Gold-catalyzed cycloisomerization of furan–ynes.

spirocyclic intermediate, followed by 1,2-alkyl shift and a subsequent gold-assisted allylic substitution. Various nucleophiles, such as alcohols and anilines were employed in this transformation. However, the reaction failed when 1,3-dicarbonyl compounds were used as nucleophiles.

In some cases, the hydroarylation of internal alkynes is followed by a fragmentation of the furan ring, which leads to the formation of a new carbonyl-containing compound. Thus, naphthyl-substituted (*Z*)-enals (and enones) are obtained from furan–ynes bearing a silyloxy group in the presence of a cationic gold(I) catalyst [JohnPhosAu(NCMe)]SbF_6 (Scheme 7.100) [169]. The mechanism most likely involves the initial furan–yne spirocyclization, followed by furan ring-opening (Scheme 7.100, path a), to provide an extended aromatic species (conjugated naphthalene derivative). Alternatively, the migration of the —CH-OTBS— group to the 3-position would result in the generation of a less-favored not fully aromatic seven-membered ring-containing structure (Scheme 7.100, path b). Later, a similar methodology was developed for 1-(2-furanyl)phenyl propargyl alcohols, which was applied in the total synthesis of wailupemycin G (Scheme 7.100) [170].

In the presence of gold catalysts, enynyl furans featuring 2-atom-tether between furan and alkyne undergo a gold-catalyzed cycloisomerization, leading to fulvene derivatives rather than arenes (Scheme 7.101) [171]. Mechanistically, this transformation was proposed to involve intramolecular hydroarylation of furan in a 6-*endo*-dig manner with the formation of cyclopropyl gold(I) carbene. Rearrangement via furan ring-opening, deprotonation and demetalation finally furnished the fulvene derivatives with an enone moiety.

The dependency of the cyclization pathway on the length of alkyne–furan tether is illustrated in the synthesis of *N*-containing heterocycles from arylalkynylamides. Highly substituted cyclopentadienes bearing a ketone substituent were obtained in case of 4-atom-tether between furan and alkyne by the fragmentation of the cyclopropyl gold(I) carbene intermediate, followed by electrocyclization (Scheme 7.102, Eq. 1). Substrates with three atoms in the tether were converted into benzoannulated heterocycles via a Friedel–Crafts-type

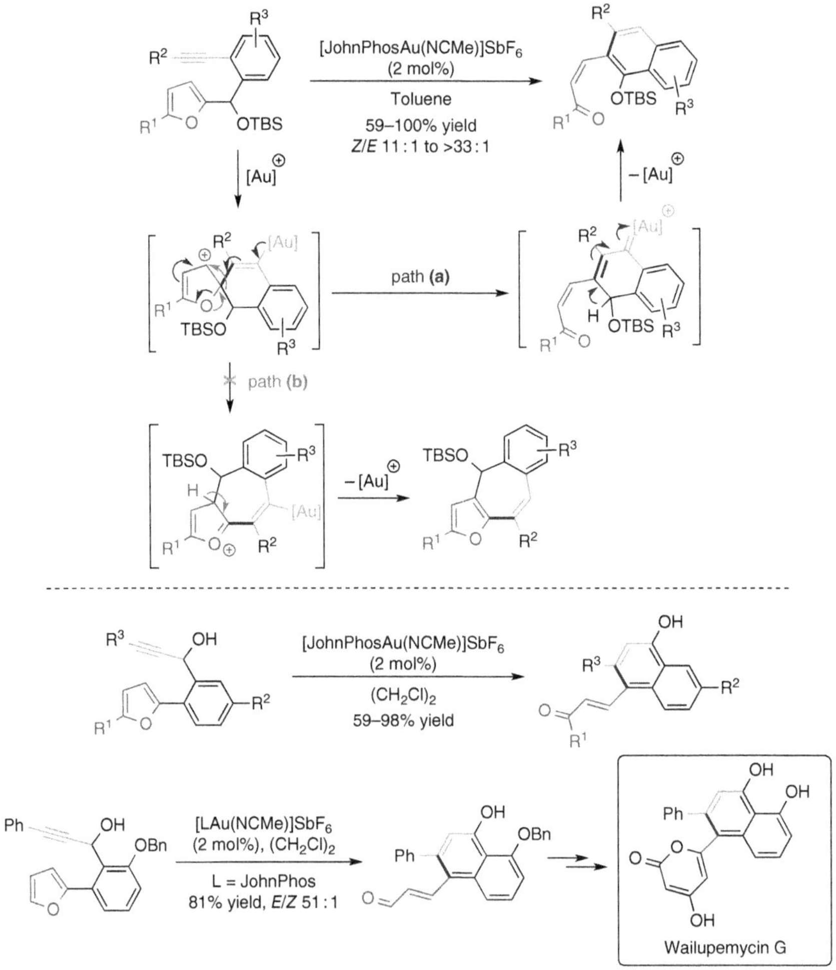

Scheme 7.100 Gold-catalyzed hydroarylation of furans in the synthesis of naphthols and its application in the total synthesis of wailupemycin G.

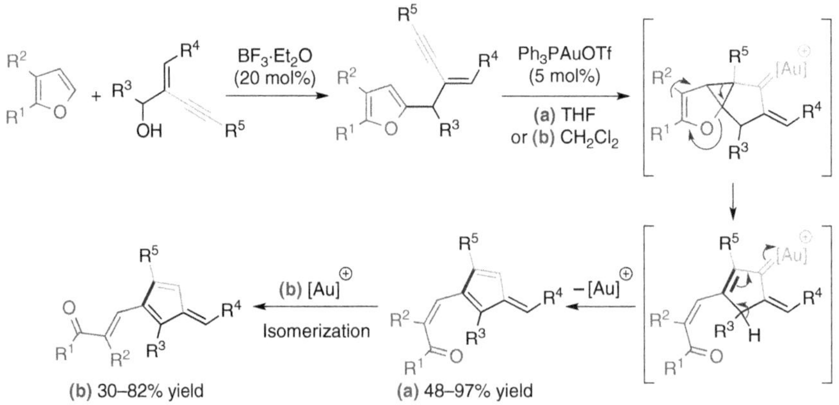

Scheme 7.101 Gold-catalyzed cycloisomerization of enynyl furans in the synthesis of fulvene derivatives.

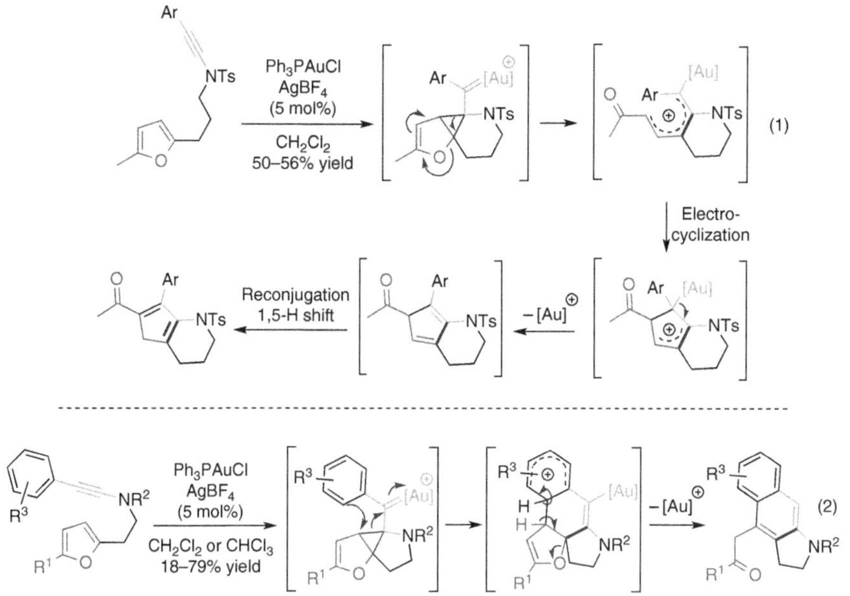

Scheme 7.102 Dependence of the outcome of gold-catalyzed cycloisomerization of furan–ynes on the tether length.

reaction of the cyclopropyl gold(I) carbene intermediate (Scheme 7.102, Eq. 2) [172].

A hydroarylation/Friedel–Crafts-type double cyclization cascade that gave rise to tetracyclic systems was uncovered when furans bearing aryl alkynyl ethers moiety, tethered through a protected amine linker, were treated with a gold(I) catalyst [146]. The mechanistic proposal for this transformation involved an initial 6-*endo*-dig cyclization leading to the formation of a stabilized furan cation intermediate. Subsequent Friedel–Crafts-like cyclization with the electron-rich arene furnishes the final tetracycles after protodeauration (Scheme 7.103).

TMS-protected (*o*-alkynyl)phenyl (2-furyl)carbinols suffer a complex rearrangement in the presence of gold(I) source, which includes either 1,4- or 1,5-furanyl migrations, yielding indene derivatives or trisubstituted olefins, respectively (Scheme 7.104) [145]. In the absence of molecular sieves, the *exo*-cyclization of furan–ynes occurs with concomitant 1,4-furanyl migration and cyclization leading to densely functionalized indene derivatives (path b). By addition of unactivated molecular sieves as a water source, the initial hydroarylation changed to the *endo*-pathway (path a).

Transformations Involving Friedel—Crafts-Type Reactions A synthesis of arene-substituted (*Z*)-enones/enals was achieved via a cascade transformation, consisting of intermolecular Friedel–Crafts reaction of 2-substituted furans with enynols, followed by intramolecular gold-catalyzed hydroarylation and subsequent fragmentation (Scheme 7.105) [173]. A similar tandem transformation featuring alkyne hydroarylation with furans was developed for the synthesis of phenanthrene derivatives. In a single step, acylated

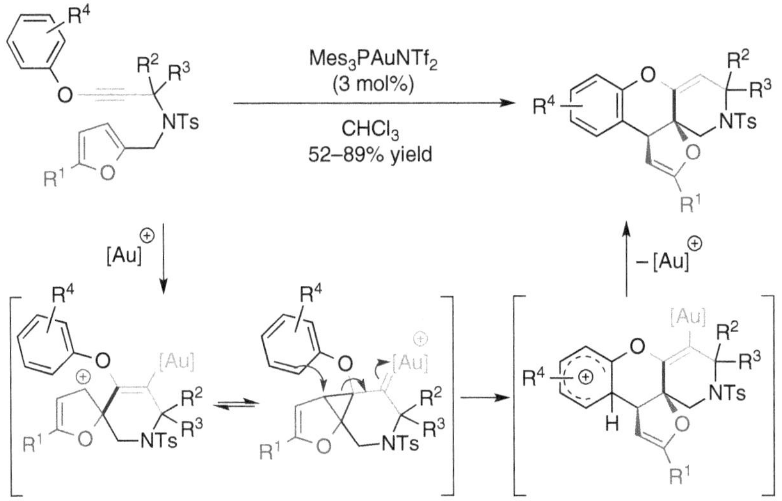

Scheme 7.103 Gold-catalyzed cycloisomerization of furans tethered to aryl alkynyl ether leading to the formation of complex tetracylcles.

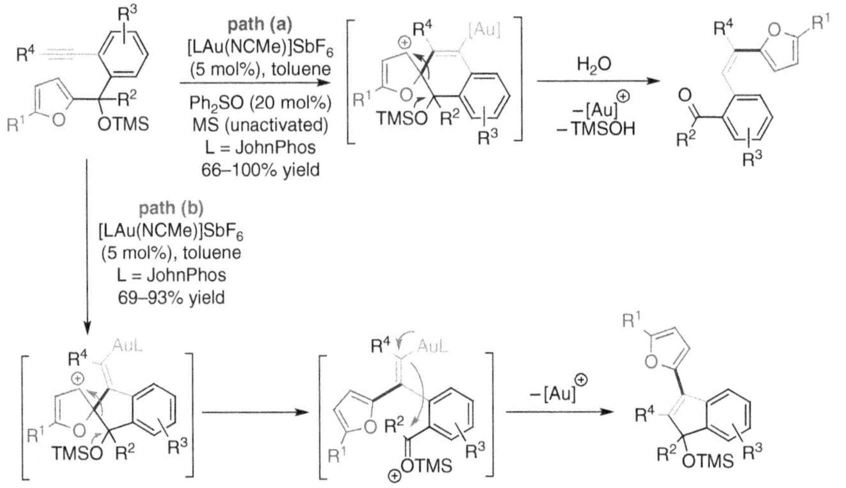

Scheme 7.104 Gold-catalyzed intramolecular hydroarylation of furan–ynes in the synthesis of trisubstituted alkenes and indenes.

polyarenes and heteroarenes were prepared from furans and allyl alcohols, bearing two alkyne chains. This gold-catalyzed cascade probably proceeds via Friedel–Crafts reaction between furans and ene–diyne–ols leading to arylated ene–diynes, followed by furan–yne hydroarylation/fragmentation and heteroenyne rearrangement (Scheme 7.106) [174].

7.2.5.4 The Furan–Yne Cycloisomerization to Phenols

In addition to the aforementioned transformations involving the alkyne hydroarylation with furans, an intriguing reactivity of α-yne-furans holding

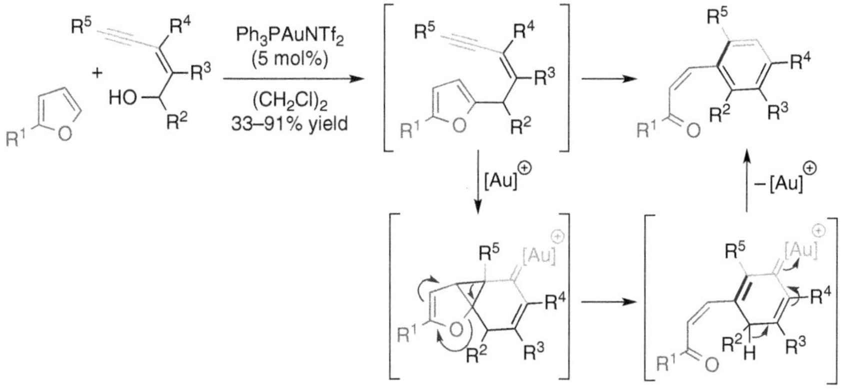

Scheme 7.105 Gold-catalyzed Friedel–Crafts reaction/cycloisomerization cascade in the synthesis of (*Z*)-enones and -enals.

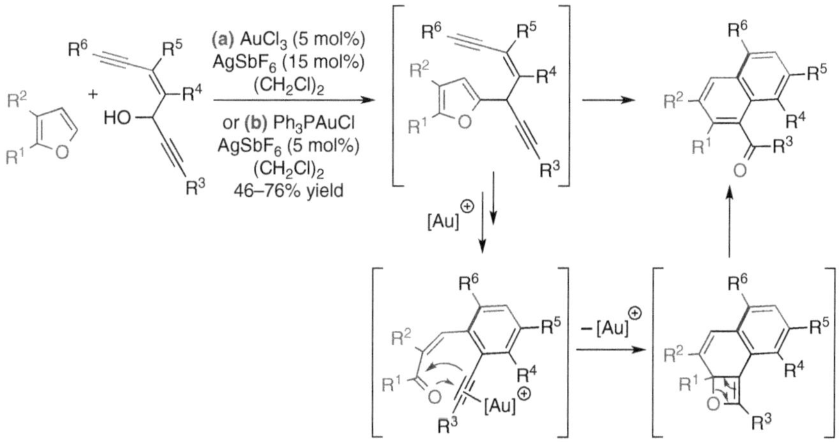

Scheme 7.106 Gold-catalyzed Friedel–Crafts reaction/cycloisomerization cascade in the synthesis of phenanthrene derivatives.

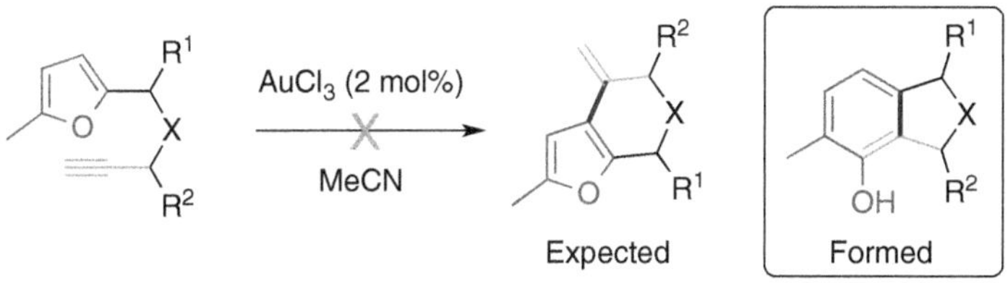

Scheme 7.107 Gold-catalyzed intramolecular hydroarylation of furan–ynes leading to phenols.

terminal alkynes was uncovered. The cycloisomerization of these α-yne-furans in the presence of gold(III) salts was shown to result in the formation of phenols (Scheme 7.107) [161].

Initially, mechanistic studies of the furan–yne cycloisomerization to phenols were carried out using Pt(II)-catalysts [5d, e]. Later, a similar mechanistic

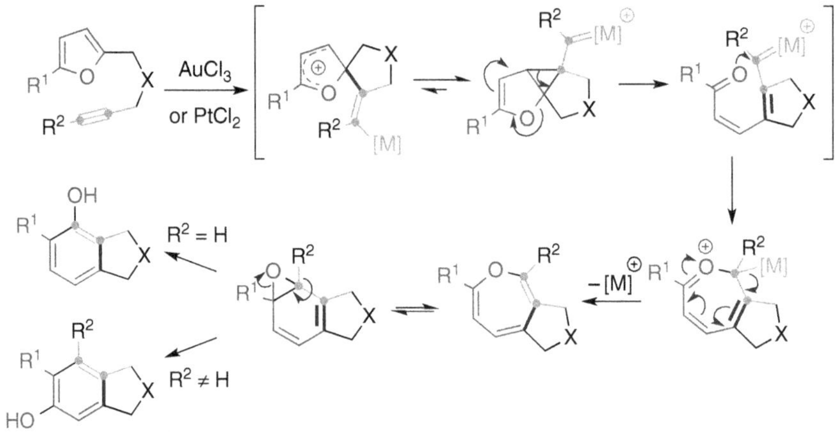

Scheme 7.108 Mechanism of platinum- and gold-catalyzed formation of phenols from furan–ynes.

pathway was suggested for gold catalysts [175]. Presumably, an initial hydroarylation of alkyne by furan results in the formation of spirocycle that evolves to cyclopropyl metal carbene. Subsequent furan ring-opening followed by cyclization and demetalation generates an oxepin. An arene oxide produced by electrocyclization undergoes rearomatization via epoxide-opening (Scheme 7.108) [5d, e]. The proposed mechanism of the gold-catalyzed cycloisomerization was also supported by addition computational [176] and experimental [177] evidences, which indicated that the phenols are formed through an oxepin and an arene oxide intermediate. It is worth to mention that the cycloisomerization of furans unsubstituted at the 5-position usually results in mixtures of regioisomeric phenols [161, 178] via a mechanism known in the literature as the "oxygen walk" (Scheme 7.109) [177b, 179].

The gold-catalyzed cycloisomerization of furan–ynes to phenols displays excellent chemoselectivity. Attempts to trap early intermediates of furan–yne cycloisomerization were unsuccessful and in all cases phenols were obtained in good yields [180]. The formation of phenols also was shown to be competitive with a simple alkyne hydroalkoxylation pathway (ca. 3:2 ratio of hydroalkoxylation vs phenol synthesis) [181]

A variety of gold salts, from simple $AuCl_3$ [5e, 161] to electrophilic complexes of Au(I) [182] and Au(III) [182c, 183], have proved to be competent in promoting

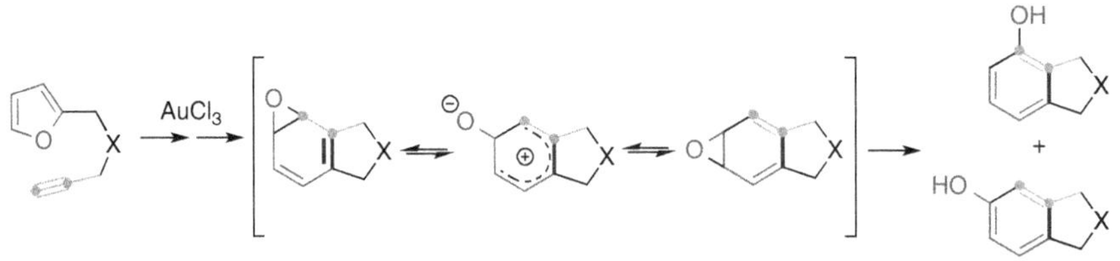

Scheme 7.109 "Oxygen walk" mechanism for the formation of regioisomeric phenols in hydroarylation of furan–ynes.

the synthesis of phenols from furans. Remarkably, in some cases the preferential reaction pathway, either the phenol formation or simple hydroarylation, strongly depends on the nature of the catalyst used and in particular on the oxidation state of the gold center: $AuCl_3$ generally leads to the formation of phenols whereas electrophilic Au(I) catalysts predominantly favor the 6-*exo*-dig hydroarylation [184]. The reaction is limited to electron-rich furans with alkyne chains tethered to the 2-position, while with electron-deficient furans the reaction is unsatisfactory. In addition, 3-alkynyl furans and other aromatic heterocycles do not give rise to similar reactivity [185].

Furans bearing an alkynyl substituent tethered at the 2-position by a 3- or 4-atom chain are generally good substrates for the phenol synthesis. A wide range of heteroatoms or functionalities at the tether is well tolerated [186]. The substrates featuring all-carbon tethers [161], ethers [161, 178, 187], and protected amines [161, 188], are converted to substituted phenols (**F4-I** and **F4-II**), dihydroisobenzofurans (**F4-III**), dihydrobenzofurans (**F4-IV**)/isochromanes (**F4-X** and **F4-XI**)/chromanes [189] (**F4-XII**), and indolines (**F4-VI**)/dihydroisoindoles (**F4-V**)/tetrahydroquinolines (**F4-IX**)/tetrahydroisoquinolines (**F4-VII** and **F4-VIII**)(Figure 7.4). However, the selectivity of the reaction toward the formation of phenol derivatives versus simple furan alkenylation proved to be affected

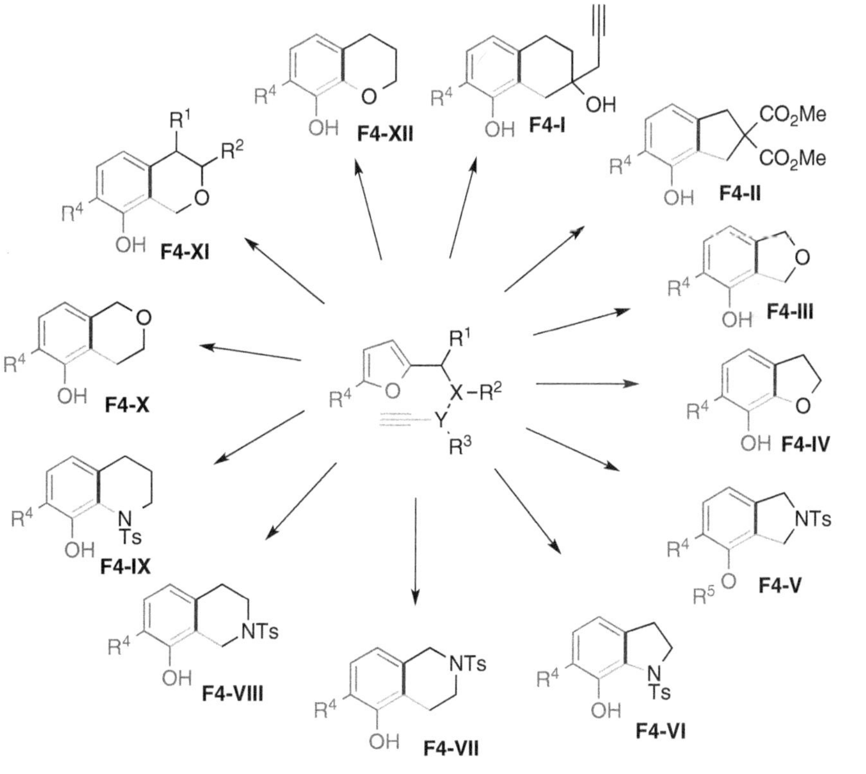

Figure 7.4 Scope of the phenol synthesis via furan–ynes intramolecular alkyne hydroarylation.

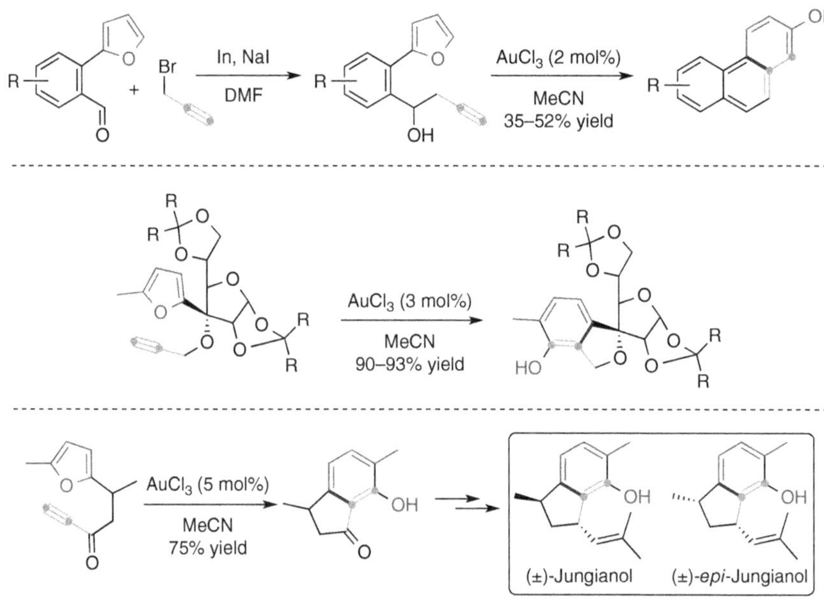

Scheme 7.110 Gold (III)-catalyzed cycloisomerization of furan–ynes in the preparation of polyarenes and biologically relevant molecules.

by the nature of the catalyst used, especially for substrates featuring longer tethers [190].

The efficiency of the gold-catalyzed cycloisomerization of furan–ynes to phenols is not substantially affected by the steric hindrance of the substituent at the 5-position of the furan moiety. Furans bearing sterically demanding substituents (mesityl, adamantyl), evolved to phenol derivatives in the presence of gold(III) source. However, the rate of the reaction depends on the electronic nature of the substituents [191]. The gold(III)-catalyzed phenol synthesis methodology was extended to more complex substrates. This transformation was applied for the preparation of the polycyclic scaffolds, such as hydroxyphenanthrene derivatives [192]. Similarly, the synthesis of spirocyclic dihydrobenzofuran-substituted monosaccharides was developed taking advantage of cycloisomerization of carbohydrate-derived furanyl propargyl ethers (Scheme 7.110) [193]. In addition, the cycloisomerization of furan–ynes was used as a key step in the synthesis of the natural product jungianol and its analogs [194]. Also, the furan–ynes cycloisomerization has been applied within a cascade transformation starting with a hydroarylation step (Scheme 7.111, Eq. 1) [195]. This transformation includes the formation of five new bonds, albeit in low yields. Alternatively, the furan moiety can be assembled *in situ* prior to the phenol synthesis cascade. Such a process that includes the hydroalkoxylation/isomerization sequence was demonstrated for en-diynols (Scheme 7.111, Eq. 2) [196].

The substrates featuring a stereogenic center and two diastereotopic alkyne groups undergo cycloisomerization to phenols with high level of diastereoselectivity (Scheme 7.112) [197]. However, the attempt to extend this method

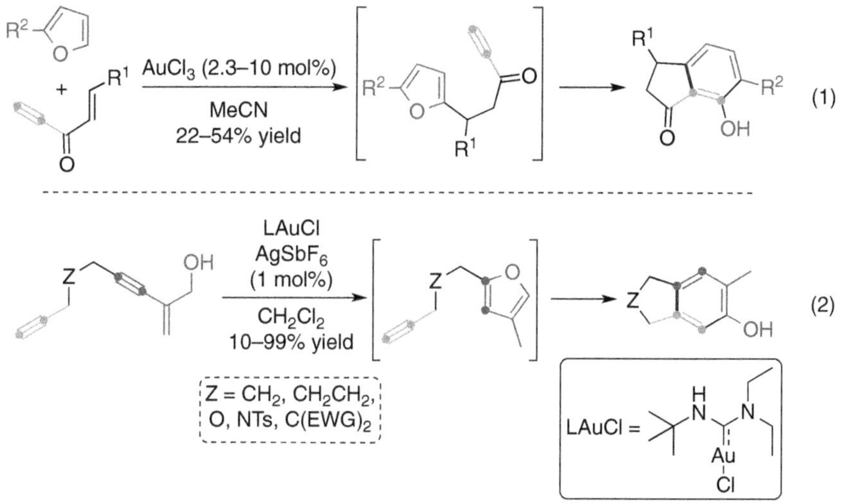

Scheme 7.111 Cascade transformations involving gold(III)-catalyzed cycloisomerization of furan–ynes.

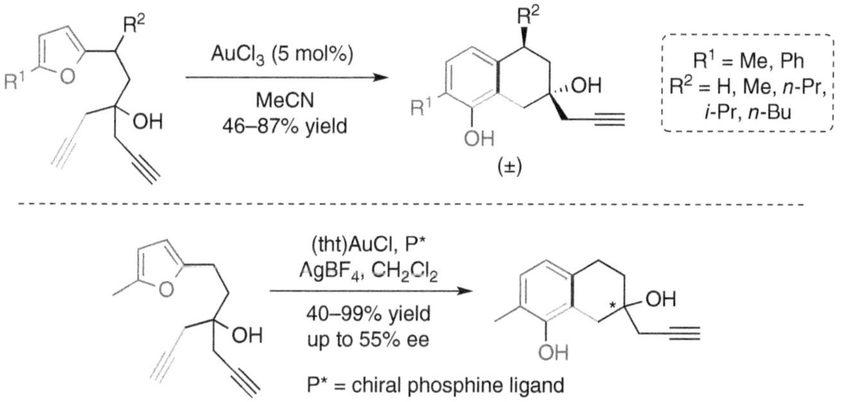

Scheme 7.112 Induction of diastereo- and enantioselectivity in gold-catalyzed cycloisomerization of furan–ynes.

for the preparation of enantioenriched tertiary alcohols, employing a chiral phosphine gold catalyst, was unsuccessful. In the absence of a stereogenic center at β-carbon of the tertiary alcohol, the desired phenol derivatives were obtained in high yields but with low *ee* (37–55% ee) [198].

Gold-catalyzed cycloisomerization of furan–ynes to phenols was used for the preparation of various enantioenriched 8-hydroxytetrahydroisoquinolines from propargyl derivatives of the non-natural amino acid furylalanine [199] (Scheme 7.113, Eq. 1) and other chiral amine tethers (Scheme 7.113, Eq. 2) [200]. In a similar fashion, diverse furan–ynes featuring chiral sulfoxide *N*-protecting groups on the tether were converted into enantioenriched hydroxylated dihydroisoindoles (Scheme 7.113, Eq. 3) [201].

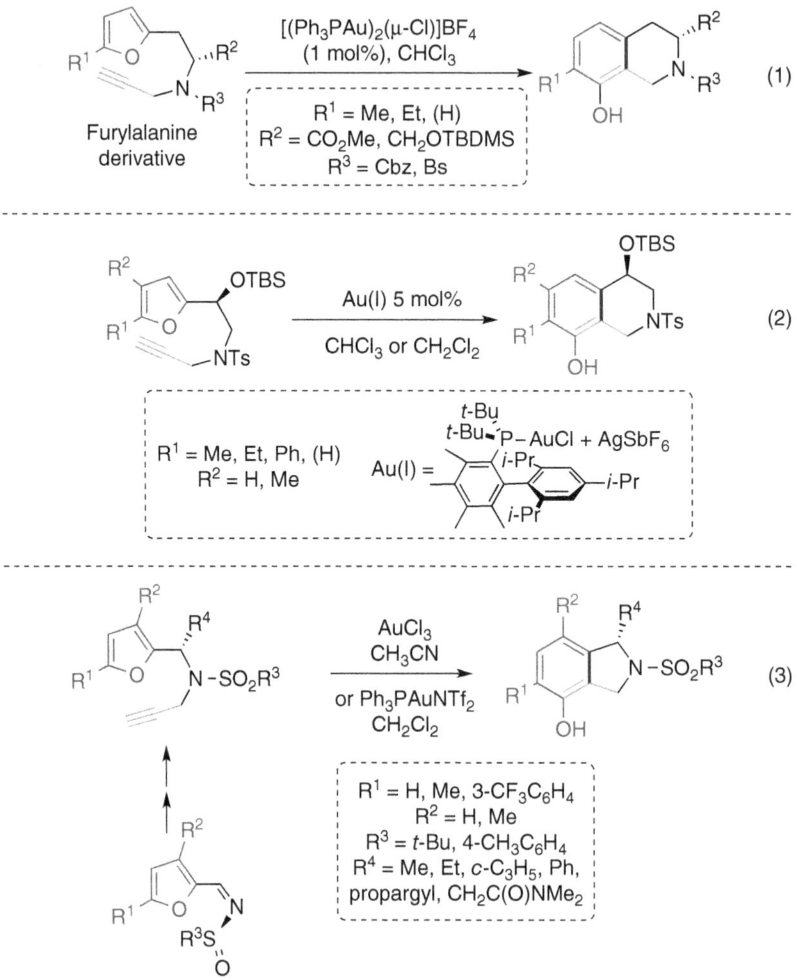

Scheme 7.113 Gold-catalyzed cycloisomerization of furan–ynes for the preparation of enantioenriched compounds.

7.2.6 Alkyne Hydroarylation with Thiophenes and Benzothiophenes

The use of thiophenes and benzothiophenes for the intramolecular hydroarylation of alkynes is much less common and usually reported as an extension of the methodology elaborated for *O*-, or *N*-containing heterocycles. The cyclization of alkynyl thiophenes, was reported to produce seven-membered ring-fused thiophenes [84, 124, 162]. Similarly to (2-alkynyl)phenyl-(2-furyl)carbinol, thiophene derivatives suffer endo-hydroarylation/1,5-heteroaryl migration, when treated with a gold(I) catalyst, furnishing trisubstituted alkenes [145]. In the presence of gold(I), 2-homopropargylic thiophene and benzothiophene derivatives undergo intramolecular hydroarylation followed by silanol elimination, yielding benzothiophenes and dibenzothiophenes, respectively (Scheme 7.114) [102d].

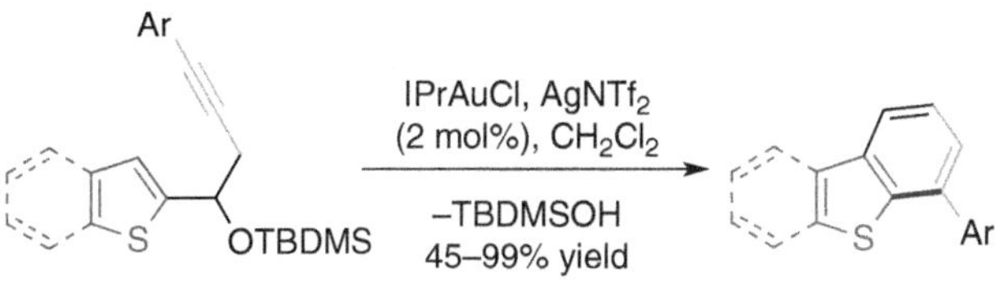

Scheme 7.114 Gold-catalyzed rearrangement of thiophene- and benzothiophene-derived homopropargyl silanols.

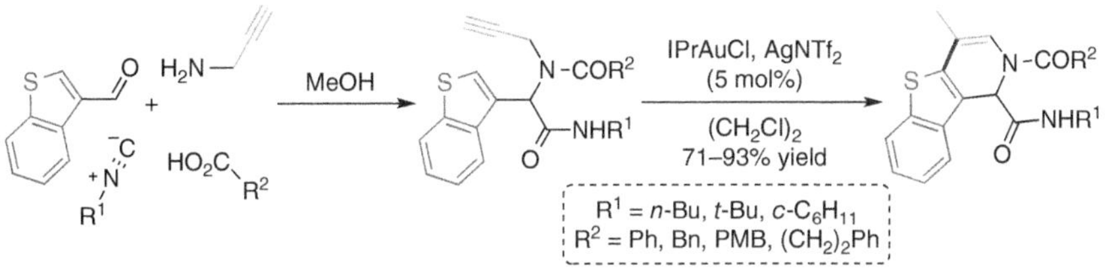

Scheme 7.115 Gold-catalyzed alkyne hydroarylation reaction of products of Ugi four-component condensation between 3-formyl-benzothiophene, 2-carboxylic acid, isonitrile, and propargylamine.

As in the case of furans and benzofurans [167], the reaction proceeds by a formal migration of an aromatic substituent.

A two-step protocol for one-pot copper(I)-mediated formation of ynamides/silver(I)-catalyzed hydroarylation, described for pyrroles and indoles, has been shown to be equally efficient for the preparation of naphtho[2,1-*b*]thiophenes [131]. Rapid synthesis of diversely substituted [6,5,6]-fused and [6,5,7]-fused heterocycles was developed through π-acid-catalyzed post-Ugi hydroarylation of alkynes [155b, 157] (Scheme 7.115).

7.3 Intermolecular Hydroarylation of Alkynes

7.3.1 Intermolecular Hydroarylation of Alkynes with Arenes

The intermolecular hydroarylation of alkynes with arenes proceeds in the presence of various metal triflates as catalysts [4a, c], and usually requires harsh conditions. Gold and copper π-Lewis acids are also competent catalysts for this transformation under milder conditions. Thus, $AuCl_3$ (1.5 mol%) in combination with silver salts (3 mol% of $AgSbF_6$) was reported to be an optimal catalyst for the hydroarylation of aryl alkynes providing 1,1-diarylethylene derivatives (Scheme 7.116, conditions **a**) [202]. Similarly, 1,1-disubstituted olefins could be obtained in good yields via copper-mediated hydroarylation of aryl alkynes with electron-rich arenes. In this case, the use of triflic acid as a solvent was essential to achieve the desired reactivity (Scheme 7.116, conditions **b**) [203].

In the presence of gold(I) catalysts, electron-poor alkynes (ethyl propiolate and 3-butyn-2-one) were converted into *anti*-Markovnikov addition products with high *Z*-selectivity. A gold(III)-derived catalytic system ($AuCl_3$/3AgOTf) was

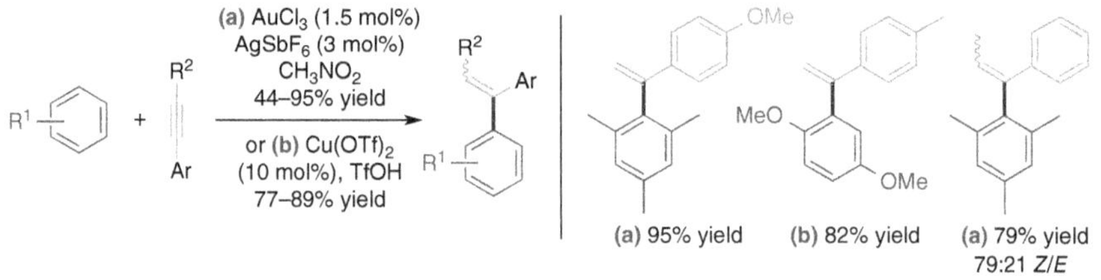

Scheme 7.116 Gold- and copper-catalyzed protocols for the intermolecular hydroarylation of aryl alkynes with electron-rich arenes.

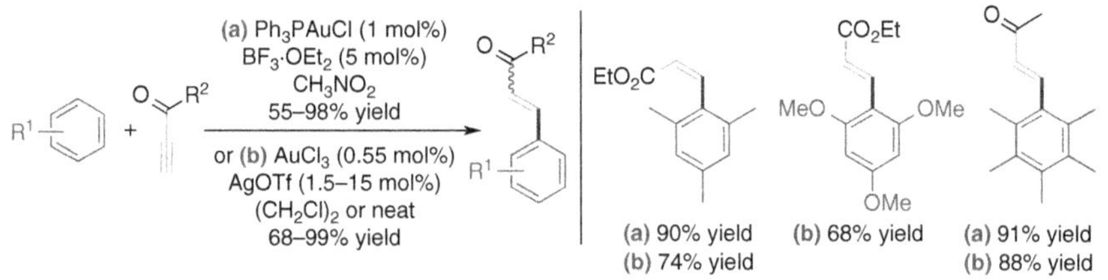

Scheme 7.117 Gold-catalyzed intermolecular hydroarylation of electron-poor alkynes with electron-rich arenes.

also developed for these transformations (Scheme 7.117) [59]. *N*-heterocyclic carbene- (NHC) [204], *N*-acyclic carbene- (NAC) [205] as well as carbodiphosphorane-ligated gold(I) catalysts [206] are alternative catalytic systems providing satisfactory chemical yields of monoalkenylated arenes. The ligand effect on π-acid-catalyzed hydroarylation has been computationally studied [207].

In addition, formation of dihalogenated aromatic compounds was achieved by cyclization of 1,2-bis(2-haloethynyl)benzenes combined with incorporation of an arene molecule that was used as a solvent (Scheme 7.118) [208]. Two possible mechanisms have been proposed to explain this reactivity. The first possible pathway involves the formation of a gold vinylidene intermediate by 1,2-iodine migration (path a). Upon the addition of benzene onto the alkyne the cyclization occurs, which was followed by protodeauration to give the substituted naphthalene. According to the second proposal, the transformation could be initiated by the intermolecular hydroarylation of the iodo-substituted alkyne by benzene followed by enyne-type cyclization to provide a gold(I) carbene intermediate (path b). The diiodonaphthalene would be then obtained by 1,2-iodine shift and demetalation. This method is limited to symmetrical 1,2-bis(2-haloethynyl)benzenes since in case of unsymmetrically substituted substrates an inseparable mixture of regioisomers was obtained.

7.3.2 Intermolecular Hydroarylation of Alkynes with Heteroarenes

The intermolecular alkyne hydroarylation with *N*- and *O*-containing heteroarenes, such as furans, indoles, and pyrroles has been less studied, although some interesting transformations were uncovered.

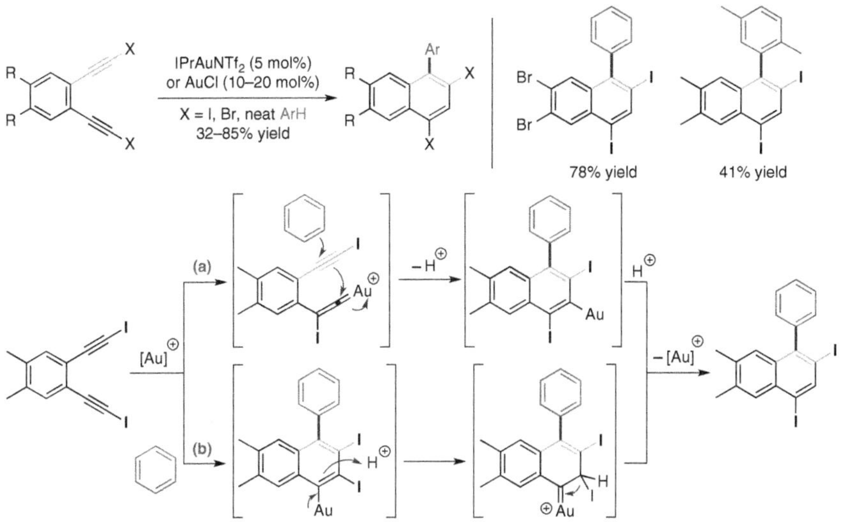

Scheme 7.118 Gold-catalyzed intermolecular hydroarylation/cycloisomerization of 1,2-bis(2-haloethynyl)benzenes.

7.3.2.1 *N*-Heterocycles

The intermolecular hydroarylation of alkynes with indoles was developed using $GaCl_3$ [3c] and a rhenium complex [10] as catalysts. The formation of bis-indole from ethyl propiolate and *N*-methyl indole was considered one of the first examples of intermolecular gold(III)-catalyzed hydroarylation of alkynes with indoles (Scheme 7.119, Eq. 1) [209]. In a somewhat related transformation, copper(II) bromide under solvent-free conditions was found to promote the *anti*-Markovnikov hydroarylation of aryl alkynes, resulting in the formation of bis(indolyl)alkanes (Scheme 7.119, Eq. 2) [210].

In contrast to these reactions, the double addition of indole to terminal alkynes occurs at the internal carbon atom of a triple bond in the presence of electrophilic gold(I) complexes (Scheme 7.120, Eq. 1) [91]. The same reactivity was observed with 2-ethylpyrrole as an arylating agent. Interestingly, the outcome of the reaction between bis(trifluoromethyl)phenylacetylene and

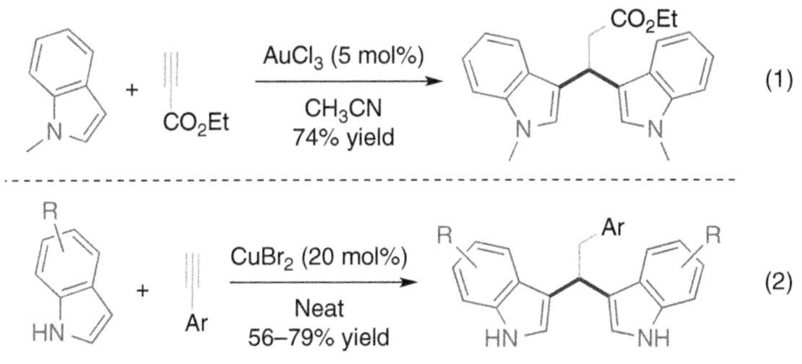

Scheme 7.119 Intermolecular double hydroarylation of alkynes with indoles at the terminal alkyne position with the formation of bis-indoles.

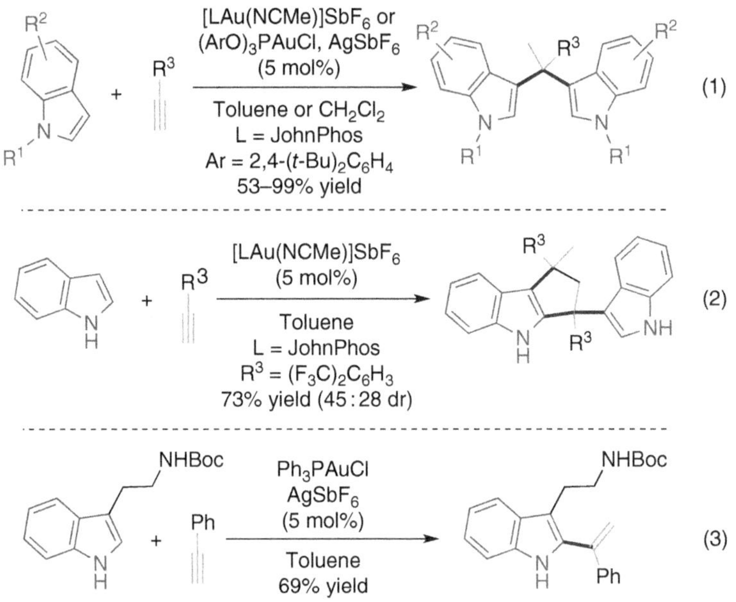

Scheme 7.120 Different outcome of the gold-catalyzed intermolecular alkyne hydroarylation with indoles.

indole was found to be solvent dependent. While the bis-indole product was isolated from the reaction conducted in CH_2Cl_2 in excellent 98% yield, the same reaction in toluene gave dimeric tetrahydrocyclopenta[*b*]indole adducts (Scheme 7.120, Eq. 2). Similar dimerization was observed when 3-methyl indole (skatole) was used. However, the alkyne hydroarylation with *N*-Boc-tryptamine led exclusively to the 2-alkenylated indole derivative (Scheme 7.120, Eq. 3).

The scope of alkyne double hydroarylation was later extended to pyrroles as arylating agent employing gold(I) catalyst and acetic acid as a solvent (Scheme 7.121, Eq. 1) [211]. The outcome of the reaction depends on the substitution pattern of alkyne. The hydroarylation of terminal alkynes results in the formation of bis(pyrrolyl)alkanes, whereas the use of alkyl-substituted internal alkynes leads to α-alkenylated pyrroles (Scheme 7.121, Eq. 2). Other heteroarenes, such as indole, furan, and thiophene, also gave bis(heretoaryl)alkanes [212]. Furthermore, acetylenic sulfones undergo hydroarylation with various indoles and pyrroles in the presence of copper(II) triflate to produce α,α-adducts (Scheme 7.121, Eq. 3) [213].

A synthesis of 2-vinylindole derivatives was developed via hydroarylation of ynamides with 3-substituted indoles (Scheme 7.122) [214]. This method provides high levels of stereoselectivity and broad substrate scope. In the presence of gold catalysts, 1,3-diynes undergo formal [4+2]-type reaction with pyrrole leading to indole derivatives. An initial intermolecular alkenylation of pyrrole results in the formation of an enyne intermediate that is perfectly set to undergo a gold(I)-catalyzed intramolecular hydroarylation in a 6-*endo*-dig fashion to form 4,7-disubstituted indoles (Scheme 7.123) [215]. The use of *N*-substituted

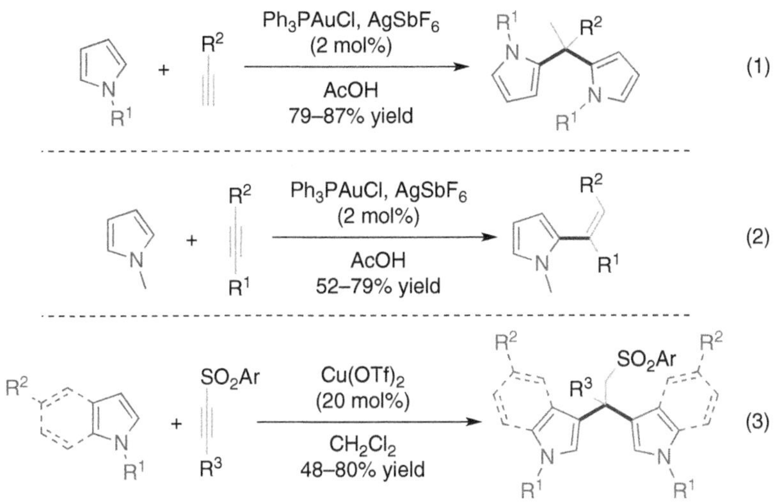

Scheme 7.121 Gold- and copper-catalyzed intermolecular hydroarylation of alkynes with pyrroles.

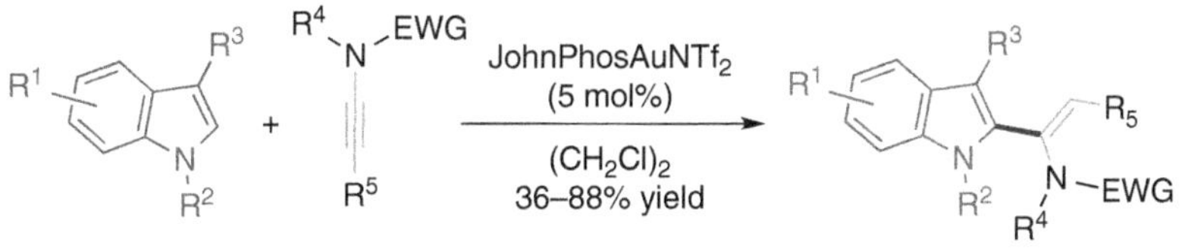

Scheme 7.122 Gold-catalyzed intermolecular hydroarylation of ynamides with 3-substituted indoles.

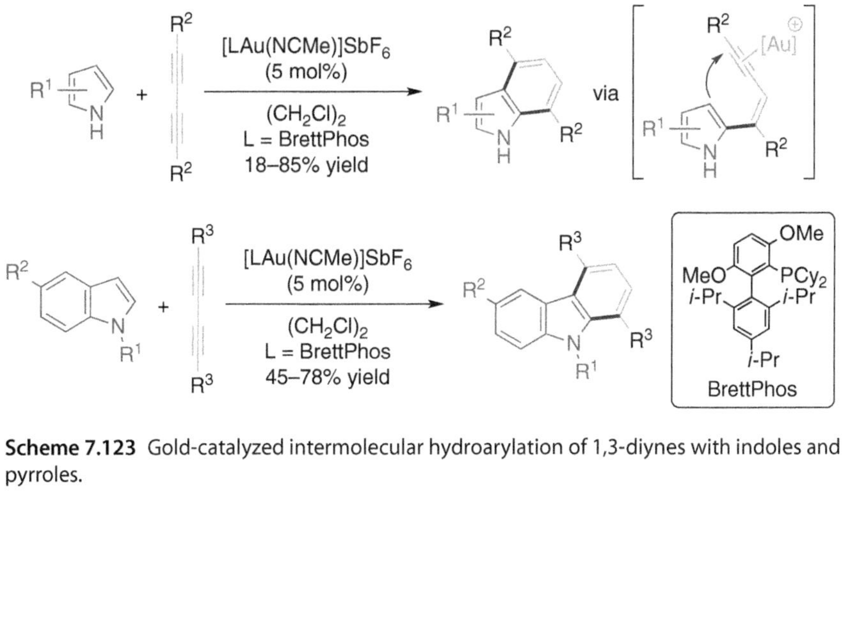

Scheme 7.123 Gold-catalyzed intermolecular hydroarylation of 1,3-diynes with indoles and pyrroles.

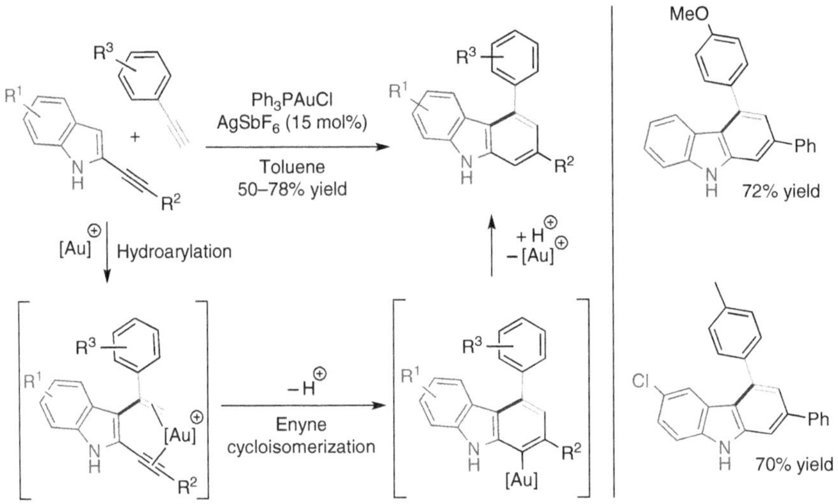

Scheme 7.124 Gold-catalyzed intermolecular hydroarylation/enyne cycloisomerization cascade in the synthesis of NH-carbazoles.

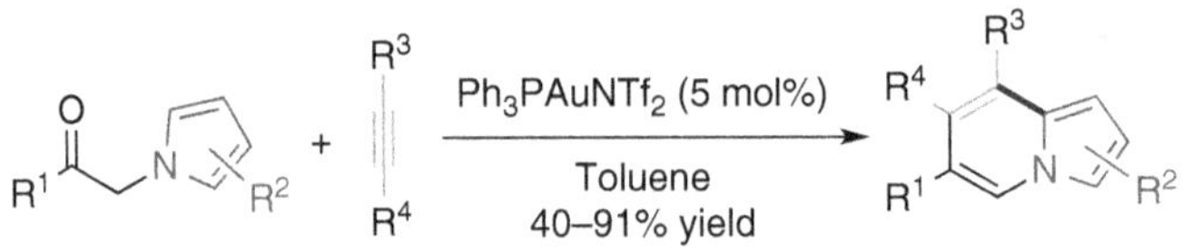

Scheme 7.125 Gold-catalyzed intermolecular hydroarylation/cycloaromatization cascade in the synthesis of indolizines.

indoles as a nucleophile in this transformation gave access to the corresponding carbazole derivatives in moderate yields.

Several complex cascade transformations, featuring intermolecular hydroarylation of alkynes by *N*-containing heterocycles, have also been disclosed. Thus, an efficient one-pot procedure for the synthesis of NH-carbazoles was developed via an alkenylation of 2-alkynylindole derivatives leading to 2-alkynyl-3-alkenyl indoles, followed by 6-*endo*-dig enyne cycloisomerization (Scheme 7.124) [216]. A simple approach for the preparation of various indolizines was developed through a one-pot gold(I)-catalyzed hydroarylation/cycloaromatization of α-(*N*-pyrrolyl)ketones (Scheme 7.125) [217]. This method provides an access to diverse indolizines decorated with a wide variety of the functional groups on the pyridine unit.

7.3.2.2 *O*-Heterocycles

As discussed earlier, alkyne hydroarylation reactions of furan analogs can either give simple alkenylation products or undergo complex rearrangements resulting in the formation of phenols, enones, and other compounds. Similar behavior was observed in intermolecular hydroarylation with furans. In one of the earliest reports describing the intermolecular hydroarylation of alkynes with furans, it was observed that 2-substituted furans reacted with alkynes in the presence

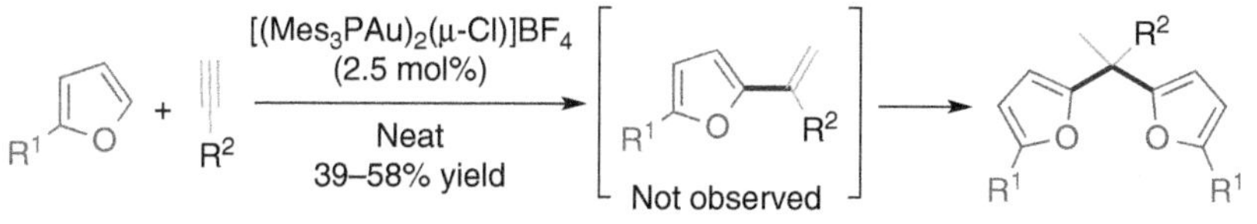

Scheme 7.126 Intermolecular double hydroarylation of terminal alkynes with furans.

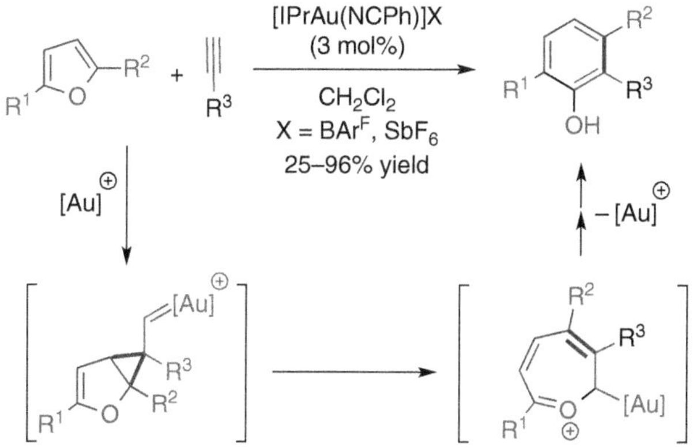

Scheme 7.127 Intermolecular gold-catalyzed hydroarylation of alkynes with furans in the synthesis of phenols.

of Schmidbaur–Bayler salt $[(Mes_3PAu)_2Cl]BF_4$, providing bisfuryl adducts of a twofold hydroarylation (Scheme 7.126) [211, 218]. Analogously to the twofold reaction of benzofuran [209], the vinyl furan derivative was not observed presumably due to faster addition of a second molecule of furan to the 2-alkenylated intermediate. Further elaboration of this strategy resulted in the development of a new catalytic system for the bis-hydroarylation of alkynes with various heterocycles, including furan derivatives [211]. When 2,5-disubstituted furans were used, the 3-alkenylation product was observed together with the rearranged phenol [190]. A practical method for the synthesis of highly substituted phenols was later developed based on this hydroarylation/rearrangement approach (Scheme 7.127). This complex transformation most likely proceeds via the initial formation of a cyclopropyl gold(I) carbene intermediate. A subsequent rearrangement and demetalation provide an oxepin, which is in equilibrium with the arene oxide that undergoes epoxide-opening to yield the phenols [219]. The extension of this reaction to 1,3-diphenylisobenzofuran resulted in the synthesis of 2,3-disubstituted indenes (Scheme 7.128) [219]. Mechanistically, this transformation is similar to the aforementioned phenol synthesis and also involves the generation of a cyclopropyl gold(I) carbene intermediate followed by the ring opening. A Friedel–Crafts-type reaction furnishes the indene after rearomatization and protodeauration.

In the presence of gold catalysts, 2,5-disubstituted furans react with alkynyl ethers leading to 2-phenoxy phenol derivatives in moderate yields (Scheme 7.129, Eq. 1) [220]. The use of sterically hindered gold complexes bearing non-coordinating anions is essential to promote the hydroarylation

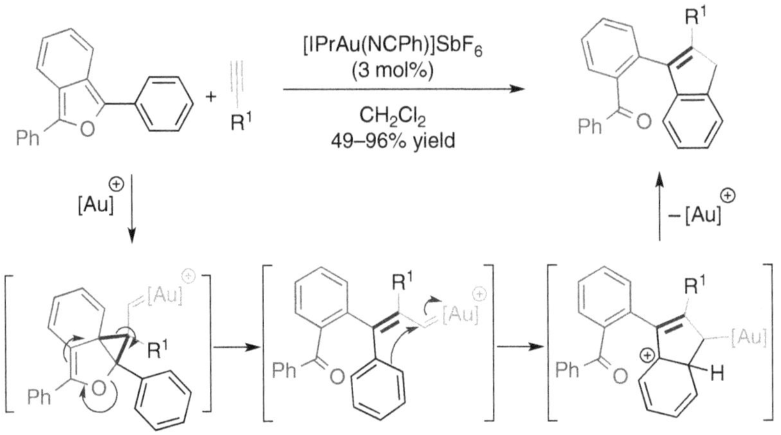

Scheme 7.128 Gold-catalyzed intermolecular hydroarylation of alkynes with 1,3-diphenylisobenzofuran.

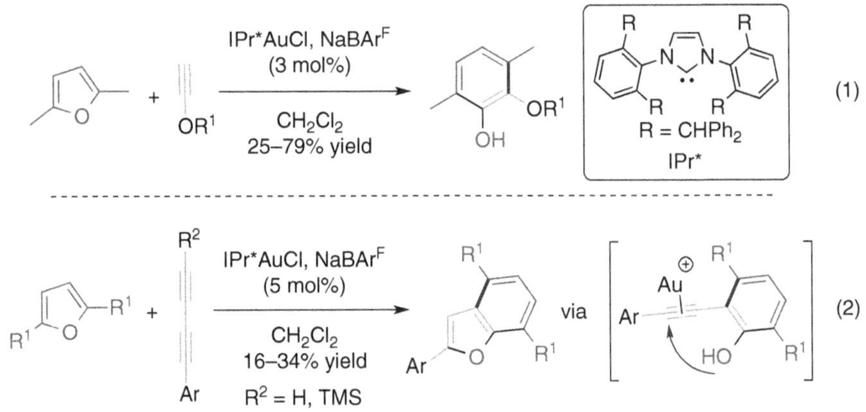

Scheme 7.129 Gold-catalyzed hydroarylation of ethynyl ethers and 1,3-diynes with 2,5-disubstituted furans.

process. This transformation is limited to terminal alkynes as internal alkynes do not provide the desired products. However, benzofurans can be obtained from 1,3-diynes and 2,5-disubstituted furans via sequential intermolecular hydroarylation, rearrangement, and intramolecular hydroalkoxylation of the alkyne by the phenol, although in low yields (Scheme 7.129, Eq. 2).

7.4 Metal-Supported Catalysts and Their Applications in Hydroarylation of Alkynes

Heterogeneous or solid supported molecular catalysts offer some advantages over the traditional soluble catalysts: no ligand is required in some cases, pure product can be isolated by simple filtration, and often the catalyst can be

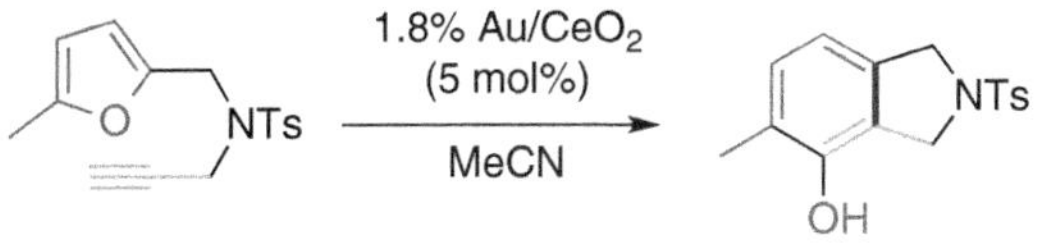

Cycle	Conversion (24 h, %)
First	100
Second	45
Third	35

Scheme 7.130 Intramolecular furan–yne cycloisomerization catalyzed by nanoparticles of gold supported on nanocrystalline CeO_2.

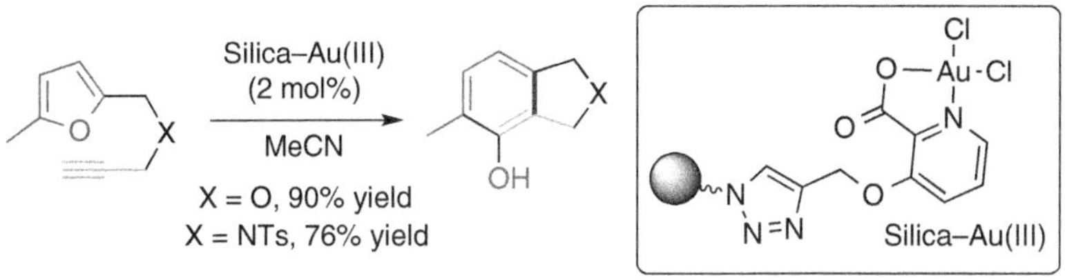

Scheme 7.131 Intramolecular furan–yne cycloisomerization catalyzed by gold(III) immobilized silica nanoparticles.

recycled and reused. Some supported catalysts have proved to be competent in promoting the hydroarylation of alkynes or related cycloisomerization reactions. To our knowledge, the first example of a supported gold catalyst in alkyne hydroarylation reaction was reported in the context of the phenol synthesis from alkynyl furans (Scheme 7.130) [221]. Presumably, cationic gold species are responsible for the catalytic activity of this system because gold(III) ions were shown to be stabilized on cerium oxide [222]. Gold nanoparticles supported on nanocrystalline cerium oxide demonstrated a good reactivity, although this catalytic system lost its initial activity within a few cycles [221].

Gold(III) supported on silica was prepared taking advantages of "click chemistry" between picolinic acid derivative (bearing an alkyne side-chain) and azide grafted silica nanoparticles [223]. In comparison to gold(III) chloride, a gold(III) catalyst immobilized on the silica nanoparticles was shown to promote phenol synthesis with an increased rate of the reaction during the first cycle, presumably, due to accelerating effect of picolinic acid anchor (Scheme 7.131). However, the catalytic activity of this system was lost after the second cycle, although no significant leaching was detected. The catalyst deactivation most likely was caused by an undesired reduction of gold(III) to gold(0). A polystyrene–supported gold(I) complex (PS-Au(I)) was also applied to the phenol synthesis and showed good efficiency (Scheme 7.132). However, a decrease in reactivity was observed leading to longer reaction times in each consecutive cycle (from 5 min for the first cycle to >35 h for the fourth and fifth cycles) [224]. Computational studies were performed to clarify the selectivity and mechanism of the cycloisomerization of ω-alkynylfuranes catalyzed by planar gold clusters [225] and subnanometer gold clusters [226] as well as core–shell nanoclusters [227].

In addition to phenol synthesis, supported gold catalysts have been shown to trigger other typical hydroarylation reactions. Gold nanoparticles supported on titanium oxide proved to be competent in promoting a hydroarylation of propargyl aryl ethers yielding chromene derivatives and, in some cases,

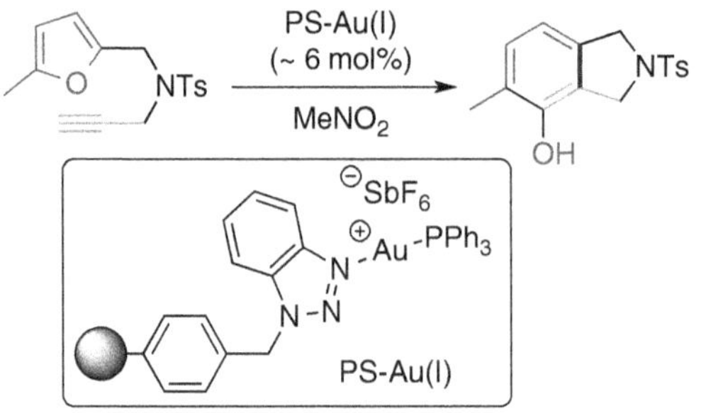

Cycle	Time (min)	Yield (%)
First	5	88
Second	30	86
Third	180	82
Fourth	2160	85
Fifth	2160	80

Scheme 7.132 Polystyrene-gold(I) (PS = polystyrene) catalyst applied to the furan–yne cycloisomerization.

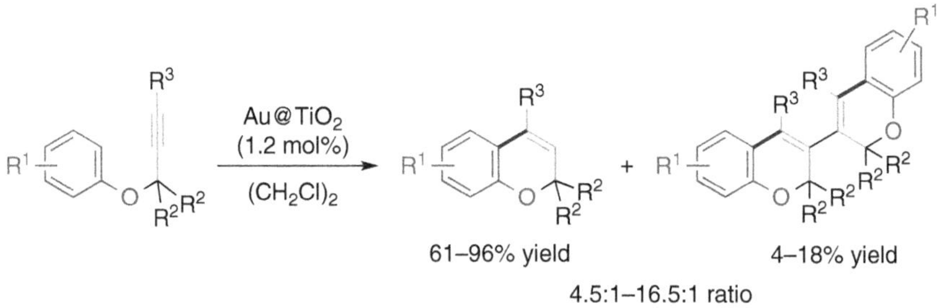

Scheme 7.133 Cycloisomerization of aryl propargyl ethers catalyzed by gold nanoparticles supported on titanium oxide.

resulting in a mixture with dimeric 2*H*-chromenes (Scheme 7.133) [228]. DFT calculations addressing the mechanism, product distribution, and substituent effect revealed that the amount of dimeric 2*H*-chromenes formed in the reaction is influenced by the substitution pattern of the substrate and the electronic properties of the active site of the catalyst [229]. The same supported catalyst and, alternatively, gold nanoparticles immobilized on alumina have been used for the synthesis of fused pyridocoumarine and quinoline derivatives via an intramolecular alkenylation of propargyl anilines (Scheme 7.134) [230].

A simple approach for the preparation of various complex-fused and spiro-heterocyclic scaffolds was developed through a one-pot two-step process

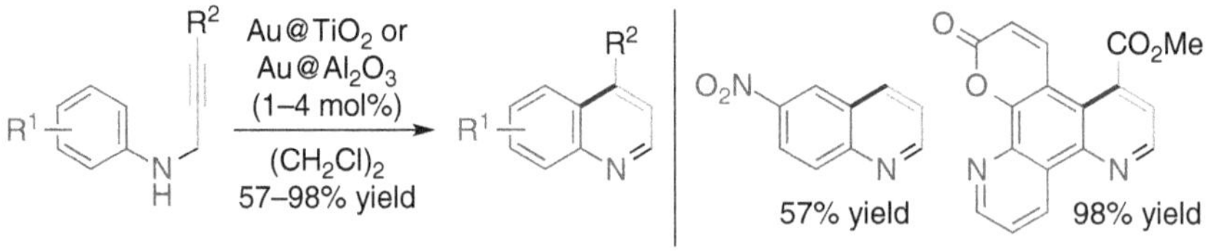

Scheme 7.134 Cycloisomerization of propargylic anilines catalyzed by gold nanoparticles supported on titanium oxide or aluminum oxide.

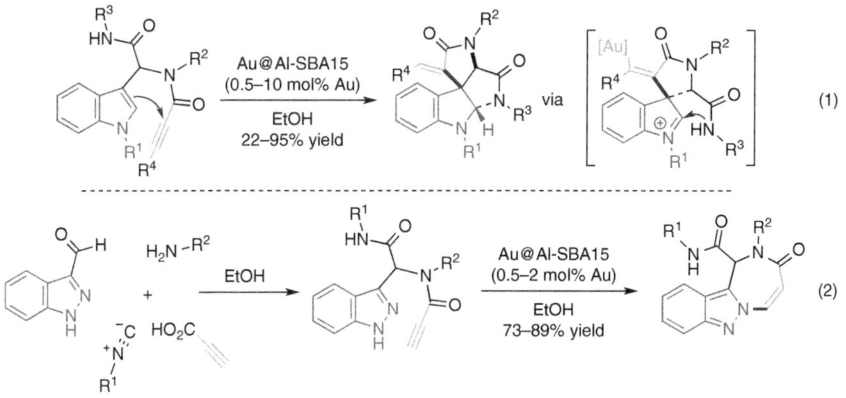

Scheme 7.135 A post-Ugi intramolecular hydroarylation of internal alkynes tethered to indole 3-position catalyzed by supported gold nanoparticles.

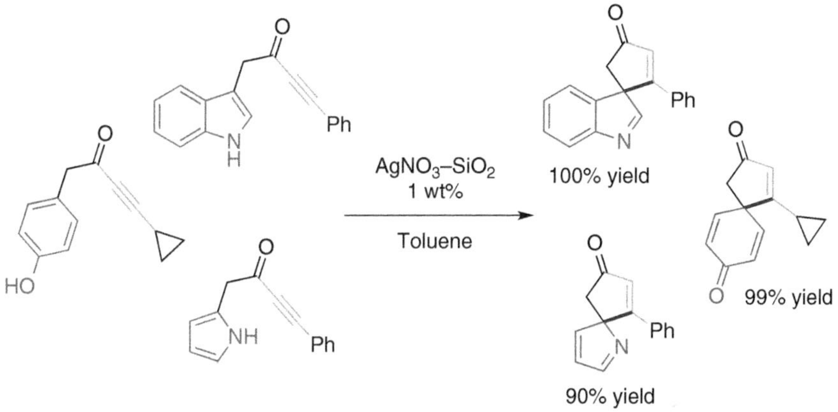

Scheme 7.136 Spirocyclization catalyzed by silica-supported silver nitrate.

by employing gold nanoparticles immobilized on Al-SBA15 [231]. Adducts of Ugi-4CR between indole-3-carboxaldehyde, primary amine, 2-alkynoic acid and isocyanide (Scheme 7.77), undergo gold-catalyzed spirocyclization with *exo*-selectivity (Scheme 7.135, Eq. 1). However, the cyclization of indazole-derived substrates proceeds in an endo-fashion, leading exclusively to a diazepino[1,2-*b*]indazole skeleton (Scheme 7.135, Eq. 2).

Dearomatizing spirocyclization of alkyne-tethered (hetero)aromatics was developed by employing a silica-supported $AgNO_3$ ($AgNO_3$–SiO_2) catalyst (Scheme 7.136). The high efficiency of this catalytic system presumably comes from the formation of silver nanoparticles and the synergistic effect caused by the silica itself. No significant catalyst leaching was detected and the system retained its activity after five cycles [232].

7.5 Hydroarylation of Alkynes in Total Synthesis

The alkyne hydroarylation promoted by π-acid catalysts is an efficient approach for the rapid generation of various structural motifs including simple alkenylated (hetero)arenes to functionalized polycyclic scaffolds, which are present in the complex architectures of some natural products. Although the gold-catalyzed hydroarylation has been widely used in the synthesis of natural products, silver and copper π-acid catalysts have been much less applied in target-oriented synthesis. The benefits of this strategy have been well illustrated by its application in target-oriented synthesis (Figure 7.5). Thus, the syntheses of diverse natural products were accomplished taking advantage of the π-acid-catalyzed

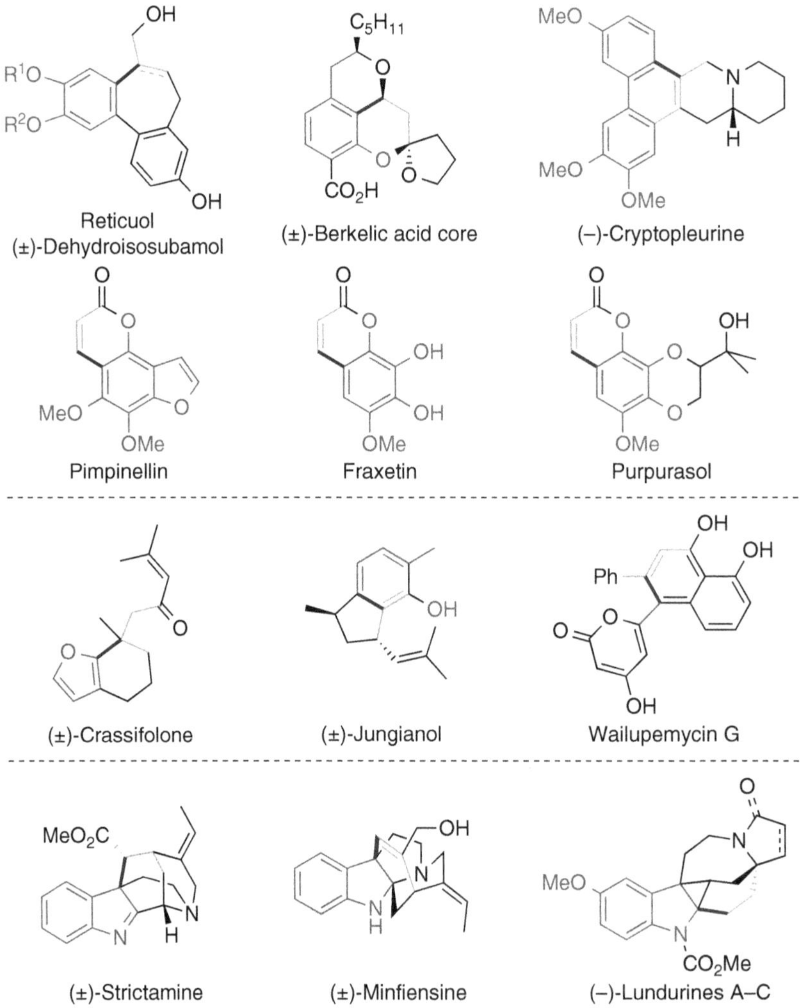

Figure 7.5 Examples of natural compounds, synthesized by gold-catalyzed hydroarylation of alkynes.

alkyne hydroarylation with arenes (reticuol [83], (±)-dehydroisosubamol [85], (±)-berkelic acid's tetracyclic core [62], (−)-cryptopleurine [76], pimpinellin, fraxetin and purpurasol [61]) as well as furans ((±)-crassifolone [163], (±)-jungianol [194] and its analogs, wailupemycin G [170]). Furthermore, the intramolecular hydroarylation of alkynes with indoles was used as a key step in formal syntheses of akuammiline alkaloids, (±)-strictamine [111], and (±)-minfiensine [114b], as well as in the total synthesis of the *Kopsia* alkaloids (−)-lundurines A–C [95].

The advantages of alkyne hydroarylation for the generation of functionalized polycyclic systems have been also illustrated in the total synthesis of the coumarin-containing natural products. The late-stage cyclization of the aryl propiolate mediated by a cationic gold(I) complex bearing JohnPhos ligand provided the natural product pimpinellin in a good yield (Scheme 7.137) [61]. The same approach was used for the assembling of the coumarin fragment presented in fraxetin and purpurasol.

The synthetic utility of the gold(I)-mediated intramolecular hydroarylation of alkyne with furan was demonstrated in the synthesis of wailupemycin G (Scheme 7.138) [170]. The gold(I)-catalyzed cyclization of furan–ynes coupled with a fragmentation of the furan ring was employed for the construction of the 1-naphthol intermediate in good yield and with high stereoselectivity (*E*/*Z* 53:1). Further functional group modification afforded the β-ketoester that was perfectly set to undergo a gold(I)-catalyzed alkyne hydrocarboxylation. The employment of $XPhosAuNTf_2$ as the catalyst provided the desired 4-hydroxy-2-pyrone, which was debenzylated to provide wailupemycin G. The benefits of the hydroarylation strategy for the generation of synthetically diverse scaffolds were demonstrated in the context of the formal total synthesis of akuammiline alkaloid (±)-strictamine (Scheme 7.139) [111]. The tetracyclic core of the natural product has been assembled via π-acid-mediated cyclization of *N*-nosylated propargyl-β-tetrahydrocarboline derivative. The *in situ* generated cationic gold(I) complex bearing a bulky electron-donating ligand ($SPhosAuCl/AgNTf_2$)

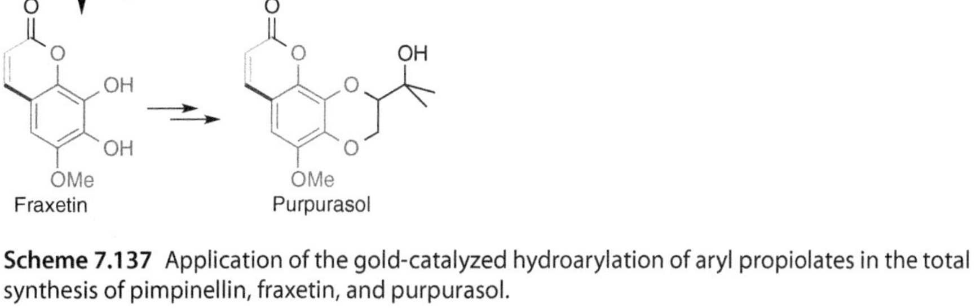

Scheme 7.137 Application of the gold-catalyzed hydroarylation of aryl propiolates in the total synthesis of pimpinellin, fraxetin, and purpurasol.

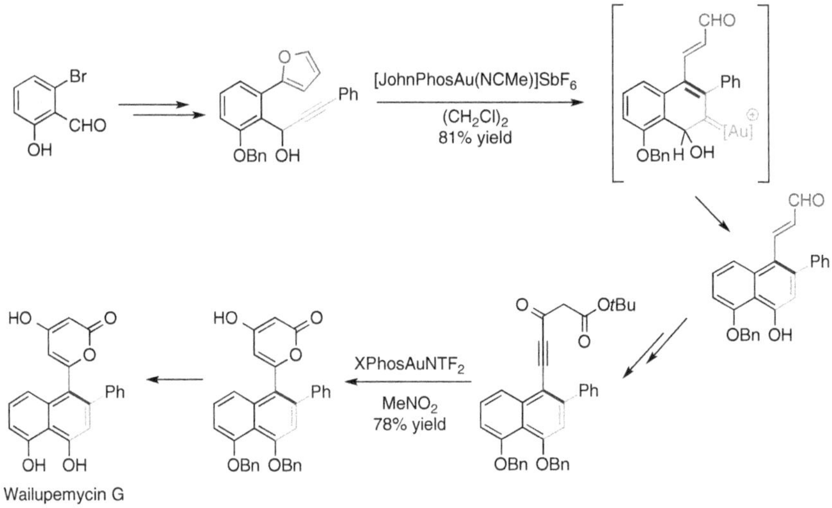

Scheme 7.138 Application of the gold-catalyzed hydroarylation of furan–ynes in the total synthesis of wailupemycin G.

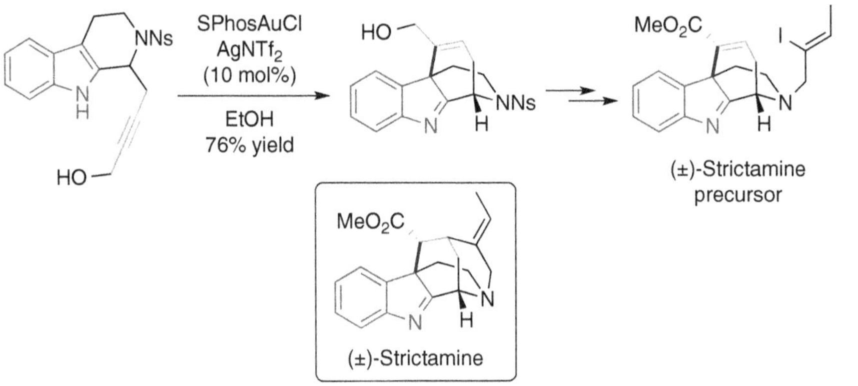

Scheme 7.139 Gold-catalyzed alkyne hydroarylation using indole–yne in the formal synthesis of (±)-strictamine.

was the most efficient in promoting the desired transformation. The multistep functional group modification provided the known enoate precursor for a Heck-type reaction to complete the formal total synthesis of (±)-strictamine.

The advantages of hydroheteroarylation strategy was illustrated by a highly regioselective formation of the lundurine tetracyclic core via gold(I)-catalyzed intramolecular alkenylation of the indole during the synthesis of *Kopsia* alkaloids (Scheme 7.140) [94, 95]. The indole–yne precursor was prepared by a tandem condensation/lactamization/[3,3]-sigmatropic Claisen rearrangement and the resulting aldehyde was homologated under the conditions of the Ohira-Bestmann reaction to give the substrate that underwent the selective gold(I)-mediated cycloisomerization to form the desired eight-membered ring-containing tetracycle. Further elaboration of the azocine intermediate

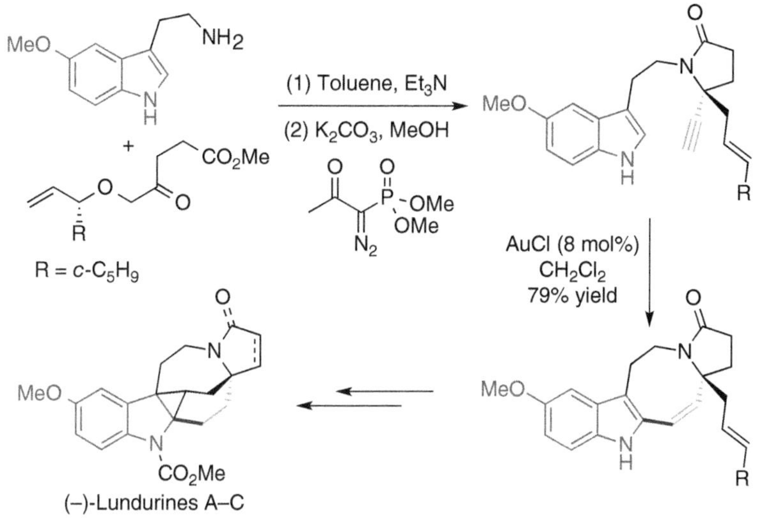

Scheme 7.140 The application of gold(I)-catalyzed hydroarylation in propargylated tryptamines in the total synthesis of (−)-lundurines A–C.

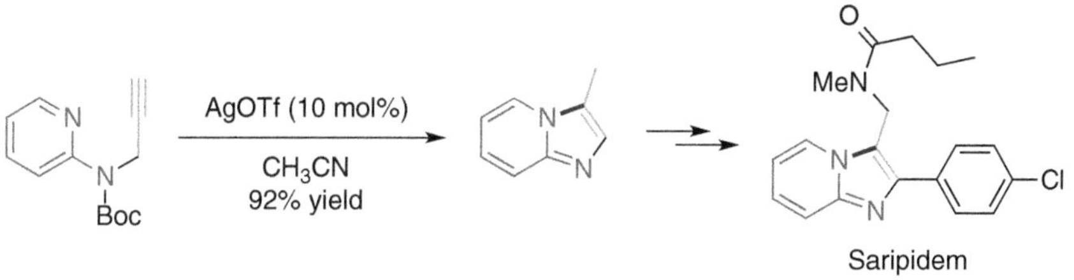

Scheme 7.141 Silver-catalyzed hydroarylation of propargylated 2-aminopyridines in the synthesis of saripidem.

completed the syntheses of the lundurines. The synthesis of anxiolytic saripidem was accomplished by employing the silver-mediated hydroarylation of alkynes as a key step (Scheme 7.141) [34]. The required imidazolopyridine skeleton was constructed by the highly efficient cycloisomerization of *N*-propargylaminopyridines.

References

1 For recent examples see:(a) Wang, H., Zhao, J., Zhang, J., and Zhu, Q. (2011) *Adv. Synth. Catal.*, **353**, 2653–2658; (b) Eom, D., Park, S., Park, Y., Lee, K., Hong, G., and Lee, P.H. (2013) *Eur. J. Org. Chem.*, 2672–2682.

2 (a) Yamamoto, Y., Gridnev, I.D., Patil, N.T., and Jin, T. (2009) *Chem. Commun.*, 5075–5087; (b) de Mendoza, P. and Echavarren, A.M. (2010) *Pure Appl. Chem.*, **82**, 801–820; For more examples see:(c) Tanaka, K. (2013) *Transition-Metal-Mediated Aromatic Ring Construction*, John

Wiley & Sons, Inc, Hoboken, New Jersey; (d) de Haro, T. and Nevado, C. (2014) Hydroarylation reactions, in *Comprehensive Organic Synthesis*, 2nd, Vol. 5 edn (eds P. Knochel and G.A. Molander), Elsevier Ltd, Zürich, pp. 1621–1659; (e) Yamamoto, Y. (2014) *Chem. Soc. Rev.*, **43**, 1575–1600.

3 Selected examples: (a) Inoue, H., Chatani, N., and Murai, S. (2002) *J. Org. Chem.*, **67**, 1414–1417; (b) Fürstner, A. and Mamane, V. (2003) *Chem. Commun.*, 2112–2113; (c) Yadav, J.S., Reddy, B.V.S., Padmavani, B., and Gupta, M.K. (2004) *Tetrahedron Lett.*, **45**, 7577–7579; (d) Li, R., Wang, S.R., and Lu, W. (2007) *Org. Lett.*, **9**, 2219–2222; (e) Li, H.-J., Guillot, R., and Gandon, V. (2010) *J. Org. Chem.*, **75**, 8435–8449; (f) Kwon, Y., Cho, H., and Kim, S. (2013) *Org. Lett.*, **15**, 920–923.

4 Selected examples: (a) Tsuchimoto, T., Maeda, T., Shirakawa, E., and Kawakami, Y. (2000) *Chem. Commun.*, 1573–1574; (b) Nishigawa, M., Takao, H., Yadav, V.K., Imagawa, H., and Sugihara, T. (2003) *Org. Lett.*, **5**, 4563–4565; (c) Yoon, M.Y., Kim, J.H., Choi, D.S., Shin, U.S., Lee, J.Y., and Song, C.E. (2007) *Adv. Synth. Catal.*, **349**, 1725–1737; (d) Donets, P.A., Van Hecke, K., Van Meervelt, L., and Van der Eycken, E.V. (2009) *Org. Lett.*, **11**, 3618–3621.

5 For selected examples see: (a) Jia, C., Piao, D., Oyamada, J., Lu, W., Kitamura, T., and Fujiwara, Y. (2000) *Science*, **287**, 1992–1995; (b) Jia, C., Lu, W., Oyamada, J., Kitamura, T., Matsuda, K., Irie, M., and Fujiwara, Y. (2000) *J. Am. Chem. Soc.*, **122**, 7252–7263; (c) Jia, C., Kitamura, T., and Fujiwara, Y. (2001) *Acc. Chem. Res.*, **34**, 633–639; (d) Martín-Matute, B., Cárdenas, D.J., and Echavarren, A.M. (2001) *Angew. Chem., Int. Ed.*, **40**, 4754–4757; (e) Martín-Matute, B., Nevado, C., Cárdenas, D.J., and Echavarren, A.M. (2003) *J. Am. Chem. Soc.*, **125**, 5757–5766; (f) Pastine, S.J., Youn, S.W., and Sames, D. (2003) *Org. Lett.*, **5**, 1055–1058; (g) Fürstner, A. and Kennedy, J.W.J. (2006) *Chem. Eur. J.*, **12**, 7398–7410; (h) Bosdet, M.J.D., Piers, W.E., Sorensen, T.S., and Parvez, M. (2007) *Angew. Chem., Int. Ed.*, **46**, 4940–4943; (i) Gruit, M., Michalik, D., Tillack, A., and Beller, M. (2009) *Angew. Chem., Int. Ed.*, **48**, 7212–7216; (j) Gruit, M., Pews-Datyan, A., and Beller, M. (2011) *Org. Biomol. Chem.*, **9**, 1148–1159; (k) Carreras, J., Patil, M., Thiel, W., and Alcarazo, M. (2012) *J. Am. Chem. Soc.*, **134**, 16753–16758.

6 Nakao, Y. (2011) *Chem. Record*, **11**, 242–251.

7 For a recent example see: (a) Eom, D., Mo, J., Lee, P.H., Gao, Z., and Kim, S. (2013) *Eur. J. Org. Chem.*, 533–540; (b) Greenhalgh, M.D., Jones, A.S., and Thomas, S.P. (2015) *Chem. Cat. Chem.*, 7, 190–222.

8 (a) Chatani, N., Inoue, H., Ikeda, T., and Murai, S. (2000) *J. Org. Chem.*, **65**, 4913–4918; (b) Manikandan, R. and Jeganmohan, M. (2015) *Org. Biomol. Chem.*, **13**, 10420–10436.

9 Maeyama, K. and Iwasawa, N. (1999) *J. Org. Chem.*, **64**, 1344–1346.

10 Xia, D., Wang, Y., Du, Z., Zheng, Q.-Y., and Wang, C. (2012) *Org. Lett.*, **14**, 588–591.

11 Easton, N.R. and Cassady, D.R. (1962) *J. Org. Chem.*, **27**, 4713–4714.

12 (a) Bandini, M. (2011) *Chem. Soc. Rev.*, **40**, 1358–1367; (b) Muratore, M.E., Homs, A., Obradors, C., and Echavarren, A.M. (2014) *Chem. Asian J.*, **9**, 3066–3082; (c) Dorel, R. and Echavarren, A.M. (2015) *Chem. Rev.*, **115**,

9028–9072; (d) Muratore, M.E. and Echavarren, A.M. (2014) Gold-catalyzed hydroarylation of alkynes, in *The Chemistry of Organogolds Compounds, Part 2, PATAI's Chemistry of Functional Group* (eds Z. Rappoport, J.F. Liebman, and I. Marek), John Wiley & Sons, Ltd, Chichester, UK, Chapter 16.

13 (a) Jia, M. and Bandini, M. (2015) *ACS Catal.*, **5**, 1638–1652; (b) Ranieri, B., Escofet, I., and Echavarren, A.M. (2015) *Org. Biomol. Chem.*, **13**, 7103–7118.

14 (a) Perchellet, E.M., Crow, K.R., Gakhar, G., Nguyen, T.A., Shi, A., Hua, D.H., and Perchellet, J.-P.H. (2010) *Int. J. Oncol.*, **36**, 673–688; (b) Li, Y.-M., Shi, J., Cai, R., Chen, X.-Y., Guo, Q.-X., and Liu, L. (2010) *Tetrahedron Lett.*, **51**, 1609–1612; (c) Dejon, L. and Speicher, A. (2013) *Tetrahedron Lett.*, **54**, 6700–6702.

15 Ryu, J.-S. (2006) *Bull. Korean Chem. Soc.*, **27**, 631–632.

16 Symeonidis, T.S., Kallitsakis, M.G., and Litinas, K.E. (2011) *Tetrahedron Lett.*, **52**, 5452–5455.

17 Boyarskiy, V.P., Ryabukhin, D.S., Bokach, N.A., and Vasilyev, A.V. (2016) *Chem. Rev.*, **116**, 5894–5986.

18 (a) Nieto-Oberhuber, C., Muñoz, M.P., Buñuel, E., Nevado, C., Cárdenas, D.J., and Echavarren, A.M. (2004) *Angew. Chem., Int. Ed.*, **43**, 2402–2406; (b) Nevado, C. and Echavarren, A.M. (2005) *Chem. Eur. J.*, **11**, 3155–3164; (c) Menon, R.S., Findlay, A.D., Bissember, A.C., and Banwell, M.G. (2009) *J. Org. Chem.*, **74**, 8901–8903.

19 Alfonsi, M., Arcadi, A., Chiarini, M., and Marinelli, F. (2011) *Tetrahedron Lett.*, **52**, 5145–5148.

20 Zhu, M., Fu, W., Zou, G., Xun, C., Deng, D., and Ji, B. (2012) *J. Fluorine Chem.*, **135**, 195–199.

21 Kuninobu, Y., Inoue, Y., and Takai, K. (2007) *Chem. Lett.*, **36**, 1422–1423.

22 Xiao, F., Chen, Y., Liu, Y., and Wang, J. (2008) *Tetrahedron*, **64**, 2755–2761.

23 (a) Huang, H., Jiang, H., Chen, K., and Liu, H. (2009) *J. Org. Chem.*, **74**, 5476–5480; (b) Meyet, C.E. and Larsen, C.H. (2014) *J. Org. Chem.*, **79**, 9835–9841.

24 Zang, X., Liu, B., Shu, X., Gao, Y., Lv, H., and Zhu, J. (2012) *J. Org. Chem.*, 77, 501–510.

25 (a) Huma, H.Z.S., Halder, R., Kalra, S.S., Das, J., and Iqbal, J. (2002) *Tetrahedron Lett.*, **43**, 6485–6488; (b) Mondal, R.R., Khamarui, S., and Maiti, D.K. (2016) *ACS Omega*, **1**, 251–263.

26 The formation of quinolines and phenanthrenes is proposed to proceed via allene-intermediates:Saifuddin, M., Agarwal, P.K., and Kundu, B. (2011) *J. Org. Chem.*, **76**, 10122–10128.

27 Periyaraja, S., Shanmugam, P., Mandal, A.B., Kumar, T.S., and Ramamurthy, P. (2013) *Tetrahedron*, **69**, 2891–2899.

28 (a) Williamson, N.M., March, D.R., and Ward, A.D. (1995) *Tetrahedron Lett.*, **36**, 7721–7724; (b) Williamson, N.M. and Ward, A.D. (2005) *Tetrahedron*, **61**, 155–165; (c) Holman, M.A., Williamson, N.M., and Ward, A.D. (2005) *Aust. J. Chem.*, **58**, 368–374; (d) Suzuki, M., Yu, D., Morris-Natschke, S.L., Smith, P.C., and Lee, K.-H. (2007) *Bioorg. Med. Chem.*, **15**, 6852–6858; (e) Fotie, J., Kaiser, M., Delfin, D.A., Manley, J., Reid, C.S., Paris, J.-M.,

Wenzler, T., Maes, L., Mahasenan, K.V., Li, C., and Werbovetz, K.A. (2010) *J. Med. Chem.*, **53**, 966–982.
29 Su, J., Ju, J., and Hua, R. (2012) *Curr. Org. Synth.*, **9**, 273–277.
30 Ding, D., Mou, T., Feng, M., and Jiang, X. (2016) *J. Am. Chem. Soc.*, **138**, 5218–5221.
31 Bala, B.D., Muthusaravanan, S., and Perumal, S. (2013) *Tetrahedron Lett.*, **54**, 3735–3739.
32 Vallerotto, S., Le Douaron, G., Bernadat, G., Ferrié, L., and Figadère, B. (2016) *Synthesis*, **48**, 3232–3240.
33 Majumbar, K.C., Nandi, R.K., Ganai, S., and Taher, A. (2011) *Synlett*, 116–120.
34 Chioua, M., Soriano, E., Infantes, L., Jimeno, M.L., Marco-Contelles, J., and Samadi, A. (2013) *Eur. J. Org. Chem.*, 35–39.
35 Shibuya, T., Nakamura, K., and Tanaka, K. (2011) *Beilstein J. Org. Chem.*, **7**, 944–950.
36 Tanaka, M., Shibata, Y., Nakamura, K., Teraoka, K., Uekusa, H., Nakazono, K., Takata, T., and Tanaka, K. (2016) *Chem. Eur. J.*, **22**, 9537–9541.
37 Gronnier, C., Odabachian, Y., and Gagosz, F. (2011) *Chem. Commun.*, **47**, 218–220.
38 Jurberg, I.D. and Gagosz, F. (2011) *J. Organomet. Chem.*, **696**, 37–41.
39 Morán-Poladura, P., Suárez-Pantiga, S., Piedrafita, M., Rubio, E., and González, J.M. (2011) *J. Organomet. Chem.*, **696**, 12–15.
40 Liu, X.-Y., Ding, P., Huang, J.-S., and Che, C.-M. (2007) *Org. Lett.*, **9**, 2645–2648.
41 Luo, Y., Li, Z., and Li, C.-J. (2005) *Org. Lett.*, **7**, 2675–2678.
42 Liu, X.-Y. and Che, C.-M. (2008) *Angew. Chem., Int. Ed.*, **47**, 3805–3810.
43 Zhou, Y., Feng, E., Liu, G., Ye, D., Li, J., Jiang, H., and Liu, H. (2009) *J. Org. Chem.*, **74**, 7344–7348.
44 Reddy, M.R., Kumar, G.R., Yarlagadda, S., Reddy, C.R., Yadav, J.S., Sridhar, B., and Reddy, S.B.V. (2016) *RSC Adv.*, **6**, 113390–113394.
45 Qiu, W.-W., Surendra, K., Yin, L., and Corey, E.J. (2011) *Org. Lett.*, **13**, 5893–5895.
46 (a) Nicolaou, K.C., Sasmal, P.K., and Xu, H. (2004) *J. Am. Chem. Soc.*, **126**, 5493–5501; (b) Garcia, J., Barluenga, S., Beebe, K., Neckers, L., and Winssinger, N. (2010) *Chem. Eur. J.*, **16**, 9767–9771.
47 Mézailles, N., Ricard, L., and Gagosz, F. (2005) *Org. Lett.*, **7**, 4133–4136.
48 Kumar, M., Hammond, G.B., and Xu, B. (2014) *Org. Lett.*, **16**, 3452–3455.
49 Curtis, N.R., Prodger, J.C., Rassias, G., and Walker, A.J. (2008) *Tetrahedron Lett.*, **49**, 6279–6281.
50 Rassias, G., Stevenson, N.G., Curtis, N.R., Northall, J.M., Gray, M., Prodger, J.C., and Walker, A.J. (2010) *Org. Proc. Res. Dev.*, **14**, 92–98.
51 Lykakis, I.N., Efe, C., Gryparis, C., and Stratakis, M. (2011) *Eur. J. Org. Chem.*, 2334–2338.
52 Sharif, S.A.I., Calder, E.D.D., Harkiss, A.H., Maduro, M., and Sutherland, A. (2016) *J. Org. Chem.*, **81**, 9810–9819.

53 Tumma, N., Jacolot, M., Jean, M., Chandrasekhar, S., and van de Weghe, P. (2012) *Synlett*, **23**, 2919–2922.
54 Arcadi, A., Blesi, F., Cacchi, S., Fabrizi, G., Goggiamani, A., and Marinelli, F. (2012) *Org. Biomol. Chem.*, **10**, 9700–9708.
55 Wang, Y.-L., Zhang, W.-M., Dai, J.-J., Feng, Y.-S., and Xu, H.-J. (2014) *RSC Adv.*, **4**, 61706–61710.
56 Mo, J., Eom, D., Lee, E., and Lee, P.H. (2012) *Org. Lett.*, **14**, 3684–3687.
57 Morán-Poladura, P., Rubio, E., and González, J.M. (2013) *Beilstein J. Org. Chem.*, **9**, 2120–2128.
58 Kim, C.-E., Ryu, T., Kim, S., Lee, K., Lee, C.-H., and Lee, P.H. (2013) *Adv. Synth. Catal.*, **355**, 2873–2883.
59 Shi, Z. and He, C. (2004) *J. Org. Chem.*, **69**, 3669–3671.
60 Vadola, P.A. and Sames, D. (2012) *J. Org. Chem.*, **77**, 7804–7814.
61 Cervi, A., Aillard, P., Hazeri, N., Petit, L., Chai, C.L.L., Willis, A.C., and Banwell, M.G. (2013) *J. Org. Chem.*, **78**, 9876–9882.
62 Brimble, M.A., Haym, I., Sperry, J., and Furkert, D.P. (2012) *Org. Lett.*, **14**, 5820–5823.
63 Do, J.H., Kim, H.N., Yoon, J., Kim, J.S., and Kim, H.-J. (2010) *Org. Lett.*, **12**, 932–934.
64 Shaikh, A.C., Shalini, S., Vaidhyanathan, R., Mane, M.V., Barui, A.K., Patra, C.R., Venkatesh, Y., Bangal, P.R., and Patil, N.T. (2015) *Eur. J. Org. Chem.*, 4860–4867.
65 Aparece, M.D. and Vadola, P.A. (2014) *Org. Lett.*, **16**, 6008–6011.
66 Nemoto, T., Matsuo, N., and Hamada, Y. (2014) *Adv. Synth. Catal.*, **356**, 2417–2421.
67 Wu, W.-T., Xu, R.-Q., Zhang, L., and You, S.-L. (2016) *Chem. Sci.*, **7**, 3427–3431.
68 James, M.J., Cuthbertson, J.D., O'Brien, P., Taylor, R.J.K., and Unsworth, W.P. (2015) *Angew. Chem., Int. Ed.*, **54**, 7640–7643.
69 Storch, J., Bernard, M., Sýkora, J., Karban, J., and Čermák, J. (2013) *Eur. J. Org. Chem.*, 260–263.
70 Tang, S., Monot, J., El-Hellani, A., Michelet, B., Guillot, R., Bour, C., and Gandon, V. (2012) *Chem. Eur. J.*, **18**, 10239–10243.
71 Mamane, V., Hannen, P., and Fürstner, A. (2004) *Chem. Eur. J.*, **10**, 4556–4575.
72 (a) Yamamoto, K., Okazumi, M., Suemune, H., and Usui, K. (2013) *Org. Lett.*, **15**, 1806–1809; (b) Oyama, H., Nakano, K., Harada, T., Kuroda, R., Naito, M., Nobusawa, K., and Nozaki, K. (2013) *Org. Lett.*, **15**, 2104–2107.
73 Fürstner, A. and Mamane, V. (2002) *J. Org. Chem.*, **67**, 6264–6267.
74 Soriano, E. and Marco-Contelles, J. (2006) *Organometallics*, **25**, 4542–4553.
75 Carreras, J., Gopakumar, G., Gu, L., Gimeno, A., Linowski, P., Petuškova, J., Thiel, W., and Alcarazo, M. (2013) *J. Am. Chem. Soc.*, **135**, 18815–18823. Addition/Correction: (2014) *J. Am. Chem. Soc.*, **136**, 3319–3319..
76 Stoye, A. and Opatz, T. (2015) *Eur. J. Org. Chem.*, 2149–2156.
77 (a) Mandadapu, A.K., Dathi, M.D., Arigela, R.K., and Kundu, B. (2012) *Tetrahedron*, **68**, 8207–8215; (b) Jin, R., Chen, J., Chen, Y., Liu, W., Xu, D., Li, Y., Ding, A., and Guo, H. (2016) *J. Org. Chem.*, **81**, 12553–12558.

78 Urbano, A., Hernández-Torres, G., del Hoyo, A.M., Martínez-Carrión, A., and Carreño, M.C. (2016) *Chem. Commun.*, **52**, 6419–6422.

79 Shu, C., Chen, C.-B., Chen, W.-X., and Ye, L.-W. (2013) *Org. Lett.*, **15**, 5542–5545.

80 Shu, C., Li, L., Chen, C.-B., Shen, H.-C., and Long-Wu, Y. (2014) *Chem. Asian J.*, **9**, 1525–1529.

81 Pascual, S., Bour, C., de Mendoza, P., and Echavarren, A.M. (2011) *Beilstein J. Org. Chem.*, 7, 1520–1525.

82 Ito, M., Kawasali, R., Kanyiva, K.S., and Shibata, T. (2016) *Eur. J. Org. Chem.*, 5234–5237.

83 Pflästerer, D., Rettenmeier, E., Schneider, S., de Las Heras Ruiz, E., Rudolph, M., and Hashmi, A.S.K. (2014) *Chem. Eur. J.*, **20**, 6752–6755.

84 Pflästerer, D., Schumacher, S., Rudolph, M., and Hashmi, A.S.K. (2015) *Chem. Eur. J.*, **21**, 11585–11589.

85 Pflästerer, D., Rudolph, M., Yates, B.F., Ariafard, A., and Hashmi, A.S.K. (2017) *Adv. Synth. Catal.*, **359**, 866–874.

86 Nakae, T., Ohnishi, R., Kitahata, Y., Soukawa, T., Sato, H., Mori, S., Okujima, T., Uno, H., and Sakaguchi, H. (2012) *Tetrahedron Lett.*, **53**, 1617–1619.

87 Huang, G., Cheng, B., Xu, L., Li, Y., and Xia, Y. (2012) *Chem. Eur. J.*, **18**, 5401–5415.

88 Lim, W. and Rhee, Y.H. (2013) *Eur. J. Org. Chem.*, 460–464.

89 Wang, H., Denton, J.R., and Davies, H.M.L. (2011) *Org. Lett.*, **13**, 4316–4319.

90 Gorin, D.J., Dube, P., and Toste, F.D. (2006) *J. Am. Chem. Soc.*, **128**, 14480–14481.

91 Ferrer, C., Amijs, H.M., and Echavarren, A.M. (2007) *Chem. Eur. J.*, **13**, 1358–1373.

92 Zhang, L., Wang, Y., Yao, Z.-J., Wang, S., and Yu, Z.-X. (2015) *J. Am. Chem. Soc.*, **137**, 13290–13300.

93 Ferrer, C. and Echavarren, A.M. (2006) *Angew. Chem., Int. Ed.*, **45**, 1105–1109.

94 Ferrer, C., Escribano-Cuesta, A., and Echavarren, A.M. (2009) *Tetrahedron*, **65**, 9015–9020.

95 Kirillova, M.S., Muratore, M.E., Dorel, R., and Echavarren, A.M. (2016) *J. Am. Chem. Soc.*, **138**, 3671–3674.

96 Peshkov, V.A., Pereshivko, O.P., and Van der Eycken, E.V. (2012) *Adv. Synth. Catal.*, **354**, 2841–2848.

97 James, M.J., Clubley, R.E., Palate, K.Y., Procter, T.J., Wyton, A.C., O'Brien, P., Taylor, R.J.K., and Unsworth, W.P. (2015) *Org. Lett.*, **17**, 4372–4375.

98 Liddon, J.T.R., James, M.J., Clarke, A.K., O'Brien, P., Taylor, R.J.K., and Unsworth, W.P. (2016) *Chem. Eur. J.*, **22**, 8777–8780.

99 (a) England, D.B. and Padwa, A. (2008) *Org. Lett.*, **10**, 3631–3634; (b) Verniest, G., England, D., de Kimpe, N., and Padwa, A. (2010) *Tetrahedron*, **66**, 1496–1502; (c) Verniest, G., Wang, X., de Kimpe, N., and Padwa, A. (2010) *J. Org. Chem.*, **75**, 424–433.

100 Ramesh, S. and Nagarajan, R. (2015) *Synthesis*, **47**, 3573–3582.

101 Praveen, C. and Perumal, P.T. (2011) *Synlett*, **4**, 521–524.

102 (a) Qiu, Y., Kong, W., Fu, C., and Ma, S. (2012) *Org. Lett.*, **14**, 6198–6201; (b) Qiu, Y., Ma, D., Kong, W., Fu, C., and Ma, S. (2014) *Org. Chem. Front.*, **1**, 62–67; (c) Shao, J., He, R., Shen, W., and Li, M. (2014) *J. Mol. Model*, **20**, 2239–2252; (d) Hashmi, A.S.K., Yang, W., and Rominger, F. (2012) *Chem. Eur. J.*, **18**, 6576–6580.

103 (a) Jakubec, P., Hawkins, A., Felzmann, W., and Dixon, D.J. (2012) *J. Am. Chem. Soc.*, **134**, 17482–17485; (b) Eguchi, K., Fujiwara, Y., Hayashida, A., Horlad, H., Kato, H., Rotinsulu, H., Losung, F., Mangindaan, R.E.P., de Voogd, N.J., Takeya, M., and Tsukamoto, S. (2013) *Bioorg. Med. Chem.*, **21**, 3831–3838.

104 (a) Zhu, D., Chen, M., Li, M., Luo, B., Zhao, Y., Huang, P., Xue, F., Rapposelli, S., Pi, R., and Wen, S. (2013) *Eur. J. Med. Chem.*, **68**, 81–88; (b) Rongved, P., Kirsh, G., Bouaziz, Z., Jose, J., and Le Borgne, M. (2013) *Eur. J. Med. Chem.*, **69**, 465–479.

105 (a) Zhang, Y., Wada, T., and Sasabe, H. (1998) *J. Mater. Chem.*, **8**, 809–828; (b) Grazulevicius, J.V., Strohriegl, P., Pielichowski, J., and Pielichowski, K. (2003) *Prog. Polym. Sci.*, **28**, 1297–1353; (c) Thomas, K.R.J., Lin, J.T., Tao, Y.-T., and Ko, C.-W. (2001) *J. Am. Chem. Soc.*, **123**, 9404–9411; (d) Díaz, J.L., Dobarro, A., Villacampa, B., and Velasco, D. (2001) *Chem. Mater.*, **13**, 2528–2536.

106 Wang, L., Li, G., and Liu, Y. (2011) *Org. Lett.*, **13**, 3786–3789.

107 (a) Yang, J.M., Zhu, C.-Z., Tang, X.-Y., and Shi, M. (2014) *Angew. Chem. Int. Ed.*, **53**, 5142–5146; (b) Yang, J.M., Li, P.-H., Wei, Y., Tang, X.-Y., and Shi, M. (2016) *Chem. Commun.*, **52**, 346–349.

108 Gao, Y., Xu, Q., and Shi, M. (2015) *ACS Catal.*, **5**, 6608–6614.

109 Zhang, L., Chang, L., Hu, H., Wang, H., Yao, Z.-J., and Wang, S. (2014) *Chem. Eur. J.*, **20**, 2925–2932.

110 Xu, W., Wang, W., and Wang, X. (2015) *Angew. Chem. Int. Ed.*, **54**, 9546–9549.

111 Nishiyama, D., Ohara, A., Chiba, H., Kumagai, H., Oishi, S., Fujii, N., and Ohno, H. (2016) *Org. Lett.*, **18**, 1670–1673.

112 Zhang, Y.-Q., Zhu, D.-Y., Jiao, Z.-W., Li, B.-S., Zhang, F.-M., Tu, Y.-Q., and Bi, Z. (2011) *Org. Lett.*, **13**, 3458–3461.

113 Cheng, B., Huang, G., Xu, L., and Xia, Y. (2012) *Org. Biomol. Chem.*, **10**, 4417–4423.

114 (a) Podoll, J.D., Liu, Y., Chang, L., Walls, S., Wang, W., and Wang, X. (2013) *Proc. Natl. Acad. Sci. USA*, **110**, 15573–15578; (b) Liu, Y., Xu, W., and Wang, X. (2010) *Org. Lett.*, **12**, 1448–1451; (c) Chang, L., Podoll, J.D., Wang, W., Walls, S., O'Rourke, C.P., and Wang, X. (2014) *J. Med. Chem.*, **57**, 3803–3817; (d) Barbour, P.M., Podoll, J.D., Marholz, L.J., and Wang, X. (2014) *Bioorg. Med. Chem. Lett.*, **24**, 5602–5605.

115 Noey, E.L., Wang, X., and Houk, K.N. (2011) *J. Org. Chem.*, **76**, 3477–3483.

116 Cera, G., Crispino, P., Monari, M., and Bandini, M. (2011) *Chem. Commun.*, **47**, 7803–7805.

117 Cera, G., Crispino, P., Chiarucci, M., Mazzanti, A., Mancinelli, M., and Bandini, M. (2012) *Org. Lett.*, **14**, 1350–1353.

118 Zheng, N., Chang, Y.-Y., Zhang, L.-J., Gong, J.-X., and Yang, Z. (2016) *Chem. Asian J.*, **11**, 371–375.
119 Hirano, K., Inaba, Y., Watanabe, T., Oishi, S., Fujii, N., and Ohno, H. (2010) *Adv. Synth. Catal.*, **352**, 368–372.
120 Hirano, K., Inaba, Y., Takasu, K., Oishi, S., Takemoto, Y., Fujii, N., and Ohno, H. (2011) *J. Org. Chem.*, **76**, 9068–9080.
121 Hirano, K., Inaba, Y., Takahashi, N., Shimano, M., Oishi, S., Fujii, N., and Ohno, H. (2011) *J. Org. Chem.*, **76**, 1212–1227.
122 Ferrara, G., Jin, T., Oniwa, K., Zhao, J., Asiri, A.M., and Yamamoto, Y. (2012) *Tetrahedron Lett.*, **53**, 914–918.
123 Boiaryna, L., El Mkaddem, M.K., Taillier, C., Dalla, V., and Othman, M. (2012) *Chem. Eur. J.*, **18**, 14192–14200.
124 Boiaryna, L., Azizi, S.M., El Bouakher, A., Picard, B., Taillier, C., Othman, M., Trabelsi-Ayadi, M., and Dalla, V. (2015) *Org. Lett.*, **17**, 2130–2133.
125 Lu, Y., Du, X., Jia, X., and Liu, Y. (2009) *Adv. Synth. Catal.*, **351**, 1517–1522.
126 Xie, X., Du, X., Chen, Y., and Liu, Y. (2011) *J. Org. Chem.*, **76**, 9175–9181.
127 Heffernan, S.J., Tellam, J.P., Queru, M.E., Silvanus, A.C., Benito, D., Mahon, M.F., Hennessy, A.J., Andrews, B.I., and Carbery, D.R. (2013) *Adv. Synth. Catal.*, **355**, 1149–1159.
128 Xu, S., Zhou, Y., Xu, J., Jiang, H., and Liu, H. (2013) *Green Chem.*, **15**, 718–726.
129 Loh, C.C.J., Badorrek, J., Raabe, G., and Enders, D. (2011) *Chem. Eur. J.*, **17**, 13409–13414.
130 Kumar, R., Arigela, R.K., Samala, S., and Kundu, B. (2015) *Chem. Eur. J.*, **21**, 18828–18833.
131 Kiruthika, S.E., Nandakumar, A., and Perumal, P.T. (2014) *Org. Lett.*, **16**, 4424–4427.
132 Kumar, A., Vachhani, D.D., Modha, S.G., Sharma, S.K., Parmar, V.S., and Van der Eycken, E.V. (2013) *Eur. J. Org. Chem.*, 2288–2292.
133 Kumar, A., Vachhani, D.D., Modha, S.G., Sharma, S.K., Parmar, V.S., and Van der Eycken, E.V. (2013) *Beilstein J. Org. Chem.*, **9**, 2097–2102.
134 Modha, S.G., Kumar, A., Vachhani, D.D., Jacobs, J., Sharma, S.K., Parmar, V.S., Van Meervelt, L., and Van der Eycken, E.V. (2012) *Angew. Chem., Int. Ed.*, **51**, 9572–9575.
135 Modha, S.G., Vachhani, D.D., Jacobs, J., Van Meervelt, L., and Van der Eycken, E.V. (2012) *Chem. Commun.*, **48**, 6550–6552.
136 Vachhani, D.D., Kumar, A., Modha, S.G., Sharma, S.K., Parmar, V.S., and Van der Eycken, E.V. (2015) *Synthesis*, **47**, 1337–1347.
137 Kumar, A., Li, Z., Sharma, S.K., Parmar, V.S., and Van der Eycken, E.V. (2013) *Chem. Commun.*, **49**, 6803–6805.
138 Hashmi, A.S.K., Yang, W., and Rominger, F. (2012) *Adv. Synth. Catal.*, **354**, 1273–1279.
139 Wang, M.-Z., Zhou, C.-Y., and Che, C.-M. (2011) *Chem. Commun.*, **47**, 1312–1314.
140 Huang, L., Yang, H.-B., Zhang, D.-H., Zhang, Z., Tang, X.-Y., Xu, Q., and Shi, M. (2013) *Angew. Chem., Int. Ed.*, **52**, 6767–6771.
141 Li, G. and Liu, Y. (2010) *J. Org. Chem.*, **75**, 3526–3528.

142 Zhang, Z., Tang, X., Xu, Q., and Shi, M. (2013) *Chem. Eur. J.*, **19**, 10625–10631.
143 Sanz, R., Miguel, D., and Rodríguez, F. (2008) *Angew. Chem., Int. Ed.*, **47**, 7354–7357.
144 Sanz, R., Miguel, D., Gohain, M., García-García, P., Fernández-Rodríguez, M.A., González-Pérez, A., Nieto-Faza, O., de Lera, Á.R., and Rodríguez, F. (2010) *Chem. Eur. J.*, **16**, 9818–9828.
145 Wang, C., Xie, X., Liu, J., Liu, Y., and Li, Y. (2015) *Chem. Eur. J.*, **21**, 559–564.
146 Hashmi, A.S.K., Rudolph, M., Huck, J., Frey, W., Bats, J.W., and Hamzić, M. (2009) *Angew. Chem., Int. Ed.*, **48**, 5848–5852.
147 Borsini, E., Broggini, G., Fasana, A., Baldassarri, C., Manzo, A.M., and Perboni, A.D. (2011) *Beilstein J. Org. Chem.*, **7**, 1468–1474.
148 Gruit, M., Michalik, D., Krüger, K., Spannenberg, A., Tillack, A., Pews-Davtyan, A., and Beller, M. (2010) *Tetrahedron*, **66**, 3341–3352.
149 Bhandari, M.R., Yousufuddin, M., and Lovely, C.J. (2011) *Org. Lett.*, **13**, 1382–1385.
150 Li, Y. and Lin, Z. (2015) *Organometallics*, **34**, 3538–3545.
151 Pan, B., Lu, X., Wang, C., Hu, Y., Wu, F., and Wan, B. (2014) *Org. Lett.*, **16**, 2244–2247.
152 Sarkar, S., Bera, K., Jalal, S., and Jana, U. (2013) *Eur. J. Org. Chem.*, 6055–6061.
153 Hack, D., Loh, C.C.J., Hartmann, J.M., Raabe, G., and Enders, D. (2014) *Chem. Eur. J.*, **20**, 3917–3921.
154 Hu, Y., Li, Y., Zhang, S., Li, C., Li, L., Zha, Z., and Wang, Z. (2015) *Org. Lett.*, **17**, 4018–4021.
155 (a) Modha, S.G., Kumar, A., Vachhani, D.D., Sharma, S.K., Parmar, V.S., and Van der Eycken, E.V. (2012) *Chem. Commun.*, **48**, 10916–10918; (b) Kumar, A., Vachhani, D.D., Modha, S.G., Sharma, S.K., Parmar, V.S., and Van der Eycken, E.V. (2013) *Synthesis*, **45**, 2571–2582.
156 Fang, R., Wei, X., and Yang, L. (2014) *Org. Biomol. Chem.*, **12**, 8433–8441.
157 Li, Z., Kumar, A., Sharma, S.K., Parmar, V.S., and Van der Eycken, E.V. (2015) *Tetrahedron*, **71**, 3333–3342.
158 Li, Z., Kumar, A., Vachhani, D.D., Sharma, S.K., Parmar, V.S., and Van der Eycken, E.V. (2014) *Eur. J. Org. Chem.*, 2084–2091.
159 Akiyama, R. and Kobayashi, S. (2002) *Angew. Chem., Int. Ed.*, **41**, 2602–2604.
160 Yamamoto, H., Sasaki, I., Imagawa, H., and Nishizawa, M. (2007) *Org. Lett.*, **9**, 1399–1402.
161 Hashmi, A.S.K., Frost, T.M., and Bats, J.W. (2000) *J. Am. Chem. Soc.*, **122**, 11553–11554.
162 Dong, Z., Liu, C.-H., Wang, Y., Lin, M., and Yu, Z.-X. (2013) *Angew. Chem. Int. Ed.*, **52**, 14157–14161.
163 Menon, R.S. and Banwell, M.G. (2010) *Org. Biomol. Chem.*, **8**, 5483–5485.
164 Hashmi, A.S.K. and Wölfle, M. (2009) *Tetrahedron*, **65**, 9021–9029.
165 (a) Grisé, C.M. and Barriault, L. (2006) *Org. Lett.*, **8**, 5905–5908; (b) Grisé, C.M., Rodrigue, E.M., and Barriault, L. (2008) *Tetrahedron*, **64**, 797–808.

166 Byers, P.M., Rashid, J.I., Mohamed, R.K., and Alabugin, I.V. (2012) *Org. Lett.*, **14**, 6032–6035.
167 Hashmi, A.S.K., Yang, W., and Rominger, F. (2011) *Angew. Chem., Int. Ed.*, **50**, 5762–5765.
168 Sun, N., Xie, X., and Liu, Y. (2014) *Chem. Eur. J.*, **20**, 7514–7519.
169 Wang, C., Chen, Y., Xie, X., Liu, J., and Liu, Y. (2012) *J. Org. Chem.*, **77**, 1915–1921.
170 Chen, Y., Wang, L., Sun, N., Xie, X., Zhou, X., Chen, H., Li, Y., and Liu, Y. (2014) *Chem. Eur. J.*, **20**, 12015–12019.
171 Chen, Y. and Liu, Y. (2011) *J. Org. Chem.*, **76**, 5274–5282.
172 Hashmi, A.S.K., Pankajakshan, S., Rudolph, M., Enns, E., Bander, T., Rominger, F., and Frey, W. (2009) *Adv. Synth. Catal.*, **351**, 2855–2875.
173 Chen, Y., Lu, Y., Li, G., and Liu, Y. (2009) *Org. Lett.*, **11**, 3838–3841.
174 Chen, Y., Li, G., and Liu, Y. (2011) *Adv. Synth. Catal.*, **353**, 392–400.
175 Echavarren, A.M., Méndez, M., Muñoz, M.P., Nevado, C., Martín-Matute, B., Nieto-Oberhuber, C., and Cárdenas, D.J. (2004) *Pure Appl. Chem.*, **76**, 453–463.
176 Rabaâ, H., Engels, B., Hupp, T., and Hashmi, A.S.K. (2007) *Int. J. Quant. Chem.*, **107**, 359–365.
177 (a) Hashmi, A.S.K., Rudolph, M., Weyrauch, J.P., Wölfle, M., Frey, W., and Bats, J.W. (2005) *Angew. Chem., Int. Ed.*, **44**, 2798–2801; (b) Hashmi, A.S.K., Kurpejović, E., Wölfle, M., Frey, W., and Bats, J.W. (2007) *Adv. Synth. Catal.*, **349**, 1743–1750; (c) Hashmi, A.S.K., Rudolph, M., Siehl, H.-U., Tanaka, M., Bats, J.W., and Frey, W. (2008) *Chem. Eur. J.*, **14**, 3703–3708; (d) For a comparison of Au(III) and Pt(II) catalysis in phenol synthesis see:Hashmi, A.S.K., Kurpejović, E., Frey, W., and Bats, J.W. (2007) *Tetrahedron*, **63**, 5879–5885.
178 Hashmi, A.S.K., Wölfle, M., Ata, F., Hamzic, M., Salathé, R., and Frey, W. (2006) *Adv. Synth. Catal.*, **348**, 2501–2508.
179 Kasperek, G.J., Bruice, P.Y., Bruice, T.C., Yagi, H., and Jerina, D.M. (1973) *J. Am. Chem. Soc.*, **95**, 6041–6046.
180 Rudolph, M., McCreery, M.Q., Frey, W., and Hashmi, A.S.K. (2011) *Beilstein J. Org. Chem.*, **7**, 794–801.
181 Hashmi, A.S.K., Bührle, M., Wölfle, M., Rudolph, M., Wieteck, M., Rominger, F., and Frey, W. (2010) *Chem. Eur. J.*, **16**, 9846–9854.
182 (a) Hashmi, A.S.K., Loos, A., Littmann, A., Braun, I., Knight, J., Doherty, S., and Rominger, F. (2009) *Adv. Synth. Catal.*, **351**, 576–582; (b) Hashmi, A.S.K., Hengst, T., Lothschütz, C., and Rominger, F. (2010) *Adv. Synth. Catal.*, **352**, 1315–1337; (c) Pažický, M., Loos, A., Ferreira, M.J., Serra, D., Vinokurov, N., Rominger, F., Jäkel, C., Hashmi, A.S.K., and Limbach, M. (2010) *Organometallics*, **29**, 4448–4458; (d) Chen, Y., Yan, W., Akhmedov, N.G., and Shi, X. (2010) *Org. Lett.*, **12**, 344–347; (e) Hashmi, A.S.K., Loos, A., Doherty, S., Knight, J.G., Robson, K.J., and Rominger, F. (2011) *Adv. Synth. Catal.*, **353**, 749–759; (f) Hashmi, A.S.K., Yu, Y., and Rominger, F. (2012) *Organometallics*, **31**, 895–904; (g) Cauteruccio, S., Loos, A., Bossi, A., Blanco Jaimes, M.C., Dova, D., Rominger, F., Prager, S., Dreuw, A., Licandro, E., and Hashmi, A.S.K. (2013) *Inorg. Chem.*, **52**, 7995–8004; (h) Manzano,

R., Rominger, F., and Hashmi, A.S.K. (2013) *Organometallics*, **32**, 2199–2203; (i) Hashmi, A.S.K., Bechem, B., Loos, A., Hamzic, M., Rominger, F., and Rabaa, H. (2014) *Aust. J. Chem.*, **67**, 481–499.

183 Muuronen, M., Perea-Buceta, J.E., Nieger, M., Patzsche, M., and Helaja, J. (2012) *Organometallics*, **31**, 4320–4330.

184 Hashmi, A.S.K., Hofmann, J., Shi, S., Schütz, A., Rudolph, M., Lothschütz, C., Wieteck, M., Bührle, M., Wölfle, M., and Rominger, F. (2013) *Chem. Eur. J.*, **19**, 382–389.

185 Hashmi, A.S.K., Weyrauch, J.P., Kurpejović, E., Frost, T.M., Miehlich, B., Frey, W., and Bats, J.W. (2006) *Chem. Eur. J.*, **12**, 5806–5814.

186 Hashmi, A.S.K. (2010) *Pure Appl. Chem.*, **82**, 1517–1528.

187 Hashmi, A.S.K., Wölfle, M., Teles, J.H., and Frey, W. (2007) *Synlett*, 1747–1752.

188 (a) Hashmi, A.S.K., Frost, T.M., and Bats, J.W. (2001) *Org. Lett.*, **3**, 3769–3771; (b) Hashmi, A.S.K., Weyrauch, J.P., Rudolph, M., and Kurpejović, E. (2004) *Angew. Chem., Int. Ed.*, **43**, 6545–6547; (c) Hashmi, A.S.K., Enns, E., Frost, T.M., Schäfer, S., Frey, W., and Rominger, F. (2008) *Synthesis*, 2707–2718; (d) Hashmi, A.S.K., Ghanbari, M., Rudolph, M., and Rominger, F. (2012) *Chem. Eur. J.*, **18**, 8113–8119; (e) Hashmi, A.S.K., Häffner, T., Yang, W., Pankajakshan, S., Schäfer, S., Schultes, L., Rominger, F., and Frey, W. (2012) *Chem. Eur. J.*, **18**, 10480–10486.

189 Hashmi, A.S.K., Rudolph, M., Bats, J.W., Frey, W., Rominger, F., and Oeser, T. (2008) *Chem. Eur. J.*, **14**, 6672–6678.

190 Hashmi, A.S.K., Blanco, M.C., Kurpejović, E., Frey, W., and Bats, J.W. (2006) *Adv. Synth. Catal.*, **348**, 709–713.

191 Hashmi, A.S.K., Salathé, R., and Frey, W. (2006) *Chem. Eur. J.*, **12**, 6991–6996.

192 Samanta, K., Kar, G.K., and Sarkar, A.K. (2012) *Tetrahedron Lett.*, **53**, 1376–1379.

193 Maurya, S.K. and Hotha, S. (2006) *Tetrahedron Lett.*, **47**, 3307–3310.

194 Hashmi, A.S.K., Ding, L., Bats, J.W., Fischer, P., and Frey, W. (2003) *Chem. Eur. J.*, **9**, 4339–4345.

195 Hashmi, A.S.K. and Grundl, L. (2005) *Tetrahedron*, **61**, 6231–6236.

196 Hashmi, A.S.K., Häffner, T., Rudolph, M., and Rominger, F. (2011) *Chem. Eur. J.*, **17**, 8195–8201.

197 Hashmi, A.S.K., Hamzić, M., Rudolph, M., Ackermann, M., and Rominger, F. (2009) *Adv. Synth. Catal.*, **351**, 2469–2481.

198 Hashmi, A.S.K., Hamzić, M., Rominger, F., and Bats, J.W. (2009) *Chem. Eur. J.*, **15**, 13318–13322.

199 Hashmi, A.S.K., Haufe, P., Schmid, C., Nass, A.R., and Frey, W. (2006) *Chem. Eur. J.*, **12**, 5376–5382.

200 (a) Hashmi, A.S.K., Ata, F., Kurpejovic, E., Huck, J., and Rudolph, M. (2007) *Top. Catal.*, **44**, 245–251; (b) Hashmi, A.S.K., Ata, F., Haufe, P., and Rominger, F. (2009) *Tetrahedron*, **65**, 1919–1927.

201 Hashmi, A.S.K., Schäfer, S., Bats, J.W., Frey, W., and Rominger, F. (2008) *Eur. J. Org. Chem.*, 4891–4899.

202 Reetz, M.T. and Sommer, K. (2003) *Eur. J. Org. Chem.*, 3485–3496.

203 Bhilare, S.V., Darvatkar, N.B., Deorukhar, A.R., Raut, D.G., Trivedi, G.K., and Salunkhe, M.M. (2009) *Tetrahedron Lett.*, **50**, 893–896.

204 Tubaro, C., Baron, M., Biffis, A., and Basato, M. (2013) *Beilstein J. Org. Chem.*, **9**, 246–253.

205 Bartolomé, C., García-Cuadrado, D., Ramiro, Z., and Espinet, P. (2010) *Organometallics*, **29**, 3589–3592.

206 El-Hellani, A., Bour, C., and Gandon, V. (2011) *Adv. Synth. Catal.*, **353**, 1865–1870.

207 García-Rodeja, Y. and Fernández, I. (2017) *Organometallics.*, **36**, 460–466.

208 Nösel, P., Müller, V., Mader, S., Moghimi, S., Rudolph, M., Braun, I., Rominger, F., and Hashmi, A.S.K. (2015) *Adv. Synth. Catal.*, **357**, 500–506.

209 Li, Z., Shi, Z., and He, C. (2005) *J. Organomet. Chem.*, **690**, 5049–5054.

210 Srivastava, A., Patel, S.S., Chandna, N., and Jain, N. (2016) *J. Org. Chem.*, **81**, 11664–11670.

211 Luo, C., Yang, H., Mao, R., Lu, C., and Cheng, G. (2015) *New J. Chem.*, **39**, 3417–3423.

212 Mehrabi, T. and Ariafard, A. (2016) *Chem. Commun.*, **52**, 9422–9425.

213 Xie, M.-H., Xie, F.-D., Lin, G.-F., and Zhang, J.-H. (2010) *Tetrahedron Lett.*, **51**, 1213–1215.

214 Pirovano, V., Negrato, M., Abbiati, G., Dell'Acqua, M., and Rossi, E. (2016) *Org. Lett.*, **18**, 4798–4801.

215 Matsuda, Y., Naoe, S., Oishi, S., Fujii, N., and Ohno, H. (2015) *Chem. Eur. J.*, **21**, 1463–1467.

216 Samala, S., Mandadapu, A.K., Saifuddin, M., and Kundu, B. (2013) *J. Org. Chem.*, **78**, 6769–6774.

217 Li, X., Xie, X., and Liu, Y. (2016) *J. Org. Chem.*, **81**, 3688–3699.

218 Hashmi, A.S.K. and Blanco, M.C. (2006) *Eur. J. Org. Chem.*, 4340–4342.

219 Huguet, N., Lebœuf, D., and Echavarren, A.M. (2013) *Chem. Eur. J.*, **19**, 6581–6585.

220 Zeiler, A., Ziegler, M.J., Rudolph, M., Rominger, F., and Hashmi, A.S.K. (2015) *Adv. Synth. Catal.*, **357**, 1507–1514.

221 Carrettin, S., Blanco, M.C., Corma, A., and Hashmi, A.S.K. (2006) *Adv. Synth. Catal.*, **348**, 1283–1288.

222 (a) Carrettin, S., Guzman, J., and Corma, A. (2005) *Angew. Chem., Int. Ed.*, **44**, 2242–2245; (b) Guzman, J., Carrettin, S., and Corma, A. (2005) *J. Am. Chem. Soc.*, **127**, 3286–3287.

223 Ganai, A.K., Bhardwaj, R., Hotha, S., Gupta, S.S., and Prasad, B.L.V. (2010) *New J. Chem.*, **34**, 2662–2670.

224 Cao, W. and Yu, B. (2011) *Adv. Synth. Catal.*, **353**, 1903–1907.

225 Yang, M., Chen, Z., Luo, Y., Znang, J., Tang, D., He, R., Shen, W., and Li, M. (2016) *RSC Adv.*, **6**, 22709–22721.

226 Tang, D., Yang, M., Zhang, J., He, R., Shen, W., and Li, M. (2016) *Chem. Cat. Chem.*, **8**, 461–470.

227 Yang, M., Chen, Z., Luo, Y., Zhang, J., He, R., Shen, W., Tang, D., and Li, M. (2016) *Chem. Cat. Chem.*, **8**, 2367–2375.

228 Efe, C., Lykakis, I.N., and Stratakis, M. (2011) *Chem. Commun.*, **47**, 803–805.

229 Tang, D., Chen, Z., Zhang, J., Tang, Y., and Xu, Z. (2014) *Organometallics*, **33**, 6633–6642.

230 Symeonidis, T.S., Lykakis, I.N., and Litinas, K.E. (2013) *Tetrahedron*, **69**, 4612–4616.

231 Schröder, F., Ojeda, M., Erdmann, N., Jacobs, J., Luque, R., Noël, T., Van Meervelt, L., Van der Eycken, J., and Van der Eycken, E.V. (2015) *Green Chem.*, **17**, 3314–3318.

232 Clarke, A.K., James, M.J., O'Brien, P., Taylor, R.J.K., and Unsworth, W.P. (2016) *Angew. Chem. Int. Ed.*, **55**, 13798–13802.

8

Catalytic Alkyne Hydroarylation Using Arylboron Reagents, Aryl Halides, and Congeners

Yoshihiko Yamamoto

Nagoya University, Department of Basic Medicinal Sciences, Graduate School of Pharmaceutical Sciences, Furo-cho, Chikusa, Nagoya 464-8601, Japan

8.1 Introduction

The addition of organometallic reagents across the multiple bonds of unsaturated organic compounds is known as carbometallation, which enables access to organometallic species with higher molecular complexity [1, 2]. Among them, carbometallation of alkynes is of particular importance: it proceeds in syn fashion to produce valuable vinylmetal species, which can be trapped by electrophiles to stereoselectively afford tri- or tetra-substituted alkenes (Scheme 8.1). Because the stereoselective synthesis of highly substituted alkenes is a formidable challenge [3, 4], carbometallation of alkynes has played a significant role in organic synthesis [5].

Alkenylation of aromatic compounds has been a central subject for research because arylalkenes are found not only in natural products but also in pharmaceuticals and functional materials, and also because they are used as versatile starting materials [6]. Therefore, transition-metal-catalyzed addition of arene C—H bonds across the triple bonds of alkynes, hydroarylation of alkynes, has been extensively investigated as a straightforward method to synthesize arylalkenes [7–11]. However, harsh reaction conditions and/or unsatisfactory regio- and stereoselectivity are often disadvantages encountered in alkyne hydroarylations. Using arylmetal reagents for carbometallation and subsequent protonation of the resultant vinylmetal intermediates also produces arylalkenes. Thus, this formal hydroarylation protocol provides complementary access to arylalkenes with controllable regio- and stereoselectivity under mild reaction conditions (Scheme 8.2).

Alkyne hydroarylation via carbometallation/protonation has been underdeveloped compared to direct hydroarylation reactions because common organometallic reagents such as organolithium, organomagnesium, and organozinc reagents are generally unreactive toward unactivated alkynes except for those bearing a directing group such as a hydroxyl group [12]. Thus, organocopper reagents have been mainly used for this purpose [13], and, later,

Catalytic Hydroarylation of Carbon-Carbon Multiple Bonds, First Edition.
Edited by Lutz Ackermann, T. Brent Gunnoe and Laurel Goj Habgood.

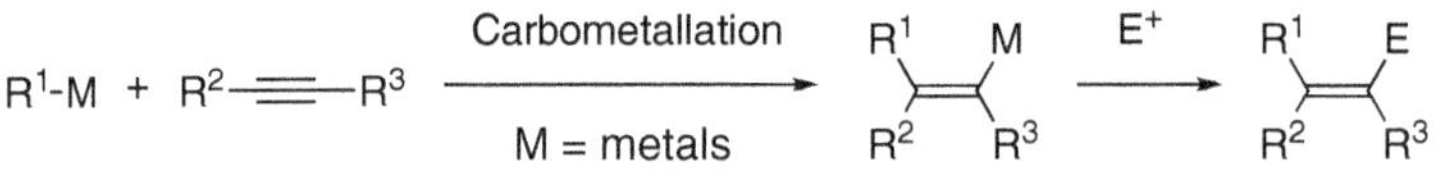

Scheme 8.1 Carbometallation of alkynes and subsequent reaction with electrophiles.

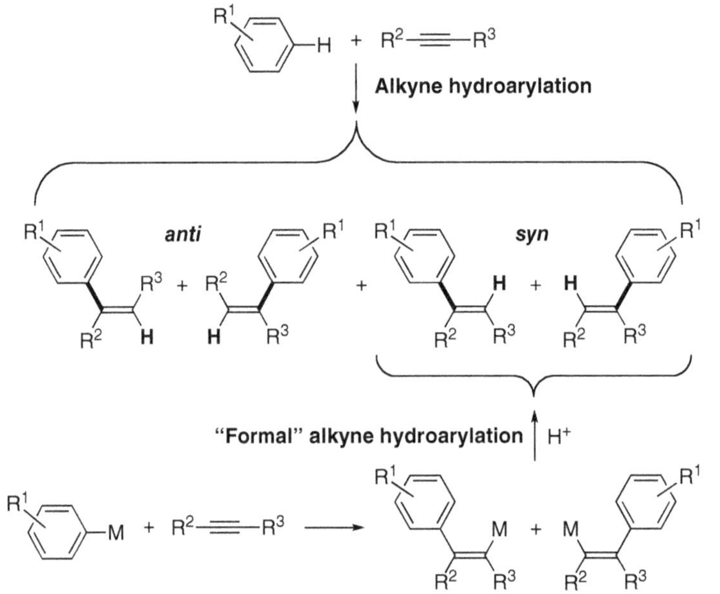

Scheme 8.2 Alkyne hydroarylation and "formal" alkyne hydroarylation via carbometallation/protonation.

transition-metal catalysts have been identified to promote carbometallation of alkynes using organolithium, organomagnesium, and organozinc reagents [14, 15]. However, the use of pyrophoric organometallic reagents in stoichiometric amounts is notable in the carbometallation approach. In addition, the lack of functional-group compatibility when using highly nucleophilic organometallic reagents severely limits the scope of this method. Therefore, alternative carbometallation approaches that utilize bench-top stable and readily accessible arylmetal surrogates are highly desirable for synthetic applications to functional molecules. In this vein, arylboron reagents fit this requirement: they are not moisture sensitive and can be handled in the air [16, 17]. Arylboron reagents have been widely used for palladium-catalyzed cross-coupling with aryl halides to produce biaryls (Suzuki–Miyaura coupling) in both industrial and laboratory settings [18, 19]. Therefore, diverse arylboron reagents are now commercially available. Alkyne hydroarylation using arylboron reagents was actually realized in the pioneering studies independently reported by the research groups of Hayashi and Shirakawa in 2001 [20, 21]. After these seminal reports, various combinations of transition-metal catalysts, arylboron reagents, and alkyne substrates have been investigated to expand the scope of alkyne hydroarylation via carbometallation to a large extent.

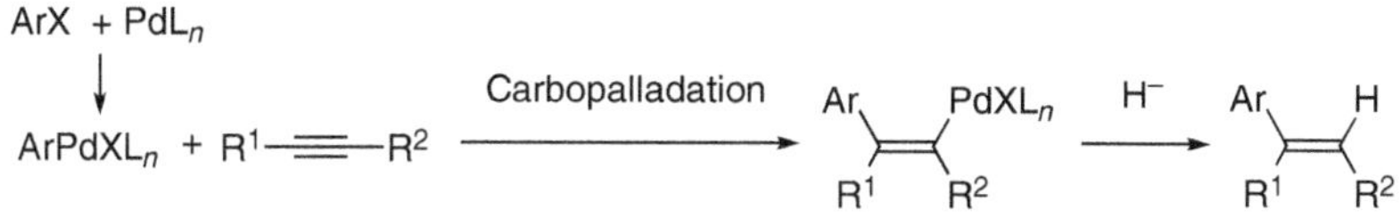

Scheme 8.3 Carbopalladation of alkynes and subsequent reaction with hydride.

Another approach to arylalkenes via formal alkyne hydroarylation is the palladium-catalyzed reaction of aryl iodides under modified Mizoroki–Heck reaction conditions (Scheme 8.3). Cacchi and co-workers performed the palladium-catalyzed reaction of aryl iodides and alkynes in the presence of formic acid as a hydride donor to obtain hydroarylation products [22]. Although, similarly to Mizoroki–Heck reaction, this reaction proceeds via carbopalladation, the resultant vinylpalladium species undergoes reductive elimination with hydride derived from formic acid (see Section 8.3). Such a formal addition of an aryl fragment and hydride across alkyne triple bonds is called reductive Heck reaction. An intramolecular variant of this reaction, reductive Heck cyclization, was also developed by Grigg and co-workers [23]. Thereafter, various reductive Heck protocols have been successfully implemented to synthesize multi-substituted arylalkenes.

Despite differences in aryl precursors involved and the underlying reaction mechanisms, alkyne hydroarylation using arylboron reagents and reductive Heck reactions shares the scope of alkyne substrates and applicability in organic synthesis. Thus, the aim of this chapter is to provide an overview of alkyne hydroarylations using arylborons, aryl halides, and their congeners. In the following sections, the scope and limitations of these two formal alkyne hydroarylations will be discussed, and then, sequential processes consisting of alkyne hydroarylation and subsequent cyclization as well as synthetic applications to bioactive compounds and natural products will be outlined. A review of reductive Heck reactions was published by Cacchi and co-workers [24]. The sequential processes leading to heterocycles were also reviewed [25, 26]. A review covering heterocycles synthesis using alkyne hydroarylations was also published recently [27].

8.2 Catalyzed Alkyne Hydroarylations Using Arylboron and Arylsilicon Reagents

In this section, the scope, regio- and stereoselectivity, experimental conditions, and underlying mechanisms of transition-metal-catalyzed alkyne hydroarylation using arylboron reagents will be discussed. Related alkyne hydroarylation reactions using arylsilanes will also be included, although it has been scarcely investigated so far. The most widely used transition-metal elements for this purpose are rhodium and palladium, and less expensive and more abundant first-low elements such as nickel, cobalt, and copper are also employed for effective catalysts. In the following subsections, rhodium- and palladium-catalyzed reactions will be overviewed followed by the survey of other protocols using first-low elements.

8.2.1 Rhodium-Catalyzed Reactions

Rhodium-catalyzed 1,4-addition of arylboronic acids to activated alkenes was developed by Miyaura and co-workers [28], and later enantioselective protocols were developed by the Hayashi group and others [29]. Hayashi and co-workers extended rhodium-catalyzed addition of arylboronic acids to internal alkynes in 2001 [20]. They used a rhodium catalyst, which is produced *in situ* from 3 mol% Rh(acac)(C_2H_4)$_2$ (acac = acetylacetonate) and 6.6 mol% 1,4-bis(diphenylphosphino)butane (dppb), in dioxane/H_2O (10:1). The reaction of symmetrical internal alkynes **1** with arylboronic acids **2** (5 equiv.) proceeded at 100 °C for 3 h to stereoselectively afford syn hydroarylation products **3** in 89–96% yields (Scheme 8.4). As ligands, 1,2-bis(diphenylphosphino)ethane (dppe), 2,2′-bis(diphenylphosphino)-1,1′-binaphthyl (binap), and 1,1′-bis(diphenyl phosphino)ferrocene (dppf) could be used in place of dppb. Lower amounts of arylboronic acids decreased the yield. Because *p*-anisylboronic acid readily underwent protodeboration under the reaction conditions, the corresponding boroxine **4** was used at 60 °C to afford **3** in 97% yield. The reaction of unsymmetrical alkyne **5** with phenylboronic acid resulted in a 3:1 mixture of regioisomers **6** and **6′**, albeit in a high yield. In contrast, unsymmetrical alkynes **7** bearing electron-withdrawing groups (EWGs) regioselectively produced the corresponding products **8**, in which the phenyl group was introduced at the β position to the EWGs.

The vinylic protons of hydroarylation products **3** were assumed to be derived from H_2O via protonolysis of putative vinylrhodium intermediates. However,

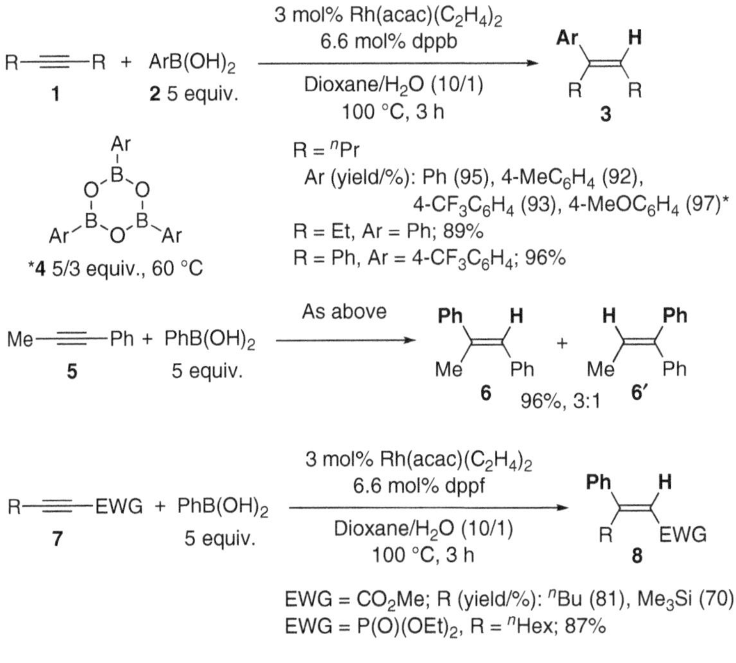

Scheme 8.4 Rh-catalyzed hydroarylation of alkynes using arylboron reagents.

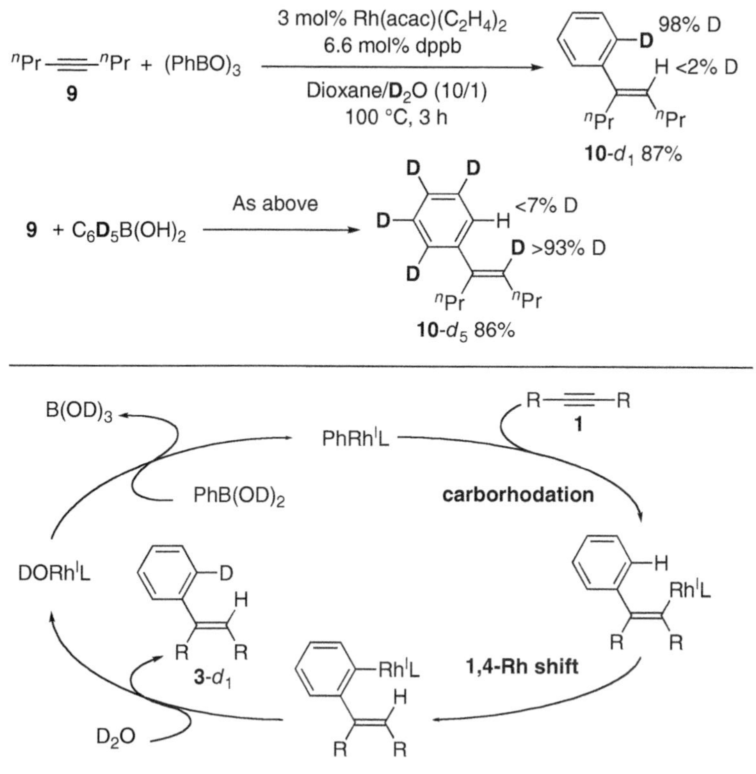

Scheme 8.5 Deuterium-labeling experiments and proposed mechanism for Rh-catalyzed hydroarylation of alkynes using arylboronic acids.

deuterium-labeling experiments shown in Scheme 8.5 revealed that the vinylic proton was transferred from the introduced phenyl group: the reaction of **9** with triphenylboroxine, $(PhBO)_3$, in the presence of D_2O, produced **10-d_1**. On the other hand, the same reaction using $C_6D_5B(OH)_2$ afforded **10-d_5**. Based on these observations, a catalytic cycle involving a 1,4-rhodium shift was proposed. The 1,4-rhodium shift generates a reactive arylrhodium species, which can be involved in further reactions. Thus, sequential processes initiated by alkyne hydroarylation have been developed (see Section 8.4.1.3).

Lautens and Yoshida discovered that internal alkynes bearing aza-heteroaryl groups underwent regio- and stereoselective hydroarylation under aqueous conditions [30, 31]. In the presence of 2 mol% $[RhCl(cod)]_2$ (cod = 1,5-cyclooctadiene), 8 mol% water-soluble pyridylphosphine ligand **L1**, sodium dodecyl sulfate (SDS) as a surfactant, and Na_2CO_3, 2-alkynylpyridines **11** reacted with arylboronic acids **2** in H_2O at 80 °C for 1–3 h to afford hydroarylation products **12** in 51–86% yields (Scheme 8.6). The aryl groups were introduced at the β position to the 2-pyridyl group. Arylboronic acids bearing ortho substituents gave higher yields, and arylboronic acids bearing 4-methoxy groups required increased catalyst loadings. The use of water-soluble ligand **L2** without a pyridyl group instead of **L1** significantly decreased reaction efficiency,

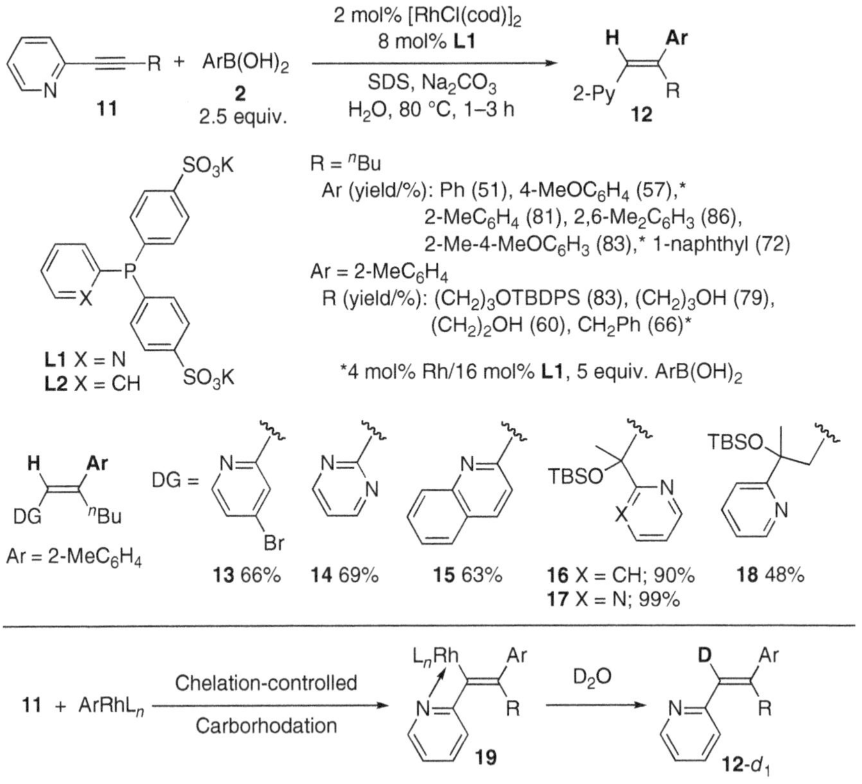

Scheme 8.6 Rh-catalyzed hydroarylation of aza-heteroaryl-substituted alkynes using arylboronic acids.

and no reaction occurred when 3- or 4-alkynylpyridines were used as substrates. Protected or unprotected hydroxyl and phenyl groups on the alkyl substituents R were tolerated, although these groups at the propargylic position inhibited the reaction. As aza-heterocyclic groups, 4-bromo-2-pyridyl (**13**), 2-pyrimidyl (**14**), and 2-quinolinyl (**15**) groups were also effective. Moreover, the insertion of one quaternary carbon between the alkyne and 2-pyridyl or 2-pyrimidyl groups (**16** and **17**) gave higher yields, while the further elongation of the tether connecting the alkyne and the 2-pyridyl group (**18**) decreased the yield. According to these observations, it was concluded that hydroarylation of aza-heteroaryl-substituted alkynes proceeds via chelation-controlled carborhodation: the aza-heteroaryl groups function as directing groups (DGs) that control regioselectivity to favorably generate vinylrhodium intermediates **19**. A reaction performed in D_2O afforded hydroarylation product **12-d_1** with the vinylic carbon adjacent to the 2-pyridyl group being deuterated, supporting the involvement of the proposed vinylrhodium intermediate **19**.

The Genêt group demonstrated that a rhodium catalyst derived from 1.5 mol% $[RhOH(cod)]_2$ and 6 mol% water-soluble ligand **L3** was highly effective for hydroarylation of symmetrical alkyne **9** using a wide range of arylboronic acids **2**

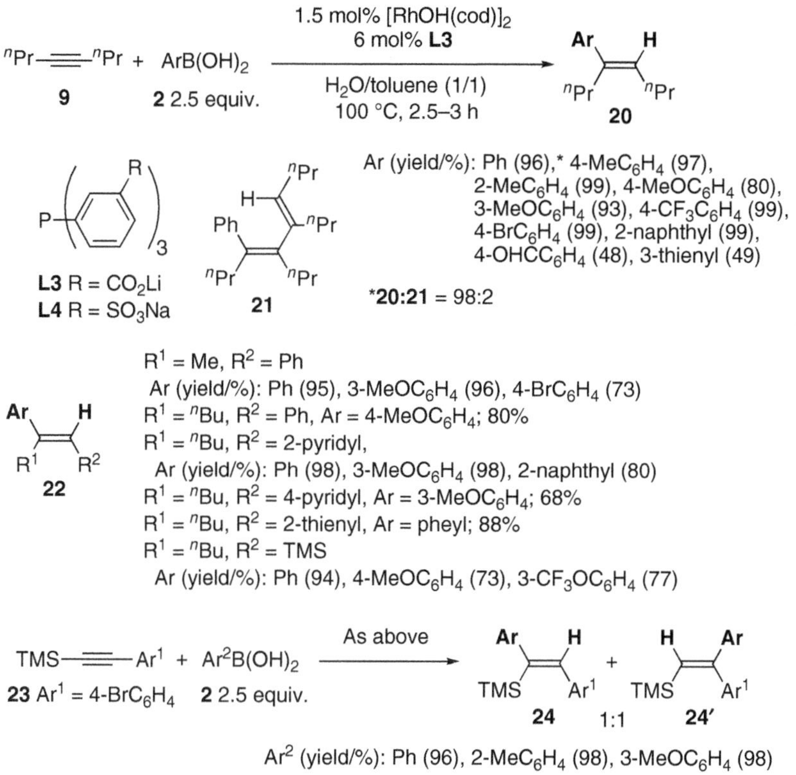

Scheme 8.7 Rh-catalyzed hydroarylation of alkynes using arylboronic acids under biphasic conditions.

under biphasic conditions in H_2O/toluene (1:1) (Scheme 8.7) [32, 33]. The reaction could be performed at 100 °C without a surfactant and a base to afford hydroarylation products **20** in high yields, except for those from 4-formylphenyl and 3-thienylboronic acids. Small amounts of **21,** which were derived from two molecules of alkyne **9** and phenylboronic acid, was also observed. In contrast, product selectivity was considerably lowered to **20:21** = 85:15 when **L4** was used as a ligand. Although Hayashi's study observed that unsymmetrical alkyne **5** bearing methyl and phenyl terminal groups gave a mixture of regioisomers [20], similar phenylalkynes could be regioselectively converted into 1,2-diarylalkenes **22** (R^2 = Ph) in good yields. Moreover, unsymmetrical alkynes bearing heteroaryl or trimethylsilyl (TMS) terminal groups (R^2) also afforded the corresponding products **22** as single regioisomers. These results indicate that the (hetero)aryl and TMS groups functioned as directing groups. In contrast to the reports of Lautens and Yoshida, a 4-pyridyl group also showed a similar directing effect. However, when unsymmetrical alkyne **23** possessed both TMS and 4-bromophenyl terminal groups, regioselectivity was totally lost, affording 1:1 mixtures of regioisomers **24** and **24′**. The additional advantage of this biphasic system is the recyclability of the expensive rhodium catalyst: the reaction of **9** with phenylboronic acid could be repeated at least four times without notable loss of efficiency and selectivity.

As discussed earlier, the regioselectivity of rhodium-catalyzed hydroarylation of unsymmetrical alkynes depends on the substrate structure as well as on reaction conditions. Chelation-controlled regioselectivity was also reported by several groups (Scheme 8.8). Moderate regioselectivity in favor of **26** bearing a phenyl group at the β carbon to the alcohol substituent was observed when hydroarylation of propargylic alcohol derivative **25** (R = OH) was performed using a rhodium catalyst, $[RhCl(cod)]_2/P(OEt)_3$, in alcohol solvents at 70 °C [34]. The regioselectivity was diminished by replacing the hydroxyl group with a methoxy group. Unsymmetrical alkyne **27** possessing a *p*-toluenesulfonyl (Ts) group at the propargylic position also underwent regioselective hydroarylation in the presence of a rhodium catalyst, $[RhOH(cod)]_2$/binap, in dioxane at 75 °C, affording allylsulfone **28** as predominant regioisomers with the ratio of **28**:**28′** = 9:1∼10:1 [35]. Furthermore, regioselective hydroarylation of ynamides **29** using arylboronic acids **2** was achieved by Lam and co-workers [36]. They

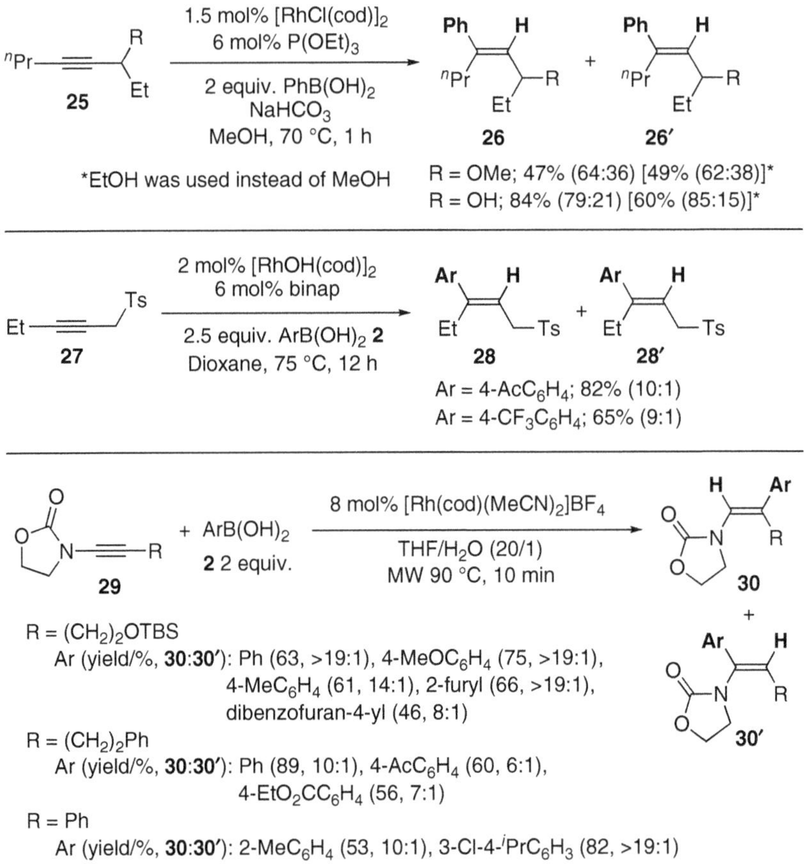

Scheme 8.8 Rh-catalyzed hydroarylation of propargylic alcohol, propargyl sulfone, and ynamides using arylboronic acids.

used a cationic rhodium complex, $[Rh(cod)(MeCN)_2]BF_4$, as a catalyst in THF/H_2O (20/1) under microwave (MW) irradiation conditions at 90 °C. As a result, hydroarylation products **30** and **30′** were obtained in 46–89% combined yields with regioselectivity ranging from 6:1 to >19:1. In the major regiosiomers **30,** the aryl groups were introduced to the β carbon to the oxazolidin-2-one moiety, which is assumed to form a chelate with a cationic arylrhodium species.

The electronic directing effect of aryl terminal groups overrides the chelation-controlled directing effect of alcohols and amines (Scheme 8.9). As disclosed by Lautens and co-workers, hydroarylation of propargylic alcohols **31** bearing aryl terminal groups with phenylboronic acid (2 equiv.) was performed using $[RhOH(cod)]_2$/binap as a catalyst in dioxane/H_2O (10:1) at 60 °C to afford **32** as major products in good yields with regioselectivity ranging

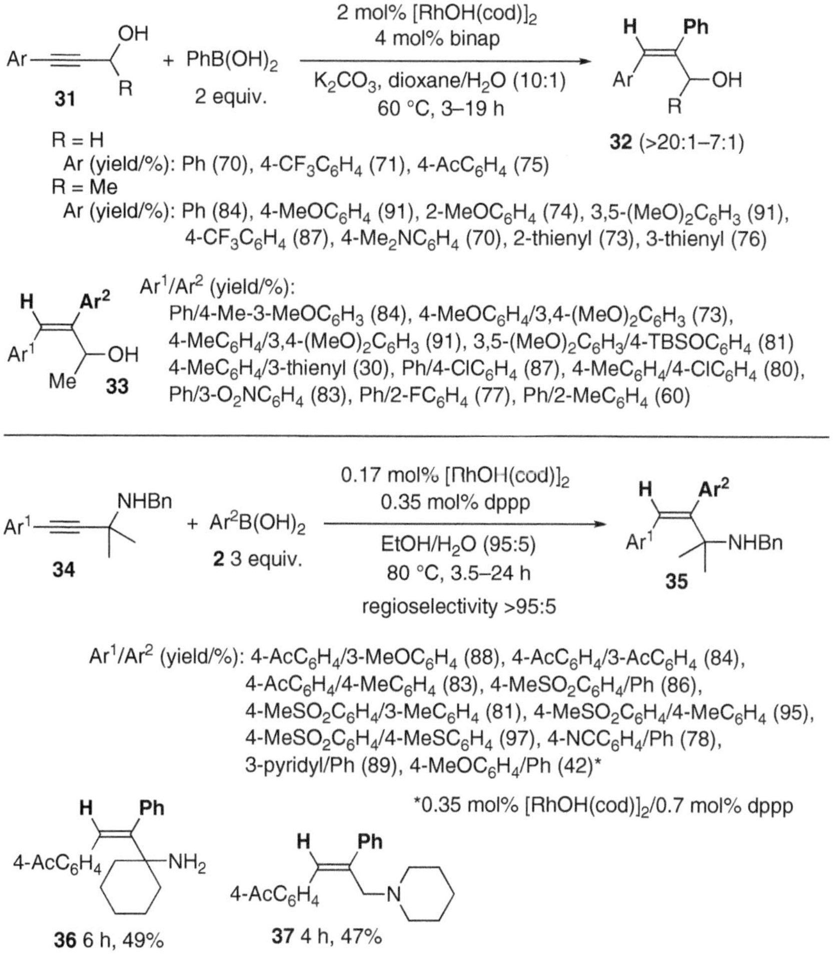

Scheme 8.9 Rh-catalyzed hydroarylation of propargylic alcohols and amines using arylboronic acids.

from 7:1 to >20:1 [37]. The newly introduced phenyl group was placed on the α carbon to the hydroxymethyl substituents. Hydroarylation was inefficient when the corresponding methyl ether was used as a substrate. The same directing effect was also observed for a homopropargylic alcohol substrate bearing a phenyl terminal group, albeit with diminished efficiency, although no reaction occurred in the absence of the hydroxyl group. The scope of this protocol was extensively investigated using diverse propargyl alcohol substrates and arylboronic acids **2**. Hydroarylation products **33** were obtained in 30–91% yields with similar regioselectivity.

Marinelli and co-workers implemented hydroarylation of secondary propargylic amines **34** bearing aryl terminal groups including a 3-pyridyl group using the same rhodium complex with 1,3-bis(diphenylphosphino)propane (dppp) as a ligand in aqueous ethanol at 80 °C [38]. When the terminal aryl groups have electron-withdrawing substituents, the corresponding products **35** are obtained in high yields with excellent regioselectivity (>95:5). However, the reaction of an electron-rich *p*-anisyl-substituted substrate was much less efficient. Similar products **36** and **37** possessing primary and tertiary amines, respectively, were also obtained, albeit in moderate yields.

Phosphine-free conditions using $Rh(acac)(CO)_2$ in toluene/H_2O at 110 °C [39] and a recyclable catalyst system using $Rh(acac)(CO)_2$ in combination with a polymer-supported bisphosphine ligand [40] were also applied to alkyne hydroarylation using arylboronic acids.

Despites recent great progresses in the Hiyama–Denmark cross-coupling of organosilanes [41], the development of alkyne hydroarylation using arylsilanes as arylating agents has lagged behind. Different types of arylsilanes were used for rhodium-catalyzed alkyne hydroarylation as outlined in Scheme 8.10. Mori and co-workers investigated the reaction of symmetrical alkyne **9** with aryl(ethyl)silanediol **38** using $[RhOH(cod)]_2$ as a catalyst without a phosphine ligand [42]. Syn hydroarylation proceeded in toluene/H_2O (10/1) at 100 °C to afford trisubstituted alkene **20** in 65–73% yields. They also used an arylstannane in the presence of a phenol additive to obtain a similar result. The use of unsymmetrical alkyne **39** bearing phenyl and TMS terminal groups resulted in the almost unselective formation of regioisomers **40** and **40′**. Nakao, Hiyama, and co-workers applied their aryl[2-(hydroxymethyl)phenyl]dimethylsilanes **41** to rhodium-catalyzed alkyne hydroarylation [43]. They performed hydroarylation of **9** under unhydrous conditions using $[RhOH(cod)]_2$ in combination with 1,2-bis(diphenylphosphino)benzene (dppBz) as a ligand at 100 °C in toluene to obtain **20** in 26–85% yields. The yield was higher for electron-rich arylsilanes. These silanes undergo transmetallation via a four-centered transition state (**42**), and the resultant silicon residue **43** can be reused [44]. Inexpensive aryl(trimethoxy)silanes **45** also could be used as demonstrated by Cheng and co-workers [45]. They developed a Rh/Cu binary catalyst system that works in air. Triarylalkenes **46** were obtained in moderate to high yields from symmetrical diarylalkynes **44**. Two examples using unsymmetrical phenylalkyne substrates were also described to give **47** with regioselectivity of 88:12–95:5.

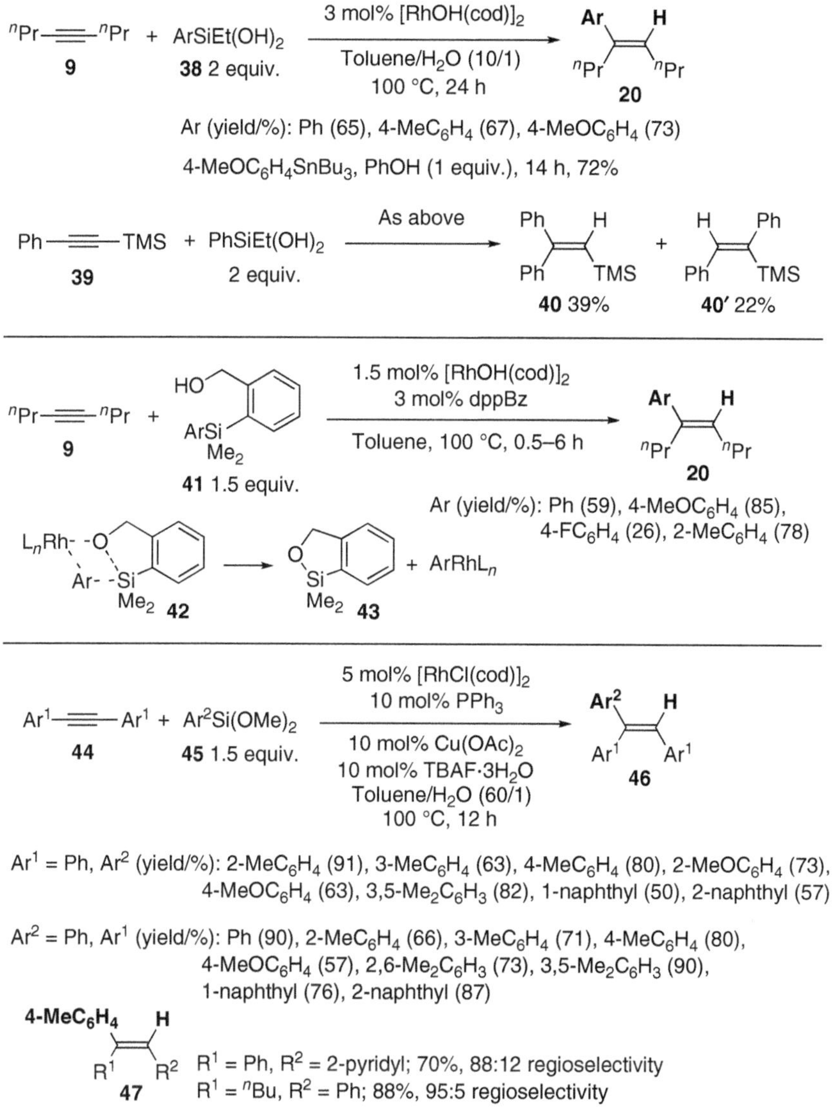

Scheme 8.10 Rh-catalyzed hydroarylation of alkynes using arylsilanes.

8.2.2 Palladium-Catalyzed Reactions

Palladium catalysts have been widely used for the cross-coupling reaction of organoboron reagents (Suzuki–Miyaura coupling) [18, 19]. Therefore, palladium-catalyzed alkyne hydroarylation using arylboron reagents has been actively investigated so far. The first example reported by Oh and co-workers in 2003 is outlined in Scheme 8.11 [46]. They performed the reaction of terminal alkynes **48** with arylboronic acids **2** in the presence of 3 mol%

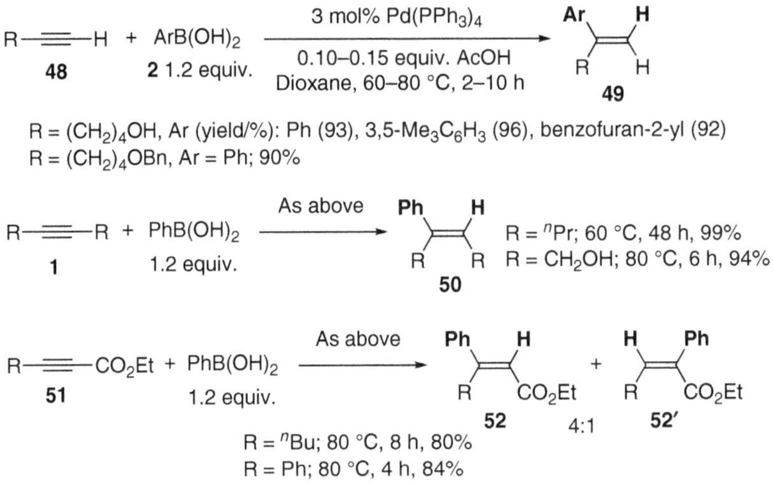

Scheme 8.11 Pd-catalyzed hydroarylation of alkynes using arylboronic acids.

$Pd(PPh_3)_4$ and acetic acid (0.10–0.15 equiv.) in dioxane at 60–80 °C. As a result, 1,1-disubstituted alkenes **49** were selectively obtained in high yields. Symmetrical internal alkynes **1** also underwent hydroarylation to afford trisubstituted alkenes **50** in high yields. When unsymmetrical propiolates **51** were used as the substrates, cinnamate derivatives **52** and regioisomers **52′** were formed with the ratio of **52**:**52′** = 4:1.

Hydroarylation of terminal alkynes **48** was repeated with excess AcOD, resulting in the formation of (*Z*)-**53**-d_1 and (*E*)-**53**-d_1 in favor of the (Z) isomer (Scheme 8.12). Under the same conditions except for the absence of phenylboronic acid, **48**-d_1 was observed with 50% deuterium incorporation. On the other hand, **48**-d_1 was subjected to the standard hydroarylation conditions to exclusively afford (*E*)-**53**-d_1. These deuterium-labeling experiments suggest that the vinyl proton introduced via syn hydroarylation comes from acetic acid and/or the arylboronic acids. Accordingly, a plausible catalytic cycle involving hydropalladation of alkynes with $HPd^{II}OAc$ species and subsequent transmetallation/reductive elimination was proposed as shown in Scheme 8.12.

The palladium-catalyzed regioselective hydroarylation of propiolates **51** was reinvestigated [46], and reversal of regioselectivity, depending on reaction conditions, was found (Scheme 8.13) [47]. Regiosiomers **54**, which have the aryl groups at the β carbon to the ethoxycarbonyl groups, were major products when a new catalyst system consisting of $Pd(OAc)_2$ and dppe was used in dioxane or chloroform. However, a bulky *tert*-butyl terminal group of **51** (R = ^{t}Bu) favored the contrasting regioisomer **54′** with the high regioisomer ratio of **54**:**54′** = 1:50. In contrast, the use of bulky and electron-donating P^tBu_3 as a ligand in THF produced **54′** as major products with the ratio of **54**:**54′** ranging from 1:2 to 1:50. Later, Ogilvie and co-workers reported that compounds **54** were exclusively formed under different catalytic conditions [48]. They used PEt_3 or P^tBu_2Me as ligands under basic conditions using aqueous Cs_2CO_3 at 23 °C in toluene. Diverse

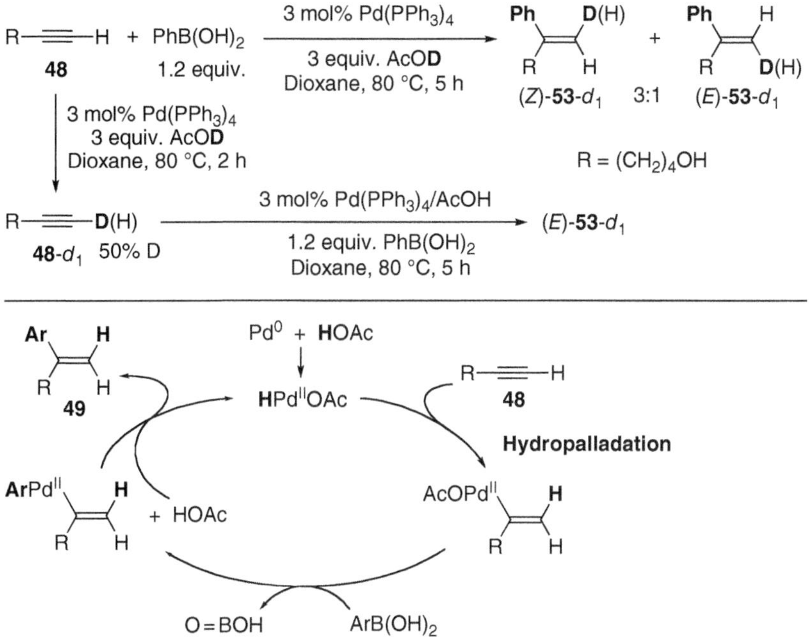

Scheme 8.12 Deuterium-labeling experiments and proposed mechanism for Pd-catalyzed hydroarylation of terminal alkynes using arylboronic acids.

cinnamate derivatives **54** were obtained in 49–85% yields. Weinreb amides **55** could also be used for the synthesis of acrylic amides **56.**

The effect of pendant hydroxyl groups and amines on regioselectivity was investigated by Oh, Marinelli, and co-workers as outlined in Scheme 8.14. When propargylic alcohol substrates **57** ($n = 1$) were subjected to the previously optimized conditions, the regioisomer ratio was altered depending on the bulkiness of the terminal groups R from **58:58′** = 100:0 (R = Me) to **58:58′** = 0:100 (R = ^{t}Bu) [49]. The regioselectivity decreased as the distance between the hydroxyl groups and the alkyne moieties increased. To explain the directing effect of the hydroxyl group observed for the substrates bearing small terminal groups R, the oxidative addition of arylboronic acids coordinated by alcohols **57** to Pd^0 to generate arylpalladium intermediates **59** was assumed to occur, followed by regioselective carbopalladation to afford vinylpalladium intermediates **60.** In contrast to alkynyl alcohols **57,** propargylic acetates **61** afforded regioisomers **62** and **62′** with the constant ratio of **62:62′** = 5:1. A similar directing effect was observed for hydroarylation of propargylic amines **35** [38]. Hydroarylation of **35** with arylboronic acids **2** (2 equiv.) was performed using $Pd(OAc)_2$ in combination with a bulky and electron-donating ligand, PCy_3, and AcOH in EtOH at 80 °C, affording secondary allylic amines **63,** regioisomers that were in contrast to those obtained via the rhodium-catalyzed hydroarylation of the same substrates (Scheme 8.9). The regioselectivity was determined to be >96:4. Primary and tertiary allylic amines **64** and **65** were also obtained in 54%

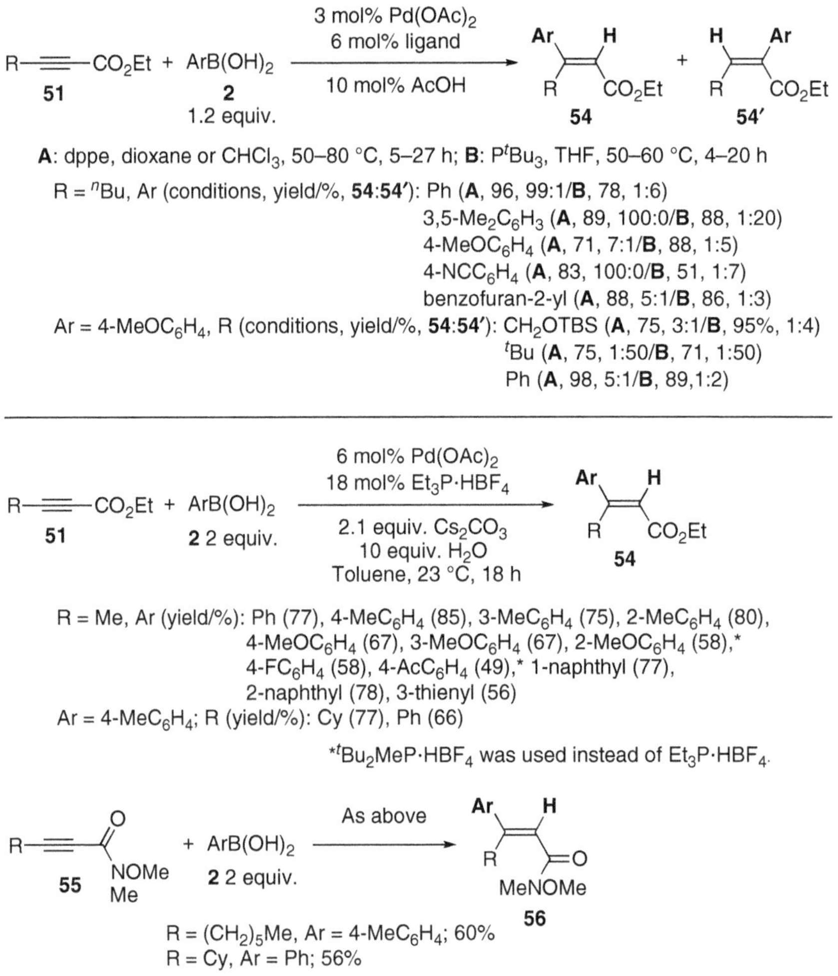

Scheme 8.13 Pd-catalyzed hydroarylation of propiolates and propiolamides using arylboronic acids.

and 78% yields, respectively. The origin of the high regioselectivity was ascribed to the electronic effect of the aryl terminal groups.

Heteroatom-substituted alkynes have been investigated as substrates for palladium-catalyzed hydroarylation. Aryloxyalkynes **66** were subjected to hydroarylation using $Pd(OAc)_2/2PPh_3$ as a catalyst in the presence of Na_2CO_3 in dioxane at 60 °C (Scheme 8.15) [50]. As a result, arylated vinyl ethers **67** were obtained in moderate to high yields with regioselectivity in favor of arylation at the α carbon to the alkoxy group. Diverse electron-rich and electron-deficient arylboronic acids could be used and substituted aryl, branched alkyl, and allyl terminal groups were tolerated. In addition to the aryloxyalkynes, ethoxyalkynes exhibited comparable efficiency.

Similarly, ynamides **29** were efficient substrates under similar conditions except for using P(*m*-tolyl)$_3$, which is a bulkier ligand than PPh_3 (Scheme 8.15) [51]. The

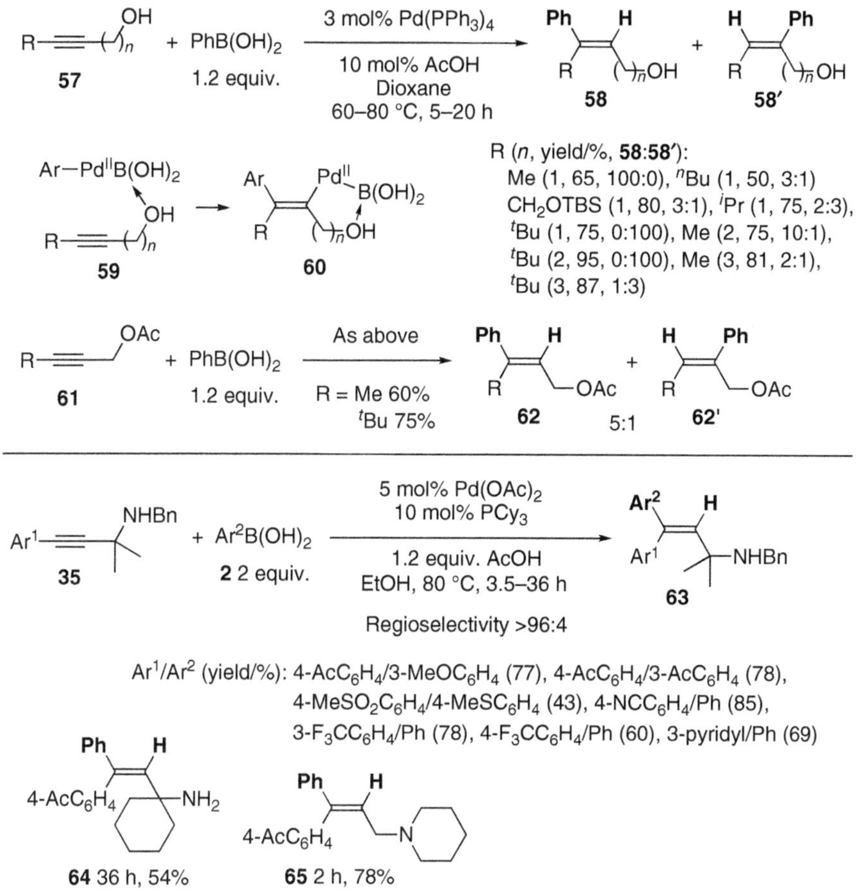

Scheme 8.14 Pd-catalyzed hydroarylation of alkynyl alcohols and propargylic amines using arylboronic acids.

obtained enamides **30′** have the newly introduced aryl groups on the α carbon to the oxazolidinone moiety. This regioselectivity is the same with the above vinyl ether formation but contrasts with that of the rhodium-catalyzed hydroarylation of similar ynamides (Scheme 8.8) [36]. This method tolerated various substituted aryl and alkyl terminal groups. Moreover, ynamines bearing imidazolyl or indolyl groups could be used to afford the corresponding enamines **68–70** in high yields. The electron-donating ability of the oxygen and nitrogen terminal groups obviously plays an important role in these regioselective hydroarylations.

As discussed earlier, rhodium- and palladium-catalyzed alkyne hydroarylations using arylboron reagents uniformly afforded syn addition products. However, anti-selective hydroarylation was reported by Zhu and co-workers (Scheme 8.16) [52]. They performed hydroarylation of ynamide **71** bearing an *N*-sulfonyl group with arylboronic acids **2** (1.5 equiv.) under conditions using $Pd(OAc)_2/2P(m\text{-tolyl})_3$ as a catalyst in ethanol at 70 °C to selectively obtain enamides **72** as a result of anti-hydroarylation. Diverse arylboronic acids **2**

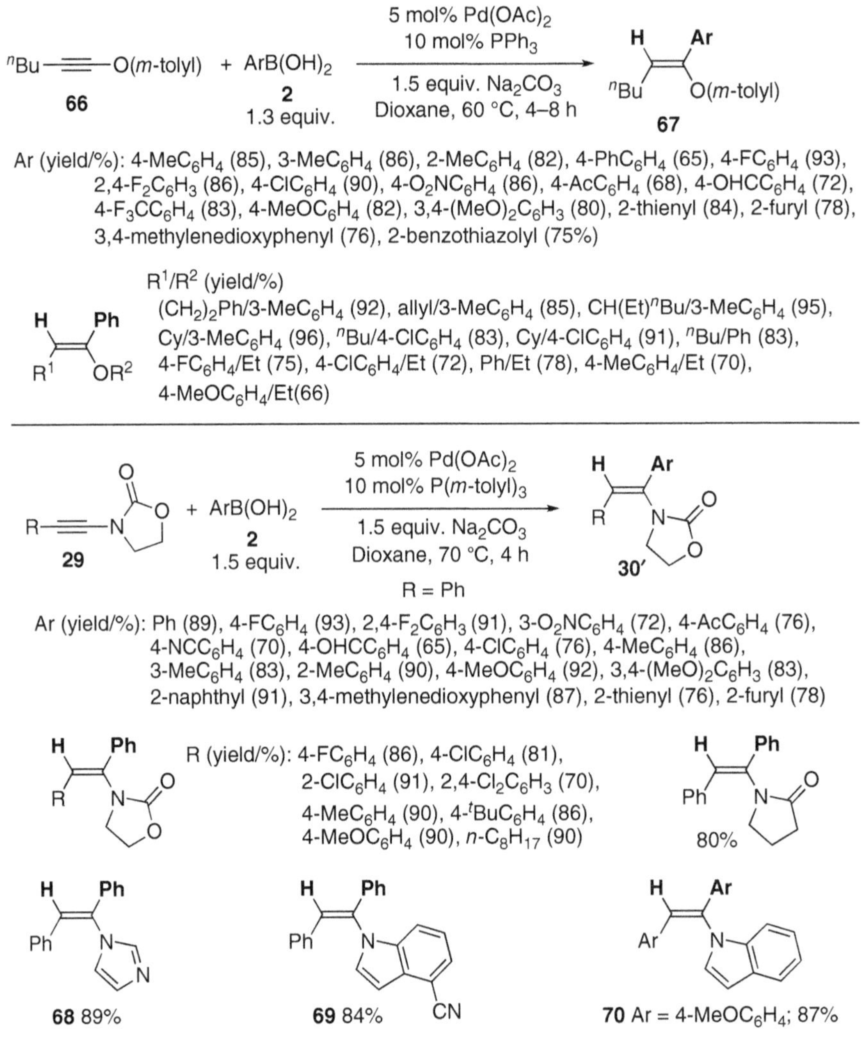

Scheme 8.15 Pd-catalyzed hydroarylation of alkoxyalkynes, ynamides, and ynamines using arylboronic acids.

substituted with electron-donating or electron-withdrawing groups afforded high yields. The use of other solvents such as dioxane, tetrahydrofuran (THF), N,N-dimethylformamide (DMF), and MeOH lowered the stereoselectivity. The anti-stereoselectivity was also dependent on ligands' identity: the use of PPh_3 and PCy_3 led to an E/Z ratio lower than 4:1, and an inverse stereoselectivity was observed using dppe. Moreover, the electron-withdrawing group on the nitrogen terminal groups proved to be important as the use of a ynamide substrate bearing a benzyloxycarbonyl group resulted in a lower stereoselectivity with $E/Z = 83:17$. To explain the observed stereoselectivity, it was assumed that arylpalladium species underwent normal syn carbopalladation

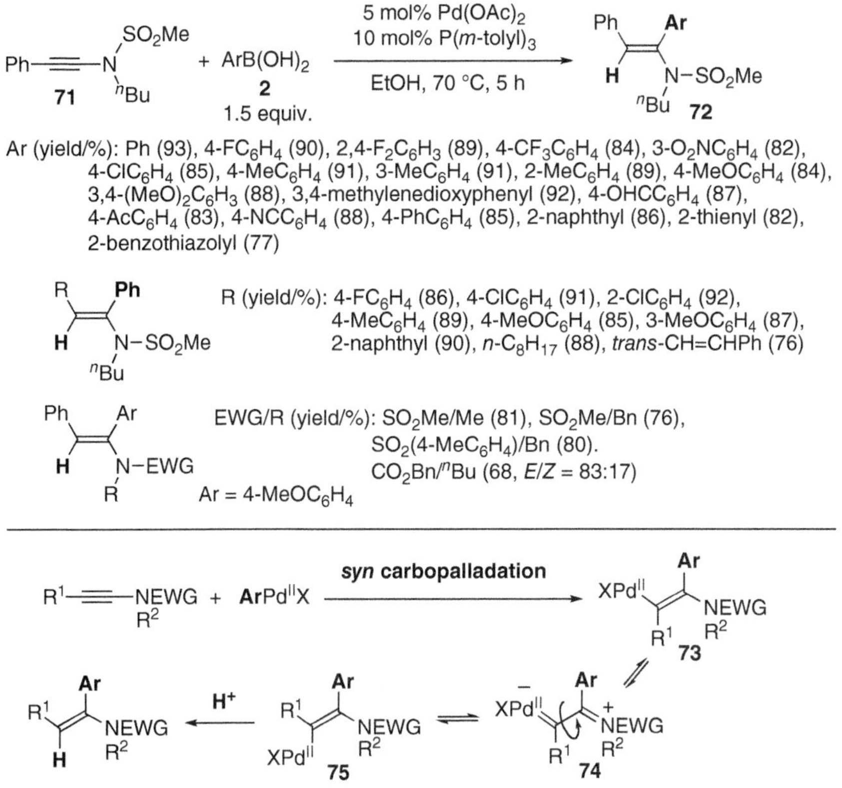

Scheme 8.16 Pd-catalyzed anti hydroarylation of ynamides using arylboronic acids.

to generate vinylpalladium intermediates **73**, which isomerized to formal anti-carbopalladation intermediares **75** via zwitterionic intermediates **74**.

After extensive investigations, various regio- and stereoselective protocols for palladium-catalyzed alkyne hydroarylation using arylboronic acids have been established as described earlier. Furthermore, other methods utilizing elaborated catalysts (PCP pincer complex and $PdCl_2/^iPr_2NPPh_2$) and stable and inexpensive $NaBPh_4$ as an arylboron reagent have also been developed [53–55].

8.2.3 Reactions Catalyzed by First Row Transition Metals

Although a majority of alkyne hydroarylations using arylboron reagents have been implemented using rhodium and palladium catalysts, the utilization of abundant and inexpensive first row transition metals as catalysts is desirable. One of the earliest examples of alkyne hydroarylation with arylboron reagents was achieved using a nickel catalyst as reported by Shirakawa and co-workers [21]. They reacted arylboronic acid propane-1,3-diol esters **76** with symmetrical alkynes **1** (3 equiv.) in the presence of 5 mol% $Ni(cod)_2$ and H_2O (1 equiv.) in dibutyl ether at 80 °C for 5–24 h (Scheme 8.17). When 4-octyne (**9**) was used with phenylboronate, a mixture of hydroarylation product **3** and 1:2 adduct **21** was obtained in 98% combined yield with the ratio of **3**:**21** = 80:20 as estimated by gas chromatography (GC). Similar product selectivity was observed for other electron-rich arylboronates and an almost 1:1 mixture of **3** and **21** was formed from *p*-(trifluoromethyl)phenylboronate. In contrast, the use of diphenylacetylene as a substrate led to the exclusive formation of **3**, albeit in 44% yield. Under different catalytic conditions using 5 mol% dppf as a ligand and phenol (3 equiv.) as a proton source, triphenylborane could be used to obtain **10** and **21** in 240% GC yield based on Ph_3B with the ratio of **10:21** = 91:9. On the other hand, 1:2 adduct **21** was exclusively formed when using phenylboronic acid.

Later, Robbins and Hartwig carried out high-throughput screening of transition-metal-catalyzed reactions, leading to the discovery of a new nickel catalyst system [56]. Their protocol used PPh_3 as a ligand and arylboronic acids, although $Ni(cod)_2$ was the same nickel source with the above example (Scheme 8.17). As such, the reaction of symmetrical alkynes **1** with arylboronic acids **2** (3 equiv.) proceeded in THF at 100 °C to selectively produce hydroarylation products **3** in 55–95% yields. Interestingly, compounds **3** were obtained as *E*/*Z*-stereoisomer mixtures in favor of *Z*-isomers with various ratios (>20:1–2.9:1). Therefore, this example is of importance as rare anti-hydroarylation, although semi-catalytic amounts of $Ni(cod)_2$ were required.

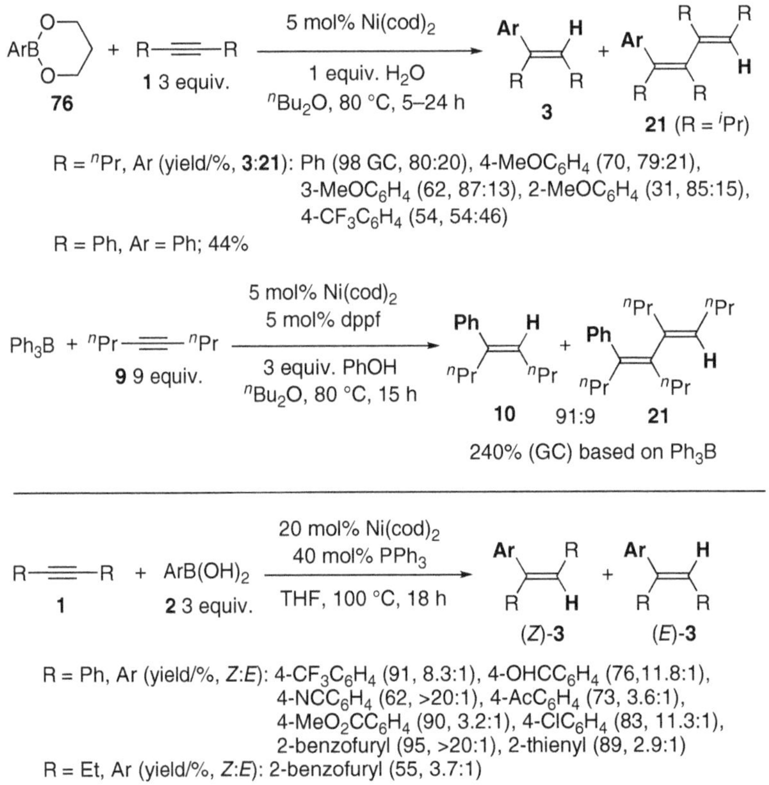

Scheme 8.17 Ni-catalyzed hydroarylation of alkynes using arylboron reagents.

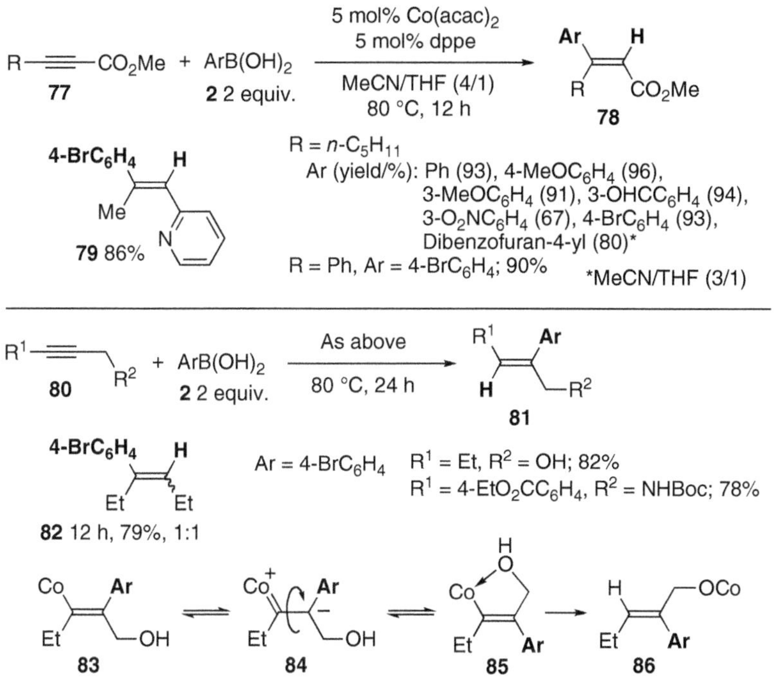

Scheme 8.18 Co-catalyzed hydroarylation of alkynes using arylboronic acids.

Cobalt-catalyzed alkyne hydroarylation using arylboronic acids was reported by Cheng and co-workers [57] (Scheme 8.18). In the presence of 5 mol% $Co(acac)_2$ and 5 mol% dppe, propiolates **77** reacted with arylboronic acids **2** (2 equiv.) in MeCN/THF (4/1) at 80 °C to afford syn hydroarylation products **78** in 67–96% yields. The aryl groups were regioselectively introduced to the β carbon to the methoxycarbonyl groups. In addition, a 2-pyridyl group also behaved similarly to afford **79** as a single isomer in 86% yield. On the other hand, propargylic alcohol and amine substrates **80** (R^2 = OH or NHBoc) reacted with *p*-bromophenylboronic acid to produce anti-hydroarylation products **81,** in which *p*-bromophenyl groups were connected to the α carbon to the CH_2R^2 groups. In the absence of a directing group, 3-hexyne afforded **82** as a 1:1 stereoisomer mixture. To explain the observed regio- and stereoselectivity, it was assumed that normal syn carbocobaltation occurs to generate vinylcobalt intermediates **83,** which then undergo isomerization via zwitterionic intermediates **84** into cyclic chelate complexes **85.** Subsequently, **85** evolve into the final products via internal proton transfer and ligand exchange of resultant alkoxycobalt complexes **86.**

Hydroarylation of **77** was repeated using $PhB(OD)_2$, which was derived from triphenylboroxine and D_2O (3 equiv.), to obtain **78-*d*$_1$** in 90% yield with 50% deuterium incorporation (Scheme 8.19). Moreover, hydroarylation did not proceed when $Co(acac)_2$ was replaced with $Co(OH)_2$, although the same reaction smoothly occurred by the addition of acetylacetone (acacH). Based on these observations, a plausible catalytic cycle was proposed as outlined in

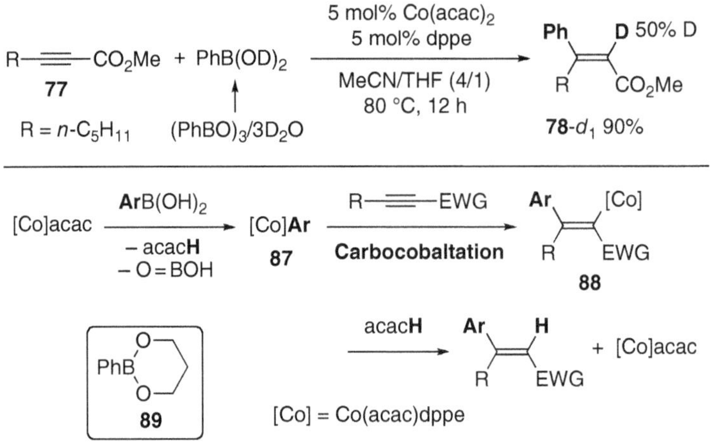

Scheme 8.19 Deuterium-labeling experiments and proposed mechanism for Co-catalyzed hydroarylation of alkynes using arylboronic acids.

Scheme 8.19. Co(acac)$_2$dppe undergoes transmetallation with arylboronic acids to generate ArCo(acac)dppe (**87**) along with acacH and B(O)OH. Subsequent syn carbocobaltation occurs from arylcobalt species **87** and the alkyne substrates to generate vinylcobalt intermediates **88,** which are protonated with acacH to ultimately afford syn hydroarylation products with concomitant restoration of the starting cobalt catalyst. Thus, acetylacetonate plays a dual role as a ligand and a proton transfer agent. The protons of arylboronic acids also proved to be important: the hydroarylation did not proceed with arylboronate **89.**

Regio- and stereoselective hydroarylation of propiolates using arylboronic acids was also achieved using copper catalysis [58]. The reaction of propiolates **77** with various arylboronic acids **2** (3 equiv.) were performed in the presence of 1–10 mol% CuOAc without any ligand and additive in methanol at 28 °C to afford syn hydroarylation products **78** in 60–94% yields (Scheme 8.20). Other propiolates bearing functionalized alkyl and aryl terminal groups could be used for this protocol. Copper-catalyzed hydroarylation of propiolates could be performed on a 30 mmol scale [59]. The reaction of **77** with phenylboronic acid in MeOD afforded **78**-d_1 in 74% yield with 86% deuterium incorporation (Scheme 8.20). The use of neopentyl glycol ester **90** led to complete deuterium incorporation. Accordingly, the methanol solvent and the arylboronic acids are the proton sources in this reaction.

The copper-catalyzed hydroarylation protocol could be applied to arylpropynenitriles **91** (Scheme 8.21) [60]. Thus, diverse 3,3-diarylacrylonitriles **92** were stereoselectively synthesized in good yields. This method proved to be effective for an alkynylnitrile possessing an alkyl terminal. 3,3-Diarylacrylonitiles can be used as building blocks in the synthesis of 3,3-diarylpropylamines, which are important components found in pharmaceuticals. Thus, Yun and co-workers prepared 3,3-diarylacrylonitriles using the copper-catalyzed hydroarylation and used them as substrates for copper-catalyzed asymmetric conjugate reduction [61].

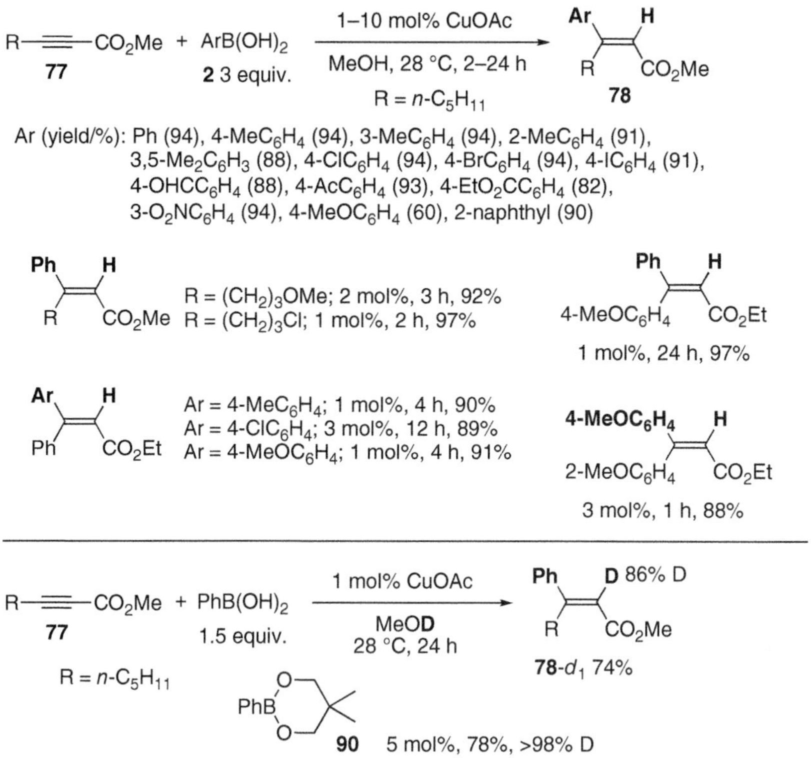

Scheme 8.20 Cu-catalyzed hydroarylation of propiolates using arylboronic acids and labeling experiments.

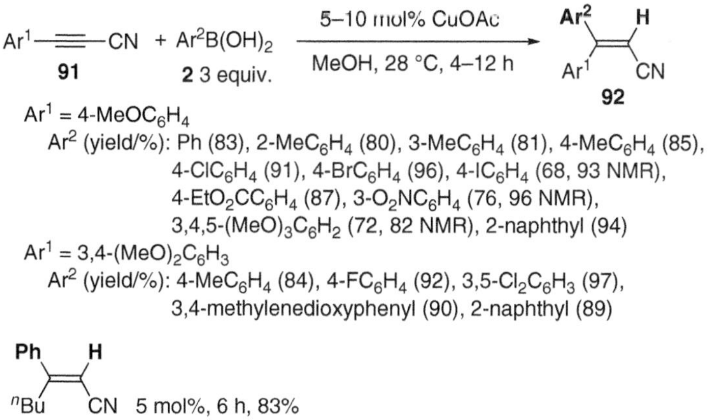

Scheme 8.21 Cu-catalyzed hydroarylation of propynenitriles using arylboronic acids.

8.3 Catalyzed Alkyne Hydroarylations Using Aryl Halides and Arenediazonium Compounds

The Mizoroki–Heck reaction, which involves palladium-catalyzed arylation of alkenes using aryl halides, is one of the most widely used methods to synthesize arylalkenes [62, 63]. However, the synthesis of highly substituted alkenes using this method is problematic as multi-substituted alkenes are usually inefficient substrates. Thus, this reaction was modified using internal alkynes as substrates by Cacchi and co-workers [22]. In the classic Mizoroki–Heck reaction, alkenes are used as substrates and after carbopalladation the palladium(0) catalyst is restored via β-H elimination and subsequent removal of HX with a base (Scheme 8.22). On the other hand, when alkynes are used instead of alkenes, vinylpalladium intermediates **93** are generated and thus β-hydride elimination is not feasible. Thus, Cacchi and co-workers used formic acid as a hydride donor, which captures vinylpalladium intermediates **93** to generate hydride complexes **94.** Subsequent reductive elimination affords trisubstituted alkenes **3** as the final

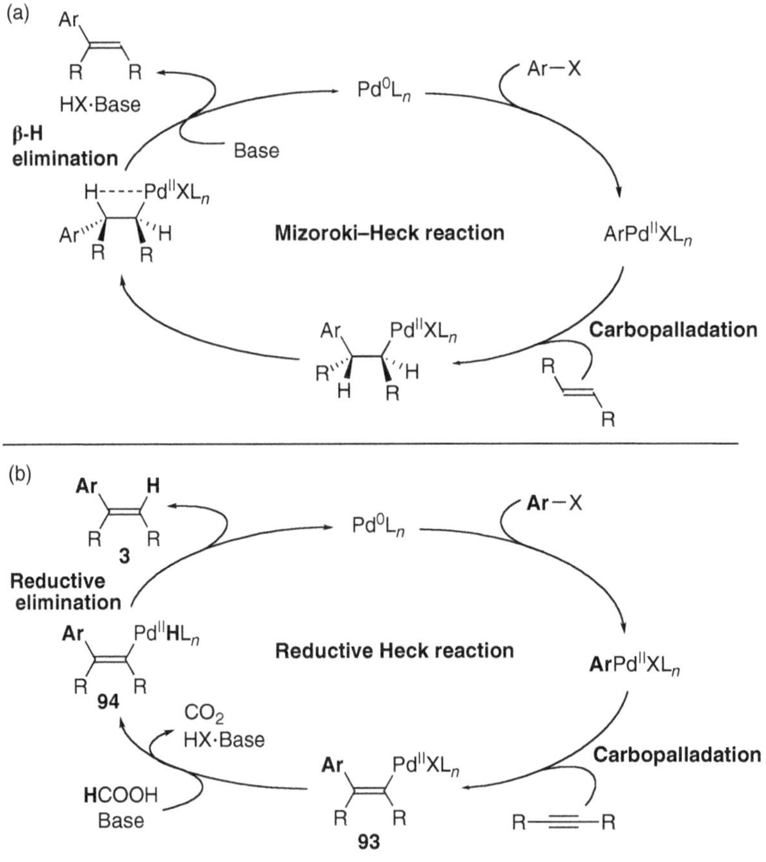

Scheme 8.22 Catalytic cycles of (a) Mizoroki–Heck reaction and (b) reductive Heck reaction.

product. This variant of Mizoroki–Heck reaction is now known as a reductive Heck reaction. In the following subsections, advances in reductive Heck reactions will be overviewed by classifying them into two types: intermolecular reactions and intramolecular versions (reductive Heck cyclizations).

8.3.1 Intermolecular Reductive Heck Reactions

In the first report of Cacchi and co-workers, the reaction of diphenylacetylene with aryl iodides (2.4 equiv.) was performed using 0.5 mol% $Pd(OAc)_2(PPh_3)_2$ and HCO_2H (2.6 equiv.) in Et_3N/MeCN (1.34/1) at 80 °C, resulting in the formation of triarylethenes in 40–96% yields [22]. However, stereochemistry of the products was not determined at this stage. Deuterium incorporation was observed when using DCO_2H. Similar reactions of unsymmetrical diarylalkynes exhibited almost no regioselectivity. Reductive Heck reactions of various unsymmetrical alkyne substrates were further investigated to reveal regioselectivity. Tertiary propargylic alcohols **95** and aryl iodides **96** (2.4 equiv.) were subjected to reductive Heck conditions to selectively afford 1,1-diarylallylic alcohols **97** in varied yields, although stereochemistry was not determined (Scheme 8.23) [64].

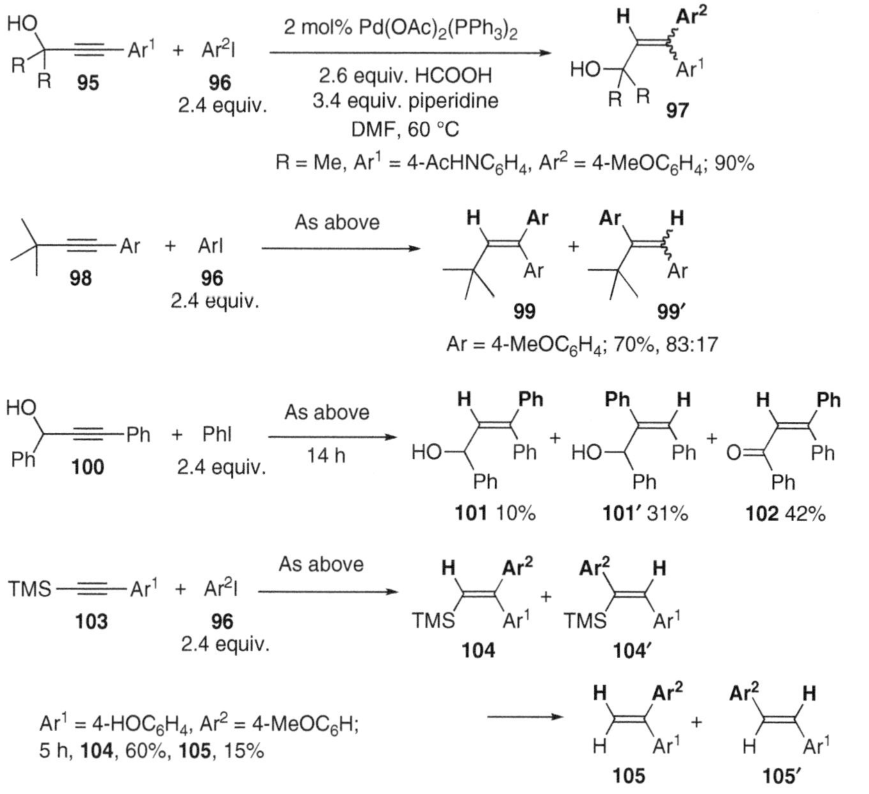

Scheme 8.23 Pd-catalyzed reductive Heck reaction of propargylic alcohols or trimethylsilylalkynes using aryl iodides.

In the absence of a hydroxyl group, unsymmetrical alkyne **98** afforded a mixture of regioisomers **99** and **99′** still in favor of 1,1-diarylalkene **99.** These results indicate that bulky terminal substituents control regioselectivity. In contrast, the reaction of secondary propargylic alcohol **100** with iodobenzene resulted in the unselective formation of 1,1-diaryl **101,** 1,2-diaryl **101′,** and **102,** which was derived from the oxidation of **101,** in 10%, 31%, and 42% yields, respectively. Later, the stereochemistry of 1,2-diphenylallylic alcohol **101′** was unambiguously confirmed by X-ray crystallography, as reductive Heck reaction proceeded in syn fashion [65]. Tandem Sonogashira coupling/reductive Heck reaction was also reported to form diarylallylic alcohols from terminal propargylic alcohols. A similar tandem reaction was later revisited by Antunes and co-workers [66, 67]. Arylalkyne substrates **103** bearing bulky TMS terminal groups were used for the reductive Heck reaction, resulting in 2,2-diarylvinylsilanes **104** as major products together with regioisomers **104′** and desilylated byproducts **105** and **105′** in various yields and selectivity [68]. Typically, **103** ($Ar^1 = 4\text{-}HOC_6H_4$) and *p*-iodoanisole (2.4 equiv.) were subjected to the standard conditions for 5 h to afford the corresponding vinylsilane **104** and its protodesilylation product **105** in 60% and 15% yields, respectively. When the crude reaction mixture was treated with 1 mol% I_2 in benzene/H_2O (2/1) at 75 °C for 7.5 h it afforded **105** as a sole product in 76% yield.

The influence of electron-withdrawing groups was investigated under ligand-free conditions. Cacchi and co-workers performed reductive Heck reaction of alkynyl ketone **106** with *p*-iodoanisole (2.4 equiv.) in the presence of 5 mol% $Pd(OAc)_2$ and HCOOK (2.4 equiv.) in DMF at 40 °C (Scheme 8.24) [69]. As a result, syn and anti-hydroarylation products **107** and **107′** were formed in 20% and 13% yields, respectively. Anti-adduct **107′** was assumed to be generated by isomerization of **107** via the zwitterionic intermediate **108.** The *p*-anisyl group was regioselectively placed on the α carbon to the ketone carbonyl group, and this regioselectivity contrasts with that observed in the Heck reaction of alkenones. Such a discrepancy in regioselectivity was ascribed to the different electronic environment of alkenones and alkynyl ketones on the basis of the density functional theory study: NBO analyses of *trans*-cinnamaldehyde and 3-phenylpropynal revealed that the coefficient in the conjugated π* orbital is larger on the carbon β to the formyl group for both compounds, while the coefficient is substantially larger on the α-carbon in the nonconjugated π* orbital of 3-phenylpropynal [70]. The corresponding acetals were also used as substrates. Thus, the reaction of cyclic acetal **109** under similar conditions afforded syn hydroarylation products **110** and **110′** in 73% combined yield with no regioselectivity. The loss of regioselectivity was ascribed to the bulky cyclic acetal moiety. In fact, syn hydroarylation product **112** was predominantly obtained in 48% yield when less sterically demanding acyclic acetal **111** was used as a substrate. Similarly, the reductive Heck reaction of 3-phenylpropiolate **113** with various aryl iodides **96** (2.4 equiv.) also afforded regioisomer mixtures of **114** and **114′** in moderate combined yields with moderate regioselectivity in favor of α-arylated isomers **114** [71]. The Cacchi group later extended this ligand-free reductive Heck reaction to tertiary propargylic amides [72]. However,

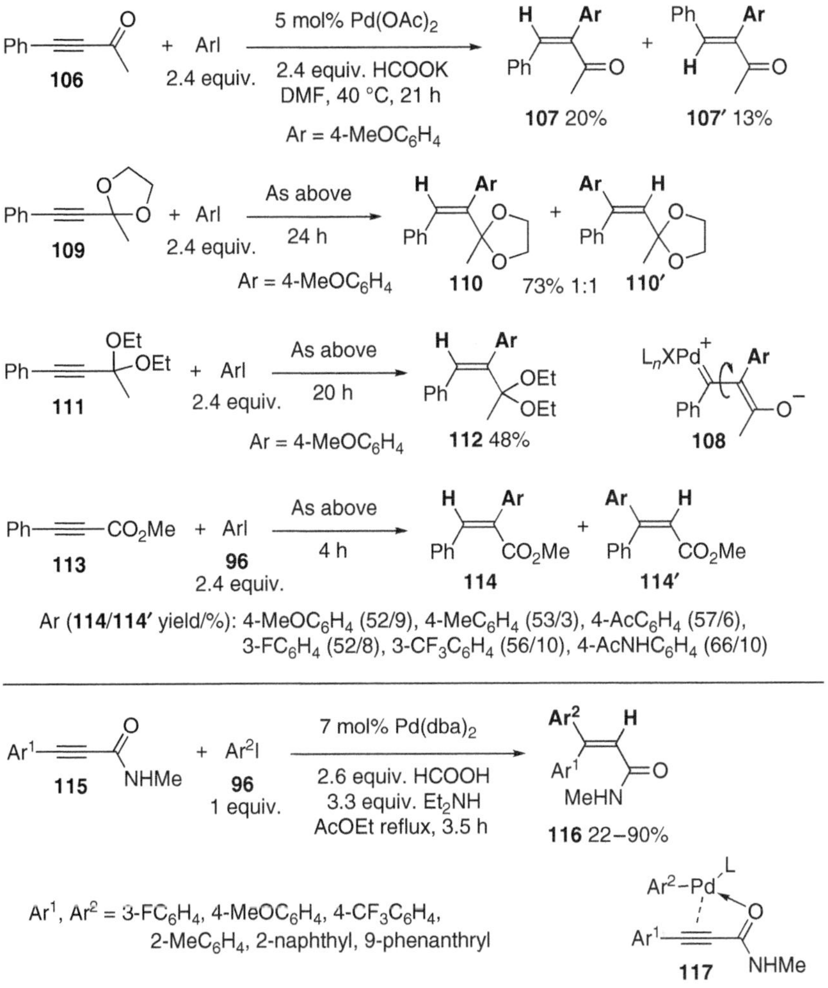

Scheme 8.24 Pd-catalyzed reductive Heck reaction of alkynyl ketone, acetals or amides using aryl iodides.

the formation of allenes via β-N elimination competed with reductive Heck reaction.

Mitchell and co-workers applied different ligand-free conditions to the reaction of alkynylamides **115** with aryl iodides **96** (Scheme 8.24) [73, 74]. In the presence of 7 mol% $Pd(dba)_2$ (dba = dibenzylideneacetone), HCO_2H (2.6 equiv.), and Et_2NH (3.3 equiv.), syn hydroarylation selectively proceeded in AcOEt under reflux to afford 3,3-diarylacrylamides **116** in 22–90% yields. It was assumed that the Ar^2 groups were introduced to the β carbon to the amide carbonyl group via chelation complexes **117**.

Reductive Heck reaction was also implemented in an ionic liquid solvent, $[bmim]BF_4$ (bmim = 3-methylimidazolium). Cacchi and co-workers demonstrated that the reductive Heck reaction of diphenylacetylene (**118**) with

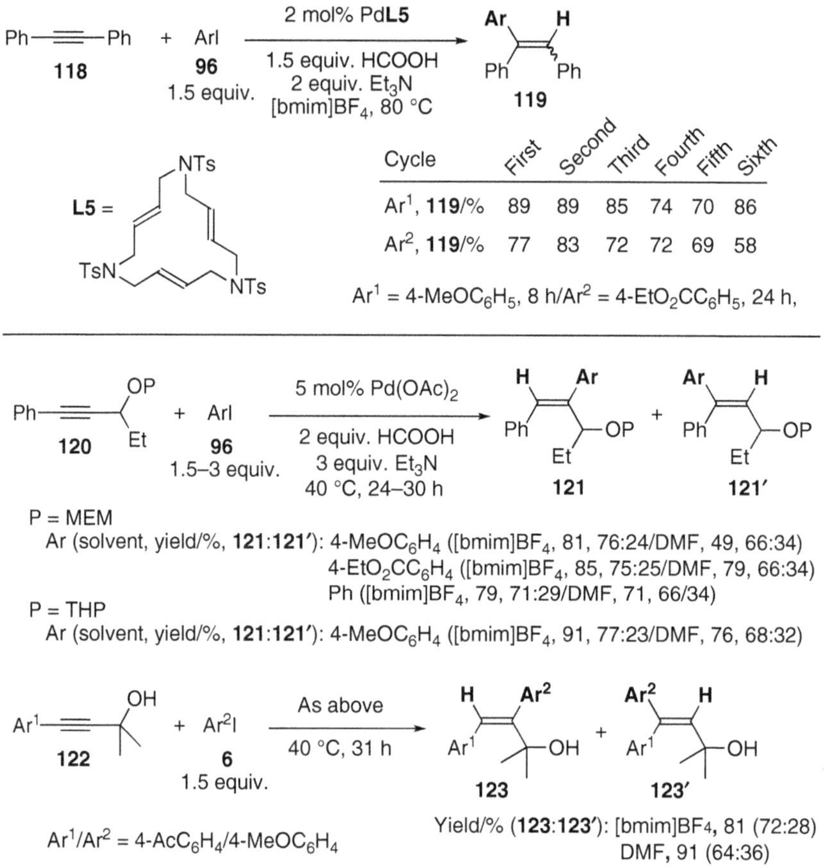

Scheme 8.25 Pd-catalyzed reductive Heck reaction of alkynes using aryl iodides in ionic liquid and DMF solvents.

p-iodoanisole or ethyl *p*-iodobenzoate (**96**, Ar = 4-$MeOC_6H_4$ and 4-$EtO_2CC_6H_4$, respectively) using a palladium catalyst with cyclic triene ligand **L5** could be performed at 80 °C in [bmim]BF_4 (Scheme 8.25) [75]. Other ionic liquids were less efficient than [bmim]BF_4. The catalyst could be reused in at least 6 cycles with compounds **119** obtained in moderate to high yields. This solvent was further used for ligand-free reductive Heck reaction of propargylic alcohol derivatives [76]. Although protecting groups such as 2-methoxyethoxymethyl (MEM) or 2-tetrahydropyranyl (THP) groups were required to suppress undesirable oxidation of secondary alcohols, propargylic alcohol derivative **120** and **122** reacted with aryl iodides **96** (1.5–3 equiv.) at 40 °C in [bmim]BF_4 to afford regioisomer mixtures of syn hydroarylation products **121**/**121′** or **123**/**123′** in higher selectivity than those in DMF. Notably, the newly introduced aryl groups were placed on the α carbon to the propargylic alcohol or ether moieties in the major regioisomers. This regioselectivity contrasts with those observed for previous examples where PPh_3 was used as a ligand (Scheme 8.23).

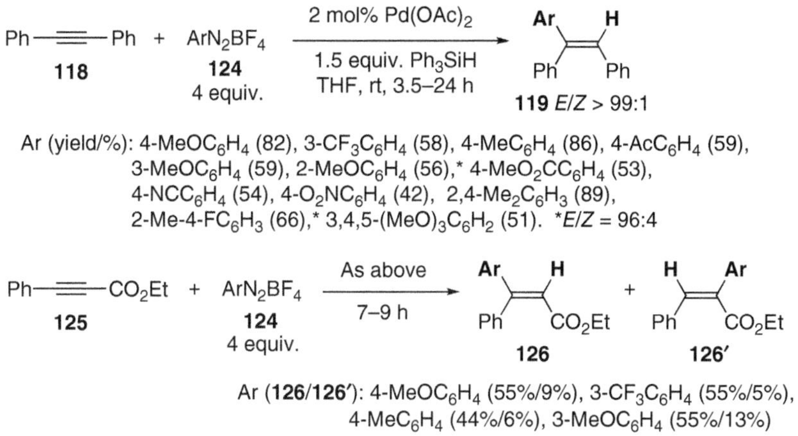

Scheme 8.26 Pd-catalyzed reductive Heck reaction of alkynes using arenediazonium salts.

In addition to formic acid and congeners, other hydride sources were used for reductive Heck reactions. For example, Wu and co-workers used NaOMe as a hydride donor instead of formic acid [77]. Cacchi and co-workers used triphenylsilane to achieve a reductive Heck reaction using arenediazonium salts **124**, which are readily available from the corresponding anilines (Scheme 8.26) [78]. In the presence of 2 mol% $Pd(OAc)_2$ and Ph_3SiH (1.5 equiv. relative to the alkyne), the reaction of diphenylacetylene (**118**) with various arenediazonium salts **124** (4 equiv. relative to the alkyne) in THF at room temperature afforded triarylalkenes **119** in high syn selectivity. The product yields were higher for electron-rich arenediazonium salts than for electron-deficient ones. In contrast to aryl iodides, ortho-substituted arenediazonium salts could be used. The use of the silane was essential: the hydroarylation product was not obtained when HCOOK was used as a hydride source. Similarly, the reaction of propiolate **125** with **124** afforded regioisomer mixtures of syn hydroarylation products **126** and **126′** in moderate combined yields. The regioselectivity in favor of β-arylation products **126** was similar to that observed for the reductive Heck reaction of propiolamides **115** (Scheme 8.24).

As the first nickel-catalyzed intermolecular reductive Heck reaction, hydroarylation of propiolates with *o*-iodobenzyl alcohols was investigated by Rayabarapu and Chen and seven-membered lactones were obtained as the final product after intramolecular transesterification (see Section 8.4.1.1) [79]. Robbins and Hartwig also identified new nickel-catalyzed hydroarylations of alkynes using arylboronic acids and aryl bromides using high-throughput reaction studies [56]. They established semi-catalytic (turnover number of <5) anti-hydroarylation of diphenylacetylene (**118**) using aryl bromides **127** as limiting reagents (Scheme 8.27). This protocol required 20 mol% $Ni(cod)_2$, 40 mol% P^nBu_3, and Et_3SiH as the hydride source, and a reaction temperature of 100 °C. The yields were moderate and the stereoselectivity ranged from 1:1.1 to 1:17.5.

Weix and co-workers also reported the nickel-catalyzed reductive Heck reaction of propiolates **77** using aryl iodides **96** [80]. The combination of

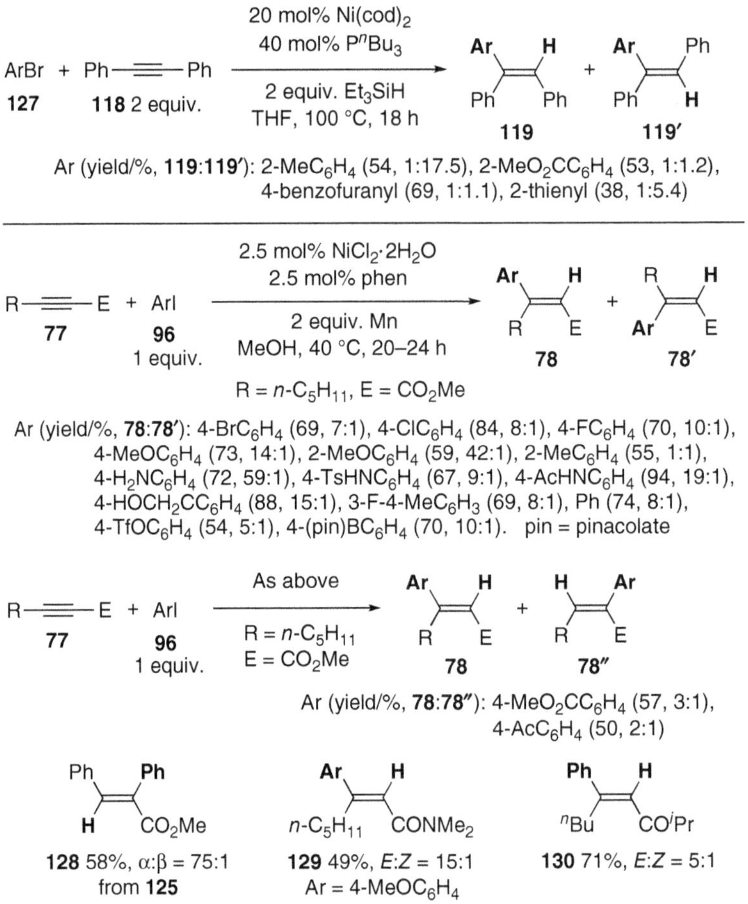

Scheme 8.27 Ni-catalyzedreductive Heck reaction of alkynes using aryl iodides.

inexpensive $NiCl_2{\cdot}2H_2O$ (2.5 mol%) and phenanthroline (phen, 2.5 mol%) was used as a catalyst with manganese powder required as a stoichiometric reductant. Using various electron-rich aryl iodides, syn hydroarylation products **78** were obtained along with stereoisomers **78′**. The combined yields were 54–94% and the *E*:*Z* ratio was found to range from 1:1 to 59:1. In contrast, the use of electron-deficient aryl iodides such as methyl *p*-iodobenzoate and *p*-iodoacetophenone led to regioisomer mixtures of **78** and **78″** in moderate yields and regioselectivity. The product selectivity varied depending on the substrate structures: ethyl 3-phenylpropiolate (**125**) underwent β selective anti-hydroarylation to afford **128** with concomitant transesterification, whereas 3-arylacrylic amide **129** and 3-phenylethenyl isopropyl ketone **130** were obtained as a result of syn hydroarylation. As for the reaction mechanism, this reaction was proposed to proceed in a similar manner with hydroarylation of propiolates using arylboronic acids rather than palladium-catalyzed reductive Heck reaction: arylnickel species ($ArNi^{II}I$) undergo syn hydroarylation and subsequent protonation of the resultant vinylnickel intermediates generates $INi^{II}(OMe)$, which is then reduced by Mn^0 to restore the Ni^0 active catalyst.

8.3.2 Intramolecular Reductive Heck Reactions

In their pioneering studies, Grigg and co-workers achieved an intramolecular reductive Heck reaction (reductive Heck cyclization) of several iodobenzene derivatives bearing tethered alkynes [23, 81]. They further optimized the reaction conditions by screening additives and hydride sources (Scheme 8.28) [82]. In the presence of 10 mol% $Pd(OAc)_2$ and 20 mol% PPh_3, *N*-propargyl-*o*-iodoanilines **131** were treated with HCOOH/piperidine or HCOONa as hydride donors to obtain 5-*exo* cyclization products **132** in 50–60% yields. Different additives such as Et_4NCl and Ag_2CO_3 were found to be optimal depending on alkyne terminal groups. Similarly, 6-*exo* cyclization of *N*-propargyl-*o*-iodobenzamides **133** with a methyl or benzyl protecting group (PG) successfully afforded **134** in comparable yields. Although a substrate bearing *N*-propargylpiperidine as a terminal group generated a mixture of two products, the use of $TlNO_3$ as an additive led to the exclusive formation of the corresponding 6-*exo* cyclization product. Moreover, 7-*exo* cyclization was also possible: *o*-iodobenzyl homopropargyl ether **135** was transformed into oxepane derivative **136** in 62% yield. The alkylidene moieties of the obtained products were stereoselectively constructed as a

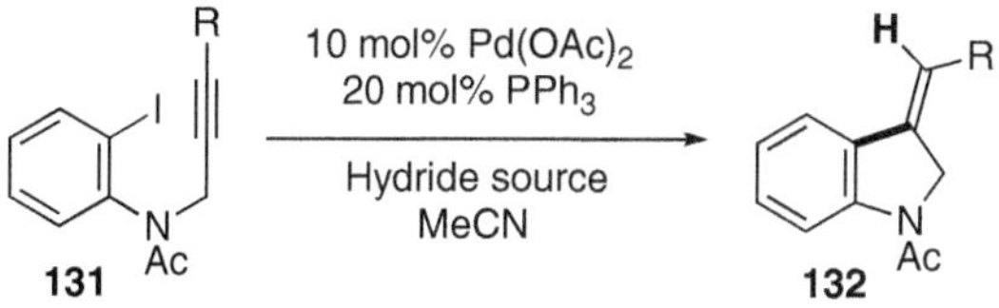

R = Me; HCOOH (3 equiv.), Ag_2CO_3 (1 equiv.), rt, 16 h, 60%
R = TMS; HCOOH (3 equiv.), piperidine (4 equiv.), Et_4NCl (1 equiv.), 60 °C, 1 h, 50%
R = CH_2NEt_2; HCOONa (1.1 equiv.), Ag_2CO_3 (1 equiv.), 60 °C, 15 h, 56%

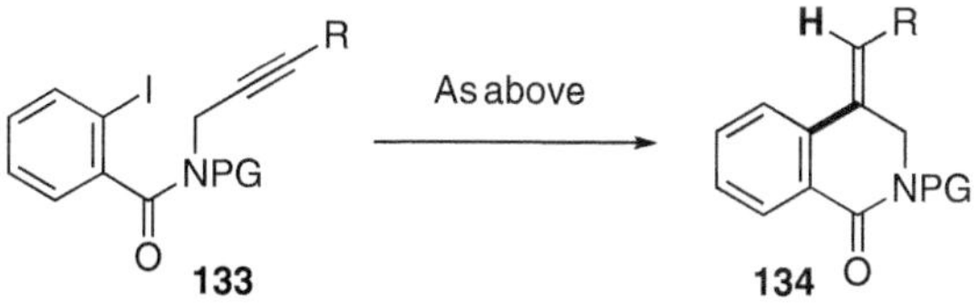

PG/R = Me/H; HCOOH (3 equiv.), piperidine (4 equiv.), Et_4NCl (1 equiv.), 60 °C, 2 h, 60%
PG/R = Bn/Me; HCOOH (3 equiv.), piperidine (4 equiv.), Et_4NCl (1 equiv.), 60 °C, 8 h, 50%
PG/R = Me/CH_2Pip; HCOONa (1.1 equiv.), $TlNO_3$ (1 equiv.), 60 °C, 12 h, 53%

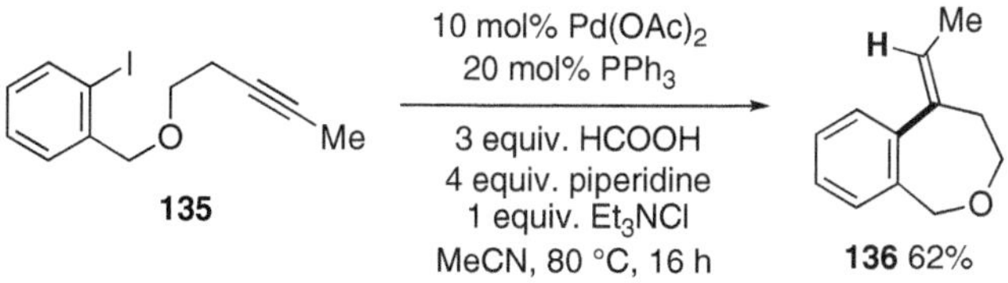

Scheme 8.28 Pd-catalyzed reductive Heck cyclization of *o*-(ω-alkynyl)iodobenzenes.

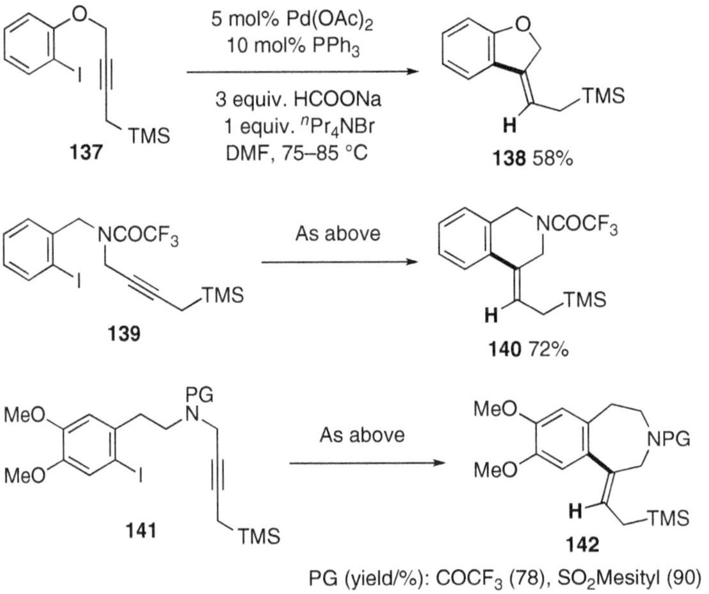

Scheme 8.29 Synthesis of heterocyclic allylsilanes via Pd-catalyzed reductive Heck cyclization.

result of syn hydroarylation. Similar 5-*exo* and 6-*exo* reductive Heck cyclizations also proceeded under Nozaki–Hiyama–Kishi coupling conditions using $NiCl_2$ as a catalyst in the presence of $CrCl_2$ as a stoichiometric reductant [83].

Reductive Heck cyclization proved to be effective for stereoselective synthesis of exocyclic allylsilanes from propargylic silanes (Scheme 8.29) [84]. Under modified reductive Heck conditions using HCOONa and $^{n}Pr_4NBr$ in DMF at 75–85 °C, ether-tethered propargylsilane **137** underwent 5-*exo* cyclization to afford benzofuran derivative **138** bearing an exocyclic allylsilane moiety in 58% yield. Although the cyclization of an aza analog of **137** was inefficient, 6-*exo* cyclization of **139** effectively afforded isoquinoline derivative **140** in 72% yield. In a similar manner, an interesting benzazepine derivative **142** was obtained in 78% yield via 7-*exo* cyclization of **141** (PG = $COCF_3$). The product yield was further improved to 90% using a bulky mesitylsulfonyl protecting group. Comins and co-workers demonstrated that 6-*exo* reductive Heck cyclization of pendant alkynes outweighed 5-*exo* cyclization involving an activated cyclic alkene [85].

Since these seminal studies, reductive Heck cyclizations have been applied to the synthesis of various heteocyclic compounds [86–90]. In particular, synthetically challenging medium-ring heterocycles have been successfully synthesized using reductive Heck protocols. The group of Van der Eycken applied microwave-assisted conditions to the construction of seven- and eight-membered nitrogen heterocycles (Scheme 8.30). Bromobenzenes **143** bearing pendant propiolamide moieties were subjected to reductive Heck conditions using 3 mol% $Pd(PPh_3)_4$ and HCOONa (1.5 equiv.) in DMF/H_2O (3/1) with microwave irradiation at 110 °C for 15 min, affording benzazepinones **144** in 32–90% yields via 7-*exo* cyclization [91]. This protocol was also applied to 8-*exo*

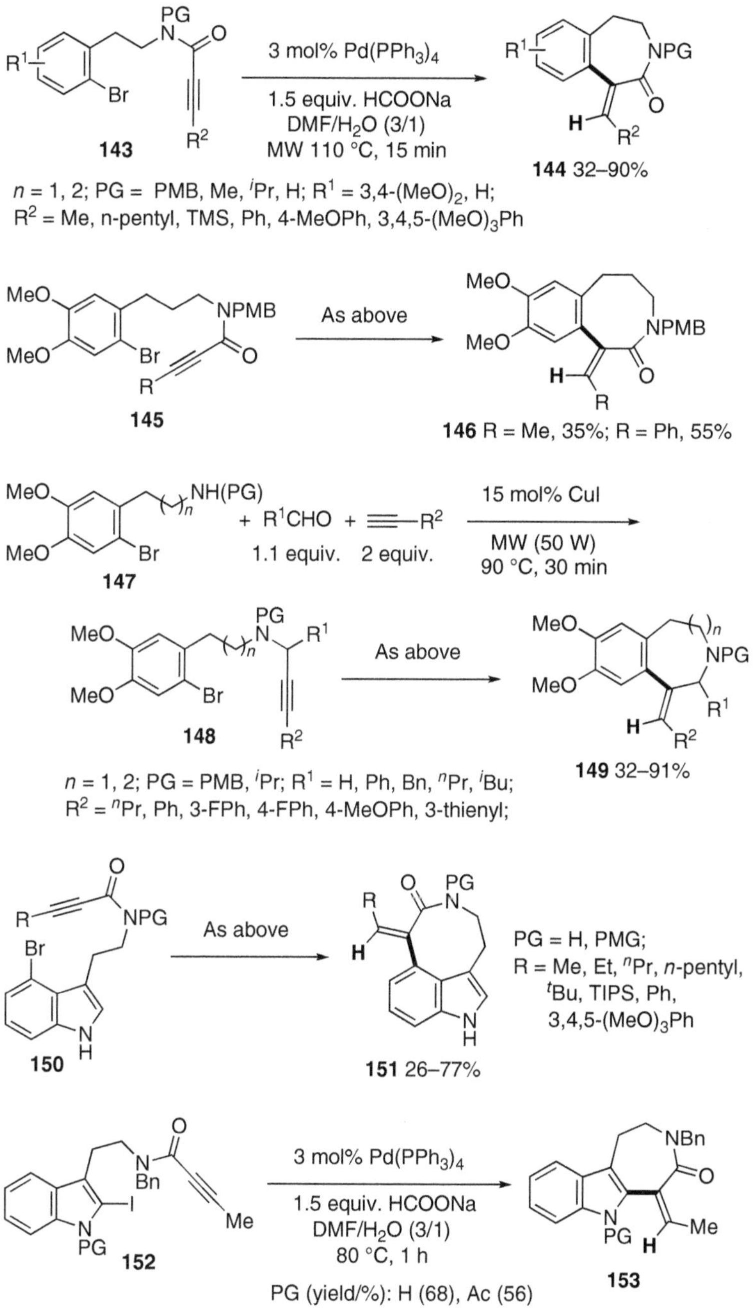

Scheme 8.30 Synthesis of medium-ring *N*-heterocycles via Pd-catalyzed reductive Heck cyclization under microwave irradiation conditions.

cyclization of **145** with longer tethers to afford benzazocinones **146,** albeit in lower yields. A diversity-oriented strategy to construct similar benzazepines and benzazocines **149** was developed by combining reductive Heck cyclization with the copper-catalyzed three-component-coupling synthesis of substrates **148** from amines **147,** aldehydes, and terminal alkynes under microwave irradiation conditions [92]. Moreover, 8-*exo* cyclization of 4-bromoindole derivatives **150** proceeded in a similar manner to deliver interesting azocino[*cd*]indoles **151** in 26–77% yields [93]. On the other hand, 7-*exo* cyclization of 2-iodoindole derivative **152** was performed under conventional heating conditions at 80 °C for 1 h to afford azepino[*b*]indoles **153** in 56–68% yields.

Majumdar and co-workers investigated reductive Heck cyclization under conventional heating conditions (Scheme 8.31). Thus, 7-*exo* cyclization of *o*-iodobenzamides **154** bearing pendant diarylalkyne moieties efficiently proceeded at 120 °C to afford dibenzoazepinones **155** in 75–90% yields [94]. Using this protocol, coumarin- and naphthalene-fused derivatives **156** and **157** were obtained in 80% and 65% yields, respectively. Furthermore, dibenzoazocinones **159** were obtained via 8-*exo* cyclization of homologous substrates **158,** albeit in lower yields [95]. Coumarin- or 2-quinolone-fused derivatives **160** and **161** were also synthesized similarly. In addition to these azepinones and azocinones, dihydrodibenzo[*b*,*e*]oxepines **163** were obtained in 67–76% yields via 7-*exo* cyclization of *o*-bromobenzylic ethers **162** [96]. Joardar and co-workers also applied this protocol to *o*-bromophenylsulfonamide **164** to obtain tricyclic sultams **165** in good yields [97].

8.4 Synthetic Applications of Alkyne Hyaroarylations Using Arylboron Reagents and Aryl Halides

This section will be devoted to the illustration of synthetic potential of alkyne hydroarylations using arylboron reagents and aryl halides. To this aim, sequential processes involving alkyne hydroarylations using arylboron reagents and aryl halides will be overviewed in the first part. Such a sequential process has gained importance because it streamlines synthetic procedures by causing multiple bond-making events in a single operation. In the second part, the synthesis of bioactive compounds and natural products will be surveyed.

8.4.1 Sequential Processes Involving Alkyne Hydroarylations Using Arylboron Reagents and Aryl Halides

8.4.1.1 Synthesis of Oxygen Heterocycles

In 1988, Cacchi and co-workers reported a sequential process comprising syn hydroarylation and subsequent lactonization (Scheme 8.32) [98]. They subjected propiolates **166,** which have secondary or tertiary propargylic alcohol moieties, to reductive Heck conditions generating syn hydroarylation products **167** and **167′.** These adducts then underwent intramolecular transesterification to afford butenolides **168** and **168′** in various yields. The hydroarylation generally proceeded with regioselectivity in favor of α-arylated adduct **167.**

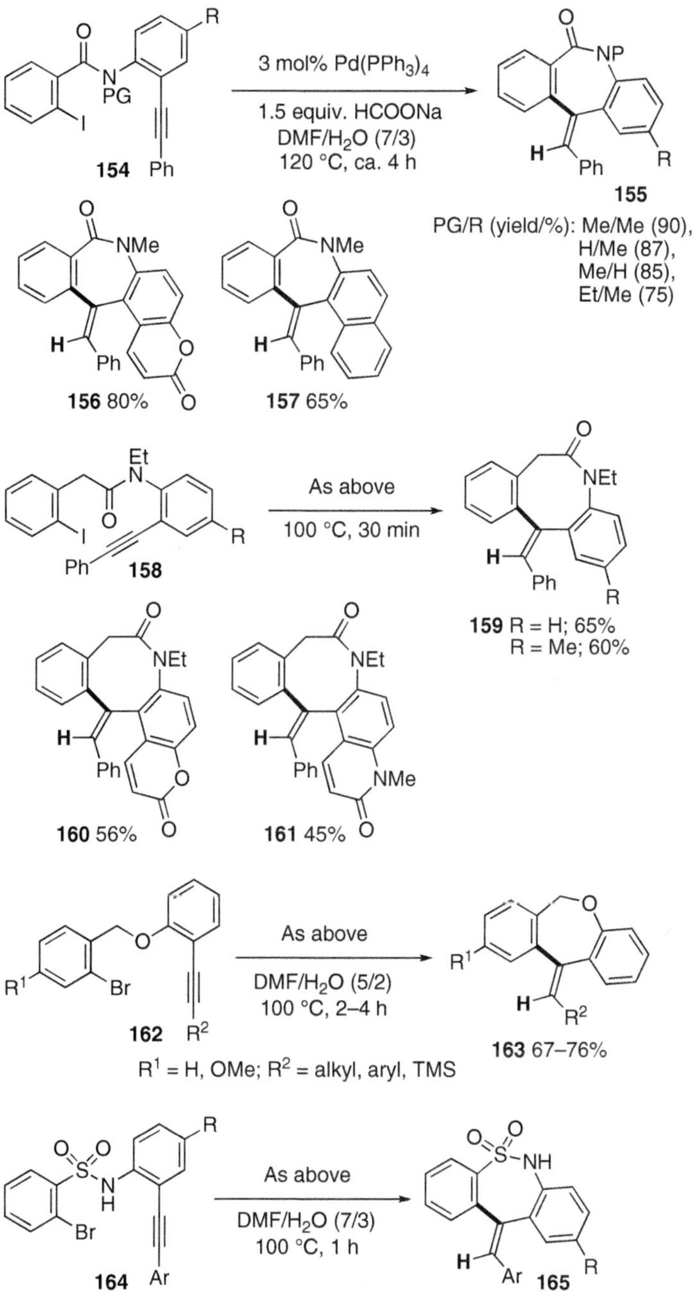

Ar/R (yield/%): Ph/H (80), Ph/Cl (76), 2-MeOC_6H_4/Cl (78), Ph/OCF_3 (70), 2-MeOC_6H_4/OCF_3 (75)

Scheme 8.31 Synthesis of medium-ring N-heterocycles via Pd-catalyzed reductive Heck cyclization under conventional heating conditions.

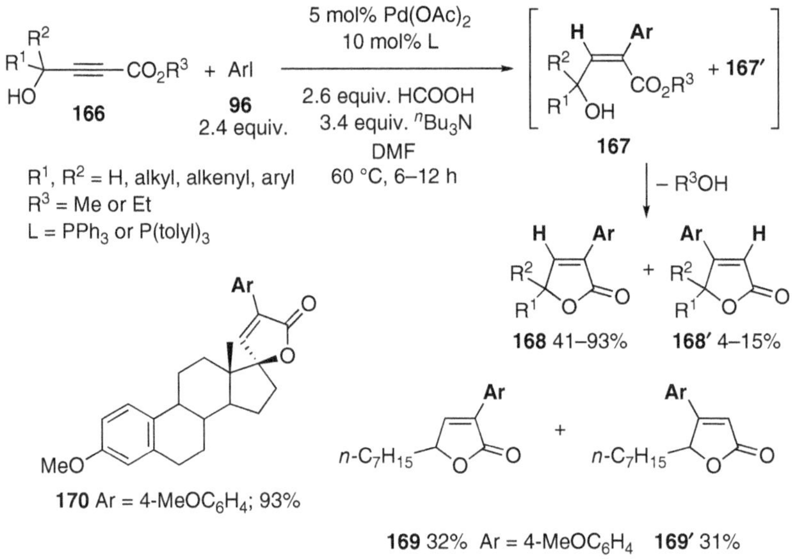

Scheme 8.32 Synthesis of butenolides via Pd-catalyzed reductive Heck reaction and subsequent lactonization.

However, butenolides **169** and **169′** were formed unselectively from ethyl 4-hydroxyundec-2-ynoate. Interestingly, butenolide **170** bearing a spirocyclic steroid framework was obtained as a sole product in the highest yield of 93%.

A similar butenolide synthesis was also achieved using arylboronic acids (Scheme 8.33). Oh and co-workers applied their palladium-catalyzed hydroarylation to 4-hydroxypropiolates **171** [99]. When 3 mol% $Pd(OAc)_2$ with 6 mol% P^tBu_3 was used, 3-arylbutenolides **168** were exclusively formed in high yields irrespective of the substitution patterns at the propargylic positions. In contrast, the use of bidentate dppe instead of P^tBu_3 resulted in the formation of regioisomer mixtures of **168** and **168′**. The regioselectivity in favor of 4-arylbutenolides **168′** decreased when the steric bulk of the propargylic substituents increased. On the other hand, Arcadi and co-workers developed the selective synthesis of 4-arylbutenolides using rhodium-catalyzed alkyne hydroarylation [100]. Hydroarylation of 4-hydroxypropiolate **172** with phenylboronic acid (2 equiv.) was performed using 3 mol% $[Rh(OH)(cod)]_2$ and 6 mol% dppb in dioxane/H_2O (10/1) at 100 °C to afford **173** in 90% yield. Although this selective butenolide formation was accompanied by a side reaction producing small amounts of dimeric products such as bisbutenolide **174,** various 4-arylbutenolides were formed in moderate to good yields using several modified catalysts.

Copper-catalyzed hydroarylation of 4-hydroxypropiolates **175** using phenylboronic acid (3 equiv.) proceeded under mild conditions to directly deliver 4-arylbutenolides **176,** albeit in moderate yields (Scheme 8.34) [101]. To improve the yield, TBS-protected 4-hydroxypropiolate **177** was used as a substrate for copper-catalyzed hydroarylation. After acidic deprotection, the desired products **178** and bisbutenolide **179** were obtained in 56% and 31% yields, respectively. In contrast, a similar reaction of homologous substrate

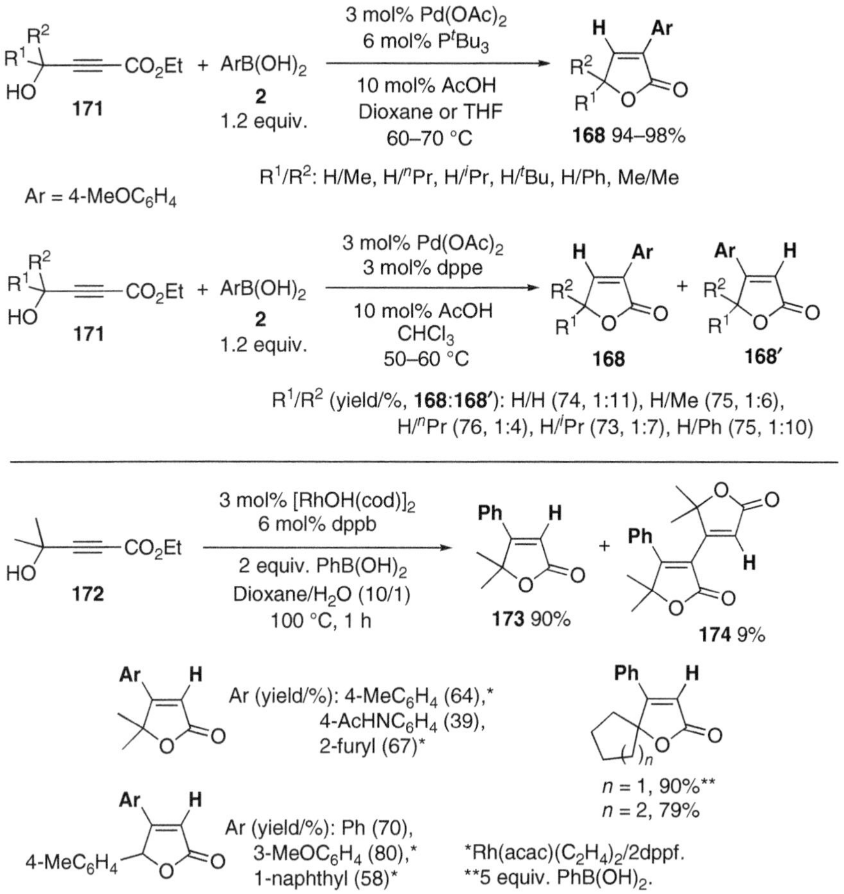

Scheme 8.33 Synthesis of butenolides via alkyne hydroarylation using arylboronic acids and subsequent lactonization.

180 with arylboronic acids **2** exclusively afforded 4-arylpentenolides **181** in 61–88% yields. These results suggest that the formation of the dimeric product was enhanced by coordination of the propargylic ether moiety. Moreover, (*o*-hydroxyphenyl)propiolate **182,** which has a hydroxyl group protected by MOM (methoxymethyl) group, underwent copper-catalyzed hydroarylation, affording 4-arylcoumarins **183** in 75–93% yields after deprotection [102].

4-Arylcoumarin **183** was also synthesized, albeit in a lower yield, via the reductive Heck reaction of alkynyl acetal **184** bearing a THP-protected phenol moiety and subsequent oxidation under acidic conditions (Scheme 8.35) [71]. Cacchi, Marinelli and co-workers also developed sequential processes involving reductive Heck reaction of propargylic alcohols **185** and **188** toward the synthesis of 4-arylchromenes **187.** In the first approach, syn hydroarylation products **186,** which were derived from phenyl acetate **185,** was deacetylated under basic conditions and subsequent Lewis-acid-promoted cation cyclization of the resultant phenol intermediates afforded 4-arylchromenes **187** in good yields

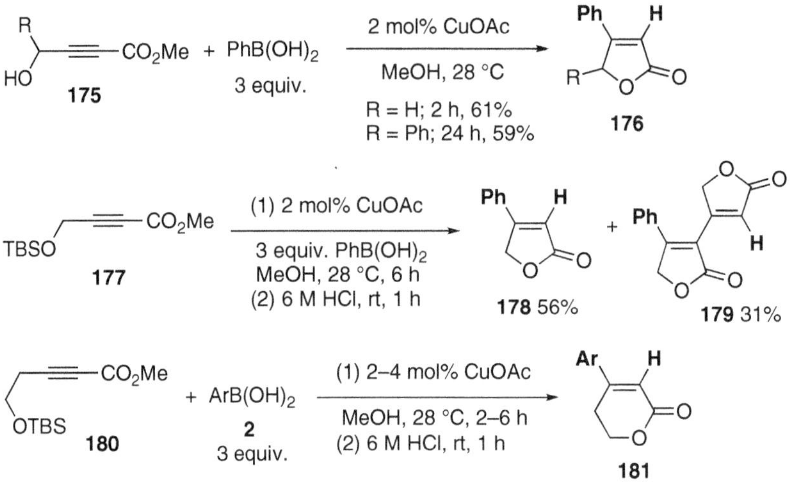

Ar (yield/%): Ph (86), 4-MeC_6H_4 (88), 4-$MeOC_6H_4$ (86), 4-ClC_6H_4 (88), 4-$OHCC_6H_4$ (61)

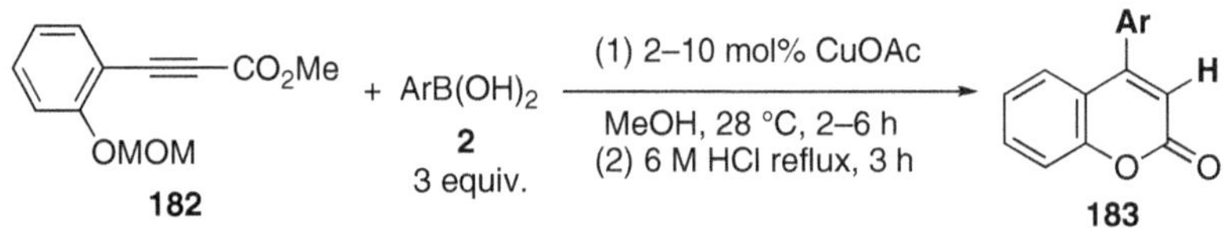

Ar (yield/%): Ph (89), 4-MeC_6H_4 (90), 3-MeC_6H_4 (93), 2-MeC_6H_4 (89), 4-$MeOC_6H_4$ (90), 4-ClC_6H_4 (88), 4-IC_6H_4 (89), 4-$OHCC_6H_4$ (89), 3-$O_2NC_6H_4$ (86), 2-naphthyl (75)

Scheme 8.34 Synthesis of butenolides, pentenolides, and coumarins via Cu-catalyzed alkyne hydroarylation using arylboronic acids and subsequent lactonization.

[103]. The other approach utilized the reductive Heck reaction of bromobenzene **188** under ligand-free conditions and subsequent palladium-catalyzed C—O bond-forming cyclization [104]. After the hydroarylation step, 2.5 mol% $Pd(OAc)_2$, 5 mol% dppf, and ^{t}BuONa were added and the second cyclization step was performed at 110 °C for 6 h, affording **187** in 57% overall yield. In a similar manner, spirocyclic chromene **189** was also obtained in a higher yield of 77%.

The synthesis of seven-membered lactones was accomplished by several different hydroarylation methods (Scheme 8.36). Marinelli and co-workers applied palladium-catalyzed reductive Heck reaction using aryl iodides **96** or rhodium-catalyzed hydroarylation using arylboronic acids **2** to benzoates **190** and **192,** which have propargylic alcohol substituents at the ortho positions [105]. Hydroarylation of these substrates with different regioselectivity, depending on the conditions, was followed by intramolecular transesterification, affording 5-arylbenzo[*c*]oxepin-1(3*H*)-ones **191** or 4-arylbenzo[*c*]oxepin-1(3*H*)-ones **193** in moderate yields. On the other hand, Rayabarapu and Cheng performed the reaction of *o*-iodobenzylic alcohols **194** with propiolates **195** in the presence of 5 mol% $NiBr_2$(dppe) as a catalyst and zinc powder in MeCN at 80 °C, to obtain benzo[*c*]oxepin-3(1*H*)-ones **196** in 44–81% yields [79]. They confirmed that

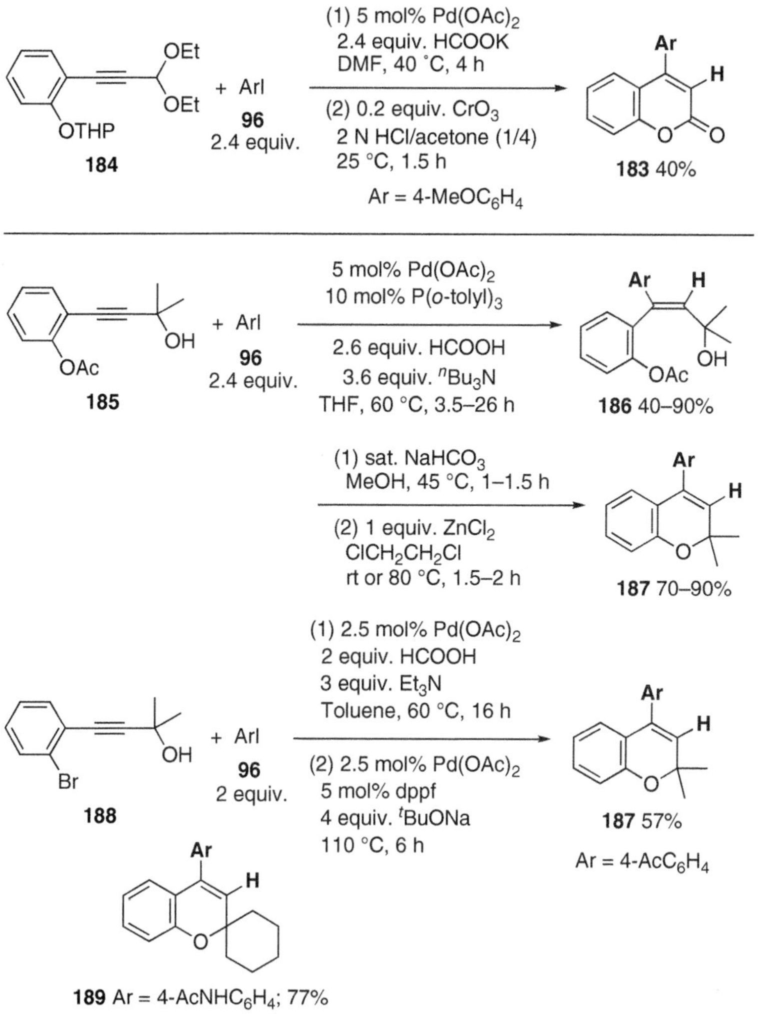

Scheme 8.35 Synthesis of coumarin and chromenes via reductive Heck reaction and subsequent cyclization.

hydroarylation proceeded at room temperature to generate *syn*-**197**, which was converted into the corresponding benzoxepinone in the presence of zinc salts (ZnX_2) at 80 °C probably via isomerization to *anti*-**197**.

8.4.1.2 Synthesis of Nitrogen and Phosphorous Heterocycles

Cacchi and co-workers synthesized quinolines via the reductive Heck reaction of acetanilides **198** bearing an alkynyl acetal substituent at the *ortho* position (Scheme 8.37) [69]. Ligand-free reductive Heck reaction of **198** with various aryl iodides **96** (2.4 equiv.) selectively generated syn hydroarylation products **199** along with small amounts of regioisomers **199′**, which were subsequently treated with 5 mol% *p*-toluenesulfonic acid (TsOH) in EtOH at 70 °C to afford 3-arylquinolines **200** as major products in 41–65% overall yields.

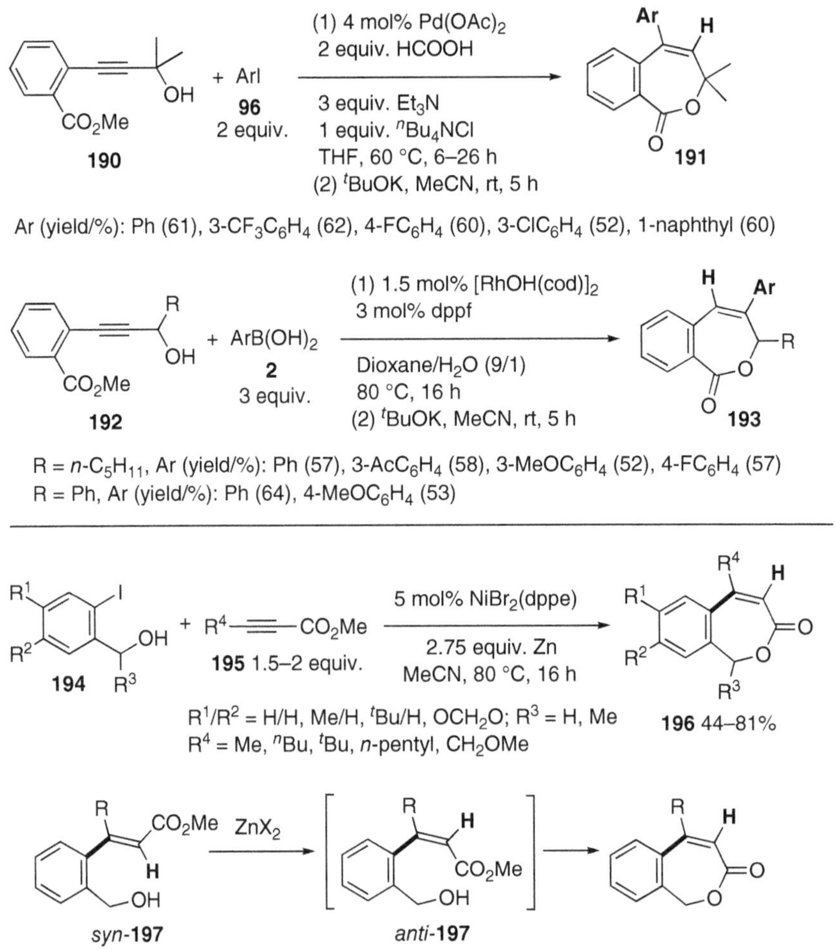

Scheme 8.36 Synthesis of benzoxepinones via alkyne hydroarylation and subsequent lactonization.

Moreover, reductive Heck reaction of **198** could be performed in an ionic liquid ([bmim]BF_4) at 60 °C in the presence of 2.5 mol% Pd**L5**, HCOOH (3 equiv.), and Et_3N (2 equiv.), directly affording quinolines **200** and **200′** in favor of the former [75].

Cacchi and co-workers attempted the synthesis of diarylquinolines via the reductive Heck reaction of (*o*-aminophenyl)ethynyl *p*-anisyl ketone (**201**) with *p*-iodotoluene (Scheme 8.38) [106]. After performing the reaction at 40 °C for 48 h under ligand-free conditions, a mixture of 2,4-diarylquinoline **202** and 2,3-diaryl isomer **202′** was obtained in a low combined yield with an almost 1:1 ratio. Because hydroarylations of alkynes bearing electron-withdrawing groups using arylboron reagents are generally regioselective, Marinelli and co-workers investigated rhodium-catalyzed hydroarylation of **201** with arylboronic acids **2** [107]. In the presence of 3.3 mol% Rh(acac)$(C_2H_4)_2$ and 6.6 mol% dppf, **201** and **2** (3–5 equiv.) were heated in dioxane/H_2O (10/1) at 100 °C, affording

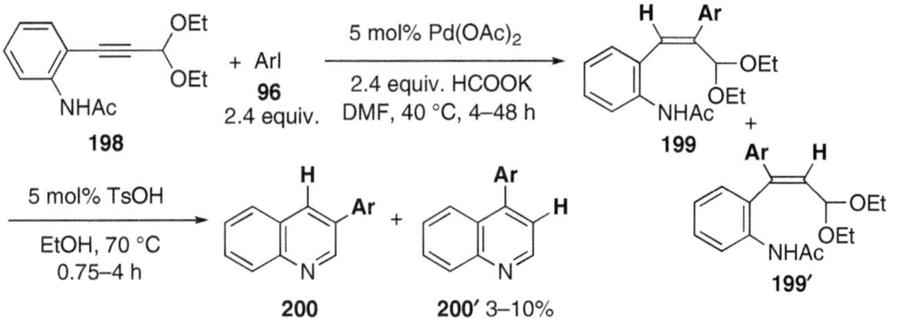

Ar (yield/%): 4-$MeOC_6H_4$ (56), Ph (51), 4-MeC_6H_4 (51), 4-AcC_6H_4 (61),* 4-$MeO_2CC_6H_4$ (49), 3-$CF_3C_6H_4$ (52), 3-FC_6H_4 (54), 3-$O_2NC_6H_4$ (46),* 4-$AcHNC_6H_4$ (50),* 3-$AcHNCH_2C_6H_4$ (41),* 4-HOC_6H_4 (63), 3-$HOCH_2C_6H_4$ (65),* 4-HO-3-MeC_6H_3 (45)*

***200′** were formed.

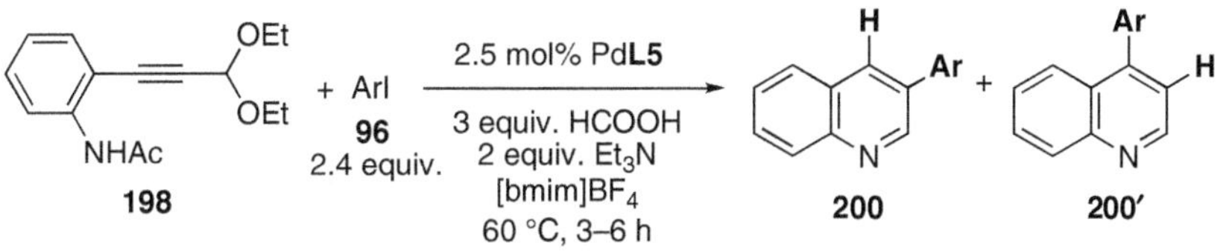

Ar (**200**/**200′** yield/%): 4-$MeOC_6H_4$ (56/9), 4-MeC_6H_4 (67/12), 4-FC_6H_4 (67/13), 4-$EtO_2CC_6H_4$ (74/21)

Scheme 8.37 Synthesis of quinolines via reductive Heck reaction and subsequent cyclization.

2,4-diarylquinolines **202** as sole products in 50–82% yields. Furthermore, an alternative protocol based on palladium-catalyzed hydroarylation using arylboronic acids **2** or potassium aryltrifluoroborate was also developed: the reaction of **201** proceeded in EtOH at 80 °C in the presence of 5 mol% $Pd(OAc)_2$, 5 mol% dppe, and AcOH (0.15 equiv.) to deliver 2,4-diarylquinolines **202** in improved yields (60–93%) [108]. Using this protocol, 2-cyclohexenyl derivatives **203** were also obtained in 60–74% yields.

2,4-Disubstituted quinolines **206** were synthesized in 56–91% yields using the nickel-catalyzed reductive Heck type reaction of *o*-iodoanilines **204** with alkynyl phenyl ketones **205** (Scheme 8.39) [109]. Cheng and Korivi proposed that arylnickel species underwent syn carbometallation across the triple bond of **205** to regio- and stereoselectively generate vinyl nickel intermediates, which were protonated by adventitious H_2O to afford *syn*-**207**. Although *syn*-**207** is unable to give the final product **206,** syn-**207** was assumed to isomerize to *anti*-**207** via reversible addition/elimination of nucleophiles such as H_2O or hydroxide. On the other hand, 2,3-disubstituted quinolines were obtained from the rhodium-catalyzed hydroarylation products from 3-arylprop-2-yn-1-ols and arylboronic acids via a three-step procedure [37].

A dual catalyst system was developed to realize a sequential process involving alkyne hydroarylation and intramolecular amidation by Lautens and co-workers [110]. They accomplished the synthesis of 3-aryl-1,2-dihydroquinolines **209** from

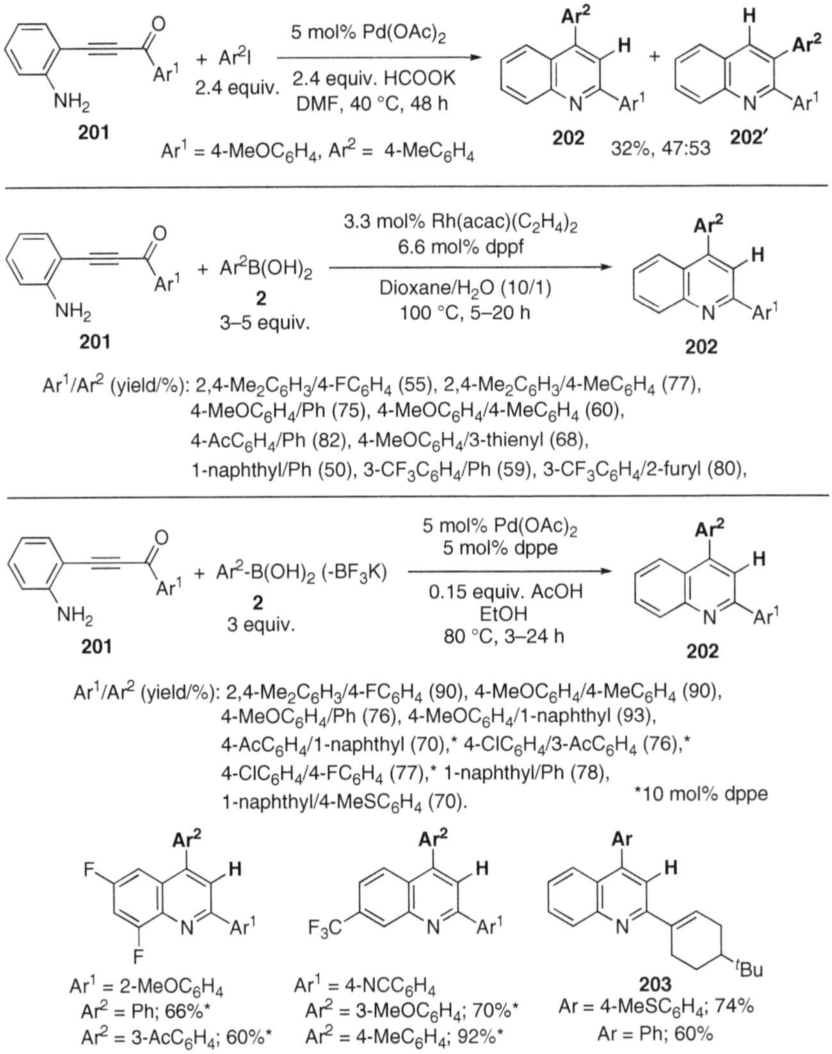

Scheme 8.38 Synthesis of diarylquinolines via alkyne hydroarylation using *p*-iodotoluene or arylboron reagents and subsequent cyclization.

N-propargyl sulfonamide **208** bearing an *o*-chlorophenyl terminal group and arylboronic acids **2** (2 equiv.) in a single process (Scheme 8.40). In this process, two individual catalytic reactions, rhodium-catalyzed alkyne hydroarylation and palladium-catalyzed intramolecular amidation of a chloroarene, proceeded without interference from each other. The addition of premixing solutions of each catalyst is important for this dual catalyst system to operate. As such, they successfully obtained a product yield comparable to that for the corresponding two-step process. Using this sequential process, various products **210,** including a 1,2-dihydro-1,5-naphthylidine derivative, were successfully synthesized in 45–78% yields.

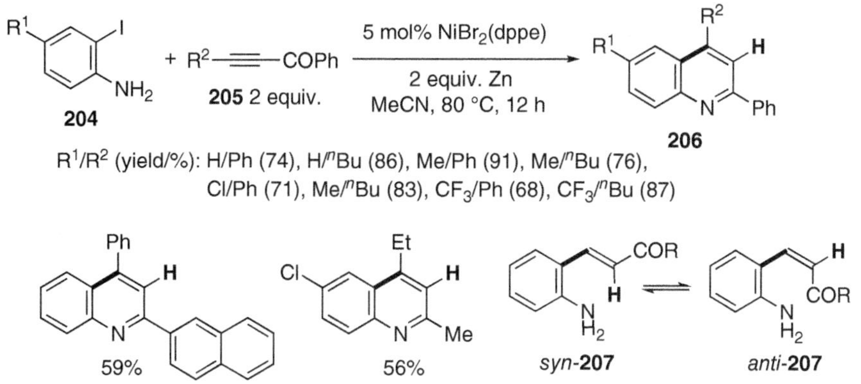

Scheme 8.39 Synthesis of 2,4-disubstituted quinolines via Ni-catalyzed reductive Heck reaction of *o*-iodoaniline with alkynones and subsequent cyclization.

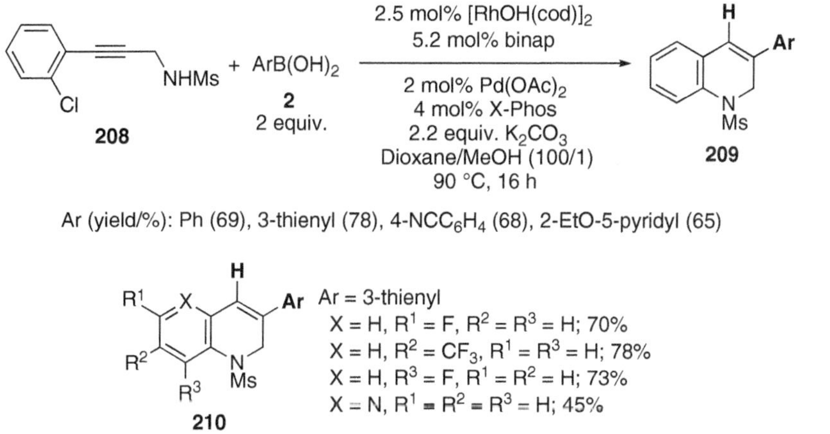

Scheme 8.40 Synthesis of 3-aryl-1,2-dihydroquinolines via Rh-catalyzed hydroarylation of propargylamines with arylboronic acids and subsequent Pd-catalyzed intramolecular amidation.

Scheme 8.41 outlines the synthesis of 3-arylindole-2-carboxylates **213** via hydroarylation product **212** [111]. Copper-catalyzed hydroarylation of (*o*-nitrophenyl)propiolate **211** using arylboronic acids **2** (2 equiv.) in MeOH at 28 °C afforded syn β-addition product **212** in 65–79% yields. Isolated **212** were then subjected to Cadogan cyclization using 5 mol% $MoCl_2O_2(dmf)_2$ and PPh_3 (2.4 equiv.) in toluene at 80 °C, affording indole-2-carboxylates **213** in 90–95% yields.

In addition to nitrogen heterocycles discussed earlier, phosphorous heterocycles, 1,2-oxaphosphole 2-oxides, were synthesized via sequential alkyne hydroarylation/cyclization (Scheme 8.42) [112]. Gao and co-workers performed the reaction of 3-hydroxypropynyl phosphonate **214** with arylboronic acids **2** (1.2 equiv.) in the presence of 3 mol% $Pd(OAc)_2$, 6 mol% dppb, and 10 mol% AcOH in dioxane at 60–70 °C to obtain 4-arylated cyclic phosphonates **215**

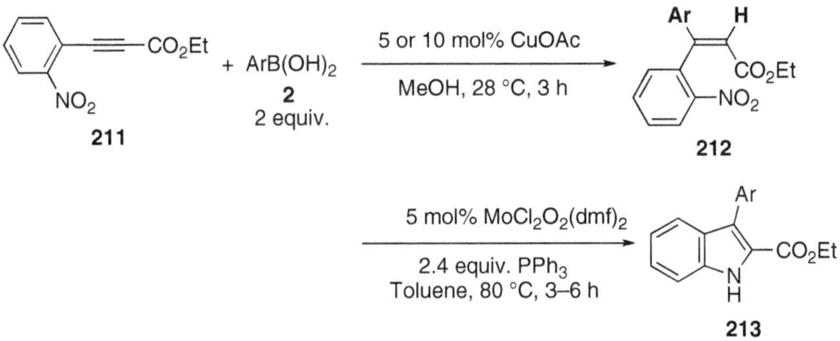

Ar (**212**/**213** yield/%): Ph (75/95), 4-MeC_6H_4 (77/94), 4-$MeOC_6H_4$ (65/92), 3-$MeOC_6H_4$ (67/94), 4-FC_6H_4 (74/90), 4-ClC_6H_4 (74/93), 4-BrC_6H_4 (78/90), 4-IC_6H_4 (79/91), 4-$EtO_2CC_6H_4$ (75/90), 4-$OHCC_6H_4$ (69/95)

Scheme 8.41 Synthesis of 3-arylindole-2-carboxylates via Cu-catalyzed hydroarylation of (*o*-nitrophenyl)propiolate with arylboronic acids and subsequent Cadogan cyclization.

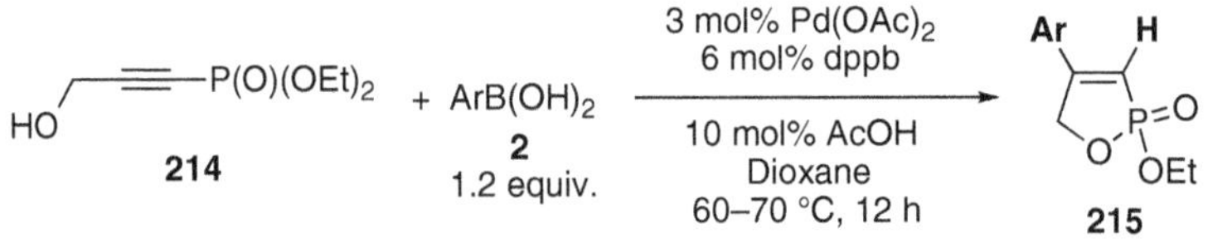

Ar (yield/%): Ph (98), 4-MeC_6H_5 (98), 4-$^nBuC_6H_5$ (91), 4-MeC_6H_5 (98), 4-$(Ph_2N)C_6H_5$ (78), 4-FC_6H_5 (81), 4-$OHCC_6H_5$ (46), 4-AcC_6H_5 (61), 4-$(HOCH_2)C_6H_5$ (52), 2-naphthyl (76), 3,4-methylenedioxyphenyl (98)

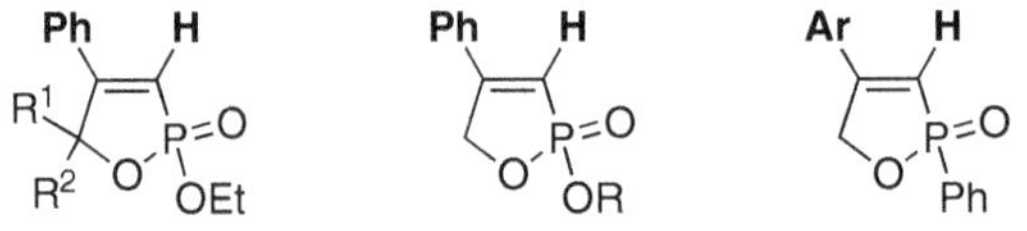

R^1 = Me, R^2 = H; 80%
R^1 = R^2 = Me; 90%

R = Me; 95%
R = ^{i}Pr; 47%

216 64–98%

Scheme 8.42 Synthesis of 4-aryl-1,2-oxaphospholene 2-oxides via Pd-catalyzed hydroarylation of 3-hydroxypropynyl phosphonates with arylboronic acids and subsequent cyclization.

in 46–98% yields. In a similar manner, cyclic phosphinates **216** were also synthesized in 64–98% yields.

8.4.1.3 Synthesis of Carbocycles

As discussed, sequential processes involving alkyne hydroarylations using arylboron reagents and aryl halides have found wide synthetic application in the preparation of oxygen, nitrogen, and phosphorous heterocycles. In contrast, sequential alkyne hydroarylation and carbocyclization has been underdeveloped, although Lautens and co-workers reported the synthesis of indenes via Friedel–Crafts cyclization of 3-arylpropenyl alcohols, which were

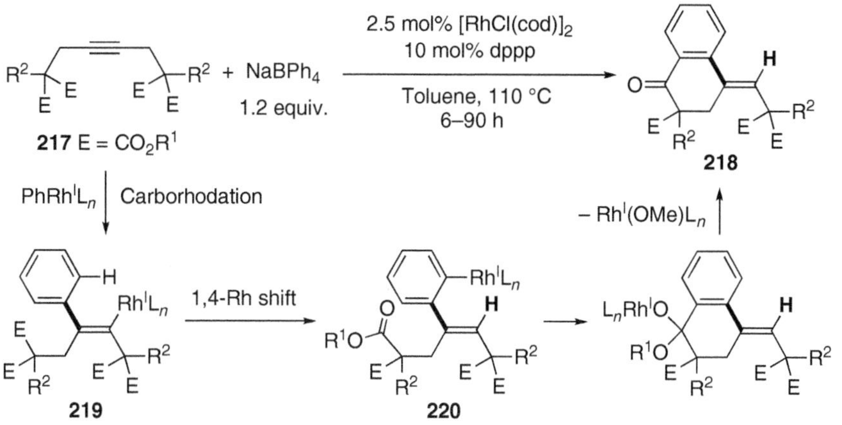

R¹/R² (yield/%): Me/Me (89), Et/Me (70), *ⁱ*Pr/Me (63), Me/*ⁱ*Bu (81), Me/Ph (69)

Scheme 8.43 Synthesis of α-tetralones via sequential process involving Rh-catalyzed alkyne hydroarylation and 1,4-Rh shift.

prepared by rhodium-catalyzed hydroarylation of 3-arylpropynyl alcohols using arylboronic acids (Scheme 8.9) [37]. Nevertheless, novel sequential processes that utilize a 1,4-rhodium shift have been described. Murakami and co-workers investigated rhodium-catalyzed reaction of internal alkynes **217** bearing two malonate moieties with phenylboron reagents (Scheme 8.43) [113]. Although a normal hydroarylation product was observed using phenylboronic acid, the use of $NaBPh_4$ (1.2 equiv.) in the presence of 2.5 mol% $[RhCl(cod)]_2$ and 10 mol% dppp in toluene at 110 °C under strictly dry conditions led to the formation of α-tetralones **218** in 63–89% yields. This interesting reaction was considered to proceed via carborhodation of **217** generating vinylrhodium intermediates **219,** which underwent a 1,4-Rh shift followed by intramolecular acylation of resultant arylrhodium intermediates **220.**

Asymmetric syntheses of carbocycles via alkyne hydroarylation involving a similar 1,4-Rh shift were also achieved by Shintani, Hayashi and co-workers. Optically active 3,3-disubstituted 1-indanones **223** were synthesized from alkynyl phenyl ketone **221** bearing a bulky triethylsilyl terminal group (Scheme 8.44) [114]. They used the combination of 2.5 mol% $[RhOH(cod)]_2$ and 5 mol% (*R*)-segphos (5,5′-bis(diphenylphosphino)-4,4′-bi-1,3-benzodioxole) as a chiral catalyst to obtain **223** in high yields with excellent enantioselectivity. This reaction started with regioselective syn carborhodation of **221,** resulting in the formation of vinylrhodium intermediates **224.** The subsequent 1,4-Rh shift was followed by intramolecular 1,4-addition of resultant arylrhodium intermediates **225,** affording **223.** Substituents on the phenyl ring of the substrates as well as other silyl and triethylgermyl terminal groups were tolerated.

On the other hand, cyclohexenones **226** bearing homopropargyl substituents at the β position underwent sequential reaction with $NaBPh_4$ (1.2–1.5 equiv.) in the presence of 2.5 mol% $[RhCl(cod)]_2$ and H_2O (2 equiv.) in THF at 65 °C, affording interesting spirocycloalkanones **227** in 60–74% yields

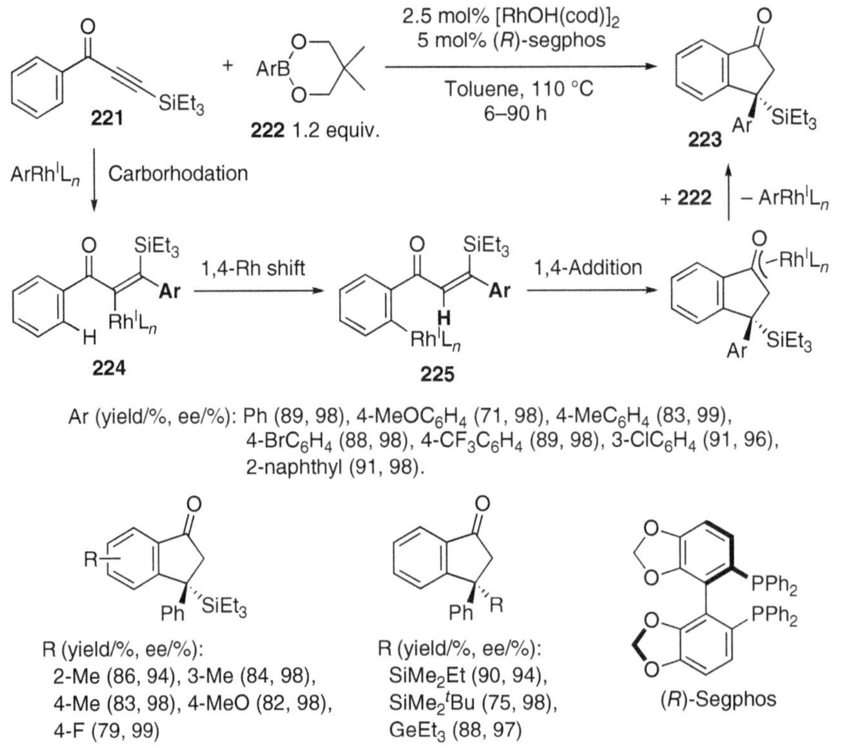

Scheme 8.44 Asymmetric synthesis of 3,3-disubstituted 1-indanones via sequential process involving Rh-catalyzed alkyne hydroarylation and 1,4-Rh shift.

(Scheme 8.45) [115]. This reaction also involved regioselective syn carborhodation and the subsequent 1,4-Rh shift. Arylrhodium intermediates **228** underwent spirocyclization via 1,4-addition to generate **229,** which were protonated to afford **227.** Similar cyclopentenone derivatives could be used as the substrates and their reaction with various tetraarylborates afforded spirocycles **230** in 62–77% yields. The asymmetric version of this reaction was further accomplished using chiral diene ligand **L6,** affording the corresponding products with the excellent enantiocontroll of quaternary spirocarbon center.

As discussed in this section, unprecedented sequential processes involving carborhodation and the subsequent 1,4-Rh shift were developed and asymmetric constructions of all carbon quaternary stereocenters were also accomplished using chiral rhodium catalysts. Other sequential processes starting from arylation of alkynes were reviewed in literatures [116, 117].

8.4.2 Synthesis of Bioactive Compounds and Natural Products via Alkyne Hydroarylations Using Arylboron Reagents and Aryl Halides

In order to illustrate the practical significance of selective alkyne hydroarylation using arylboron reagents and aryl halides, their applications to the synthesis of bioactive compounds and natural products will be surveyed in this section.

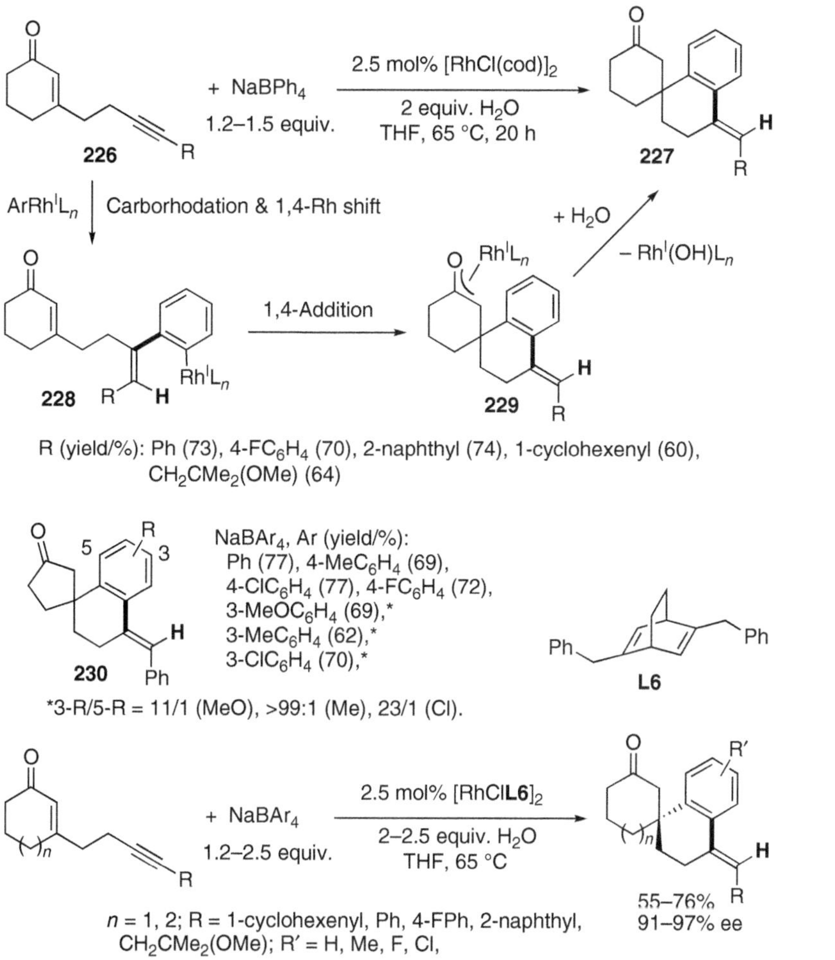

Scheme 8.45 Synthesis of spirocycles via sequential process involving Rh-catalyzed alkyne hydroarylation and 1,4-Rh shift.

The first example is the stereo-defined synthesis of conformationally restrained analogs of GR85305 (**231**), which has thromboxane antagonist and thromboxane synthase inhibitory activities (Scheme 8.46) [118]. Intramolecular Stille coupling was used for making a strategic C—C bond in the synthesis of **231.** However, this strategy proved to be inefficient. Thus, Finch and co-workers subjected alkyne substrates **232** and **233** to reductive Heck cyclization using Pd(OAc)$_2$/P(*o*-tolyl)$_3$ as a catalyst and HCOOH as a hydride source (detailed conditions are not provided in the original report). Accordingly, the desired 7-*exo* cyclization products (*Z*)-**234** and (*E*)-**234** were obtained in 56% and 60% yields, respectively. Because an initial attempt of 7-*exo* radical cyclization was not successful, the above reductive Heck cyclizations are of great significance as a stereospecific route to both stereoisomers of **234,** which were ultimately converted into the targeted (*Z*)/(*E*)-**231** in two steps.

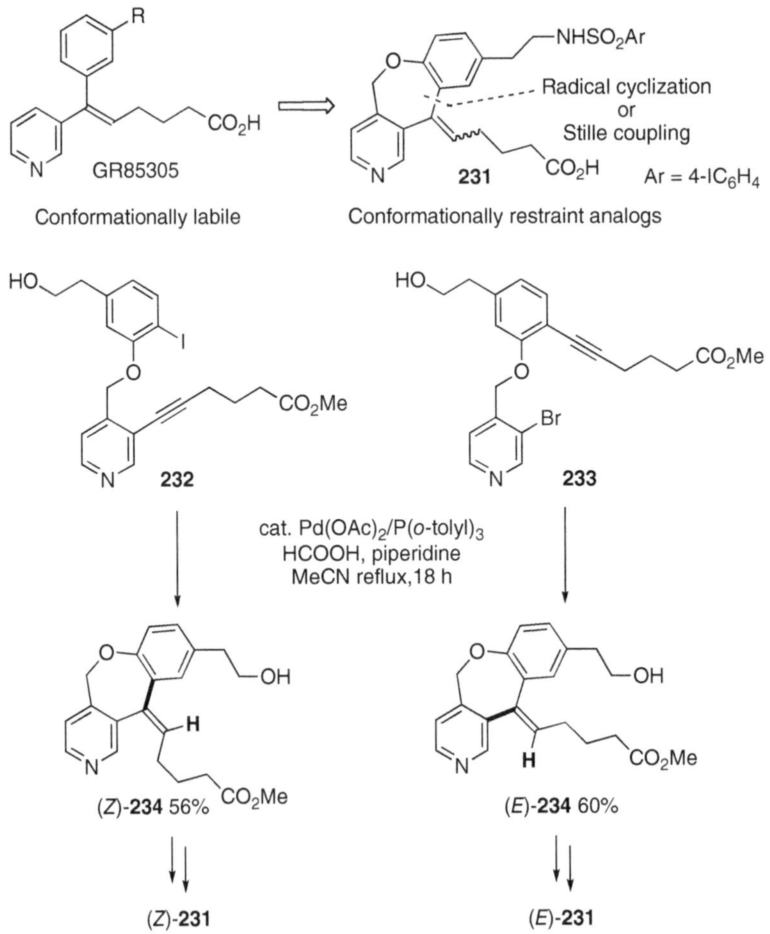

Scheme 8.46 Synthesis of conformationally restrained analogs of GR85305 via Pd-catalyzed reductive Heck cyclization.

3,3-Diarylacrylamides **235** and **236**, which possess 2-aminobenzimidazole and fluorinated benzene rings at the β positions to the amide moieties, are antirhinoviral agents (Scheme 8.47). Mitchell and co-workers investigated the synthesis of these targets using the reductive Heck reaction of iodide **237** and propiolamides **238** or **239** [74]. However, conventional protocols resulted in the formation of several unwanted products including regioisomers, a protodeiodination product from **237**, and semihydrogenation side products from the propiolamides. Therefore, they used their own ligand-free conditions using $Pd(dba)_2$ as a catalyst in ethyl acetate. Consequently, the reaction of **237** with **238** (1.3 equiv.) effectively proceeded with high regioselectivity (>17:1), affording the desired syn hydroarylation product **235** in 81% yield. Similarly, the use of **239** led to the formation of **236** in 75% yield, albeit with decreased regioselectivity (4.2:1). In addition, an attempt to synthesize a stereoisomer of **235** from propiolamide **240** and 3-fluoroiodobenzene was not successful.

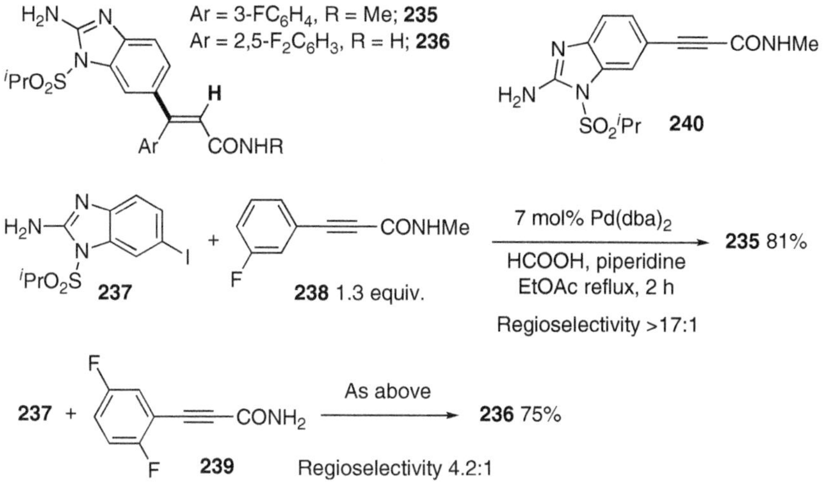

Scheme 8.47 Synthesis of antirhinoviral agents via Pd-catalyzed reductive Heck reaction.

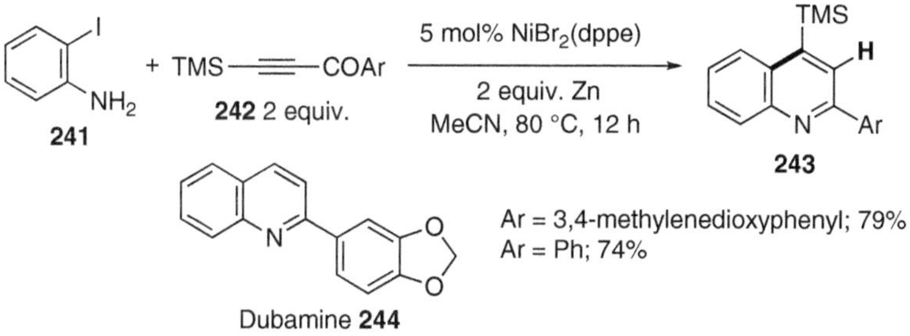

Scheme 8.48 Synthesis of 2-arylquinolines via Ni-catalyzed reductive Heck type reaction and subsequent cyclization.

Cheng and Korivi applied their sequential process via nickel-catalyzed reductive Heck reaction of *o*-iodoaniline (**241**) with TMS-protected alkynyl aryl ketones **242** (Scheme 8.48) [109]. As a result, 4-TMS-2-arylquinolines **243** were obtained in good yields, and protodesilylation using tetrabutylammonium fluoride converted **243** into biologically active 2-arylquinolines such as natural alkaloid dubamine (**244**).

The synthesis of unnatural analogs of 3-benzazepine alkaloid (−)-aphanorphine was reported by Van der Eycken and co-workers (Scheme 8.49) [119]. The desired compounds **245** were considered to be obtained from lactam precursors **246** according to the previous synthesis of (−)-aphanorphine. Thus, optically active aryl bromides **247** bearing pendant propiolamide moieties were prepared and subjected to reductive Heck cyclization under microwave irradiation conditions using palladacycle catalyst **248** in DMF/H_2O (3/1) at 110 °C. The cyclization effectively proceeded in 20 min to selectively afford 7-*exo* cyclization products **249** in 60–77% yields. Subsequently, **249** were converted into lactams **246** in two steps.

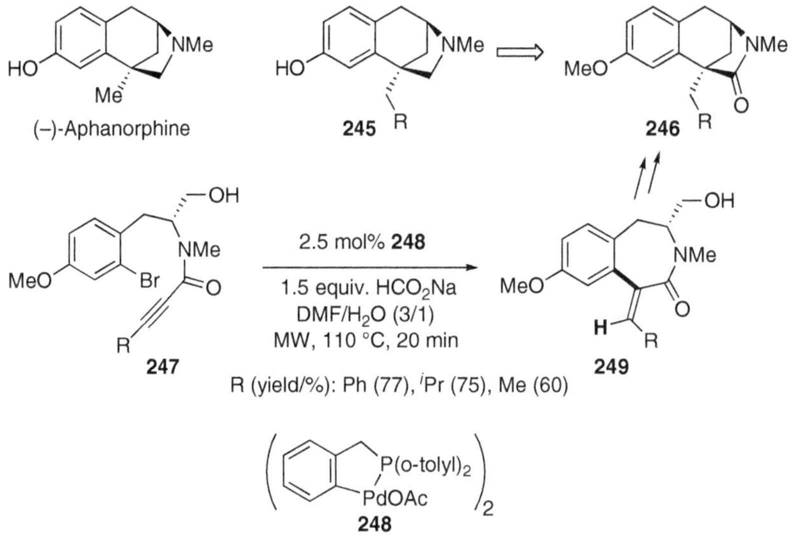

Scheme 8.49 Synthesis of analogs of (−)-aphanorphine via Pd-catalyzed reductive Heck cyclization.

Naturally occurring 4-arylcoumarins are known as neoflavones and display a wide range of biological activities (Scheme 8.50). Yamamoto and Kirai synthesized natural neoflavones via copper-catalyzed syn-selective hydroarylation of MOM-protected (*o*-hydroxyphenyl)propiolates **250** using arylboronic acids **2** and subsequent deprotection [102]. The significance of their strategy is demonstrated by the modular synthesis of seven natural products in 76–98% yields by combining three alkyne substrates with four arylboronic acids.

The synthetic utility of copper-catalyzed alkyne hydroarylation using arylboronic acids was further demonstrated in the stereo-defined synthesis of an artificial antimitotic agent CC-5079 (Scheme 8.51) [60]. Although this molecule was originally synthesized as a mixture of stereosiomers using Wittg reaction, copper-catalyzed hydroarylation of propynenitrile **251** with arylboronic acid **252** afforded (*E*)-stereoisomer of CC-5079 in 84% yield. The (*Z*)-isomer was also obtained from the reaction of propynenitrile **253** with arylboronic acid **254,** albeit in a lower yield of 71%.

In conclusion, reductive Heck reaction has been successfully applied to the synthesis of bioloactive compounds and natural products. In contrast, alkyne hydroarylation using arylboron reagents has been almost omitted in terms of such synthetic applications, although the copper-catalyzed method was effectively utilized for the modular synthesis of naturally occuring neoflavones and the stereospecific synthesis of antimimetic agent CC-5079.

8.5 Summary

As described in the preceeding discussions, formal alkyne hydroarylations catalyzed by transition metals have been extensively developed using

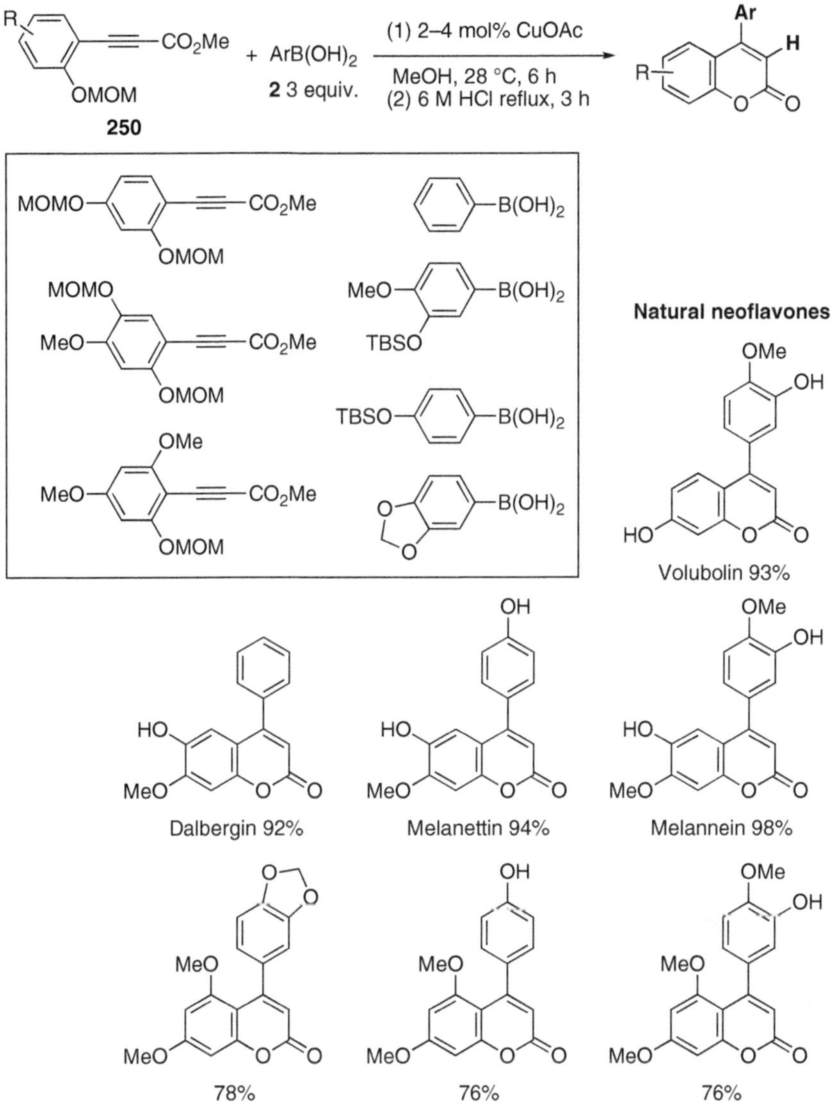

Scheme 8.50 Synthesis of natural neoflavones via Cu-catalyzed alkyne hydroarylation using arylboronic acids and subsequent cyclization.

arylboron reagents as well as aryl halides as arylating agents. Among them, palladium-catalyzed reductive Heck reaction using formic acid or congeners has been widely applied to various types of alkynes, although regioselectivity generally depends on the substrates and reaction conditions. Intramolecular reductive Heck reaction, has also proved to be highly powerful for the construction of medium-ring heterocyclic compounds. On the other hand, relatively new methods using bench-top stable and readily available arylboron reagents have rapidly progressed as no hydride donor reagent is required and

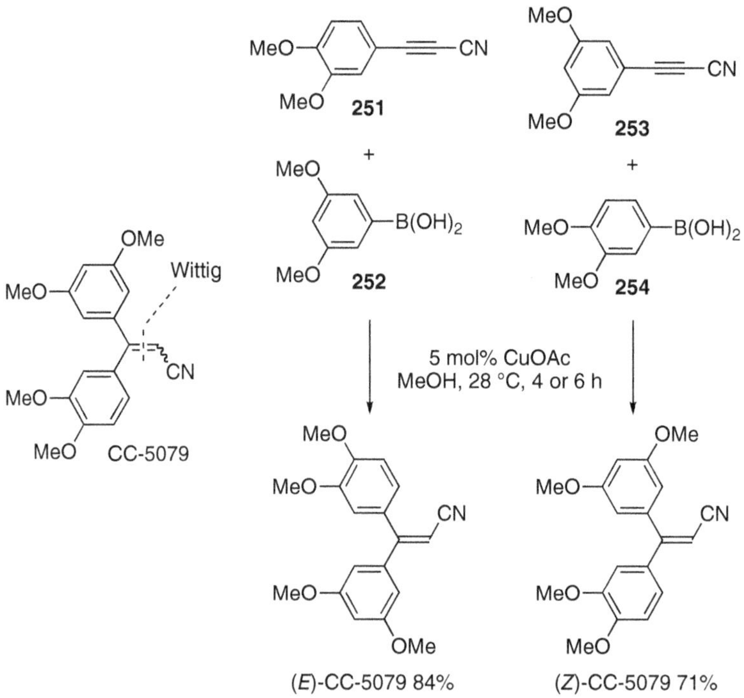

Scheme 8.51 Synthesis of CC-5079 via Cu-catalyzed hydroarylation of propynenitriles using arylboronic acids.

regioselectivity can be better controlled compared to reductive Heck reaction. Because multi-substituted alkenes can be obtained stereoselectively via a common fundamental step, syn carbometallation, both methods have found wide synthetic applications to bioactive compounds and natural products. Therefore, it is expected that selective alkyne hydroarylations using arylborons and aryl halides will be increasingly featured as key transformations in the total synthesis of complex natural and artificial molecules in the future.

References

1 Marek, I., Chinkov, N., and Banon-Tenne, D. (2004, Chapter 7) Carbometallation Reactions, in *Metal-Catalyzed Cross-Coupling Reactions, 2nd, Completely Revised and Enlarged Edition* (eds A. de Meijere and F. Diederich), Wiley-VCH, Weinheim, pp. 395–478.
2 Ding, A. and Guo, H. (2014, Vol. 4, Chapter 4.15) Carbometalation and Heterometalation Reactions of Alkenes, Alkynes, and Allenes, in *Comprehensive Organic Synthesis II* (eds P. Knochel, G.A. Molander, S. Ma, and J. Zhang), Elsevier, Amsterdam, pp. 891–938.
3 Flynn, A.B. and Ogilvie, W.W. (2007) *Chem. Rev.*, **107**, 4698–4745.

4 Wang, J. (ed.), Topics in Current Chemistry 327 (2012) *Stereoselective Alkene Synthesis*, Springer-Verlag, Berlin.
5 Normant, J.F. and Alexakis, A. (1981) *Synthesis*, 841–870.
6 Vasil'ev, A.V. (2009) *Russ. J. Org. Chem.*, **45**, 1–17.
7 Nevado, C. and Echavarren, A.M. (2005) *Synthesis*, 167–182.
8 Goj, L.A. and Gunnoe, T.B. (2005) *Curr. Org. Chem.*, **9**, 671–685.
9 Bandini, M., Emer, E., Tommasi, S., and Umani-Ronchi, A. (2006) *Eur. J. Org. Chem.*, 3527–3544.
10 Kitamura, T. (2009) *Eur. J. Org. Chem.*, 1111–1125.
11 Wang, X., Zhou, L., and Lu, W. (2010) *Curr. Org. Chem.*, **14**, 289–307.
12 Fallis, A.G. and Forgione, P. (2001) *Tetrahedron*, **57**, 589–5913.
13 Normant, J.-F. (1994, Chapter 11) Alkyne carbocupration and polyene synthesis, in *Organocopper Reagents, A Practical Approach* (ed. R.J.K. Taylor), Oxford University Press, Oxford, pp. 237–256.
14 Murakami, K. and Yorimitsu, H. (2013) *Beilstein J. Org. Chem.*, **9**, 278–302.
15 Shirakawa, E., Ikeda, D., Ozawa, T., Watanabe, S., and Hayashi, T. (2009) *Chem. Commun.*, 1885–1887.
16 Hall, D.G. (2011) *Boronic Acids: Preparation and Applications in Organic Synthesis, Medicine and Materials*, 2nd edn Wiley-VCH, Weinheim.
17 Miyaura, N. and Yamamoto, Y. (2006Vol. 9, Chapter 9.05) Boron, in *Comprehensive Organometallic Chemistry III, From Fundamentals to Applications* (eds R.H. Crabtree, M.P. Mingos, and P. Knochel), Elsevier, Amsterdam, pp. 146–244.
18 Miyaura, N. (2004, Chapter 2,) Metal-Catalyzed Cross-Coupling Reactions of Organoboron Compounds with Organic Halides, in *Metal-Catalyzed Cross-Coupling Reactions, 2nd, Completely Revised and Enlarged Edition* (eds A. de Meijere and F. Diederich), Wiley-VCH, Weinheim, pp. 41–123.
19 Suzuki, A. (2011) *Angew. Chem. Int. Ed.*, **50**, 6723–6737.
20 Hayashi, T., Inoue, K., Taniguchi, N., and Ogasawara, M. (2001) *J. Am. Chem. Soc.*, **123**, 9918–9919.
21 Shirakawa, E., Takahashi, G., Tsuchimoto, T., and Kawakami, Y. (2001) *Chem. Commun.*, 2688–2689.
22 Cacchi, S., Felici, M., and Pietroni, B. (1984) *Tetrahedron Lett.*, **25**, 3137–3140.
23 Burns, B., Grigg, R., Sridharan, V., and Worakun, T. (1988) *Tetrahedron Lett.*, **29**, 4325–4328.
24 Cacchi, S. (1990) *Pure Appl. Chem.*, **62**, 713–722.
25 Cacchi, S. (1999) *J. Organomet. Chem.*, **576**, 42–64.
26 Abbiati, G., Arcadi, A., Marinelli, F., and Rossi, E. (2014) *Synthesis*, **46**, 687–721.
27 Yamamoto, Y. (2014) *Chem. Soc. Rev.*, **43**, 1575–1600.
28 Sakai, M., Hayashi, H., and Miyaura, N. (1997) *Organometallics*, **16**, 4229–4231.
29 Hayashi, T. and Yamasaki, K. (2003) *Chem. Rev.*, **103**, 2829–2844.
30 Lautens, M. and Yoshida, M. (2002) *Org. Lett.*, **4**, 123–125.
31 Lautens, M. and Yoshida, M. (2003) *J. Org. Chem.*, **68**, 762–769.

32 Genin, E., Michelet, V., and Genêt, J.-P. (2004) *Tetrahedron Lett.*, **45**, 4157–4161.
33 Genin, E., Michelet, V., and Genêt, J.-P. (2004) *J. Organomet. Chem.*, **689**, 3820–3830.
34 Murakami, M. and Igawa, H. (2002) *Helv. Chim. Acta*, **85**, 4182–4188.
35 Tsui, G.C. and Lautens, M. (2010) *Angew. Chem. Int. Ed.*, **49**, 8938–8941.
36 Gourdet, B., Smith, D.L., and Lam, H.W. (2010) *Tetrahedron*, **66**, 6026–6031.
37 Panteleev, J., Huang, R.Y., Lui, E.K.J., and Lautens, M. (2011) *Org. Lett.*, **13**, 5314–5317.
38 Arcadi, A., Aschi, M., Chiarini, M., Ferrara, G., and Marinelli, F. (2010) *Adv. Synth. Catal.*, **352**, 493–498.
39 Zhang, W., Liu, M., Wu, H., Ding, J., and Cheng, J. (2008) *Tetrahedron Lett.*, **49**, 5214–5216.
40 Jana, R. and Tunge, J.A. (2011) *J. Org. Chem.*, **76**, 8376–8385.
41 Chang, W.-T.T., Smith, R.C., Regens, C.S., Bailey, A.D., Werner, N.S., and Denmark, S.E. (2011) *Org. React.*, **75**, 213–745.
42 Fujii, T., Koike, T., Mori, A., and Osakada, K. (2002) *Synlett*, 295–297.
43 Nakao, Y., Takeda, M., Chen, J., and Hiyama, T. (2008) *Synlett*, 774–776.
44 Nakao, Y., Chen, J., Imanaka, H., Hiyama, T., Ishikawa, Y., Duan, W.-L., Shintani, R., and Hayashi, T. (2007) *J. Am. Chem. Soc.*, **129**, 9137–9143.
45 Lin, B., Liu, M., Ye, Z., Zhang, Q., and Cheng, J. (2009) *Tetrahedron Lett.*, **50**, 1714–1716.
46 Oh, C.H., Jung, H.H., Kim, K.S., and Kim, N. (2003) *Angew. Chem. Int. Ed.*, **42**, 805–808.
47 Oh, C.H. and Ryu, J.H. (2003) *Bull. Korean Chem. Soc*, **24**, 1563–1564.
48 Bush, A.G., Jiang, J.L., Payne, P.R., and Ogilvie, W.W. (2009) *Tetrahedron*, **65**, 8502–8506.
49 Kim, N., Kim, K.S., Gupta, A.K., and Oh, C.H. (2004) *Chem. Commun.*, 618–619.
50 Bai, Y., Yin, J., Kong, W., Mao, M., and Zhu, G. (2013) *Chem. Commun.*, **49**, 7650–7652.
51 Yang, Y., Wang, L., Zhang, F., and Zhu, G. (2014) *J. Org. Chem.*, **79**, 9319–9324.
52 Yang, Y., Wang, L., Zhang, J., Jin, Y., and Zhu, G. (2014) *Chem. Commun.*, **50**, 2347–2349.
53 Gupta, A.K., Kim, K.S., and Oh, C.H. (2005) *Synlett*, 457–460.
54 Xu, X., Chen, J., Gao, W., Wu, H., Ding, J., and Su, W. (2010) *Tetrahedron*, **66**, 2433–2438.
55 Zeng, H. and Hua, R. (2008) *J. Org. Chem.*, **73**, 558–562.
56 Robbins, D.W. and Hartwig, J.F. (2011) *Science*, **333**, 1423–1427.
57 Lin, P.-S., Jeganmohan, M., and Cheng, C.-H. (2008) *Chem. Eur. J.*, **14**, 11296–11299.
58 Yamamoto, Y., Kirai, N., and Harada, Y. (2008) *Chem. Commun.*, 2010–2012.
59 Kirai, N., Yamamoto, Y., Wipf, P., and Vowell, C.L. (2010) *Org. Synth.*, **87**, 53–58.
60 Yamamoto, Y., Asatani, T., and Kirai, N. (2009) *Adv. Synth. Catal.*, **351**, 1243–1249.

61 Yoo, K., Kim, H., and Yun, J. (2009) *Chem. Eur. J.*, **15**, 11134–11138.

62 Bräse, S. and de Meijere, A. (2004, Chapter 5) Cross-Coupling of Organyl Halides with Alkenes: the Heck Reaction, in *Metal-Catalyzed Cross-Coupling Reactions, 2nd, Completely Revised and Enlarged Edition* (eds A. de Meijere and F. Diederich), Wiley-VCH, Weinheim, pp. 217–315.

63 Oestreich, M. (ed.) (2009) *The Mizoroki-Heck Reaction*, John Wiley & Sons, Chichester.

64 Arcadi, A., Cacchi, S., and Marinelli, F. (1985) *Tetrahedron*, **41**, 5121–5131.

65 Arcadi, A., Cacchi, S., Ianelli, S., Marinelli, F., and Nardelli, M. (1986) *Gazz. Chim. Ital.*, **116**, 725–729.

66 Barros, J.C., de Souza, A.L.F., de Lima, P.G., da Silva, J.F.M., and Antunes, O.A.C. (2008) *Applied Organomet. Chem.*, **22**, 249–252.

67 Barros, J.C., Souza, A.L.F., da Silva, J.F.M., and Antunes, O.A.C. (2011) *Catal. Lett.*, **141**, 549–553.

68 Arcadi, A., Cacchi, S., and Marinelli, F. (1986) *Tetrahedron Lett.*, **27**, 6397–6400.

69 Cacchi, S., Fabrizi, G., Marinelli, F., Moro, L., and Pace, P. (1996) *Tetrahedron*, **52**, 10225–10240.

70 Ahlquist, M., Fabrizi, G., Cacchi, S., and Norrby, P.-O. (2006) *J. Am. Chem. Soc.*, **128**, 12785–12793.

71 Cacchi, S., Fabrizi, G., Moro, L., and Pace, P. (1997) *Synlett*, 1367–1370.

72 Arcadi, A., Cacchi, S., Fabrizi, G., Marinelli, F., and Parisi, L.M. (2003) *J. Organomet. Chem.*, **687**, 562–566.

73 Hay, L.A. and Mitchell, D. (1997) *Tetrahedron Lett.*, **38**, 6533–6536.

74 Hay, L.A., Koenig, T.M., Ginah, F.O., Copp, J.D., and Mitchell, D. (1998) *J. Org. Chem.*, **63**, 5050–5058.

75 Cacchi, S., Fabrizi, G., Goggiamani, A., Moreno-Mañas, M., and Vallribera, A. (2002) *Tetrahedron Lett.*, **43**, 5537–5540.

76 Cacchi, S., Fabrizi, G., and Goggiamani, A. (2004) *J. Mol. Catal. A: Chem.*, **214**, 57–64.

77 Wu, M.-J., Wei, L.-M., Lin, C.-F., Leou, S.-P., and Wei, L.-L. (2001) *Tetrahedron*, **57**, 7839–7844.

78 Cacchi, S., Fabrizi, G., Goggiamani, A., and Persiani, D. (2008) *Org. Lett.*, **10**, 1597–1600.

79 Rayabarapu, D.K. and Cheng, C.-H. (2002) *J. Am. Chem. Soc.*, **124**, 5630–5631.

80 Dorn, S.C.M., Olsen, A.K., Kelemen, R.E., Shrestha, R., and Weix, D.J. (2015) *Tetrahedron Lett.*, **56**, 3365–3367.

81 Grigg, R., Loganathan, V., and Sukirthalingam, S. (1990) *Tetrahedron Lett.*, **31**, 6573–6576.

82 Grigg, R., Loganathan, V., Sridharan, V., Stevenson, P., Sukirthalingam, S., and Worakun, T. (1996) *Tetrahedron*, **52**, 11479–11502.

83 Hodgson, D.M. and Wells, C. (1994) *Tetrahedron Lett.*, **35**, 1601–1604.

84 Tietze, L.F. and Schimpf, R. (1994) *Chem. Ber.*, **127**, 2235–2240.

85 Comins, D.L., Joseph, S.P., and Zhang, Y.-m. (1996) *Tetrahedon Lett.*, **37**, 793–796.

86 Barberan, O., Alami, M., and Brion, J.-D. (2001) *Tetrahedon Lett.*, **37**, 793–796.
87 Couty, S., Liégault, B., Meyer, C., and Cossy, J. (2004) *Org. Lett.*, **6**, 2511–2514.
88 Couty, S., Liégault, B., Meyer, C., and Cossy, J. (2006) *Tetrahedron*, **62**, 3882–3895.
89 Le Strat, F., Harrowven, D.C., and Maddaluno, J. (2005) *J. Org. Chem.*, **70**, 489–498.
90 Yanada, R., Obika, S., Inokuma, T., Yanada, K., Yamashita, M., Ohta, S., and Takemoto, Y. (2005) *J. Org. Chem.*, **70**, 6972–6975.
91 Donets, P.A. and Van der Eycken, E.V. (2007) *Org. Lett.*, **9**, 3017–3020.
92 Peshkov, V.A., Pereshivko, O.P., Donets, P.A., Mehta, V.P., and Van der Eycken, E.V. (2010) *Eur. J. Org. Chem.*, 4861–4867.
93 Peshkov, V.A., Van Hove, S., Donets, P.A., Pereshivko, O.P., Van Hecke, K., Van Meervelt, L., and Van der Eycken, E.V. (2011) *Eur. J. Org. Chem.*, 1837–1840.
94 Majumdar, K.C., Chakravorty, S., Ghosh, T., and Sridhar, B. (2009) *Synlett*, 3127–3130.
95 Majumdar, K.C., Ghosh, T., and Chakravorty, S. (2010) *Tetrahedon Lett.*, **51**, 3372–3375.
96 Majumdar, K.C., Ghosh, T., and Ponra, S. (2013) *Tetrahedon Lett.*, **54**, 4661–4665.
97 Joardar, S., Chakravorty, S., and Das, S. (2015) *Synlett*, **26**, 359–362.
98 Arcadi, A., Bernocchi, E., Burini, A., Cacchi, S., Marinelli, F., and Pietroni, B. (1988) *Tetrahedron*, **44**, 481–490.
99 Oh, C.H., Park, S.J., Ryu, J.H., and Gupta, A.K. (2004) *Tetrahedon Lett.*, **45**, 7039–7042.
100 Alfonsi, M., Arcadi, A., Chiarini, M., and Marinelli, F. (2007) *J. Org. Chem.*, **72**, 9510–9517.
101 Yamamoto, Y. and Kirai, N. (2010) *Hetrocycles*, **80**, 269–279.
102 Yamamoto, Y. and Kirai, N. (2008) *Org. Lett.*, **10**, 5513–5516.
103 Arcadi, A., Cacchi, S., Fabrizi, G., Marinelli, F., and Pace, P. (2000) *Eur. J. Org. Chem.*, 4099–4108.
104 Arcadi, A., Cacchi, S., Fabrizi, G., Marinelli, F., and Verdecchia, M. (2006) *Synlett*, 909–915.
105 Arcadi, A., Fabrizi, G., Goggiamani, A., and Marinelli, F. (2015) *J. Org. Chem*, **80**, 6986–6995.
106 Cacchi, S., Fabrizi, G., and Marinelli, F. (1999) *Synlett*, 401–404.
107 Abbiati, G., Arcadi, A., Marinelli, F., Rossi, E., and Verdecchia, M. (2006) *Synlett*, 3218–3224.
108 Arcadi, A., Aschi, M., Marinelli, F., and Verdecchia, M. (2008) *Tetrahedron*, **64**, 5354–5361.
109 Korivi, R.P. and Cheng, C.-H. (2006) *J. Org. Chem.*, **71**, 7079–7082.
110 Panteleev, J., Zhang, L., and Lautens, M. (2011) *Angew. Chem. Int. Ed.*, **50**, 9089–9092.
111 Yamamoto, Y., Yamada, S., and Nishiyama, H. (2011) *Adv. Synth. Catal.*, **353**, 701–706.

112 Lv, Y., Hu, G., Ma, D., Liu, L., Gao, Y., and Zhao, Y. (2015) *J. Org. Chem.*, **80**, 6908–6914.
113 Miura, T., Sasaki, T., Nakazawa, H., and Murakami, M. (2005) *J. Am. Chem. Soc.*, **127**, 1390–1391.
114 Shintani, R., Takatsu, K., and Hayashi, T. (2007) *Angew. Chem. Int. Ed.*, **46**, 3735–3737.
115 Shintani, R., Isobe, S., Takeda, M., and Hayashi, T. (2010) *Angew. Chem. Int. Ed.*, **49**, 3795–3798.
116 Miura, T. and Murakami, M. (2007) *Chem. Commun.*, 217–224.
117 Youn, S.W. (2009) *Eur. J. Org. Chem.*, 2597–2605.
118 Finch, H., Pegg, N.A., and Evans, B. (1993) *Tetrahedron Lett.*, **34**, 8353–8356.
119 Donets, P.A., Goeman, J.L., Van der Eycken, J., Robeyns, K., Van Meervelt, L., and Van der Eycken, E.V. (2009) *Eur. J. Org. Chem.*, 793–796.

9

Transition Metal-Catalyzed Hydroarylation of Allenes

Ross A. Widenhoefer

Duke University, French Family Science Center, Department of Chemistry, Durham, NC 27708-0346, USA

9.1 Introduction

The transition metal-catalyzed addition of the C—H bond of an arene across a C=C double bond of an unactivated alkene (i.e., alkene hydroarylation) has attracted considerable attention as an atom-economical approach to the functionalization of arenes with potential application to both the large-scale alkylation of simple arenes and the synthesis of complex molecules [1]. In comparison to the numerous advances made in the catalytic hydroarylation of alkenes, the catalytic hydroarylation of allenes remains less developed. Although the transition metal-catalyzed hydroarylation of allenes was first disclosed by Panunzi in 1983 [2], this area of inquiry received little attention in the subsequent two decades. In 2006, Widenhoefer and Nelson independently disclosed gold(I)-catalyzed methods for the 6-*exo*-hydroarylation of 2-(γ-allenyl)indoles [3] and *N*-(γ-allenyl)pyrroles [4], respectively, and since this time, interest in the transition metal-catalyzed hydroarylation of allenes as a route to allylic aromatic and heteroaromatic compounds has increased dramatically. As a result of these efforts, the scope of catalytic allene hydroarylation has expanded to include the *exo*- and *endo*- intramolecular hydroarylation and intermolecular hydroarylation of allenes employing a range of heteroaromatic and heteroatom-substituted arenes. Several examples of enantioselective allene hydroarylation have likewise been disclosed.

In contrast to the hydroarylation of simple alkenes, which is typically initiated by directed C—H bond activation [1], the vast majority of transition metal-catalyzed allene hydroarylation protocols appear to involve C—C bond formation via π-activation of the allene followed by the outer-sphere addition of the arene. As such, the catalytic hydroarylation of allenes is largely restricted to electron-rich arenes such as indoles and other heteroaromatic compounds and oxygen- and nitrogen-substituted arenes. These transformations are typically catalyzed by highly electrophilic, carbophilic catalysts, most notably cationic gold(I) complexes [5] that have recently emerged as particularly effective catalysts for the π-activation of C—C multiple bonds [6]. Herein we provide a review of the transition metal-catalyzed hydroarylation of allenes with electron-rich

Catalytic Hydroarylation of Carbon-Carbon Multiple Bonds, First Edition.
Edited by Lutz Ackermann, T. Brent Gunnoe and Laurel Goj Habgood.

arenes that have appeared largely in the past decade. Related processes such as the dehydrative arylation of allenyl alcohols [7] and the formal hydroarylation of allenes with metal aryl precursors [8] are not included.

9.2 Intramolecular Hydroarylation

9.2.1 Indoles as Nucleophiles

9.2.1.1 6-*exo*-Hydroarylation

In 2006, Widenhoefer and co-workers reported a highly active gold(I) catalyst for the intramolecular *exo*-hydroarylation of 2-allenyl indoles [3]. For example, treatment of the 2-(4,5-hexadienyl)indole **1** with a catalytic 1:1 mixture of (**P1**)AuCl [**P1** = P(*t*-Bu)$_2$*o*-biphenyl] and AgOTf in dioxane at 25 °C for 30 min led to isolation of tetrahydrocarbazole **2** in 87% yield (Scheme 9.1). Gold-catalyzed hydroarylation was effective for 2-allenyl indoles that possess substitution at either the internal or terminal allenyl carbon atom and was effective for the formation of seven-membered rings (Table 9.1). Of particular interest were allenyl indoles possessing an axially chiral 1,3-disubstituted allene moiety, which underwent gold-catalyzed hydroarylation with complete (>98%) (*E*)-diastereoselectivity and axial to tetrahedral transfer of chirality (Table 9.1).

Stereochemical analysis of gold-catalyzed allene hydroarylation and the related intramolecular hydroamination and hydroalkoxylation of allenes supported an outer-sphere mechanism for gold-catalyzed hydroarylation. This mechanism involves addition of the indole to the cis-gold π-allene complex *cis*-**I**, which is in rapid equilibrium with the more stable trans diastereomer trans-**I** [9], to form the gold vinyl complex **II**. Protodemetallation of **II** with retention of configuration releases the tetrahydrocarbazole with regeneration of the active gold catalyst (Scheme 9.2). Worth noting is that the high levels of chirality transfer observed for the gold-catalyzed hydroarylation of allenyl indoles possessing an axially chiral allene moiety points to facile cyclization of *cis*-**I** owing to the rapid racemization of gold π-allene complexes [10].

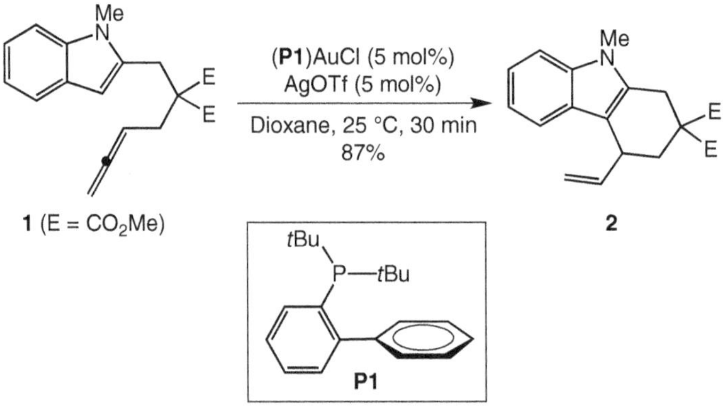

Scheme 9.1 Gold-catalyzed 6-*exo* hydroarylation of 2-allenyl indole **1**.

Table 9.1 Intramolecular hydroarylation of 2-allenyl indoles catalyzed by a 1:1 mixture of (**P1**)AuCl and AgOTf in dioxane at 25 °C for 30 min.

Allenyl indole	Product	Yield (dr)
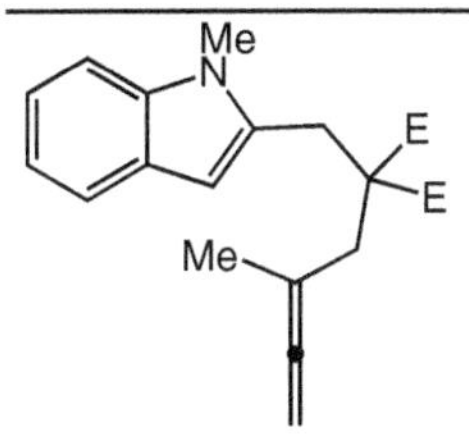	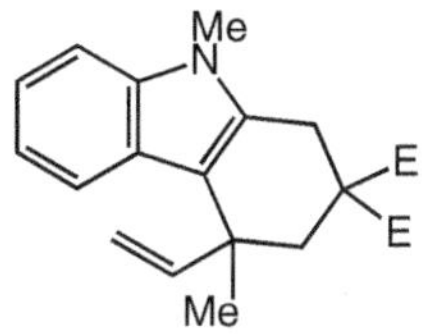	71%
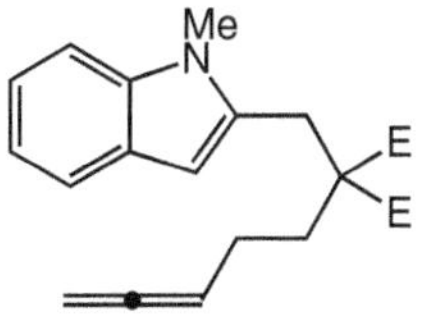	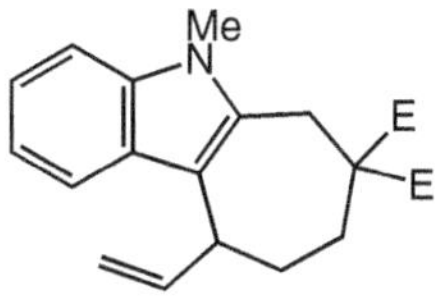	70%
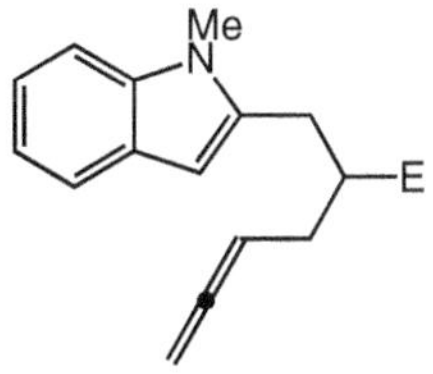	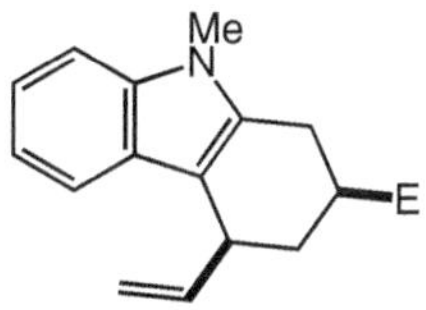	94% (5:1)
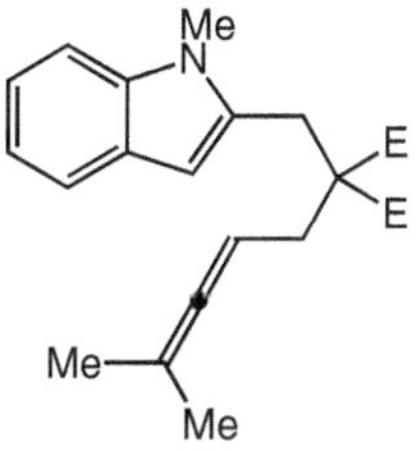	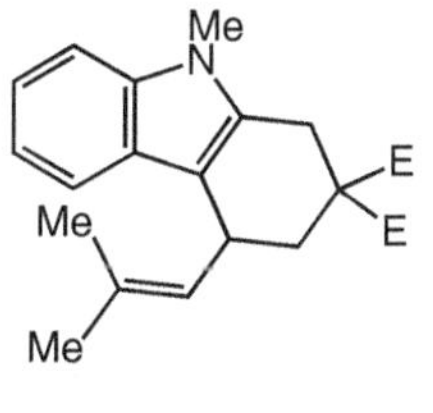	92%
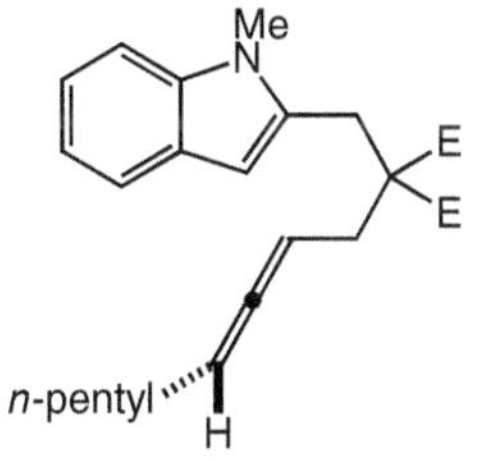 e.r. = 76 : 24	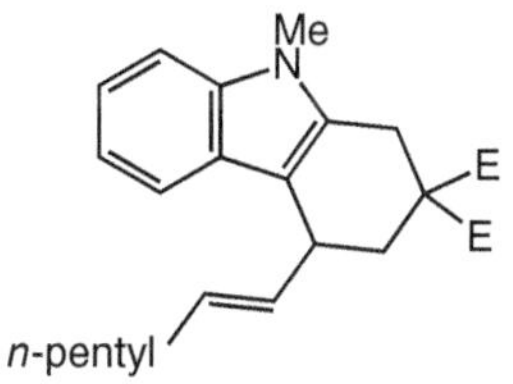	82% (≥50 : 1) e.r. = 76 : 24

9.2.1.2 5-*exo*-Hydroarylation

Toste and co-workers employed the gold-catalyzed intramolecular 5-*exo* hydroarylation of an *N*-3,4-hexadienyl indole as a key C—C bond forming step in the total synthesis of the anti-malarial agents flinderole B and C [11]. To this end, treatment of the *N*-3,4-hexadienyl indole **3** with a catalytic 1:1 mixture of

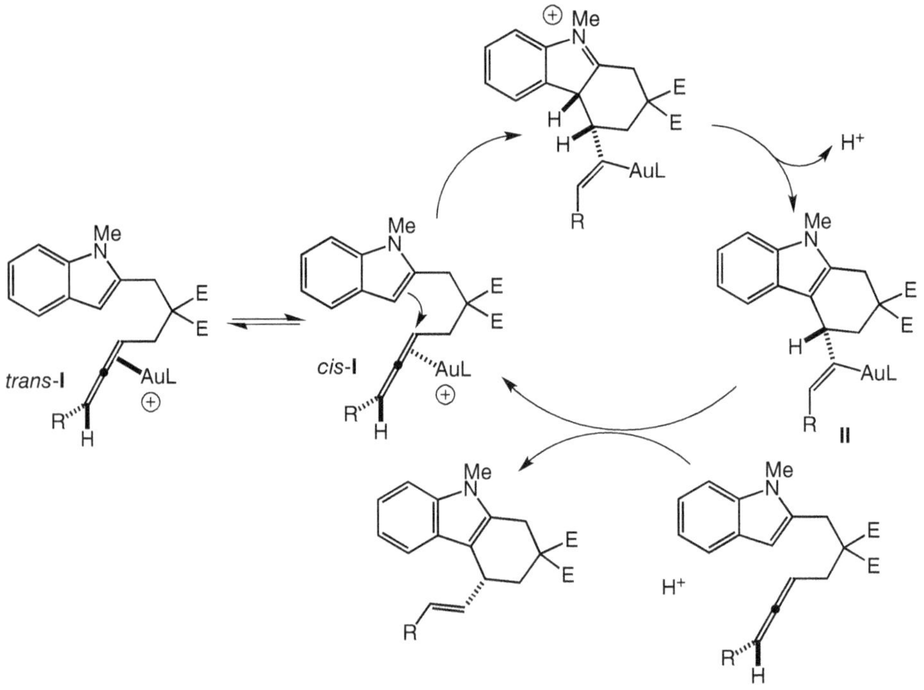

Scheme 9.2 Proposed mechanism of the gold(I)-catalyzed hydroarylation of 2-allenyl indoles.

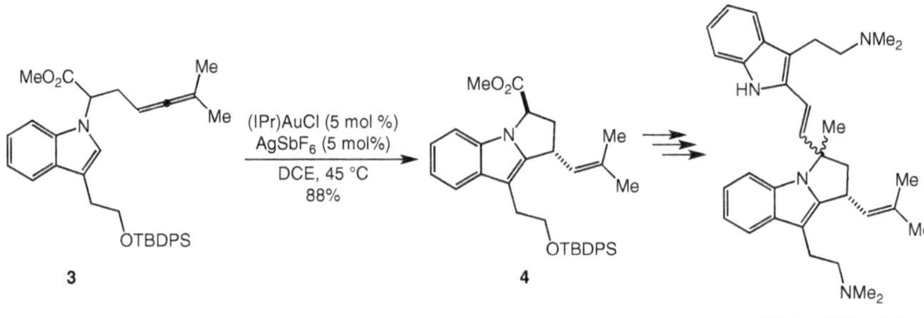

Scheme 9.3 Gold-catalyzed 5-*exo* hydroarylation of *N*-allenyl indole **3** as a route to flinderole B and C.

(IPr)AuCl [IPr = 1,3-bis(2,6-diisopropylphenyl)imidazol-2-ylidine] and $AgSbF_6$ in 1,2-dichloroethane (DCE) at 45 °C gave pyrrolidine **4** in 88% yield as a single diastereomer (Scheme 9.3). In this transformation, C—C bond formation occurs via attack of the less nucleophilic C2 position of the indole on the gold π-allene intermediate.

9.2.1.3 6-*endo*-Hydroarylation

In 2010, Barluenga and co-workers reported the gold-catalyzed intramolecular 6-*endo* hydroarylation of *N*-(2,3-butadienyl)indoles to form pyrido[1,2-*a*]indoles

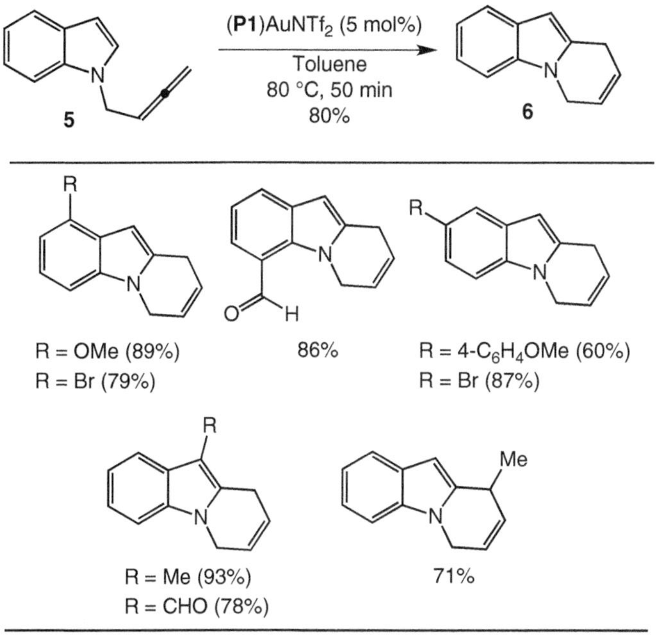

Scheme 9.4 Gold-catalyzed 6-*endo* hydroarylation of *N*-allenyl indoles.

resulting from attack of the indole C2 carbon atom on the distal allene carbon atom [12]. For example, treatment of *N*-allenyl indole **5** with a catalytic amount of (**P1**)$AuNTf_2$ in toluene at 80 °C for 50 min led to isolation of pyrido[1,2-*a*]indole **6** in 80% yield (Scheme 9.4). The gold-catalyzed intramolecular 6-*endo*-hydroarylation of *N*-allenyl indoles tolerated substitution at the 3, 4, 5, or 7 positions of the *N*-allenyl indole and was also applied to the intramolecular hydroarylation of an *N*-allenyl indole possessing an axially chiral 1,3-disubstituted allene moiety (Scheme 9.4). Gold-catalyzed cyclization of 2-deuterio-*N*-allenyl indole **5**-2-d_1 led to the formation of a 4.2:1 mixture of **6**-8-d_1 and **6**-10-d_1 with 89% retention of deuterium. This result is consistent with the initial nucleophilic attack of indole on a gold π-allene complex to generate the cationic gold vinyl intermediate **III** that isomerizes to form the iminium ion-stabilized species **IV**. Preferential loss of proton from **IV** followed by protodeauration would form **6**-8-d_1 while the less favored loss of deuteron from **IV** followed by deuterodeauration would form **6**-10-d_1 (Scheme 9.5).

9.2.1.4 5-*endo*-Hydroarylation

Ma and co-workers reported a gold-catalyzed procedure for the 5-*endo*-hydroarylation of 3-allenyl indoles as a route to dihydrocyclopenta[*b*]indole derivatives [13]. As an example of the protocol, treatment of 3-allenyl indole **7a** with a catalytic 1:1 mixture of Ph_3PAuCl and AgOTf in toluene for 110 °C for 9 h led to isolation of the dihydrocyclopenta[*b*]indole **8a** in 93% yield (Scheme 9.6). Catalyst screening experiments revealed that (**P1**)AuCl and (IPr)AuCl were similarly active catalyst precursors when paired with AgOTf, while $AuCl_3$ was

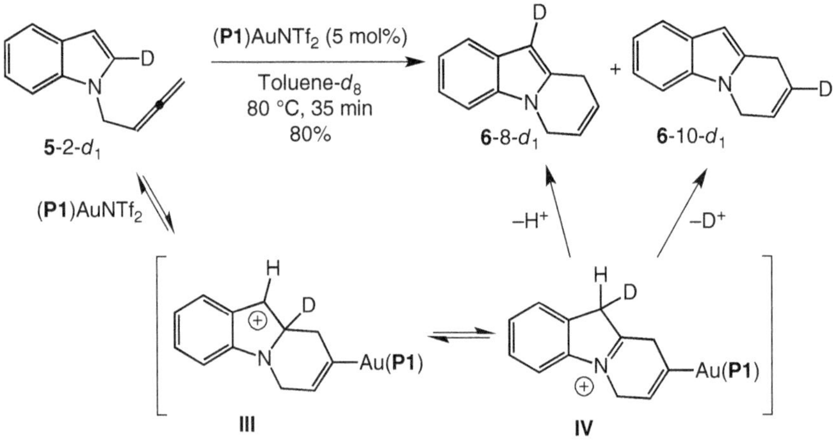

Scheme 9.5 Deuterium distribution in the gold-catalyzed 6-*endo*hydroarylation of 2-deuterio indole **5**-2-d_1.

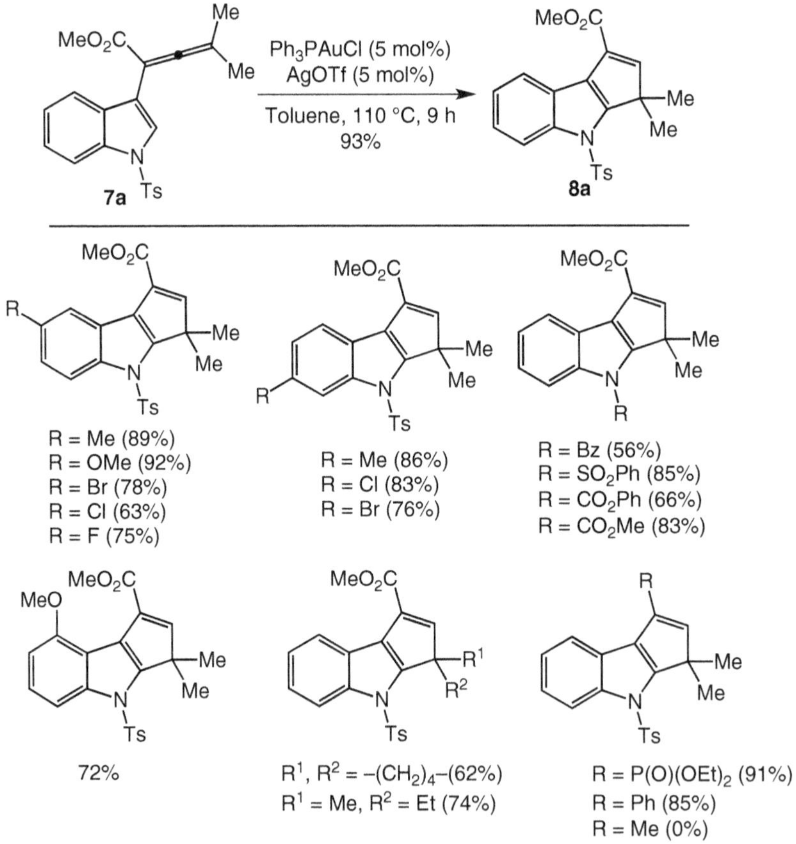

Scheme 9.6 Scope of the gold-catalyzed 5-*endo* hydroarylation of 3-allenyl indoles.

much less effective. The method was effective for 3-allenyl indoles substituted at the 1-, 4-, 5-, or 6-position of the indole ring, but required *gem*-disubstitution at the terminal allene carbon atom and an electron-withdrawing group at the allene carbon atom α to the indole ring (Scheme 9.6). Gold-catalyzed cyclization of 3-allenyl indole **7b** in the presence of 10 equiv. of acetic acid-d_4 generated the dihydrocyclopenta[*b*]indole **8b**-2-d_1 in 82% yield with 84% deuterium incorporation at the C2 position. This observation is consistent with a mechanism involving 5-*endo* addition of the indole to a gold π-allene complex followed by deuteriodeauration of the resulting gold vinyl complex **V** (Scheme 9.7).

9.2.1.5 Less Common Modes of Ring Closure

In 2015, Alcaide, Almendros and co-workers reported a gold-catalyzed process for the intramolecular hydroarylation of β-lactam-tethered allenyl indoles to form seven- and eight-membered rings [14]. For example, treatment of allenyl indole **9**, in which the allene moiety is tethered to the 3-position of the β-lactam and the indole is tethered to the 4-position of the β-lactam, with a catalytic 1:1 mixture of (IPr)AuCl and $AgSbF_6$ in DCE at 20 °C led to 7-*exo* hydroarylation and isolation of azeto-oxepino[4,5-*b*]indol-2-one **10** in 72% yield (Scheme 9.8). In comparison, $AuCl_3$, AuCl, $Ph_3PAuNTf_2$, or a 1:1 mixture of Ph_3PAuCl and AgOTf proved unsuitable as catalysts for the conversion of **9** to **10**. The protocol was similarly effective for β-lactam-tethered allenyl indoles possessing a range of *N*-protecting groups on the lactam moiety. Under more forcing conditions, **9** underwent 7-*exo* hydroarylation accompanied by C—N bond cleavage of the lactam moiety to form 1,6-dihydro-2*H*-oxepino[4,5-*b*]indole-4-carboxamide **11** in modest yield, as did a number of β-lactam-tethered allenyl indoles (Scheme 9.8).

The same authors also demonstrated the intramolecular hydroarylation of lactam-tethered allenyl indoles in which the allene moiety is tethered to the 1-position of the β-lactam and the indole is tethered to the 4-position

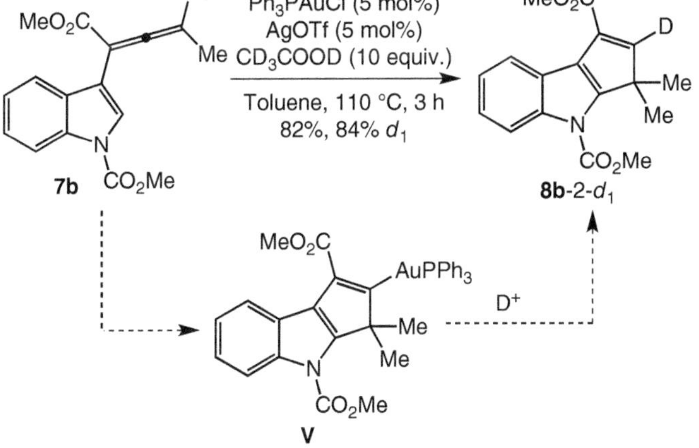

Scheme 9.7 Gold-catalyzed hydroarylation of **7b** in the presence of CD_3COOD.

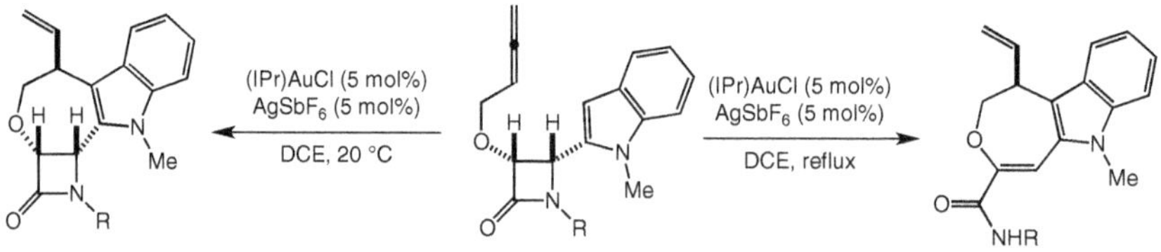

10, R = Bn (72%)
R = 4-MeOC$_6$H$_4$CH$_2$ (68%)
R = 4-MeOC$_6$H$_4$ (63%)
R = 4-BrC$_6$H$_4$CH$_2$ (82%)
R = *i*-Bu (89%)

9

11, R = Bn (60%)
R = 4-MeOC$_6$H$_4$CH$_2$ (53%)
R = 4-MeOC$_6$H$_4$ (67%)
R = 4-BrC$_6$H$_4$CH$_2$ (40%)
R = *i*-Bu (58%)

Scheme 9.8 Gold-catalyzed intramolecular 7-*exo* hydroarylation of β-lactam-tethered allenyl indoles.

of the β-lactam [14]. In the case of 1-(2,3-butadienyl) lactams **12**, reaction with a catalytic 1:1 mixture of (IPr)AuCl and AgSbF$_6$ in DCE at reflux under microwave conditions led to 8-*endo* hydroarylation to form tetrahydroazeto-azocino[3,4-*b*]indol-2-ones **13** in moderate yield (Scheme 9.9). Under similar conditions, 1-(penta-3,4-dien-1-yl) lactams **14** underwent 7-*exo* hydroarylation to form hexahydroazeto-azepino[3,4-*b*]indol-2-ones **15** in modest yield (Scheme 9.9). In these latter cases, the formation of unidentified side products, presumably resulting from isomerization or polymerization diminished the yields of compounds **15**.

9.2.2 Other Nucleophiles

9.2.2.1 6-*exo*-Hydroarylation

In 2006, Nelson and co-workers reported the diastereoselective intramolecular hydroarylation of enantiomerically enriched allenes with pyrroles as a key step in the enantioselective synthesis of (−)-rhazinilam [4]. The key pyrrole-tethered allene **16** was obtained in 60% yield as a single diastereomer (^{1}H NMR) in three steps from 2-pentynal via alkaloid-catalyzed acyl halide-aldehyde cyclocondensation with propionyl chloride followed by copper(I)-catalyzed S$_N$2′ ring opening of the resulting β-lactone with a pyrrole-substituted Grignard reagent and methylation of the free carboxylic acid. Treatment of **16** with a catalytic amount of Ph$_3$PAuOTf (5 mol%) led to 6-*exo* hydroarylation to form bicyclic pyrrole **17** containing the tetahydroindolizine ring system and stereogenic quaternary carbon atom in 92% isolated yield as a 97:3 mixture of diastereomers (Scheme 9.10).

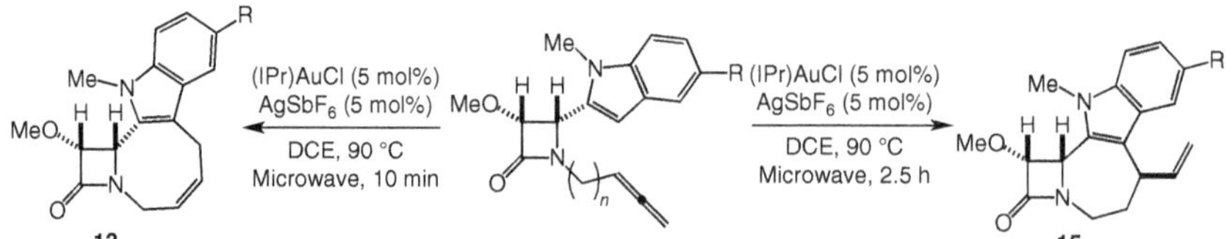

13

12 (*n* = 1), **14** (*n* = 2)

15

n = 1, R = H (59%)
n = 1, R = Cl (70%)
n = 1, R = OMe (53%)

n = 2, R = H (35%)
n = 2, R = Cl (33%)
n = 2, R = OMe (36%)

Scheme 9.9 Gold-catalyzed intramolecular hydroarylation of lactam-tethered allenyl indoles.

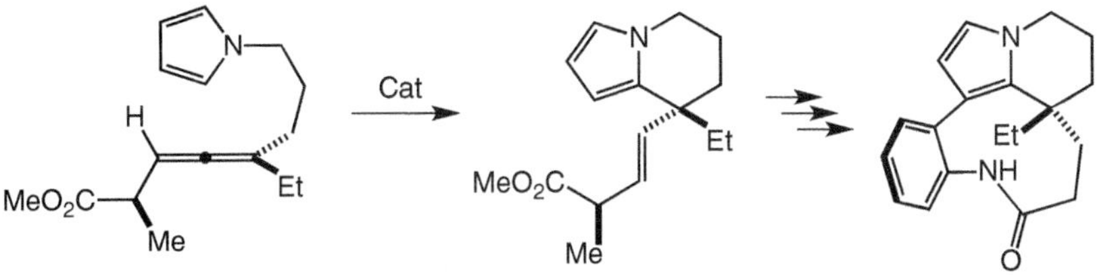

16 17

Cat	Yield	dr
$Ph_3PAuOTf$ (5 mol%)	92%	97 : 3
$AuCl_3$ (10 mol%)	27%	92 : 8
$AuCl_3/AgOTf$ (20 mol%)	82%	92 : 8
$(MeCN)_2PdCl_2$ (30 mol%)	83%	67 : 33

Scheme 9.10 Gold-catalyzed 6-*exo* hydroarylation of allenes with pyrroles as a route to (−)-rhazinilam.

Other metal complexes including $(MeCN)_2PdCl_2$, $AuCl_3$, and mixtures of $AuCl_3$ and AgOTf also catalyzed the conversion of **16** to **17**, albeit with lower efficiency and/or poor diastereoselectivity, whereas Ag(I) salts were completely ineffective. Bicyclic pyrrole was subsequently converted to (−)-rhazinilam over six steps in 39% overall yield (from **16**).

In 2007, Gagné and co-workers expanded the scope of gold-catalyzed intramolecular allene hydroarylation to include less nucleophilic arenes [15], as well as informed the development of the gold-catalyzed hydroalkenylation of allenes [16]. For example, reaction of the 4-allenyl arene **18** with a catalytic 1:1 mixture of $[P(OPh)_3]AuCl$ and $AgSbF_6$ in CH_2Cl_2 at room temperature for 6 h led to isolation of the vinyl-substituted benzocycle **19** in 85% yield (Scheme 9.11). The protocol was effective for a number of electron-rich arenes and for terminally disubstituted allenes, whereas attempted gold-catalyzed intramolecular hydroarylation of 4-allenyl arenes containing a less nucleophilic arene moiety led predominantly to hydration of the allene with adventitious moisture.

Mechanistic analysis of the gold-catalyzed conversion of **18** to **19** provided strong evidence for the involvement of *gem*-diaurated vinyl complexes in the gold-catalyzed hydroarylation of 4-allenyl arenes, and indeed, the first direct evidence for the involvement of *gem*-diaurated vinyl complexes in gold-catalyzed hydrofunctionalization [17]. For example, stoichiometric reaction of **18** with 2 equiv. of $(Ph_3P)AuNTf_2$ in the presence of 2,6-di-*tert*-butylpyridine (DTBP) led to complete consumption of **18** to form the *gem*-diaurated vinyl complex **20**, in which the $(Ph_3P)Au$ fragments were proposed to bind to the vinyl moiety through a bridging three-center two-electron mode stabilized by a Au—Au interaction (Scheme 9.12). ^{31}P NMR analysis of catalytic mixtures of **18** and $(Ph_3P)AuNTf_2$ established **20** as the only organometallic species that accumulated under reaction conditions. Attempted purification of **20** on neutral alumina or treatment with a strong donor ligand such as Br^- or PMe_3 led to deauration to form the neutral mono(gold) vinyl complex **21**; treatment of **21**

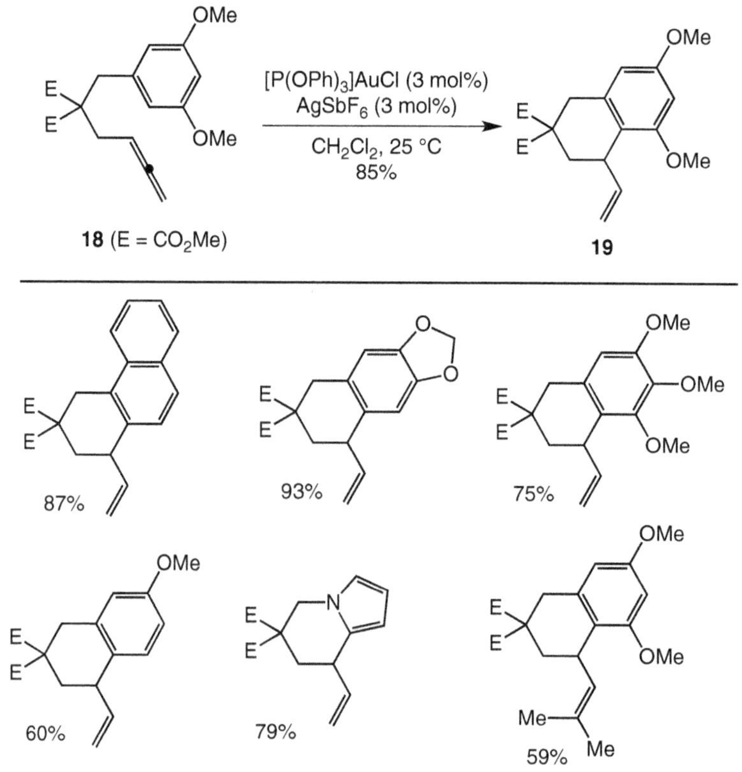

Scheme 9.11 Gold-catalyzed intramolecular 6-*exo* hydroarylation of allenes with substituted arenes.

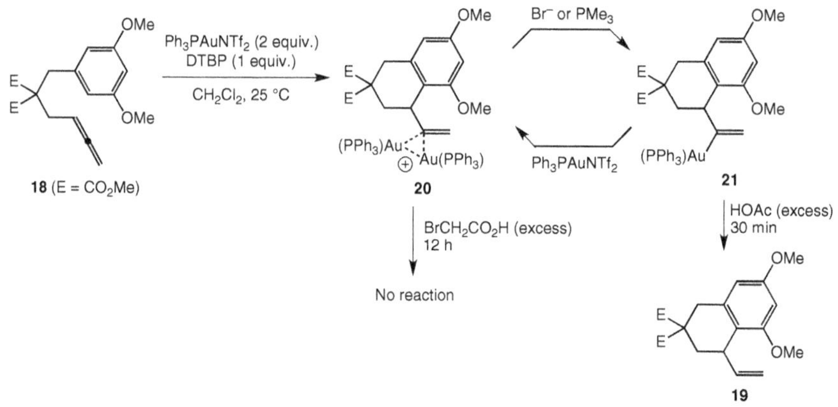

Scheme 9.12 Generation and reactivity of mono(gold) and bis(gold) vinyl complexes in the gold-catalyzed hydroarylation of allenes with substituted arenes.

with a stoichiometric amount of $(Ph_3P)AuNTf_2$ led to clean regeneration to **20**. Mono(gold) vinyl complex **21** was significantly more reactive with respect to protodeauration than was **20**; whereas treatment of **21** with excess acetic acid led to immediate protodeauration to form **19; 20** failed to react with α-bromoacetic acid after 12 h at room temperature although **20** did undergo protodeauration in the presence of $HNTf_2$ (Scheme 9.12).

Interestingly, conversion of **18** to **19** catalyzed by a 1:5 mixture of $(Ph_3P)AuCl$ (5 mol%) and $AgNTf_2$ (25 mol%) revealed no formation of **20** under reaction conditions and instead the silver/gold dimetallated vinyl complex **22** was observed. Complex **22** was synthesized independently from reaction of a 1:1:5 mixture of **18**, $(Ph_3P)AuCl$, and $AgNTf_2$ in the presence of DTBP (Scheme 9.13) [18]. The reactivity of complex **22** mirrored that of *gem*-diaurated complex **21** and was converted to **22** in the presence of excess $(Ph_3P)Au^+$. Importantly, the rate of conversion of **18** to **19** was significantly retarded by the presence of excess $AgNTf_2$, which provided direct evidence for the potential deleterious effects of silver in gold-catalyzed hydrofunctionalization reactions.

In 2014, Yu and co-workers reported the gold-catalyzed intramolecular 6-*exo* hydroarylation of allenes with furans to form furan-fused vinyl cyclohexanes [19]. For example, treatment of the sulfonamide-tethered allenyl furan **23** with a catalytic amount of $[(\mathbf{P1})Au(NCMe)]^+ \ SbF_6^-$ in dioxane at 50 °C for 6 h led to isolation of the bicyclic furan **24** in 91% yield as a single (*E*) diastereomer (Scheme 9.14). Interestingly, a catalytic amount of water was required for efficient gold-catalyzed intramolecular hydroarylation, a phenomenon that was similarly observed for the gold-catalyzed α-alkenylation of β-yne-furans [20]. The method tolerated a number of aryl and alkyl groups at the terminal allene carbon atom and a number of groups linking allene and furan functionality. The

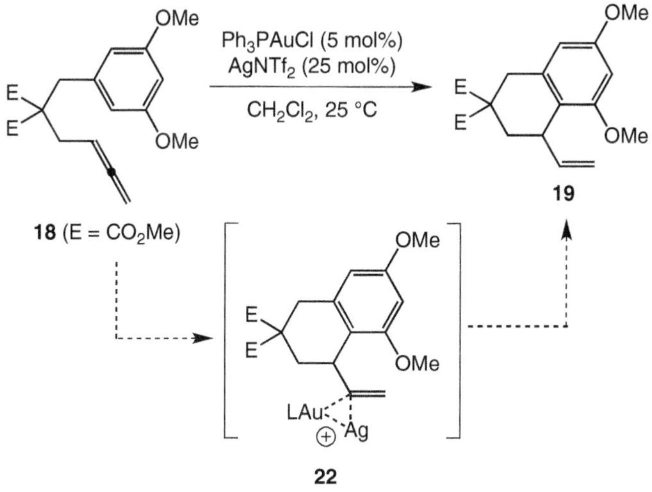

Scheme 9.13 Formation of silver/gold dimetallated vinyl complex **22** from the reaction of **18** with a mixture of gold and silver.

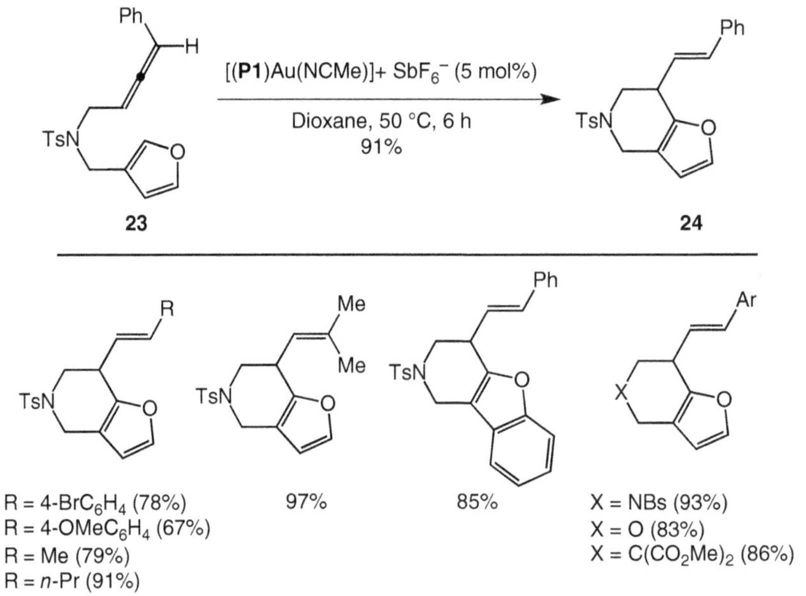

Scheme 9.14 Gold-catalyzed intramolecular 6-*exo* hydroarylation of allenes with furans.

transformation was likewise effective for the intramolecular hydroarylation of terminally disubstituted allenes and benzofuran nucleophiles (Scheme 9.14).

Yu and co-workers similarly developed a complementary platinum-catalyzed procedure for the 6-*exo*-hydroarylation of allenes with furans to form furan-fused alkylidenecyclohexanes [19]. For example, treatment of tosylate-tethered allenyl indole **25** with a catalytic amount of $PtCl_4$ in THF at 50 °C for 2 h led to isolation of alkylidenecyclohexane **26** in 92% yield as a single diastereomer (Scheme 9.15). $PtCl_2$ likewise catalyzed the conversion of **25** to **26**, albeit with longer reaction time (17 h, 73%). The $PtCl_4$-catalyzed intramolecular hydroarylation of allenyl furans tolerated a number of groups at the terminal allene carbon atom and functionality along the chain that tethered the allene and furan groups and was effective for terminally disubstituted allenes and benzofurans. Regarding the mechanism, the authors firmly ruled out the formation of **26** via platinum-catalyzed isomerization of **24** under reaction conditions, which pointed to alkene isomerization within the coordination sphere of platinum, potentially through a platinum carbene intermediate.

Hashmi and co-workers have reported a procedure for the synthesis of chromanes via the gold-catalyzed intramolecular 6-*exo* hydroarylation of aryl allenyl ethers [21]. As an example, treatment of aryl allenyl ether **27** with a catalytic 1:1 mixture of $(Ph_3P)AuCl$ and $AgNTf_2$ in chloroform at room temperature for 20 h led to isolation of 4-alkenyl chromane **28** in 98% yield (Scheme 9.16). The method was restricted to aryl allenyl ethers containing a 5-methylhexa-3,4-dienyl group, but tolerated a range of substitution on the aryl ring. The intramolecular hydroarylation of aryl allenyl ethers had previously been realized employing stoichiometric amounts of mercury trifluoroacetate [22].

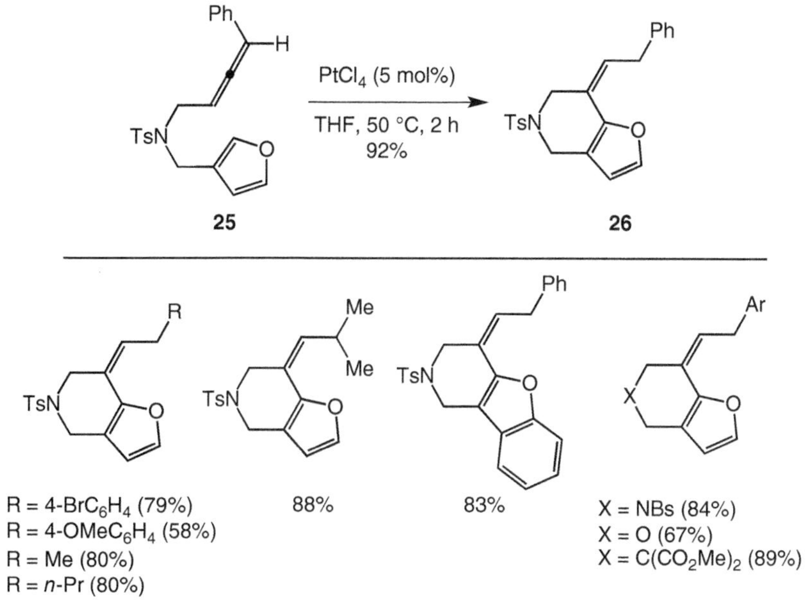

Scheme 9.15 Platinum-catalyzed 6-*exo* hydroarylation of allenes with furans.

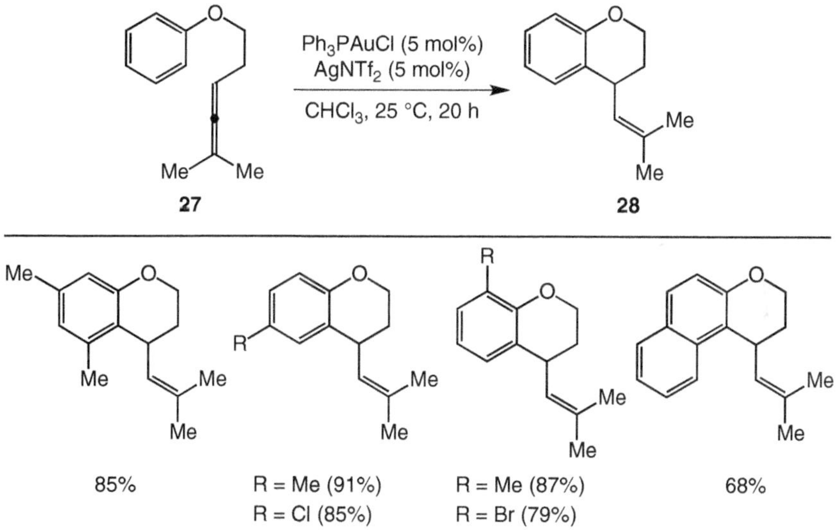

Scheme 9.16 Synthesis of chromanes via the gold-catalyzed intramolecular 6-*exo* hydroarylation of aryl allenyl ethers.

9.2.2.2 6-*endo*-Hydroarylation

The intramolecular 6-*endo* hydroarylation of an allene was first described by Ohno and co-workers in 2007 in the context of the gold-catalyzed 6-*endo* hydroarylation of allenic anilines [23]. For example, treatment of allenic aniline **29** with a catalytic 1:1 mixture of (**P1**)AuCl and AgOTf in dioxane at 25 °C

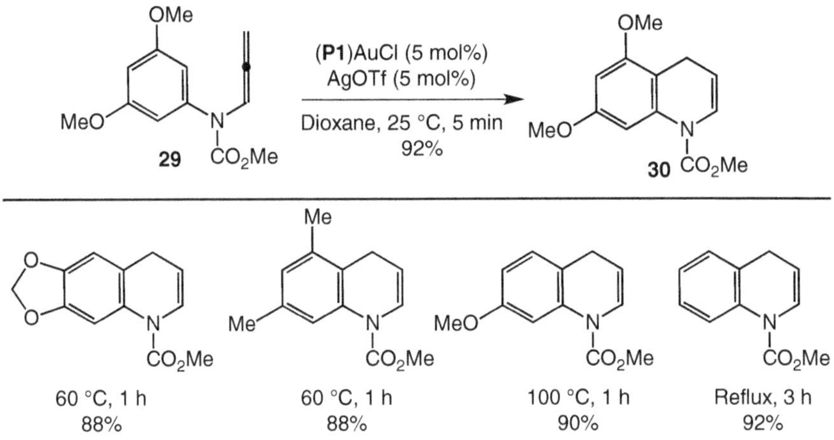

Scheme 9.17 Gold-catalyzed intramolecular 6-*endo* hydroarylation of an allene with benzene derivatives.

for 5 min led to isolation of 1,4-dihydroquinoline **30** in 92% yield as a single isomer (Scheme 9.17). The protocol was effective for a number of electron-rich and electron-neutral allenyl anilines although the rate of reaction was strongly dependent on the nature of the arene.

Lee and co-workers have reported the platinum-catalyzed 6-*endo* hydroarylation of 2-benzyl-2,3-butadienoates to form 1,4-dihydronaphthalene-2-carboxylates [24]. For example, treatment of ethyl 2-benzyl-2,3-butadienoate (**31**) with a catalytic 1:2 mixture of $PtCl_2$ and AgOTf in DCE at 80 °C for 2 h led to isolation of ethyl 1,4-dihydronaphthalene-2-carboxylate (**32**) in 93% yield (Scheme 9.18). However, a number of other potential catalysts, including $AuCl_3$, $AuCl_3$/AgOTf, (Ph_3P)AuCl/AgOTf, $AgSbF_6$, $PtCl_2$, and $PtCl_4$ were ineffective. The protocol tolerated monosubstitution at the terminal allene carbon atom and at various positions of the aryl ring and, in one case, was effective for the formation of a seven-membered ring via 7-*endo* hydroarylation of ethyl 2-phenethylbuta-2,3-dienoate.

In 2015, Ma, Wang and co-workers reported a gold(I)-catalyzed procedure for the 6-*endo* intramolecular hydroarylation of aryl allenes containing an axially chiral 1,3-disubstituted allene moiety [25]. For example, treatment of the 2-allenylbenzothiophene **33** with a catalytic 1:1 mixture of (Ph_3P)AuCl and AgOTf in dioxane at room temperature for 24 h led to isolation of the 1,4-dihydrodibenzothiophene **34** in 87% yield (Scheme 9.19). The method was likewise effective for the intramolecular 6-*endo* hydroarylation of 4-allenyl-1,2-dialkoxybenzenes and 3-(2,3-butadienyl)thiophene. The authors extended this protocol to include the highly diastereoselective 6-*endo* hydroarylation of enantiomerically enriched aryl allenes containing an axially chiral allene moiety. In an optimized protocol, treatment of the enantiomerically enriched allenyl benzo[1,3]dioxole (*R*)-**35** (99% ee) with a catalytic 1:2 mixture of the bis(gold) phosphine complex (dppm)Au_2Cl_2 [dppm = bis(diphenylphosphinomethane] and AgOTf in toluene at room temperature led to isolation of the 5,8-dihydronaphth[1,3]dioxole (*R*)-**36**

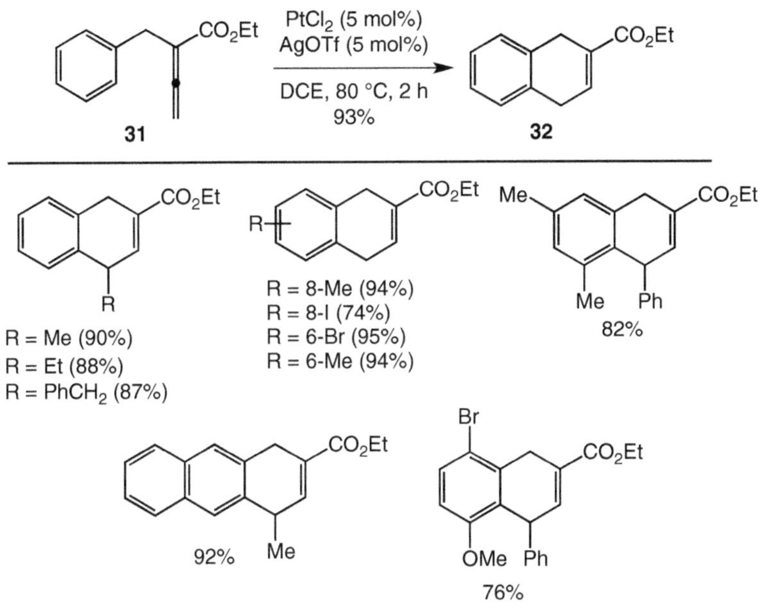

Scheme 9.18 Platinum-catalyzed 6-*endo* hydroarylation of 2-benzyl-2,3-butadienoates.

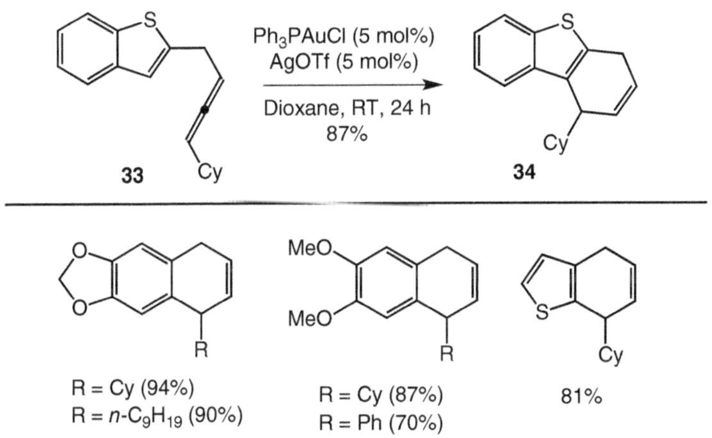

Scheme 9.19 Gold(I)-catalyzed 6-*endo* intramolecular hydroarylation of aryl allenes containing a 1,3-disubstituted allene moiety.

in 91% yield with 97% ee (Scheme 9.20). Under otherwise identical conditions, a range of other mononuclear gold precursors including (IPr)AuCl [IPr = 1,3-bis(2,6-diisopropylphenyl-imidazol-2-ylidene)], (Me_3P)AuCl, (Ph_3P)AuCl, and (*t*-Bu_3P)AuCl gave **36** in high yield, but with diminished enantiopurity (50–83% ee). The optimized protocol was applied effectively to a number of electron-rich aryl allenes with ≥ 98% transfer of chirality in all cases (Scheme 9.20).

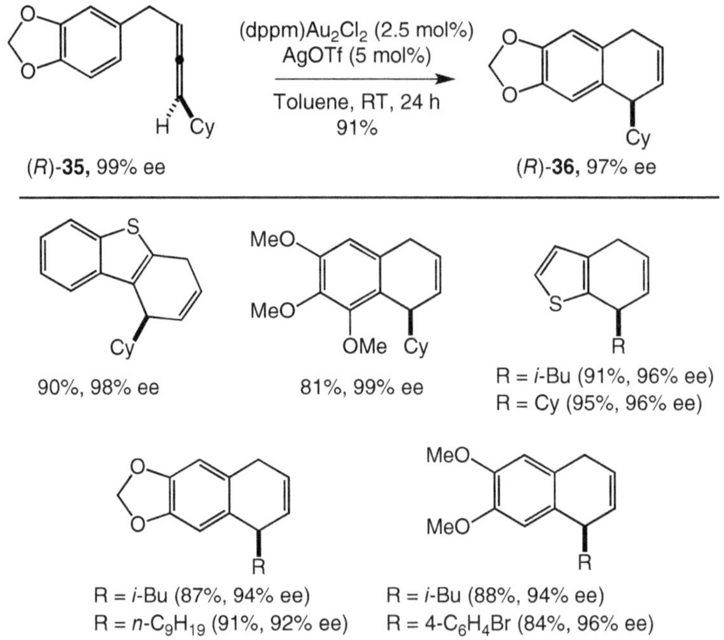

Scheme 9.20 Gold(I)-catalyzed diastereoselective 6-*endo* hydroarylation of enantiomerically enriched 1,3-disubstituted allenes.

9.2.2.3 Less Common Modes of Ring Closure

During the course of an investigation of the gold-catalyzed intramolecular 6-*endo* hydroarylation of allenic anilines, Ohno and co-workers found that the gold-catalyzed cyclization of the *N*-(buta-2,3-dienyl)-anilines led neither to 7-*endo* nor to 5-*exo* hydroarylation but instead formed 1,2-dihydroquinoline derivatives via addition of the arene to the central carbon atom of the allene moiety [23]. For example, treatment of *N*-(buta-2,3-dienyl)-aniline **37** with a catalytic 1:1 mixture of (**P1**)AuCl and AgOTf (3 mol%) in glacial acetic acid at room temperature for 1 h led to isolation of 1,2-dihydroquinoline **38** in 82% yield (Scheme 9.21). The method was similarly successful for the intramolecular hydroarylation of *N*-(buta-2,3-dienyl)-anilines possessing substitution on the internal allenyl carbon atom as well as for the cyclization of *N*-(buta-2,3-dienyl)-phenols (Scheme 9.21) [23].

During the course of an investigation of the gold-catalyzed [2+2] and [2+3] cycloaddition of *N*-allenyl indoles, Shi and co-workers demonstrated the highly efficient intramolecular 7-*exo* and 5-*endo* hydroarylation of allenes with a pyrrole and a phenyl group, respectively (Scheme 9.22) [26]. For example, treatment of the *N*-δ-allenyl pyrrole **39** with a catalytic amount of (**P1**)$AuNTf_2$ in DCE at 70 °C led to 7-*exo* hydroarylation to form 2,3,4,5-tetrahydro-1*H*-pyrrolo[1,2-*d*][1,4]diazepine **40** in 95% yield (Scheme 9.22). Under similar conditions, the 1-phenyl allene **41** underwent 5-*endo* hydroarylation to form the 1-cyclopropyl-1*H*-indene **42** in 93% isolated yield (Scheme 9.22).

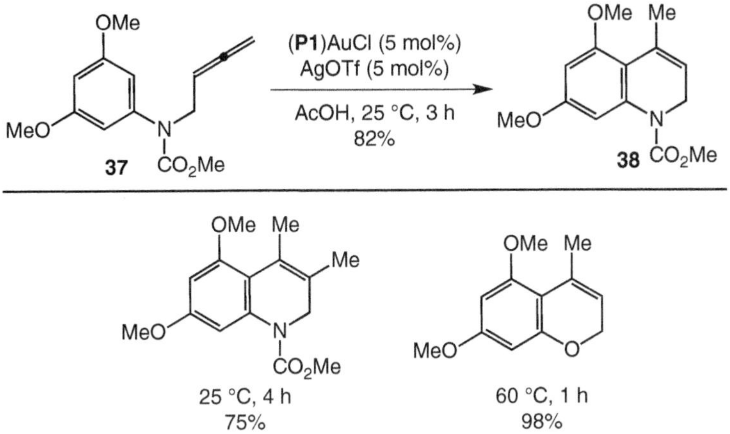

Scheme 9.21 Gold-catalyzed intramolecular hydroarylation of *N*-allenyl anilines.

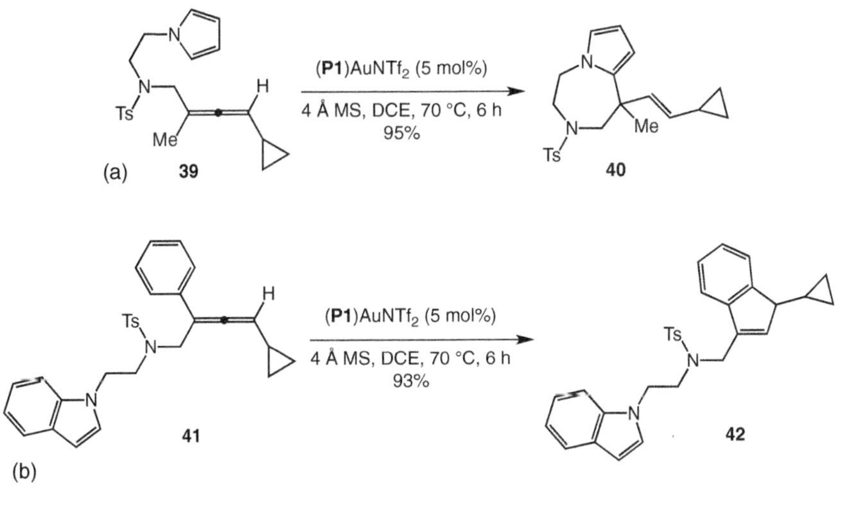

Scheme 9.22 Gold-catalyzed intramolecular 7-*exo* hydroarylation of an allene with a pyrrole (a) and the 5-*endo* hydroarylation of an allene with a phenyl group (b).

Alcaide, Almendros and co-workers demonstrated the gold-catalyzed 9-*endo* hydroarylation of (aryl)allene-tethered 2-azetidinones [27]. For example, treatment of allene **43** with a catalytic amount of $(Ph_3P)AuNTf_2$ in DCE at 110 °C under microwave irradiation led to isolation of benzo[*b*]oxonine **44** in 72% yield (Scheme 9.23). The gold-catalyzed 9-*endo* hydroarylation was applied to a number of (aryl)allene-tethered 2-azetidinones containing either a methyl or phenyl group at the internal allene carbon atom and an *N*-*i*-Bu, *i*-Pr, Bn, or methallyl group. DFT calculations of the gold-catalyzed hydroarylation of (aryl)allene-tethered 2-azetidinones supported a mechanism involving turnover-limiting, outer-sphere, 9-*endo* addition of the aryloxy group to a gold π-allene complex followed by deprotonation of the resulting oxonium ion with NTf_2^- and protodeauration of the neutral gold vinyl complex with $HNTf_2$. The

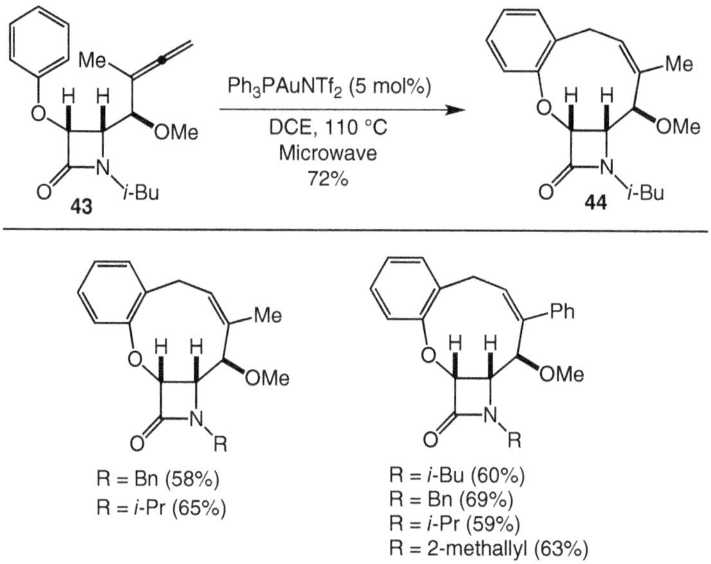

Scheme 9.23 Gold-catalyzed 9-*endo* hydroarylation of (aryl)allene-tethered 2-azetidinones.

calculated energy barrier for 9-*endo* cyclization ($\Delta G^{\ddagger} = 29.9$ kcal/mol), although high, was lower than the energy barriers calculated for seven- or eight-membered ring formation, in accordance with the experimentally observed selectivity for compounds **43**.

9.3 Intermolecular Hydroarylation

9.3.1 Indoles as Nucleophiles

9.3.1.1 Monoaddition Processes

In 2009, Widenhoefer and co-workers reported the gold-catalyzed intermolecular hydroarylation of allenes with 2-alkyl indoles as an extension of this group's efforts in the area of gold-catalyzed intramolecular allene hydroarylation [28]. For example, reaction of a 1:1.05 mixture of 1,2-dimethylindole with dimethyl (2,3-butadienyl)malonate catalyzed by a 1:1 mixture of (IPr)AuCl and AgOTf in toluene at room temperature for 48 h led to isolation of the 3-allyl indole **45** in 82% yield as a single regio- and diastereomer resulting from selective addition of indole to the less-substituted allene terminus (Scheme 9.24). The protocol was likewise effective for the intermolecular hydroarylation of 1,3-disubstituted and tetrasubstituted allenes, providing high levels of regiocontrol in the former cases, and was effective for a number of substituted indoles (Scheme 9.24).

9.3.1.2 Tandem Addition Processes

Ma and co-workers have reported the Pt(IV)-catalyzed tandem hydroarylation/hydroalkoxylation of indoles with allenyl alcohols to form indole derivatives containing a six-membered ether ring at the three-position [29].

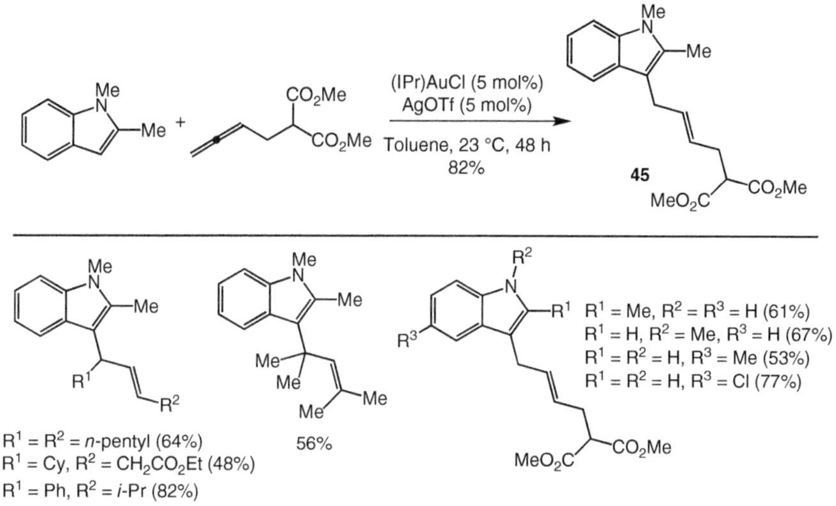

Scheme 9.24 Gold-catalyzed intermolecular hydroarylation of allenes with indoles.

For example, treatment of a 1:1.2 mixture of indole and undeca-3,4-dien-1-ol with a catalytic amount of $PtCl_4$ in THF at room temperature for 4 h led to isolation of indole ether **46** in 77% yield (Scheme 9.25). This tandem hydroarylation/hydroalkoxylation tolerated a number of indoles and allenyl alcohols. Control experiments ruled out a mechanism involving initial intramolecular hydroalkoxylation of the allenyl alcohol followed by intermolecular hydroarylation of the resulting 3,6-dihydro-2*H*-pyran. Rather, the transformation was proposed to occur via initial intermolecular hydroarylation of the allene followed by β-hydride elimination to form a 3-allenyl indole derivative and then by platinum-catalyzed intramolecular hydroalkoxylation.

Muñoz and co-workers have reported a platinum-catalyzed procedure for the intermolecular bisindolylation of allenes to form 1,1-bis(indolyl) alkanes [30]. For example, treatment of a 3:1 mixture of *N*-methylindole and propa-1,2-dien-1-ylcyclohexane with a catalytic amount of $PtCl_2$ in THF containing 3 equiv. of methanol at 70 °C for 20 h led to isolation of the 3-cyclohexyl-1,1-bis(indole) **47** in 62% yield (Scheme 9.26). The corresponding reaction in the absence of methanol formed **47** in only 23% yield; methanol presumably either stabilizes zwitterionic intermediates or functions as a proton shuttle. The protocol was effective for a number of indole derivatives and monosubstituted allenes (Scheme 9.26). Related to this process, Ma and co-workers have reported the Sc(III)-catalyzed bisindolylation of electron-deficient allenyl ketones in which the allenyl ketone functioned as a Michael acceptor [31].

9.3.2 Furans as Nucleophiles

In 2000, Hashmi and co-workers reported the gold(III)-catalyzed tandem cyclization/hydroarylation of allenyl ketones, which was the first example of a gold-catalyzed allene hydroarylation [32]. For example, treatment of 3,4-pentadien-2-one with a catalytic amount of $AuCl_3$ in acetonitrile led to

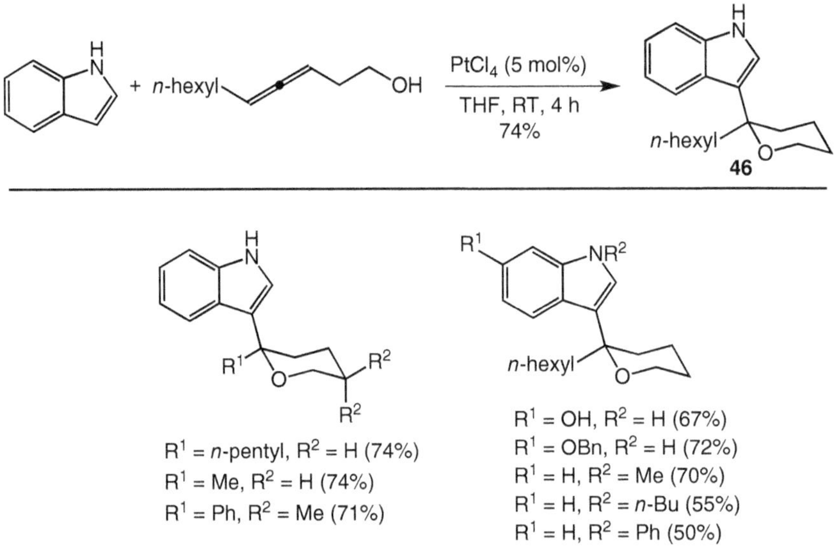

Scheme 9.25 Platinum(IV)-catalyzed tandem hydroarylation/hydroalkoxylation of indoles with allenyl alcohols.

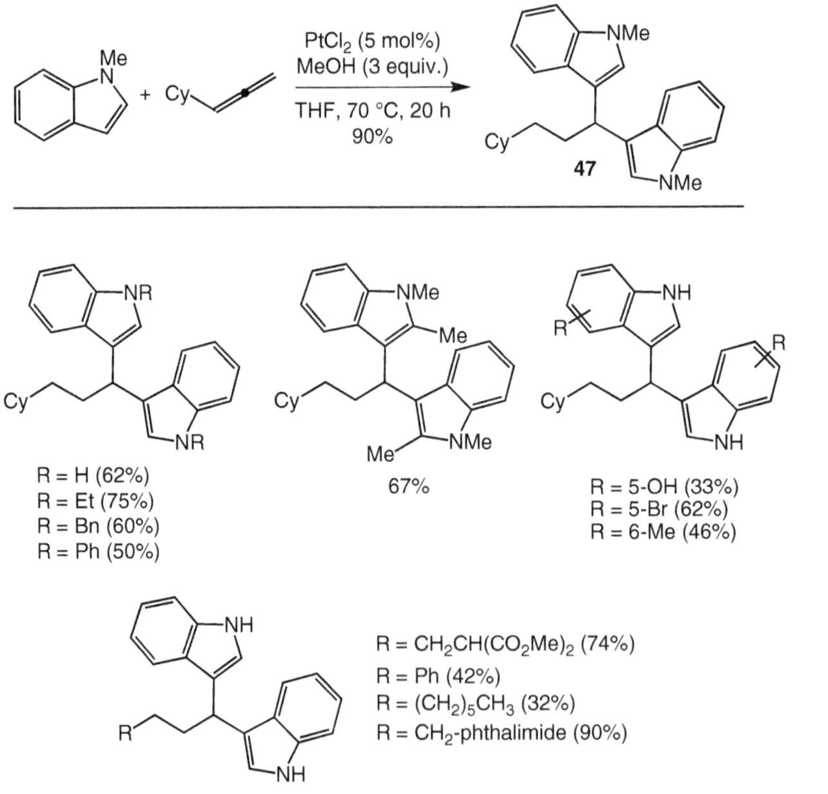

Scheme 9.26 Platinum-catalyzed intermolecular bisindolylation of allenes to form 1,1-bis (indolyl) alkanes.

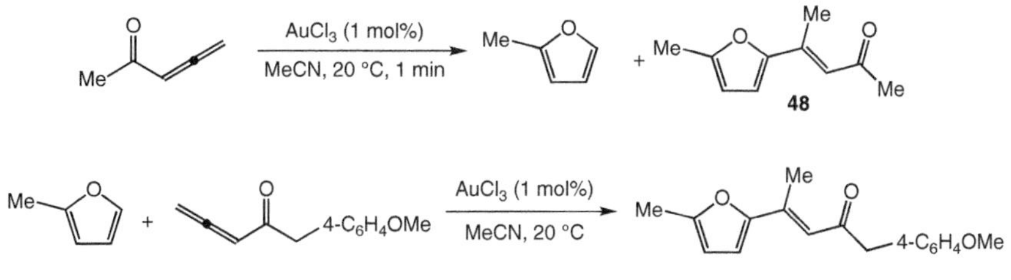

Scheme 9.27 Gold(III)-catalyzed tandem cyclization/hydroarylation of allenyl ketones.

rapid consumption of allenyl ketone to form a 1:1 mixture of 2-methylfuran and the 5-furanyl-3-penten-2-one **48** in 94% combined yield (Scheme 9.27). The gold(III)-catalyzed cycloisomerization/hydroarylation of allenyl ketones was applied to a range of alkyl and aryl ketones and, in the case of aryl ketones, the furan:5-furanyl-3-penten-2-one ratio increased with the decreasing electron donor ability of the arene, presumably due to the decreased nucleophilicity of the 2-aryl furan. That the 5-furanyl-3-penten-2-one was formed via intermolecular hydroarylation of allenyl ketone with furan was established through the gold(III)-catalyzed hydroarylation of allenyl ketone with 2-methylfuran (Scheme 9.27). It should also be noted that the hydroarylation of allenyl ketones is also catalyzed by Lewis acids [33].

9.3.3 Alkoxy Benzenes as Nucleophiles

In 1983, Panunzi and co-workers reported the platinum(II)-catalyzed intermolecular hydroarylation of phenols with 3-methyl-1,2-butadiene, which was the first example of the catalytic hydroarylation of an allene [2]. For example, treatment of a 1:1.2 mixture of 3-methylphenol and 3-methyl-1,2-butadiene with a catalytic amount of $[N(CH_2CH_2CH_3)_4]^+$ $[PtCl_3(C_2H_4)]^-$ in chloroform at 30 °C for 6 h led to 57% conversion to form a 93:4 mixture of *o*-isopentenylphenol (**49**) and 2,2-dimethylchroman (**50**), the latter product presumably resulting from secondary hydroalkoxylation of **49** (Scheme 9.28). Increasing the reaction time to 24 h led to 78% conversion to form a 58:25 mixture of **49** and **50**. Platinum-catalyzed intermolecular hydroarylation of 3-methyl-1,2-butadiene was likewise demonstrated with phenol, 4-methylphenol, and 3- and 4-methoxyphenol. Relevant to these efforts, Panunzi described the hydroarylation of *N*-acyl aniline derivatives mediated by a stoichiometric amount of the platinum π-3-methyl-1,2-butadiene complex $(PPh_3)Pt(\pi\text{-}H_2C{=}C{=}CMe_2)Cl_2$ (Scheme 9.28) [34].

In 2009, Gagné and co-workers reported a gold-catalyzed procedure for the intermolecular hydroarylation of allenes with electron-rich arenes [35]. For example, reaction of a 2:1 mixture of 1,3,5-trimethoxybenzene and 3-methyl-1,2-butadiene with a catalytic 1:1 mixture of the electron-deficient gold(I) complex $[(4\text{-}Cl\text{-}PhO)_3P]AuCl$ and $AgBF_4$ in CH_2Cl_2 at room temperature for 4 h led to isolation of prenylated arene **51** in 67% yield (Scheme 9.29). The hydroarylation protocol was applied to a number of electron-rich methoxy-substituted arenes but was restricted to terminal allenes. In the case of monosubstituted allenes, hydroarylation occurred to selectively form

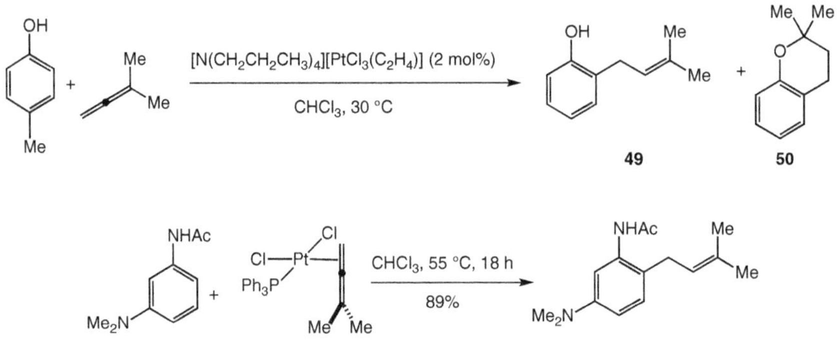

Scheme 9.28 Platinum(II)-catalyzed and Platinum(II)-mediated intermolecular hydroarylation of substituted arenes with 3-methyl-1,2-butadiene.

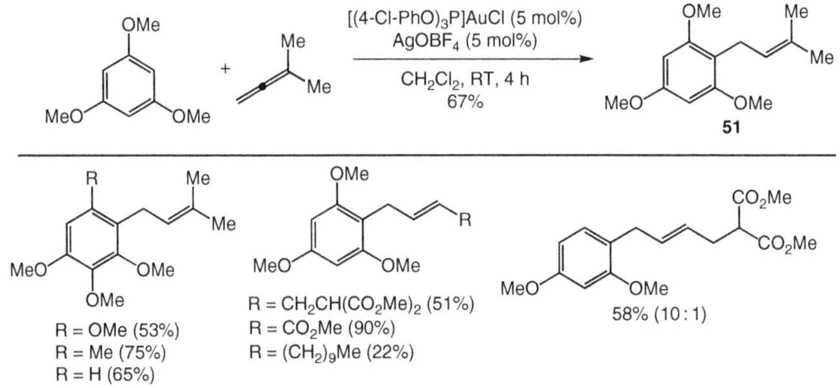

Scheme 9.29 Gold-catalyzed intermolecular hydroarylation of allenes with electron-rich arenes.

the (*E*)-allylic arene. Interestingly, electron-rich heteroaromatic compounds such as indole, pyrrole, and furan were ineffective nucleophiles for these transformations.

Ma and co-workers have reported Pd(II)/Brønsted acid-catalyzed intermolecular hydroarylation of allenoates with alkoxyl benzenes [36]. For example, treatment of a mixture of 1,3,5-trimethoxybenzene and ethyl 2-methyl-4-phenylbuta-2,3-dienoate with a catalytic amount of $PdCl_2(MeCN)_2$ (5 mol%) in a 4:1 mixture of trifluoroacetic acid (TFA) and dimethylacetamide (DMA) to TFA at 40 °C for 15 h led to isolation of the (*E*)-2-methyl-4,4-diaryl-2-butenoate **52** in 64% yield as a single regio- and diastereomer (Scheme 9.30). It should be noted, however, that arene **52** was formed in 50% yield in the absence of $PdCl_2(MeCN)_2$ under otherwise identical conditions. Nevertheless, the Pd(II)/Brønsted acid-catalyzed protocol was moderately effective for a range of ethyl 2,3-dienoates and electron-rich arenes (Scheme 9.30). Ma and co-workers subsequently expanded the scope of this procedure to include the intermolecular hydroarylation of allenoates with *N*-protected indoles, although in these cases, Pd(II) had no beneficial effect on the transformation [37].

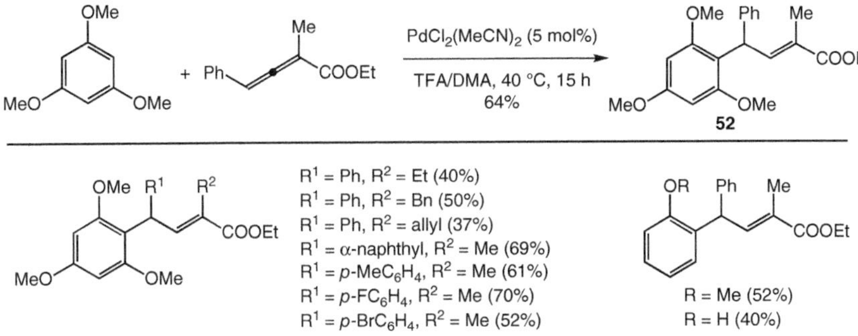

Scheme 9.30 Pd(II)/Brønsted acid-catalyzed intermolecular hydroarylation of allenoates with alkoxyl benzenes.

9.3.4 Alkyl Benzenes as Nucleophiles

In 2001, Yamaguchi and co-workers reported the hydroarylation of allenylsilanes with simple arenes to form vinyl arenes in the presence of a stoichiometric amount of $GaCl_3$ (Eq. (1)), although effective catalysis was not realized [38]. In 2007, Li and co-workers extended these efforts to realize the gold(I)-catalyzed hydroarylation of 1-phenyl-1,2-propadiene with substituted arenes [39]. In an optimized procedure, reaction of a 1:1.2 mixture of mesitylene and 1-phenyl-1,2-propadiene in dioxane that contained a trace of acetic acid at 105 °C for 18 h led to isolation of 2-cinnamyl-1,3,5-trimethylbenzene in 70% yield (Scheme 9.31). The transformation was moderately effective for a range of mono-, di-, tri-, tetra-, and pentasubstituted arenes.

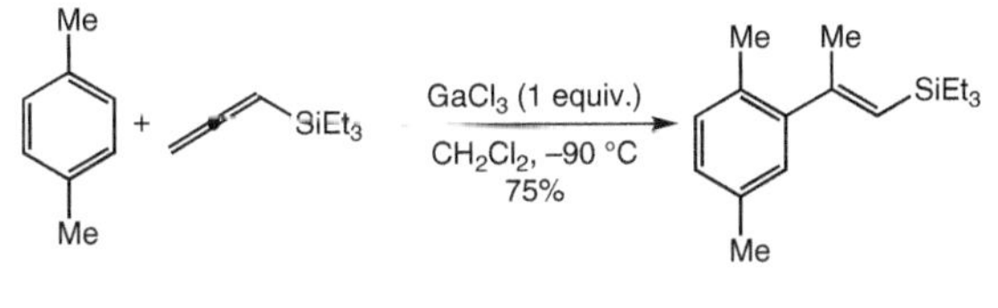

(9.1)

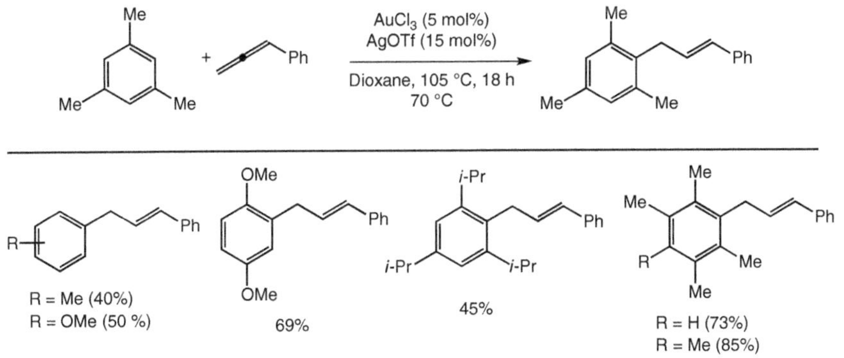

Scheme 9.31 Gold(I)-catalyzed hydroarylation of 1-phenyl-1,2-propadiene with substituted arenes.

9.4 Enantioselective Hydroarylation

9.4.1 Intramolecular Hydroarylation

Building upon their previous work in both the platinum-catalyzed enantioselective hydroarylation of 2-alkenyl indoles [40] and the gold-catalyzed hydroarylation of 2-allenyl indoles [3], Widenhoefer and Liu developed an effective gold-catalyzed protocol for the enantioselective intramolecular 6-*exo* hydroarylation of 2-allenyl indoles [41]. For example, treatment of the 2-(γ-allenyl)indole **1** with a catalytic 1:2 mixture of the axially chiral bis(gold)phosphine complex [(*S*)-**P2**]Au_2Cl_2 and $AgBF_4$ in toluene at −10 °C for 17 h led to isolation of the vinyl tetrahydrocarbazole **2** in 88% yield with 92% ee (Scheme 9.32). Enantioselective gold-catalyzed hydroarylation of 2-(γ-allenyl)indoles tolerated some substitution on the indole ring and alkyl chain and was effective for both the hydroarylation of terminally disubstituted allenes and for the formation of seven-membered rings (Scheme 9.32).

9.4.2 Intermolecular Hydroarylation

Che and co-workers subsequently reported the gold-catalyzed enantioselective intermolecular hydroarylation of 1,3-diaryl allenes with indoles [42]. As an example, treatment of a 5:1 mixture of *N*-methylindole and racemic

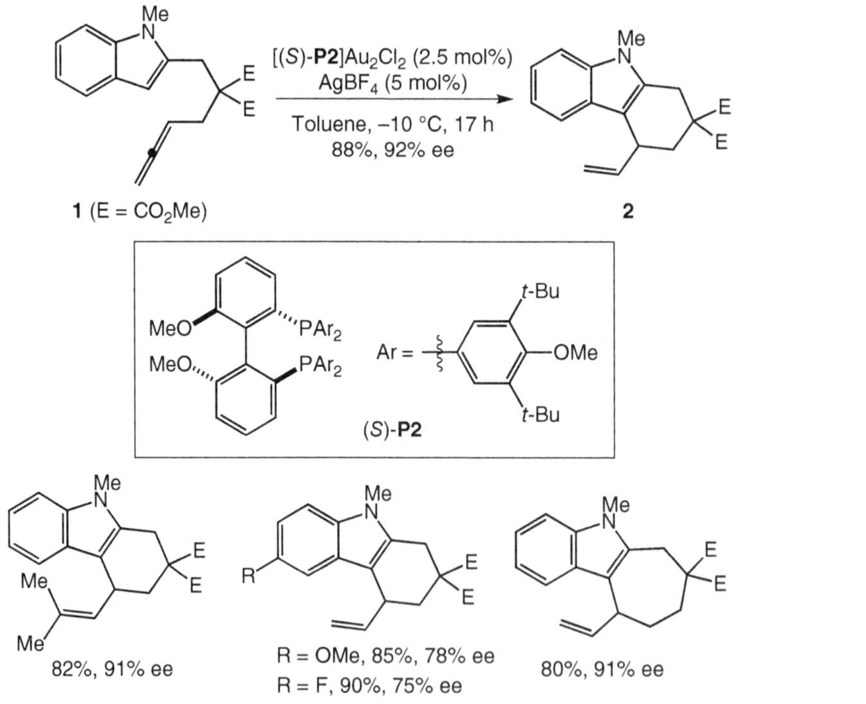

Scheme 9.32 Gold-catalyzed enantioselective intramolecular 6-*exo* hydroarylation of 2-allenyl indoles.

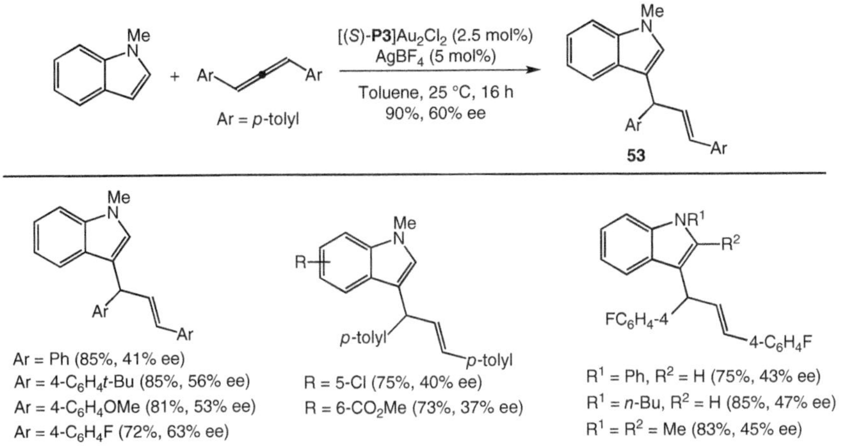

Scheme 9.33 Gold-catalyzed enantioselective intermolecular hydroarylation of 1,3-diaryl allenes with indoles.

1,3-di-*p*-tolylpropa-1,2-diene with a catalytic 1:2 mixture of [(*S*)-**P3**]Au_2Cl_2 and AgOTf in toluene at room temperature for 16 h led to isolation of 3-allyl indole **53** in 90% yield with 60% ee (Scheme 9.33). A number of *para*-substituted 1,3-diaryl allenes and indole derivatives underwent gold-catalyzed hydroarylation to give the corresponding 3-allyl indoles in good yields with moderate enantioselectivities (Scheme 9.33). Enantioselective hydroarylation of racemic 1,3-disubstituted allenes points to facile gold-catalyzed stereomutation of the allene under reaction conditions, a process known to be particularly facile in the case of 1,3-diaryl allenes owing to the stabilization of positive charge in the formation of the achiral η^1-allylic cation intermediate [10, 43]. As such, this transformation represents a rare example of the dynamic kinetic enantioselective hydrofunctionalization of an axially chiral racemic allene [44].

9.5 Summary and Outlook

Although slow to develop, following the initial disclosure of the platinum-catalyzed intermolecular hydroarylation of phenols with 3-methyl-1,2-butadiene by Panunzi and co-workers in 1983 [2], efforts directed toward the catalytic hydroarylation of allenes have increased over the past decade. Effective catalytic hydroarylation of allenes is dominated by π-activation processes involving the addition of nucleophilic, electron-rich arenes to allenes catalyzed by electrophilic and carbophilic late transition metal complexes, cationic gold(I) complexes in particular. Highlights from these efforts include the development of effective gold(I)-catalyzed methods for the intramolecular *exo*-hydroarylation of allenes with heteroaromatic nucleophiles that have found utility in target-oriented synthesis [3, 4, 11], the gold(I)-catalyzed intramolecular *exo*-hydroarylation of allenes with electron-rich arenes that has provided critical insight into the mechanisms of gold(I)-catalyzed allene hydrofunctionalization [15–18], and the enantioselective intramolecular hydroarylation of 2-allenyl indoles catalyzed by chiral bis(gold) phosphine complexes [41]. Despite these advances, many limitations regarding the catalytic hydroarylation of allenes still remain, most notably, the requirement of electron-rich arenes as coupling partners and the general dearth of effective enantioselective allene hydroarylation processes that might provide ready access to enantiomerically enriched chiral allylic arenes.

References

1 (a) Goj, L.A. and Gunnoe, T.B. (2005) *Curr. Org. Chem.*, **9**, 671–685; (b) Kakiuchi, F. and Chatani, N. (2003) *Adv. Synth. Catal.*, **345**, 1077–1101; (c) Kakiuchi, F. and Kochi, T. (2008) *Synthesis*, 3013–3039; (d) Tsurugi, H., Yamamoto, K., Nagae, H., Kaneko, H., and Mashima, K. (2014) *Dalton Trans.*, **43**, 2331–2343; (e) Colby, D.A., Bergman, R.G., and Ellman, J.A. (2010) *Chem. Rev.*, **110**, 624–655; (f) Arockiam, P.B., Bruneau, C., and Dixneuf, P.H. (2012) *Chem. Rev.*, **112**, 5879–5918; (g) Zheng, Q.-Z. and Jiao, N. (2014) *Tetrahedron Lett.*, **55**, 1121–11210.

2 De Renzi, A., Panunzi, A., Saporito, A., and Vitagliano, A. (1983) *J. Chem. Soc., Perkin Trans. II*, 993–9910.

3 Zhang, Z., Liu, C., Kinder, R.E., Han, X., Qian, H., and Widenhoefer, R.A. (2006) *J. Am. Chem. Soc.*, **128**, 9066–9073.

4 Liu, Z., Wasmuth, A.S., and Nelson, S.G. (2006) *J. Am. Chem. Soc.*, **128**, 10352–10353.

5 a Shen, H.C. (2008) *Tetrahedron*, **64**, 3885–3903; (b) Krause, N. and Winter, C. (2011) *Chem. Rev.*, **111**, 1994–2009.

6 (a) Gorin, D.J., Sherry, B.D., and Toste, F.D. (2008) *Chem. Rev.*, **108**, 3351–3378; (b) Arcadi, A. (2008) *Chem. Rev.*, **108**, 3266–3325; (c) Li, Z., Brouwer, C., and He, C. (2008) *Chem. Rev.*, **108**, 3239–3265; (d) Corma, A., Leyva-Pérez, A., and Sabater, M.J. (2011) *Chem. Rev.*, **111**, 1657–1712; (e) Liu, L.-P. and Hammond, G.B. (2012) *Chem. Soc. Rev.*, **41**, 3129–3139; (f) Bandini, M. (2011) *Chem. Soc. Rev.*, **40**, 1358–1367; (g) Hashmi, A.S.K. (2010) *Angew. Chem., Int. Ed.*, **49**, 5232–5241; (h) Shapiro, N.D. and Toste, F.D. (2010) *Synlett*, 675–691.

7 (a) Park, C. and Lee, P.H. (2008) *Org. Lett.*, **10**, 3359–3362; (b) Kong, W., Fu, C., and Ma, S. (2009) *Chem. Commun.*, 4572–4574; (c) Alcaide, B., Almendros, P., Alonso, J.M., and Fernańdez, I. (2013) *J. Org. Chem.*, **78**, 6688–6701; (d) Alcaide, B., Almendros, P., Alonso, J.M., Quirós, M.T., and Gadzinski, P. (2011) *Adv. Synth. Catal.*, **353**, 1871–1876; (e) Kong, W., Fu, C., and Ma, S. (2012) *Org. Biomol. Chem.*, **10**, 2164–2173.

8 Miller, Z.D. and Montgomery, J. (2014) *Org. Lett.*, **16**, 5486–5489.

9 (a) Brown, T.J., Sugie, A., Dickens, M.G., and Widenhoefer, R.A. (2010) *Organometallics*, **29**, 4207–4209; (b) Brown, T.J., Sugie, A., Dickens, M.G., and Widenhoefer, R.A. (2012) *Chem. - Eur. J.*, **18**, 6959–6971.

10 (a) Harris, R.J., Nokafuku, K., Li, H., and Widenhoefer, R.A. (2014) *Chem. - Eur. J.*, **20**, 12245–12254; (b) Li, H., Harris, R.J., Nokafuku, K., and Widenhoefer, R.A. (2016) *Organometallics*, **35**, 2242–2248.

11 Zeldin, R.M. and Toste, F.D. (2011) *Chem. Sci.*, **2**, 1706–1709.

12 Barluenga, J., Piedrafita, M., Ballesteros, A., Suárez-Sobrino, A.L., and González, J.M. (2010) *Chem. - Eur. J.*, **16**, 11827–11831.

13 Chen, B., Fan, W., Chai, G., and Ma, S. (2012) *Org. Lett.*, **14**, 3616–3619.

14 Alcaide, B., Almendros, P., Cembellín, S., and del Campo, T.M. (2015) *J. Org. Chem.*, **80**, 4650–4660.

15 Tarselli, M.A. and Gagné, M.R. (2008) *J. Org. Chem.*, **73**, 2439–2441.

16 Tarselli, M.A., Chianese, A.R., Lee, S.J., and Gagné, M.R. (2007) *Angew. Chem., Int. Ed.*, **46**, 6670–6673.

17 Weber, D., Tarselli, M.A., and Gagné, M.R. (2009) *Angew. Chem., Int. Ed.*, **48**, 5733–57310.

18 Weber, D. and Gagné, M.R. (2009) *Org. Lett.*, **11**, 4962–4965.

19 Liu, C.-H. and Yu, Z.-X. (2014) *Org. Chem. Front.*, **1**, 1205–1209.

20 Dong, Z., Liu, C.-H., Wang, Y., Lin, M., and Yu, Z.-X. (2013) *Angew. Chem., Int. Ed.*, **52**, 14157–14161.

21 Fazeli, A., Pflästerer, D., Rudolph, M., and Hashmi, A.S.K. (2015) *J. Organomet. Chem.*, **795**, 68–70.

22 Balasubramanian, T. and Balasubramanian, K.K. (1991) *Tetrahedron Lett.*, **32**, 6641–6644.

23 Watanabe, T., Oishi, S., Fujii, N., and Ohno, H. (2007) *Org. Lett.*, **9**, 4821–4824.

24 Park, C. and Lee, P.H. (2008) *Org. Lett.*, **10**, 3359–3362.

25 Qiu, Y., Zhou, J., Li, J., Fu, C., Guo, Y., Wang, H., and Ma, S. (2015) *Chem. - Eur. J.*, **21**, 15939–15943.

26 Mei, L.-Y., Wei, Y., Tang, X.-Y., and Shi, M. (2015) *J. Am. Chem. Soc.*, **137**, 8131–8137.

27 Alcaide, B., Almendros, P., Cembellín, S., del Campo, T.M., and Fernández, I. (2013) *Chem. Commun.*, **49**, 1282–1284.

28 Toups, K.L., Liu, G.T., and Widenhoefer, R.A. (2009) *J. Organomet. Chem.*, **694**, 571–575.

29 Kong, W., Cui, J., Yu, Y., Chen, G., Fu, C., and Ma, S. (2009) *Org. Lett.*, **11**, 1213–12110.

30 Muñoz, M.P., de la Torre, M.C., and Sierra, M.A. (2012) *Chem. - Eur. J.*, **18**, 4499–4504.

31 Ma, S. and Yu, S. (2005) *Org. Lett.*, **7**, 5063–5065.

32 Hashmi, A.S.K., Schwarz, L., Choi, J.-H., and Frost, T.M. (1990) *Angew. Chem., Int. Ed.*, **39**, 2285–2288.

33 (a) Nagao, Y., Lee, W.S., Komaki, Y., Sano, S., and Shiro, M. (1994) *Chem. Lett.*, **23**, 597–600; (b) Nagao, Y., Lee, W.S., Jeong, I.Y., and Shiro, M. (1995) *Tetrahedron Lett.*, **36**, 2799–2802; (c) Lee, W.S., Jeong, I.Y., Shiro, M., Sano, S., and Nagao, Y. (1997) *Tetrahedron Lett.*, **38**, 611–614; (d) Nagao, Y., Lee, W.S., and Kim, K. (1994) *Chem. Lett.*, **23**, 389–392.

34 De Renzi, A., Panunzi, A., Saporito, A., and Vitagliano, A. (1985) *J. Chem. Soc., Perkin Trans. I*, 1095–1098.

35 Tarselli, M.A., Liu, A., and Gagné, M.R. (2009) *Tetrahedron*, **65**, 1785–1789.

36 Fang, Z., Fu, C., and Ma, S. (2010) *Chem. - Eur. J.*, **16**, 3910–3913.

37 Fang, Z., Fu, C., and Ma, S. (2011) *Eur. J. Org. Chem.*, 1227–1231.

38 Kido, Y., Yonehara, F., and Yamaguchi, M. (2001) *Tetrahedron*, **57**, 827–833.
39 Skouta, R. and Li, C.-J. (2008) *Can. J. Chem.*, **86**, 616–620.
40 Han, X. and Widenhoefer, R.A. (2006) *Org. Lett.*, **8**, 3801–3804.
41 Liu, C. and Widenhoefer, R.A. (2007) *Org. Lett.*, **9**, 1935–1938.
42 Wang, M.-Z., Zhou, C.-Y., Guo, Z., Wong, E.L.-M., Wong, M.-K., and Che, C.-M. (2011) *Chem. - Asian J.*, **6**, 812–824.
43 Tudela, E., González, J., Vicente, R., Santamaría, J., Rodríguez, M.A., and Ballesteros, A. (2014) *Angew. Chem., Int. Ed.*, **53**, 12097–12100.
44 (a) Zhang, Z., Bender, C.F., and Widenhoefer, R.A. (2007) *J. Am. Chem. Soc.*, **129**, 14148–14149; (b) Butler, K.L., Tragni, M., and Widenhoefer, R.A. (2012) *Angew. Chem., Int. Ed.*, **51**, 5175–5178.

Index

Catalytic Hydroarylation of Carbon-Carbon Multiple Bonds, First Edition.
Edited by Lutz Ackermann, T. Brent Gunnoe and Laurel Goj Habgood.

b

c

d

g

i

S